GASEOUS DIELECTRICS VIII

GASEOUS DIELECTRICS VIII

Edited by
Loucas G. Christophorou
and
James K. Olthoff

National Institute of Standards and Technology
Gaithersburg, Maryland

KLUWER ACADEMIC / PLENUM PUBLISHERS
NEW YORK, BOSTON, DORDRECHT, LONDON, MOSCOW

Proceedings of the Eighth International Symposium on Gaseous Dielectrics,
held June 2–5, 1998, in Virginia Beach, Virginia

ISBN 0-306-46056-4

© 1998 Kluwer Academic / Plenum Publishers, New York
233 Spring Street, New York, N.Y. 10013

10 9 8 7 6 5 4 3 2 1

All rights reserved

A C.I.P. Catalogue record for this book is available from the Library of Congress

No part of this book may be reproduced, stored in a retrieval system, or transmitted in any
form ȍr by any means, electronic, mechanical, photocopying, microfilming, recording, or
otherwise, without written permission from the Publisher

Printed in the United States of America

PREFACE

The Eighth International Symposium on Gaseous Dielectrics was held in Virginia Beach, Virginia, U S A, June 2-5, 1998. The symposium continued the interdisciplinary character and comprehensive approach of the preceding seven symposia.

Gaseous Dielectrics VIII is a detailed record of the symposium proceedings. It covers recent advances and developments in a wide range of basic, applied, and industrial areas of gaseous dielectrics. It is hoped that *Gaseous Dielectrics VIII* will aid future research and development in, and encourage wider industrial use of, gaseous dielectrics.

The Organizing Committee of the Eighth International Symposium on Gaseous Dielectrics consisted of D. W. Branston (Germany), L. G. Christophorou (USA), A. H. Cookson (USA), E. Dutrow (USA), O. Farish (UK), M. E. Frechette (Canada), I. Gallimberti (Italy), A. Garscadden (USA), T. Kawamura (Japan), E. Marode (France), I. W. McAllister (Denmark), H. Morrison (Canada), A. H. Mufti (Saudi Arabia), K. Nakanishi (Japan), L. Niemeyer (Switzerland), J. K. Olthoff (USA), W. Pfeiffer (Germany), Y. Qiu (China), I. Sauers (USA), M. Schmidt (Germany), R. J. Van Brunt (USA), S. Yanabu (Japan), and J. W. Wetzer (The Netherlands). The Local Arrangements Committee consisted of members of the Conference Services Department of the National Institute of Standards and Technology (NIST) and members of the Old Dominion University (ODU). The contributions of each member of these committees, the work of the session chairmen, the interest of the participants, and the advice of many colleagues are gratefully acknowledged. I am especially indebted to Dr. James K. Olthoff of NIST and to Professor Vishnu Lakdawala of ODU for their exceptional and multiple contributions that made this such a successful meeting. Special thanks are also due to Kathy Kilmer, Tammie Grice, Sylvia Mahon, and Yicheng Wang of NIST for their assistance in organizing and running the symposium.

The Eighth International Symposium on Gaseous Dielectrics was hosted by the National Institute of Standards and Technology and the Old Dominion University. It was sponsored by the National Institute of Standards and Technology; Old Dominion University; Aero Propulsion & Power Directorate, Wright Laboratory; ABB, Switzerland; Hitachi, Ltd., Japan; Kansai Electric Power Corporation, Japan; Mitsubishi Electric Corporation, Japan; Tokyo Electric Power Company, Japan; and Toshiba Corporation, Japan. The symposium was technically sponsored by the IEEE Dielectrics and Electrical Insulation Society. The Support of the host institutions and the financial assistance of the sponsors are acknowledged with gratitude.

Loucas G. Christophorou, Symposium Chairman

Gaithersburg, Maryland

CONTENTS

SECTION 1: BASIC PHYSICS OF GASEOUS DIELECTRICS

Electron Impact Ionization and Dissociation of Molecules
Relevant to Gaseous Dielectrics.
V. Tarnovsky, H. Deutsch, S. Matt, T. D. Märk,
R. Basner, M. Schmidt, and K. H. Becker 3
Discussion 13

Radicals from Electron Impact on Fluorocarbons
S. Motlagh and J. H. Moore 15
Discussion 21

Electron Collision Cross Sections and
Transport Parameters in CHF_3
J. D. Clark, B. W. Wright, J. D. Wrbanek, and A. Garscadden 23
Discussion 29

Ion-Molecule Reactions and Ion Kinetics in DC Townsend Discharges
in Dielectric Gases
M. V. V. S. Rao and J. K. Olthoff 31
Discussion 37

Electron Drift Velocities and Electron Attachment
Coefficients in Pure CHF_3 and its Mixtures with Argon
Yicheng Wang, L. G. Christophorou, J. K. Olthoff, and J. K. Verbrugge 39

Cross Section Measurements for Various
Reactions Occurring in CF_4 and CHF_3 Discharges
B. L. Peko, I. V. Dyakov, and R. L. Champion 45

A Study of Ion-Molecule Reactions of
Methane Ions in Parent Gas at High E/N
J. de Urquijo, E. Basurto, C. Cisneros, and I. Alvarez 51

Ionization and Ion Kinetics in $c\text{-}C_4F_8$
C. Q. Jiao, A. Garscadden, and P. D. Haaland 57

The Dependence of Electron Capture Rate
Constants on Electronic Polarizability
I. Szamrej 63

New Approach to Analysis of the Data of Electron
Swarm Experiment in Ionization Chamber
A. Rosa, M. Foryś, and I. Szamrej 69

Electron Attachment to O_2 in Dense He and Ar Gases
A. F. Borghesani and M. Santini 73

Electron Mobility and Localization in High-Density
Helium Gas in an Extended Temperature Range
A. F. Borghesani and M. Santini 79

O_2^- Ion Mobility in Dense Supercritical and Near Critical Argon Gas
A. F. Borghesani and M. Santini 85

Electron Mobility Maximum in High-Density Ar Gas
A. F. Borghesani and M. Santini 91

Low-Energy Electron Collision by Excited Atoms
L. Vušković, C. H. Ying, Y. Wang, and S. Popović 97

SECTION 2: BASIC MECHANISMS

A Study of Prebreakdown in SF_6
H. H. R. Gaxiola and J. M. Wetzer 105
Discussion 111

From Corona Stabilization to Spark Breakdown in Point-to-Plane SF_6 Gaps
H. Champain, A. Goldman, and M. Lalmas 113
Discussion 123

Generation Mechanism of Partial Discharge in Different
Kind of Pure Gases and Gas Mixtures with SF_6
T. Yamada, T. Takahashi, T. Toda, and H. Okubo 125
Discussion 131

Dielectric Characteristics of SF_6/N_2 Mixtures for Very Fast Transient Voltages
W. Pfeiffer, D. Schoen, and C. Zender 133
Discussion 139

Theoretical and Experimental Analysis of a High Pressure Glow Discharge
Controlled by a Dielectric Barrier
P. Ségur, F. Massines, A. Rabehi, and M.-C. Bordage 141

Electrical and Physical Behaviour of a Point-to-Plane Barrier
Discharge in Air Governed by the Water Vapour Content
L. Parissi, E. Odic, A. Goldman, M. Goldman, and S. Koch 147

Experimental and Theoretical Investigations on the Statistics of Time Lags
to Corona Inception and Breakdown of SF_6 in Non-Uniform Electric Fields
S. K. Venkatesh and M. S. Naidu 155

Properties of the RF-Discharge in Ar-Tetraethoxylsilane
R. Foest, R. Basner, M. Schmidt, F. Hempel, and *K. Becker* 161

SECTION 3: SIMULATIONS / BREAKDOWN IN GAS MIXTURES

Thermodynamic and Electrical Properties of SF_6-N_2 Thermal Plasmas
A. Gleizes 169
Discussion 179

Leader Step Time and Low Probability Impulse Breakdown Voltage
Measured in SF_6 Gas Mixtures
W. Gu, Q. Zhang, and Y. Qiu 181
Discussion 187

The Study of SF_6 Mixtures with Buffer Gases
I. D. Chalmers, X. Q. Qiu, and P. Coventry 189
Discussion 195

Breakdown Strength of SF_6/CF_4 Mixtures in Highly Non-Uniform Fields
E. Kuffel and K. Toufani 197
Discussion 203

The Effect of Electrode Geometry on the Lightning
Impulse Breakdown in SF_6/N_2 Mixtures
D. Raghavender and M. S. Naidu 205
Discussion 211

Macroscopic Modeling of Inhomogeneous Field Breakdown in SF_6
Under Lightning Impulse Stress
H. N. Suresh and M. S. Naidu 213

A Comparative Evaluation of Cost/Benefit Analysis with Lightning
Impulse Breakdown Strength of Binary and Ternary Gas
Mixtures Containing SF_6 in Non-Uniform Field Gaps
D. Raghavender and M. S. Naidu 219

Simulation of Electrical Circuit of Plasma Opening Switch Using PSPICE
Z. Zeng, Y. Qiu, and A. Qiu 225

Power Absorbed in a RF Discharge in Oxygen
G. R. G. Raju 231

SECTION 4: PARTIAL DISCHARGES / DIAGNOSTICS

Partial Discharge Transients: The Field Theoretical Approach
I. W. McAllister and G. C. Crichton 239
Discussion 252

UHF Diagnostics and Monitoring for GIS
B. Hampton 253
Discussion 262

Dielectric Properties of Small Gas Gap in SF_6/N_2 Mixtures
M. Ishikawa, T. Kobayashi, T. Goda, T. Inoue, M. Hanai, and T. Teranishi 263
Discussion 267

Discrimination of Streamer/Leader Type Partial Discharge
in SF_6 Gas Based on Discharge Mechanism
H. Okubo, T. Takahashi, T. Yamada, and M. Hikita 269
Discussion 275

The Influence of Accelerated Partial Discharges Tests
on the Effects on Epoxy Composite Surface in SF_6
H. Słowikowska, T. Łaś, and J. Słowikowski 277

Influence of Void Geometry and Bulk Dielectric
Polarization upon PD Transients
I. W. McAllister and G. C. Crichton 283

Partial Discharge Inception and Breakdown Characteristics
in Gas Mixtures with SF_6
H. Okubo, T. Yamada, T. Takahashi, and T. Toda 289

Dependence of Partial Discharge and Breakdown Characteristics
on Applied Power Frequency in SF_6 Gas
T. Takahashi, T. Yamada, M. Hikita, and H. Okubo 295

Partial Discharge Characteristics of Small Gaps in SF_6 Gas
M. Ishikawa, T. Kobayashi, T. Goda, T. Inoue, M. Hanai, and T. Teranishi 301

Characteristics of Partial Discharges on a Dielectric Surface in SF_6-N_2 Mixtures
X. Han, Yicheng Wang, L. G. Christophorou, and R. J. Van Brunt 307

Analysis of the Degradation of Polyethylene in Air
Using Electrical and Physical Data
J. Horwarth, D. L. Schweickart, and Yicheng Wang 313

Digital Analysis of PD Sources in Gas-Insulated Switchgear Substation
A. H. Mufti, W. M. Al-Baiz, and A. O. Arafa 319

Experimental Estimation of Schwaiger Factor Limit (η_{lim}) in Atmospheric Air
R. Arora and S. Prem 325

SECTION 5: HIGH PRESSURE GAS DIELECTRICS/SF_6-N_2 MIXTURES

SF_6/N_2 Mixtures for HV Equipment and Practical Problems
T. Kawamura, S. Matsumoto, M. Hanai, and Y. Murayama 333
Discussion 343

Low SF_6 Concentration SF_6/N_2 Mixtures for GIL
X. Waymel 345
Discussion 351

Application Problems of SF_6/N_2 Mixtures to Gas Insulated Bus
H. Hama, M. Yoshimura, K. Inami, and S. Hamano 353
Discussion 359

A Search for Possible "Universal-Application" Gas Mixtures
L. G. Christophorou, J. K. Olthoff, and D. S. Green 361
Discussion 367

Some Aspects of Compressed Air and Nitrogen Insulation
M. Piemontesi, F. Koenig, L. Niemeyer, and C. Heitz 369
Discussion 375

SECTION 6: GAS DECOMPOSITION / PARTICLES

Decomposition of SF_6 under AC and DC Corona Discharges
in High-Pressure SF_6 and SF_6/N_2 (10-90 %) Mixtures
A.-M. Casanovas, L. Vial, I. Coll, M. Storer, J. Casanovas, and R. Clavreul 379
Discussion 385

Influence of a Solid Insulator on the Spark Decomposition
of SF_6 and SF_6 + 50 % CF_4 Mixtures
I. Coll, A.-M. Casanovas, C. Pradayrol, and J. Casanovas 387
Discussion 393

Chemical Reactions and Kinetics of Mixtures of SF_6 and
Fluorocarbon Dielectric Gases in Electrical Discharges
J. Castonguay 395
Discussion 402

Dielectric Coatings and Particle Movement in GIS/GITL Systems
M. M. Morcos, K. D. Srivastava, M. Holmberg, and S. Gubanski 403
Discussion 409

Breakdown Characteristics of a Short Airgap with
Conducting Particle under Composite Voltages
R. Sarathi and M. Krishnamurthi 411

SF_6 Handling and Maintenance Processes Offered
by Quadrupole Mass Spectrometry
C. T. Dervos and P. Vassiliou 417

SECTION 7: ENVIRONMENTAL ASPECTS OF GASEOUS DIELECTRICS / RECYCLING

The United States Environmental Protection Agency's SF_6 Emissions Reduction
Partnership for Electric Power Systems: An Opportunity for Industry
E. J. Dolin 425
Discussion 430

SF_6 Recycling in Electric Power Equipment
L. Niemeyer 431
Discussion 442

Study of the Decomposition of SF_6 in the Lower Atmosphere:
The Experimental Approach
J.-M. Gauthier and J. Castonguay 443
Discussion 450

Extremely Low Frequency Electric and
Magnetic Field Measurement Methods
M. Misakian 451
Discussion 457

A Systematic Search for Insulation Gases
and their Environmental Evaluation
L. Niemeyer 459

SF_6 ReUse Concept and SF_6 New Applications
M. Pittroff, A. Schütte, and A. Meier 465

SECTION 8: SURFACE DISCHARGES / DESIGN ENGINEERING

The Surface Flashover Characteristics of Spacer
for GIS in SF_6 Gas and Mixtures
Y. Hoshina, J. Sato, H. Murase, H. Aoyagi, M. Hanai, and E. Kaneko 473
Discussion 479

Generation and Investigation of Planar Surface Discharges
S. J. MacGregor, R. A. Fouracre, and S. M. Turnbull 481
Discussion 487

Electrical Surface Discharges on Wet Polymer Surfaces
H.-J. Kloes, D. Koenig, and M. G. Danikas 489
Discussion 495

Termination of Creeping Discharges on a Covered Conductor
*L. Walfridsson, U. Fromm, A. Kron, L. Ming, R. Liu,
T. Schütte, D. Windmar, and M. Sjöberg* 497
Discussion 503

Static Electrification Phenomena on Dielectric Materials
of SF_6 Gas-Insulated Transformer
*M. Ishikawa, T. Kobayashi, T. Inoue, T. Goda, A. Inui,
T. Kobayashi, T. Teranishi, and M. Meguro* 505

Prebreakdown Phenomena in Dry Air along PTFE Spacers
H. H. R. Gaxiola and J. M. Wetzer 511

SECTION 9: GAS-INSULATED EQUIPMENT I

Development of the SF_6/N_2 Gas Mixture Insulated Transformer in China
Y. Qiu and E. Kuffel 519
Discussion 527

Possibility Studies on Application of SF_6/N_2 Gas Mixtures
to a Core-Type Gas-Insulated Transformer
K. Tsuji, M. Yoshimura, T. Hoshino, T. Yoshikawa, H. Fujii, and K. Nakanishi 529
Discussion 539

Power Frequency and SIL Withstand Performance
of a GIC with 5 and 10 Percent SF_6/N_2 Mixtures
H. I. Marsden, M. D. Hopkins, and C. R. Eck, III 541
Discussion 545

Insulation Characteristics of GIS for Non-Standard
Lightning Surge Waveforms
M. Koto, S. Okabe, T. Kawashima, T. Yamagiwa, and T. Ishikawa 547
Discussion 553

SECTION 10: GAS-INSULATED EQUIPMENT II

A Utility Perspective on SF_6 Gas Management Issues
H. D. Morrison, F. Y. Chu, J.-M. Braun, and G. L. Ford 557
Discussion 564

The Comparison of Arc-Extinguishing Capability of Sulfur Hexafluoride
(SF_6) with Alternative Gases in High-Voltage Circuit-Breakers
H. Knobloch 565
Discussion 571

Insulation Characteristics of DC 500 kV GIS
M. Shikata, K. Yamaji, M. Hatano, R. Shinohara, F. Endo, and T. Yamagiwa 573
Discussion 579

Improvement of Withstand Voltage at Particle Contamination
in DC-GIS due to Dielectric Coating on Conductor
*T. Hasegawa, A. Kawahara, M. Hatano, M. Yoshimura,
H. Fujii, K. Inami, H. Hama, and K. Nakanishi* 581

SECTION 11: DISCUSSION GROUPS

Discussion Group A: 589
Other Industrial Applications of Gaseous Dielectrics and Data Bases

    PANELISTS: A. Garscadden (Chairman, WPAEB, USA),
                          D. S. Green (NIST, USA),
                          P. Haaland (Mobium Co, USA),
                          E. Marode (CNRS/ESE, France)

Discussion Group B: 601
SF_6 Substitutes

    PANELISTS: A. Cookson (Chairman, NIST, USA),
                          W. Boeck (Technical University of Munich, Germany),
                          M. Frechette (IREQ, Canada),
                          A. Sabot (EDF, France),
                          T.Yamagiwa (Hitachi, Japan)

PARTICIPANTS 615

PHOTOGRAPHS OF PARTICIPANTS 625

AUTHOR INDEX 633

SUBJECT INDEX 637

SECTION 1: BASIC PHYSICS OF GASEOUS DIELECTRICS

ELECTRON IMPACT IONIZATION AND DISSOCIATION OF MOLECULES RELEVANT TO GASEOUS DIELECTRICS

V. Tarnovsky[1], H. Deutsch[2], S. Matt[3], T.D. Märk[3],
R. Basner[4], M. Schmidt[4], and K. Becker[1]

[1]Dept. of Physics, Stevens Institute of Technology, Hoboken, NJ 07030, USA
[2]Institut für Physik, Ernst-Moritz-Arndt Universität, D-17487 Greifswald, Germany
[3]Institut für Ionenphysik, Leopold-Franzens Universität, A-6020 Innsbruck, Austria
[4]Institut für Niedertemperatur Plasmaphysik (INP), D-17489 Greifswald, Germany

INTRODUCTION

Molecules such as methane (CH_4), carbon-tetrafluoride (CF_4), the perfluoroalkanes (C_nF_{2n+2}, n = 2,3,4), and sulfur-hexafluoride (SF_6) have received much attention because of their favorable electron transport properties in various applications such as gaseous dielectrics, plasma processing, and diffuse discharge switches[1]. Ion formation processes involving SF_6 have been studied extensively in the past. A summary of available partial and total cross sections for the formation of positive ions from SF_6 can be found e.g. in the recent paper of Deutsch et al.[2]. Negative ion formation processes have been reviewed by Christophorou and co-workers[3]. By contrast, little is know about the ion formation processes involving the SF_x (x = 1-5) free radicals or about the electron-impact dissociation of SF_6 into neutral ground-state fragments even though these are processes of great relevance for the understanding of the properties of gas mixtures containing SF_6. Sugai and co-workers[4] measured the relative cross sections for the formation of SF_x (x = 1-3) radicals following electron impact on SF_6. Recently, the fullerenes C_{60} and C_{70} have been proposed as potential electron scavengers because of their large attachment cross sections at very low electron energies[5-7].

This article describes recent experimental results for the electron-impact ionization and dissociative ionization of the SF_x (x = 1-5) free radicals using the fast-neutral beam technique. We will further discuss recent developments in the measurement of cross sections for ion formation processes involving the fullerene C_{60} and we will review the status of ionization cross section measurements for H_2O and OH which are ubiquitous trace contaminants in many gaseous dielectrics applications and environments. In all cases comparisons of the experimentally determined cross sections will be made with available theoretical calculations which include semi-empirical, semi-classical, and in some cases also with more rigorous cross section calculation schemes.

EXPERIMENTAL DETAILS

All ionization studies described here were carried out in our fast - neutral - beam apparatus. The apparatus and experimental technique have been described in previous papers[8-11]. A dc discharge between a heated tungsten filament and a positively biased anode (2-4 kV) through a suitably chosen target serves as the primary ion source. The primary ions are mass selected in a Wien filter and a fraction of them is neutralized by near-resonant charge transfer in a cell filled with an appropriately chosen charge-transfer gas (often Xe). The residual ions are removed from the target beam by electrostatic deflection and most species in Rydberg states are quenched in a region of high electric field. The neutral beam is subsequently crossed at right angles by an electron beam (5 - 200 eV beam energy, 0.5 eV FWHM energy spread, 0.03 - 0.4 mA beam current). The product ions are focused in the entrance plane of an electrostatic hemispherical analyzer which separates ions of different charge-to-mass ratios. The ions leaving the analyzer are detected by a channel electron multiplier (CEM). The neutral beam density in the interaction region can be determined from a measurement of the energy deposited by the fast neutral beam into a pyroelectric crystal whose response is first calibrated by a well-characterized ion beam[8]. As an alternative, the well-established Kr or Ar absolute ionization cross sections (known to better than 5%) can be used to calibrate the pyroelectric crystal. The calibrated detector, in turn, is then used to determine the flux of the neutral target beam in absolute terms. This procedure avoids the frequent and prolonged exposure of the delicate pyroelectric crystal to fairly intense ion beams[9,11].

When studying the ionization of C_{60}, the conventional fast-beam technique was modified in order to obtain C_{60} target beams of sufficient intensity and stability. Rather than striking a discharge between the heated tungsten filament and the anode through a heated vapor of C_{60}, the voltage between the filament and the anode was kept close to zero. As a consequence, no breakdown occurs in the ion source. The heated tungsten filament simply serves as a thermionic source of near-zero-energy electrons which attach readily to the C_{60} vapor to form C_{60}^- ions because of the large attachment cross section of C_{60}. Thus, rather than using C_{60}^+ ions as the primary ions we used the negatively charged C_{60}^- ions as primary ions. These ions when accelerated to 3 kV are readily stripped of the extra electron in collisions with the background gas atoms/molecules and one obtains an intense and stable fast beam of C_{60} neutral clusters which is then crossed by the electron beam.

We also carried out extensive SIMION ion trajectory modeling calculations[12] in an effort to demonstrate to what extent our apparatus is capable of detecting energetic fragment ions with 100% efficiency and to what extent our hemispherical analyzer is capable of separating the various product ions from a given parent.

RESULTS AND DISCUSSION

In the following paragraphs we will discuss the results of recent experimental and theoretical studies relating to the ionization of the SF_x free radicals (x = 1-5), the fullerene C_{60} as well as the water molecule and the hydroxyl radical.

Electron Impact Ionization of the SF_x (x = 1-5) Free Radicals

The electron impact ionization of SF_6 has been investigated thoroughly and absolute total and partial ionization cross sections for the formation of the various positive ions have been measured by several groups and there is generally good agreement between the results obtained by different techniques. There is also good agreement between the measured total SF_6 ionization cross sections and calculations based on the binary-encounter Bethe (BEB) theory of Kim and co-workers[13] and on the modified additivity rule of Deutsch et al.[14] We have recently begun a series of ionization and dissociative ionization studies of the SF_x free radicals (x =1-5) using the fast-beam technique. Both positive and negative primary ions were used in these experiments. Only preliminary results for the absolute cross sections are available at the time this article is written. A full account of the results will be presented at the Conference. It appears that the experimental results for the total single SF_x ionization cross sections support the result of cross section calculations using the additivity rule[14] as shown in figure 1. We note in particular the unusual partial "inversion" in the magnitude of the cross sections, i.e. the fact that the cross section for SF is larger than the cross sections for SF_2 and SF_3.

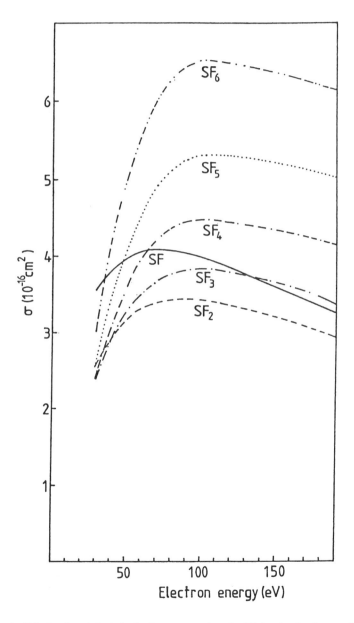

Figure 1. Calculated total single ionization cross sections for SF_x (x =1 - 6) using an additivity concept[14].

Electron Impact Ionization of C_{60}

Despite many experimental studies of the collisional interaction between electrons and fullerenes[15-17], there have only been a few absolute electron-impact ionization cross section measurements for C_{60} (and C_{70}). This is largely due to the great difficulty in normalizing measured relative cross section functions for the fullerenes. This requires among other things a quantitative knowledge of the neutral fullerene number density in the ion source.

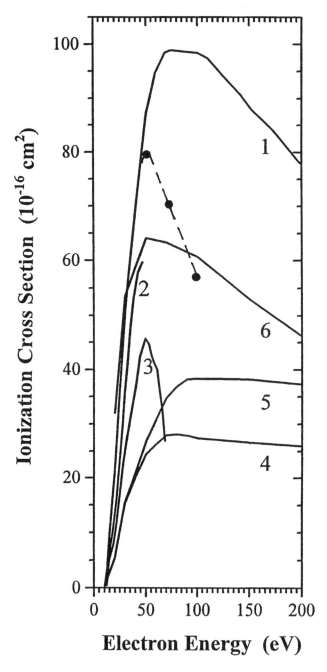

Figure 2. Various calculated and measured ionization cross sections for C_{60}. The different curves and data points are identified and explained in the text.

Until recently, the only absolute ionization cross section measurements were those of Baba et al.[18,19], Vostrikov et al.[20], Popovic[21], and Märk and co-workers[22-24]. All experiments except for those by Märk and co-workers employed a Knudsen-type technique for the absolute calibration which relies on the vapor pressure data given by Mathews et al.[25]. As discussed recently[26,27], these vapor pressure data might be in error by as much as a factor of 2.8. Matt et al.[23,24] measured a comprehensive set of cross sections for single and multiple ionization as well for dissociative ionization of C_{60} and C_{70} from threshold to 1000 eV using a mass spectrometric technique in conjunction with a novel calibration method[22] to normalize their relative cross section functions. The absolute cross section data of Baba et al.[18,19], Vostrikov et al.[20], and Popovic[21] are much larger than the corresponding values reported by Matt et al.[24]. In addition, there are significant differences in the energy dependence of the measured cross sections (see figure 2). Curves 1 and 6 in fig. 2 refer to two calculations of the total single C_{60} ionization cross section using the semi-classical DM formalism[28] (1) and an additivity concept[29] (6). Curve 2 and curve 3 represent respectively the experimental data of Baba et al.[18,19] and Vostrikov et al.[20]. The dashed line connecting the three data points at 50 eV, 75 eV and 100 eV represents the C_{60}^+ cross section reported by Popovic[21]. Curve 5 refers to the C_{60} counting cross section and curve 6 to the total C_{60} single ionization cross section of Matt et al.[24] (corrected according to the procedure discussed in the recent paper of Foltin et al.[30]). It is apparent that there are major discrepancies among the various experimental data sets and between the experimental data and the calculated cross section curves.

Very recently, Tarnovsky et al.[31] performed an absolute measurement of the C_{60}^+ electron-impact ionization cross section using the fast-neutral-beam technique which allows the absolute determination of ionization cross sections without the need to normalize the relative cross section to a previously determined "benchmark" cross section or to theory. This is accomplished by measuring all quantities that enter into the determination of an ionization cross section in absolute terms. In addition, the fast-neutral-

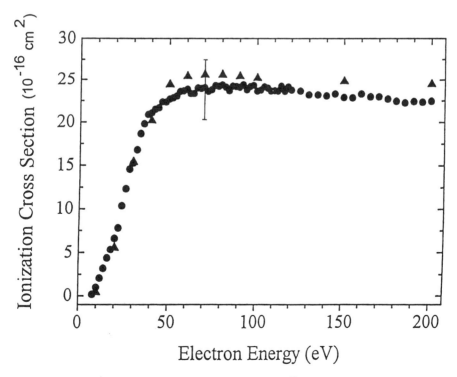

Figure 3. C_{60}^+ ionization cross section of Tarnovsky et al.[31] (●) and Matt et al.[24] (▲).

beam technique avoids the many problems associated with a direct measurement of the density of the target species in the ion source (e.g. unreliable vapor pressure data in the case of C_{60}) by determining the neutral target gas density from a measurement of the energy deposited by the fast (2-4 keV) neutral target beam into a calibrated pyroelectric detector (as described before). Figure 3 shows the result of this C_{60}^+ parent ionization cross section measurement from threshold to 200 eV in comparison with the C_{60}^+ cross section reported by Matt et al.[24]. The error bar at 70 eV indicates the statistical and all known systematic uncertainties in the absolute cross section measurement of Tarnovsky et al.[31]. The agreement between the two independent cross section measurements is excellent over the entire energy range from threshold to 200 eV.

There is an additional important conclusion that can be drawn from the excellent agreement between the two data sets shown in fig. 3. The uncertainty in the C_{60}^- attachment cross section of as much as 50% represented perhaps the largest contribution in the total uncertainty of the cross sections reported by Matt et al.[24] as discussed in ref. 32. The excellent agreement between the C_{60}^+ cross section of Tarnovsky et al.[31], which has been determined without using any benchmark cross section for calibration purposes and the cross section values of Matt et al.[32] serves as an independent confirmation of the reliability of the C_{60}^- attachment cross section used by these authors[33-35] (and which is shown in figure 4 in the energy range from zero to 15 eV).

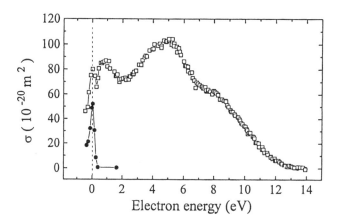

Figure 4. C_{60} electron attachment cross section in absolute units as used by Matt et al.[32] in the calibration of their C_{60} ionization cross sections (open squares) as a function of electron energy. Also shown is the relative zero-energy peak in the SF_6^- parent ion yield (full dots) which was measured under the same conditions[32].

Electron Impact Ionization of H_2O and OH

Water, H_2O, is among the most abundant molecules in the universe. It is a prominent constituent of many planetary atmospheres and it plays an important role in radiation chemistry and in biological systems. Water is also an undesirable contaminant in many gaseous dielectrics applications. The interaction of water with energetic photons and charged particles resulting in the dissociation and dissociative ionization of water are particularly important processes which lead to the formation of various neutral and ionic species. Among those, the neutral hydroxyl (OH) radicals play a prominent role due to their importance in secondary reaction processes in environments which contain H_2O in gaseous, liquid, or solid form. Electron collision processes with water vapor have received considerable attention from both experimentalists and theorists. Absolute cross sections for the electron impact ionization of H_2O (and D_2O) have been determined by 8 different groups over a period of 30 years. Those authors who reported measurements of ionization cross sections for both H_2O and D_2O found no isotope effects in the absolute partial and total ionization cross sections for the two targets. However, there are significant discrepancies among the ionization cross section data reported by the various groups, e.g. a difference of a factor of 4 in the total H_2O ionization cross section and significant differences in the partial ionization cross sections and in the cross section shapes. In contrast to the broad data base on collisions with H_2O (and D_2O), there are far fewer collisional data in the literature for the hydroxyl radical (OH, OD).

We measured absolute partial cross sections for the electron-impact ionization and dissociative ionization of the OD radical from threshold to 200 eV. We used the deuterated rather than the protonated target species to facilitate a better separation of the various product ions in our apparatus. The experimental work was complemented by extensive ion trajectory modeling calculations in an effort to quantify the loss of light D^+ fragment ions in our apparatus. In the course of this work, we also carried out partial ionization cross section measurements for the D_2O molecule, which are compared with the various previously reported H_2O and D_2O ionization cross sections. For both OD and D_2O, the total single ionization cross section was derived from the measured partial ionization cross sections. We determined the following absolute partial cross section values for D_2O at 70 eV impact energy: $1.65 \pm 0.30 \times 10^{-16}$ cm^2 (D_2O^+), $0.45 \pm 0.08 \times 10^{-16}$ cm^2 (OD^+), and less than 0.05×10^{-16} cm^2 (O^+). These cross sections are shown in figure 5. We also find a D^+ partial cross section of 0.2×10^{-16} cm^2 which we consider a lower limit of this cross section due to the loss of energetic D^+ fragment ions in our apparatus. This notion is supported by the fact that all partial D_2O ionization cross sections measured here are in very good agreement with the recent D_2O partial ionization cross sections reported by Straub et al.[36] using a time-of-flight technique with the exception of the D^+ cross section where our value is roughly 40% below their cross section. If we combine our measured partial ionization cross sections, we arrive at a total single D_2O ionization cross section of $2.35 \pm 0.5 \times 10^{-16}$ cm^2 at 70 eV. This total D_2O ionization cross section agrees very well

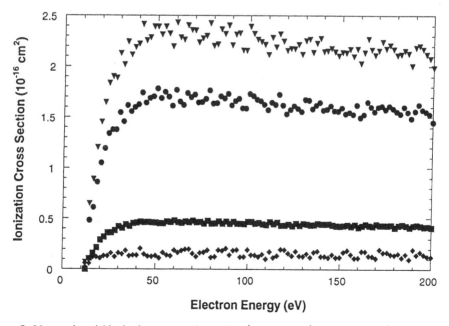

Figure 5. Measured partial ionization cross sections of D_2O^+ (●), and OD^+ (■) as well as D^+ (♦) from D_2O. Also shown is the sum of the partial cross sections (▼).

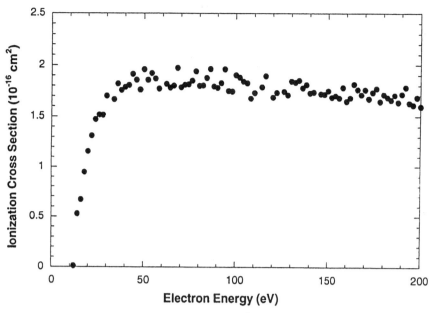

Figure 6. Absolute cross section for the formation of OD$^+$ parent ions from OD as a function of energy.

with the H$_2$O data of Djuric et al.[37] and with the D$_2$O data of Straub et al.[36]. Calculations of the total (single) ionization cross section for H$_2$O (D$_2$O) have been carried out by Deutsch et al.[14] using the modified additivity rule as well as by Kim and co-workers using their Binary-Encounter Bethe (BEB) approach[13]. In general, we find good agreement between our measured D$_2$O ionization cross sections and the calculated cross sections.

The electron-impact ionization of OD is dominated by formation of OD$^+$ parent ions. Figure 6 shows the absolute cross section for the formation of OD$^+$ ions from OD from threshold to 200 eV. This curve represents the result of a single data run. We find a cross section value at 70 eV of $1.85 \pm 0.25 \times 10^{-16}$ cm^2. The measured appearance energy of the OD$^+$ parent ion of 13.3 ± 0.5 eV is in very good agreement with the known 13.0 eV ionization energy of OD in its vibrational ground state. We found only a very small ion signal corresponding to the formation of D$^+$ fragment ions. Since ion trajectory modeling calculations indicate that our apparatus is not able to collect energetic D$^+$ fragment ions formed with excess kinetic energies of more than about 0.5 eV per fragment ion with 100% efficiency, we are not able to determine a reliable absolute value for the D$^+$ ionization cross section from OD. However, for the purpose of estimating the total single OD ionization cross section, we estimate based on trajectory modeling calculations and the measured ion signals that the combined (O$^+$ + D$^+$) ionization cross section contributes about 0.1×10^{-16} cm^2 to the total single OD ionization cross section at 70 eV. We obtained a total single OD ionization cross section by adding a "generic" cross section shape with a value at 70 eV of 0.1×10^{-16} cm^2 representing the unobserved (O$^+$ + D$^+$) cross section to the OD$^+$ partial cross section. This yields a total single OD ionization cross section of about 2.0×10^{-16} cm^2 at 70 eV. This cross section agrees well with the calculated OH total single ionization cross sections of Deutsch et al.[14] who used an additivity concept.

SUMMARY

A summary has been given of recent activities regarding the measurement and the calculation of absolute total and partial ionization cross sections for the SF$_x$ (x = 1-5) free radicals, for the fullerene C$_{60}$ as well as for the water molecule and the hydroxyl radical.

Acknowledgments

The work presented here has been supported in part by the U.S. National Science Foundation (NSF), the U.S. National Aeronautics and Space Agency (NASA), and the Division of Chemical Sciences, Office of Basic Energy Sciences, Office of Energy Research, U.S. Department of Energy.

REFERENCES

1. see e.g. L.G, Christophorou and S.R. Hunter, in "Electron-Molecule Interactions and Their Applications", Vol. 2, Chapter 5, Academic Press, New York 1984
2. H. Deutsch, K. Becker, and T.D. Märk, Int. J. Mass Spectrom. Ion Proc. **167/168**, 503-517 (1997)
3. S.R. Hunter, J.G. Carter, and L.G. Christophorou, J. Chem. Phys. **90**, 237 (1989)
4. M. Ito, M. Goto, H. Toyoda, and H. Sugai, Contr. Plasma Phys. **35**, 405 (1995)
5. R.C. Haddon, Phil. Trans. R. Soc. A **343**, 53 (1993)
6. D. Smith and P. Spanel, J. Phys. B **29**, 5199 (1996)
7. S. Matt, P. Scheier, T.D. Märk and K. Becker, in "Novel Aspects of Electron Molecule Collisions", Chapter 1, World Scientific Publ., Singapore, 1998
8. 21. R.C. Wetzel, F.A. Biaocchi, T.R. Hayes, and R.S. Freund, Phys. Rev. A **35**, 559 (1987)
9. R.S. Freund, R.C. Wetzel, R.J. Shul, and T.R. Hayes, Phys. Rev. A **41**, 3575 (1990)
10. V. Tarnovsky and K. Becker, Z. Phys. D **22**, 603 (1992)
11. V. Tarnovsky and K. Becker, J. Chem. Phys. **98**, 7868 (1993)
12. SIMION-3D (version 6.0), Energy Science and Technology Software Center
13. W. Hwang, Y.-K. Kim, and M.E. Rudd, J. Chem. Phys. **104**, 2956 (1996)
14. H. Deutsch, K. Becker, and T.D. Märk, Int. J. Mass Spectrom. Ion Proc. **167/168**, 503 (1997)
15. S.W. McElvany, M.M. Ross and J.H. Callahan, Accounts Chem.Research, **25**, 162 (1992)
16. C.Lifshitz, Mass Spectrom.Rev., **12**, 261 (1993)
17. P.Scheier, B.Dünser, R.Wörgötter, S.Matt, D.Muigg, G.Senn and T.D.Märk, Int. Rev. Phys. Chemistry, **15**, 93 (1996)
18. M.S. Baba, T.S.L. Narasimhan, R. Balasubramanian, and C.K. Mathews, Int. J. Mass Spectrom. Ion Proc., **114**, R1 (1992).
19. M.S.Baba, T.S.L.Narasimhan, R.Balasubramanian, and C.K.Mathews, Int. J. Mass Spectrom. Ion Proc., **116**, R1 (1992).
20. A.A. Vostrikov, D.Y. Dubov, and A.A.Agarkov, Tech. Phys. Lett., **21**, 715 (1995); A.A. Vostrikov, private communication, 1996
21. A. Popovic, private communication (1997); to be published in Rapid Communications in Mass Spectrometry
22. B. Dünser, M. Lezius, P. Scheier, H. Deutsch, and T.D.Märk, Phys. Rev. Lett., **74**, 3364 (1995)
23. S. Matt, O. Echt, R. Wörgötter, V. Grill, P. Scheier, C. Lifshitz, and T.D.Märk, Chem. Phys. Lett., **264**, 149 (1997)

24. S. Matt, B. Dünser, M. Lezius, H. Deutsch, K. Becker, A. Stamatovic, P. Scheier, and T.D. Märk, J. Chem. Phys., 105, 1880 (1996)
25. C.K. Mathews, M.S. Baba, T.S.L. Narasimhan, R. Balasubramanian, N. Sivaraman, T.G. Srinivasan, and P.R. Vasudeva Rao, J. Phys. Chem., 96, 3566 (1992)
26. A.L. Smith, J. Phys. B, 29, 4975 (1996)
27. Q. Gong, Y. Sun, Z. Huang, X. Zhou, Z. Gu, and D.Qiang, J. Phys. B, 29, 4981 (1996)
28. H. Deutsch, K. Becker, and T.D. Märk, J. Phys. B 29, 5175 (1996)
29. H. Deutsch, private communication (1998)
30. V. Foltin, M. Foltin, S. Matt, P. Scheier, K. Becker, H. Deutsch, and T.D. Märk, Chem. Phys. Lett. (1998), submitted for publication
31. V. Tarnovsky, P. Kurunczi, K. Becker, H. Deutsch, S. Matt, and T.D. Märk, J. Phys. B (1998), in press
32. S. Matt, P. Scheier, T.D. Märk, and K. Becker, "Positive and Negative Ion Formation in Electron Collisions with Fullerenes", in "Novel Aspects of Electron-Molecule Collisions", K. Becker (editor), World Scientific Publishing, Singapore (1998), in press
33. D.Smith, P.Spanel and T.D.Märk, Chem. Phys. Lett., 213, 202 (1993)
34. T. Jaffke, E. Illenberger, M. Lezius, S. Matejcik, D. Smith, and T.D. Märk, Chem. Phys. Lett., 226, 213 (1994)
35. S.Matejcik, T.D.Märk, P.Spanel, D.Smith, T.Jaffke and E.Illenberger, J. Chem. Phys., 102, 2516 (1995)
36. H.C. Straub, B.G. Lindsay, K.A. Smith, and R.F. Stebbings, J. Chem. Phys. 108, 109
37. N. Djuric, I.M. Cadez, and M.V. Kurepa, Int. J. Mass Spectrom. Ion Proc. 83, R7 (1988)

DISCUSSION

E. MARODE: As you have mentioned, dissociation cross sections are very much needed to evaluate subsequent chemistry in gases. My question is about how to get them. When the dissociation fragments are in an excited state, it is possible to detect the fragments, but when they are in the fundamental state the detection becomes more difficult. As a result it is difficult to obtain total dissociation cross sections for simple molecules. Can you comment on that?

K. BECKER: Neutral dissociation cross sections are difficult to measure and have been measured only for a few molecules (e.g., N_2, Cl_2 by Cosby et al. using a fast beam technique, and CF_4, CH_4 by Winters et al. using a chemical getter technique). More recent measurements have been made (e.g., on CF_4, CH_4, CHF_3) by Sugai et al. using appearance energy mass spectrometry, and by Moore et al. using the tellurium conversion technique (following paper).

G. R. G. RAJU: With regard to cross sections in SF_6 and N_2, it would be helpful to have tabulated values. Also, for electrical discharge applications the cross sections in the energy range 15 eV to 30 eV are most important. Could you please comment?

K. BECKER: Tabulated values for SF_6, N_2 and other molecules are available in many papers. Near-threshold cross sections have been measured carefully by us and by many other groups and are usually included in tables in 0.5 eV or 1 eV steps.

YICHENG WANG: I have a question about the possible influence of the internal excitation on the ionization cross section of radicals. Your results for SF_6 fragments show good agreement with theoretical calculations. Yet, SF_6 fragments prepared in the laboratory may be internally excited to some degree and the calculations usually assume that the radicals are in the ground states. Does that mean that (i) the radicals prepared in the laboratory are predominantly in their ground states or (ii) the ionization cross sections do not depend on the internal excitations significantly in the energy range investigated?

K. BECKER: Our technique can lead to vibrationally excited targets. We check for their presence by careful threshold studies. Ionization of vibrationally excited molecules/radicals results in (i) a shift of the threshold to lower values, and (ii) an enhancement of the ionization cross section in the range of up to ~ 30 eV. Generally, no effect of vibrational excitation can be found on the peak and high-energy part of the ionization cross section (electron energies greater than 50 eV).

J. H. MOORE: It's important to keep elementary chemistry in mind when thinking about plasma chemistry. For example, Professor Becker found it reasonable that SF_3^+ did not dissociate upon being ionized. A good freshman chemistry student could draw a Lewis dot diagram and predict that SF_3^+ is stable.

RADICALS FROM ELECTRON IMPACT ON FLUOROCARBONS

Safa Motlagh and John H. Moore

Department of Chemistry and Biochemistry
University of Maryland
College Park, MD 20742

INTRODUCTION

In a low-pressure, nonequilibrium plasma in a fluorocarbon gas such as that used in a processing reactor in semiconductor device fabrication, the active chemical species are all fragments of the feed gas. The fluorocarbon feed gas is itself chemically benign and the parent ions are unstable and do not persist for a sufficient time to contribute to plasma chemistry. The active fragments are produced by electron impact in the 5 to 30 eV energy range typical of the electron-energy distribution in the plasma. In general, electron-molecule collisions leading to dissociation include dissociative ionization, neutral dissociation, dissociative electron attachment, and dipolar dissociation. For fluorocarbons, only the first two are important. Dissociative attachment must be initiated by the resonant attachment of an electron to yield a negative ion that is sufficiently long-lived for dissociation to compete with autodetachment of the electron. Resonant electron attachment is not a significant feature of electron-fluorocarbon interactions. Dipolar dissociation proceeds through very high-lying electronic states. The cross sections are generally small, especially for electrons with energies less than about 30 eV. Although not a mathematical certainty, it is generally observed that the neutral fragments from both dissociative ionization and neutral dissociation are odd-electron species; they are radicals. Chemically they are very reactive. In addition, in comparison to ions, these species have relatively low sticking probabilities at surfaces. As a consequence radicals may persist in the plasma for a much longer time than ions. The concentration of radicals in a plasma can exceed the concentration of ions by as much as four orders of magnitude.

Measurements of cross sections for the production of radicals in low energy electron-molecule collisions are clearly needed to understand plasma chemistry in general, and to model and optimize the operation of particular plasma reactors. To measure a dissociation cross section one must detect one or the other of the fragments.

Experimentally it is much easier to detect a charged species so, if the choice exists, a charged fragment of a dissociation is detected rather than a neutral fragment. This is obvious from the literature describing radical production from electron-impact on fluorocarbons; there are many more reports of measurements of cross sections for dissociative ionization (e$^-$ + AB → A$^+$ + ·B + 2 e$^-$) than for neutral dissociation (e$^-$ + AB → ·A + ·B + e$^-$) since a measurement of a neutral dissociation cross section requires the detection of one or the other of the neutral products.[1]

Mass spectrometric and optical techniques are the obvious choices for the detection of radicals, but it must be borne in mind that cross section measurements should be carried out under single-collision conditions requiring essentially single-radical sensitivity. Optical methods can be very sensitive in the detection of atoms, however these methods are usually not sufficiently sensitive for the detection of molecular radicals where the absorption and emission probability is spread over many transitions. In addition, optical techniques, such as laser-induced fluorescence, suffer a lack of generality--one must know in detail the spectrum of each species to be detected. On the other hand, a mass spectrometer can be employed to detect almost any volatile species at concentrations below 10^2 cm^{-3} owing to the fact that the typical electron-impact-ionization source ionizes all species with approximately the same high efficiency; however, this lack of specificity may be either a blessing or a curse. To measure a cross section for radical production, a beam of electrons is passed through a target gas at such low densities that only a small fraction of the target is dissociated. The problem that arises with mass spectrometric detection is that ionization of both the radicals and the target gas give the same ionic species. For example, the 69 amu CF_3^+ peak is the most prominent feature in the mass spectrum of both CF_4 and CF_3. Some degree of specificity has been achieved by adjusting the ionizer electron energy to the threshold for the species of interest.[2,3]

Our approach to the specific and nearly-universal detection of radicals involves the method by which radicals in the gas phase were first identified. In these experiments a stream of dissociated gas was exposed to a mirror of lead or tellurium.[4-6] The disappearance of the mirror along with the subsequent isolation of various volatile lead and tellurium compounds in the effluent confirmed the existence of radicals as a distinct chemical entity. In our case we have the advantage of very sensitive mass spectrometric analysis to detect the volatile products of radical-metal reactions.

THE EXPERIMENT

In our apparatus a beam of electrons passes through a target gas in a collision cell that has a tellurium mirror on its inner surface. With a mass spectrometer we measure a signal that is proportional to the partial pressure of a diradical telluride (e.g., $Te(CF_3)_2$) from the reaction of the tellurium with radicals produced by electron impact on the target gas. The telluride partial pressure is proportional to the radical production rate that is in turn proportional to a cross section. The virtue of this technique is that the tellurium surface does not react with the target gas. In addition, the portion of the mass spectrum under observation is displaced by more than 128 amu (the nominal tellurium mass) from the region displaying peaks characteristic of the parent gas.

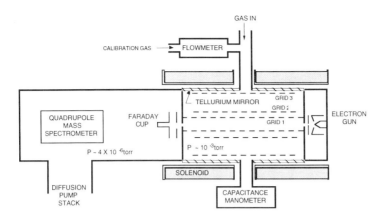

Figure 1. Schematic diagram of the apparatus.

The apparatus is shown schematically in Figure 1. It is divided into two compartments separated by an aperture: the first compartment is the dissociation region that contains the target gas, the electron gun, and the tellurium surface; the second is the analysis region containing the quadrupole mass spectrometer. The target gas is admitted to the dissociation region and evacuated from the analysis region so there is a significant pressure drop across the aperture separating the two compartments. The mass spectrometer requires pressures below 10^{-5} torr, while for these experiments the target-gas pressure in the dissociation region was of the order of 10^{-3} torr. In operation, an electron beam is directed along the axis of the dissociation region, through the aperture, and into a Faraday cup. A coaxial magnet helps collimate the electron beam. Three coaxial grids surround the region of interaction between the beam and the target gas: the innermost establishes a field free interaction region, the second is biased to prevent ions from reaching the tellurium mirror, and the outermost is in contact with, and thus establishes the electrical potential of, the mirror. Radicals produced in electron-molecule collisions are converted to volatile tellurides at the tellurium surface.

The production of a radical from electron impact on an alkane or fluoroalkane proceeds *via* neutral dissociation and dissociative ionization, so the mass spectrometer signal in our experiment is related to the sum of the corresponding partial cross sections. For the production of a radical R by electron impact on the parent XR we record the mass spectrometer signal rate S_{TeR_2} at the mass of the primary ion TeR_2^+. This signal is related to the cross section for radical production, σ_R (cm^2), (or the sum of the corresponding neutral dissociation and dissociative ionization partial cross sections, $\sigma_{nd}(R)+\sigma_{di}(R)$, by

$$\sigma_R = \sigma_{nd}(R) + \sigma_{di}(R) = \left(\frac{10^3 eRT}{N_A}\right)\left(\frac{S_{TeR_2}}{I\ell P_{XR}}\right)\left(\frac{1}{s_{TeR_2}\varepsilon}\right)$$

where s_{TeR_2} (Hz torr^{-1}) represents the mass spectrometer sensitivity, ε is the radical-to-telluride conversion efficiency at the tellurium mirror surface, I (A) is the incident electron beam current, ℓ (cm) the path length of the electron beam through the parent gas, P_{XR} (torr) the pressure of target gas in the dissociation region, the elementary charge $e = 1.60 \times 10^{-19}$ C, the gas constant $R = 62.4$ torr L K^{-1} mol^{-1}, Avogadro's constant $N = 6.02 \times 10^{23}$ mol^{-1}, and 10^3 converts L to cm^3. The quantities S_{TeR_2}, I, ℓ, and P_{XR} are measured directly with sufficient accuracy and precision that the uncertainty in the quantity $\left(S_{TeR_2}/I\ell P_{XR}\right)$ is less than 10%. The determination of the product $s_{TeR_2}\varepsilon$ is much less certain, however, since the product does not vary with incident electron energy, the precision of our cross sections as a function of electron energy is determined by the precision of the directly measured quantities. The mass spectrometer sensitivity can be measured if the appropriate calibration gas is available. The conversion efficiency has been found to be between 0.05 and 0.1 (bear in mind that 100% conversion gives $\varepsilon = 0.5$, since two radicals are required for each telluride molecule). For the moment, a comparison with related measurements is required to place our cross sections on an absolute scale.

RESULTS

Methane

We measure the partial pressure of $Te(CH_3)_2$ to obtain a relative cross section for the production CH_3 from CH_4 by electron impact at energies between 10 and 500 eV. The total dissociation cross section[7] as well as all the partial ionization cross sections[8] for electron impact on CH_4 have been reported. As shown in Figure 2, we place our results on an absolute scale by normalization of our cross section to the total dissociation cross section minus the cross sections for all the dissociative ionization channels that do not yield CH_3. Also shown on the figure is the partial dissociative ionization cross section for the production of H^+. This is equivalent to the cross section for the production of CH_3 by dissociative ionization and can be seen to be much smaller than the overall cross section for producing CH_3. Apparently neutral dissociation is the major source of CH_3. This observation reinforces Winters' conclusion that, for electron-impact dissociation of methane, neutral dissociation accounts for about half the total dissociation.[7]

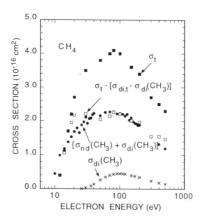

Figure 2. Cross section for production of CH_3 from CH_4 by neutral dissociation and dissociative ionization ($[\sigma_{nd}(CH_3) + \sigma_{di}(CH_3)]$) normalized to the total dissociation cross section (σ_t) minus that portion of the dissociative ionization cross section that does **not** yield CH_3 ($[\sigma_{di,t} - \sigma_{di}(CH_3)]$, reference 8).

Tetrafluoromethane

We measure the relative cross section for the electron-impact production of CF_3 from CF_4 by monitoring the intensity of the $Te(CF_3)_2^+$ peak in the mass spectrum and the cross section for F from CF_4 by monitoring the TeF_2^+ peak. The CF_3 data are shown in Figure 3 along with total neutral-dissociation cross sections in the 22 to 34 eV electron-energy range recently reported by Mi and Bonham[9] as the difference between their measurements of total inelastic electron scattering and total ion production. Also shown are cross sections for the production of CF_3 by dissociative ionization (assumed equivalent to the F^+ partial ionization cross sections of Ma, Bruce, and Bonham[10] and Poll, Winkler, Margreiter, Grill, and Mark[11]). It is apparent that the total neutral dissociation measurements of Mi and Bonham fall in the energy range below that for dissociative ionization yielding CF_3, and almost certainly below the thresholds for all neutral dissociation processes yielding radicals other than CF_3; we therefore place our measurements on an absolute scale by normalization to the total neutral dissociation cross section in the 22-34 eV range.

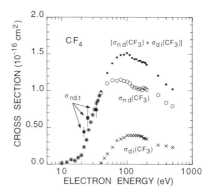

Figure 3. Cross section for production of CF_3 from CF_4 by neutral dissociation and dissociative ionization ($[\sigma_{nd}(CF_3) + \sigma_{di}(CF_3)]$) normalized to the total neutral-dissociation cross section ($\sigma_{nd,t}$, reference 9). The cross section for production of CF_3 by neutral dissociation ($\sigma_{nd}(CF_3)$) is gotten by subtracting the cross section for production of CF_3 by dissociative ionization ($\sigma_{di}(CF_3)$, references 10 and 11) from the total cross section for the production of CF_3.

CONCLUSION

Out method for radical detection is being employed in the observation of molecular radicals as well as F atoms from electron-impact dissociation of a variety of fluoroalkanes used, or proposed for use, in plasma processing in semiconductor-device fabrication. It is apparent that neutral dissociation is responsible for the bulk of the molecular radical production and a significant proportion of atomic radical production. For most fluoroalkanes there are reliable measurements of partial ionization cross sections and hence cross sections for radical production *via* dissociative ionization. Our highest priority is finding a means of placing our measurements of radical production directly on an absolute scale in order to complete the literature of electron-impact dissociation measurements.

ACKNOWLEDGMENTS

This research was supported by NSF grant CHE-95-03348. We are grateful to L. G. Christophorou, J. K. Olthoff, and M. V. V. S. Rao of NIST-Gaithersburg for many helpful discussions.

REFERENCES

1. see for example: L. G. Christophorou, J. K. Olthoff, and M. V. V. S. Rao, *J. Phys. Chem. Ref. Data*, **25**, 1341 (1996); L. G. Christophorou, J. K. Olthoff, and M. V. V. S. Rao, *J. Phys. Chem. Ref. Data*, **26**, 1 (1997); L. G. Christophorou and J. K. Olthoff, *J. Phys. Chem. Ref. Data*, **27**, 1 (1998).
2. H. Sugai and H. Toyoda, *J. Vac. Sci. Technol. A*, **10**, 1193 (1992).
3. A, Gallagher, *J. Appl. Phys.*, **63**, 2406 (1988).
4. E. Paneth and W. Hofeditz, *Ber. dt. Chem. Ges. B*, **62**, 1335 (1929).
5. F. O. Rice and A. L. Glasebrook, *J. Amer. Chem. Soc.*, **56**, 2472 (1934); F. O. Rice and M. D. Dooley, *J. Am. Chem. Soc.*, **56**, 2747 (1934).
6. L. Belchetz and E. K. Rideal, *J. Am. Chem. Soc.*, **57**, 1168 (1935).
7. H. F. Winters, *J. Chem. Phys.*, **63**, 3462-6 (1975).
8. H. C. Straub, D. Lin, B. G. Lindsay, K. A. Smith, and R. F. Stebbings, *J. Chem. Phys.*, **106**, 4430-5 (1997).
9. Mi, L. and Bonham, R.A., *J. Chem. Phys.*, **108**, 1910 (1998).
10. C. Ma, M. R. Bruce, and R. A. Bonham, *Phys. Rev. A*, **44**, 2921 (1991); Erratum, **45**, 6932 (1992).
11. H. U. Poll, C. Winkler, D. Margreiter, V. Grill, and T. D. Mark, *Int. J. Mass Spectrom. Ion Processes*, **112**, 1 (1992).

DISCUSSION

L. G. CHRISTOPHOROU: How well do the Binary-Encounter-Bethe (BEB) theory calculations of Y.-K. Kim agree with your measurements?

J. H. MOORE: Kim's soon-to-be-published calculations show better than factor-of-two agreement.

K. BECKER: A comment was made regarding the level of agreement between the BEB calculations of Kim and experimental ionization cross section data. The agreement is good for many molecules, but it is less satisfactory for some radicals, for instance, CF_3 and NF_2.

ELECTRON COLLISION CROSS SECTIONS AND TRANSPORT PARAMETERS IN CHF_3

J. D. Clark[1], B.W. Wright[1], J. D. Wrbanek[1], and A. Garscadden[2]

[1]Dept. of Physics, Wright State University, Dayton OH 45435
[2]Air Force Research Lab, WPAFB

INTRODUCTION

The modeling of gas discharges requires the accurate experimental or theoretical determinations of a large amount of atomic and molecular collision data. Among these are the interactions of electrons with CHF_3 due to its applications to replace CF_4 in low pressure plasma processing employed in the semi-conductor industry for etching and deposition. However only a modest amount has been published on low energy electron scattering cross sections of CHF_3 and nothing on its electron transport parameters. This study looks at electron drift velocities in the low E/N regime to determine vibrational excitation and total scattering cross sections in CHF_3. The total scattering cross section was estimated by **Christ**ophorou[1] using the Born approximation and as expected, it is quite large due to the polar character of CHF_3. Recently, Sanabia etal[2] have measured the total cross section for 0-20eV electrons. It is approximately constant above 2 eV at 15 x 10^{-16} cm^2 and then below 2eV increases rapidly as the electron energy tends to zero—characteristic of s-wave scattering. Other electron collision cross sections have not been reported in the literature for low electron energies, however the similarity of CHF_3 to CH_4 and CF_4 would suggest that it has large low-energy vibrational excitation cross sections.

EXPERIMENTAL APPARATUS

The experiment is a classical pulsed-Townsend type described in detail in reference 3 with only slight modification. The drift tube consisted of parallel plates enclosed in a glass vacuum housing. A fixed drift distance of 2.45 cm was chosen with 6 cm diameter electrodes used to create the uniform electric field. A photoelectron swarm is initiated by illuminating the cathode with a 1 mJ, 6 ns quadrupled NdYag laser pulse at 266 nm. This swarm drifts, under the influence of a uniform applied electric field, towards the anode where the electrons are collected. A charge-coupled current-integrating amplifier converts the electron swarm (displacement-) current into an integrated current versus time waveform, which is analyzed to determine the arrival time of the centroid of the swarm. The drift velocity is then determined as the ratio of the drift distance to the arrival time.

Drift velocity measurements (Figure 1) were made as a function of E/N in gas mixtures of 0.1% to 2% CHF_3 in Argon. The mixtures were made by volume mixing in the drift tube using pure Ar (Airco Research Grade) and CHF_3 (98% Aldrich Chemical Company) or premix of 5% CHF_3 in Ar (Matheson Semiconductor grade). The mixing ratios were determined by pressure measurements using 10 Torr and 1000 Torr capacitive manometers. The drift velocities were reproducible between gas samples and fills to within 2%. This accuracy was within the experimental uncertainties. The largest uncertainties occur when the drift velocities change rapidly with E/N or with fractional concentration of CHF_3. These uncertainties can rise as high as 10% for particular E/N points at the lowest concentrations but the average uncertainties are less than 5%.

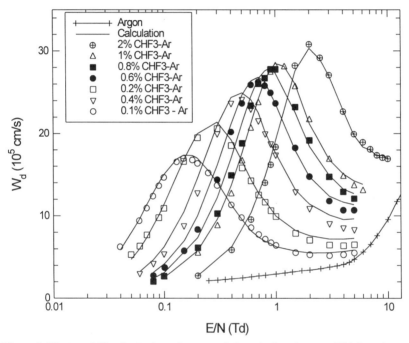

Figure 1. Electron drift velocity (experiments and theory) plotted versus E/N for mixtures of CHF_3 in Argon. The drift velocity in pure Ar is also shown for comparison.

In Figure 1, we can see that changes in the drift velocity both in magnitude and shape are affected by small changes in mixture concentration. In these mixtures a maximum in the drift velocity appears and is known to occur when the ratio of the inelastic scattering frequency to the total momentum transfer frequency changes rapidly with electron energy. This phenomenon is called "negative differential conductivity" (NDC) and is generally observed in mixtures with molecular gases with large rotational or vibrational excitation channels. Mixtures which show NDC exhibit a behavior where the peak in the drift velocity correlates with the same average electron energy and in Ramsauer gas-molecular gas mixtures, is close to the Ramsauer minimum. This result can be verified from Figure

2, where the drift velocity is plotted as a function of the average electron energy. At low molecular gas concentrations, the acute sensitivity of the drift velocity to small changes in the inelastic channels created by the addition of the molecular gas shows the usefulness of this swarm technique to determine these low energy cross sections.

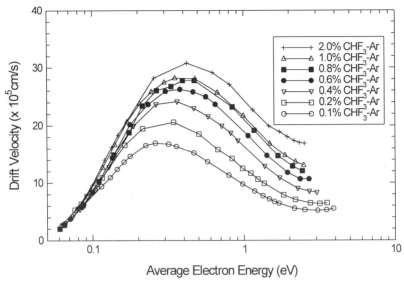

Figure 2. Drift velocity plotted versus average electron energy.

ANALYSIS OF CROSS SECTIONS

The motion of an electron swarm in a uniform electric field can be related to the scattering cross sections of the gas by the Boltzmann Transport Equation (BTE). The electron drift velocity (w_d) was obtained by a conventional "two-term" numerical solution of the BTE. The analysis proceeds by constructing a trial cross section set for the CHF_3, calculating the drift velocity and comparing to the experimental data. The comparisons and modifications of the cross sections were repeated until a satisfactory agreement between the calculated and experimental data was achieved. The collision cross sections of Ar were taken from Ref. 4, and were fixed throughout this study.

The trial cross section set was chosen based on comparisons to similar molecular species such as CH_4 and CF_4. These gases have large low energy vibrational excitation channels and Ramsauer-Townsend minima in their momentum transfer cross sections. The similarity of CHF_3 to these gases suggested that it would also have large vibrational excitation cross sections. The vibrational thresholds and relative excitation strengths were taken to be similar to the oscillator strengths from IR absorption spectra[5] shown in Table 1.

For simplicity only two vibrational excitation modes were assumed to be important, ν5, which corresponds to a C-F vibrational mode with by far, the largest oscillator strength and ν1, which corresponds to a C-H vibrational mode and to the highest threshold energy. The initial scale of the excitation cross sections was taken from the relative absorption strength. The shape of the cross sections was taken as similar to those published for CH_4 and CF_4. However, the momentum transfer cross section should differ from these gases since CHF_3 is highly polar. The momentum transfer cross section has been estimated by Christrophorou[1] using a Born approximation and was used in the initial trial cross section set. All three cross sections were modified until the calculated drift velocities converged with the experimental measurements for all of the mixture ratios. These final cross sections are shown in Figure 3.

Table 1

Modes	Frequency cm^{-1}	Threshold Energy (eV)	Absorption (arb. Units)
ν1	3051	0.378	12
ν2	1140	0.141	n/a
ν3	700	0.087	6.7
ν4	1372	0.170	44.2
ν5	1152	0.144	327
ν6	507	0.063	2.4

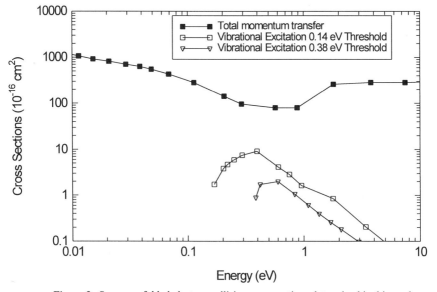

Figure 3. Swarm unfolded electron collision cross sections determined in this work.

The calculated drift velocities using the converged cross section set are shown in Figure 1 as the solid lines through the experimental data points. Sufficient agreement was obtained and no additional excitation cross sections were added to the cross section set. Comparisons of calculated to experimental drift velocities are shown in Figure 4 as a fractional difference versus the average energy of the electron swarm. The agreement of the calculated drift velocities lies within the experimental uncertainties of 5% for most of the data sets with divergence at the extremes where the sensitivities of the experiment are reduced. The agreement is good from the threshold region of the vibrational excitation toward higher energy.

The deduced vibrational excitation cross sections were consistent in size and shape to similar molecules. The total momentum cross section was exceptionally large and exceeded recent measurements[2] by at least an order of magnitude. The origin of this difference is not known; changes in the inelastic cross sections with a lower momentum cross section did not give agreement with experimental data.

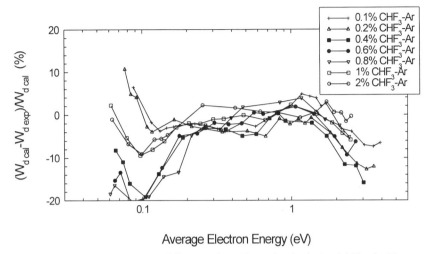

Figure 4. Fractional difference of experimental and calculated drift velocities

ELECTRON ATTACHMENT

In this study of electron drift in CHF_3- Ar mixtures the formation of negative ions was observed. This was evidenced by the continued growth in the collected charge from the drift tube after the electron swarm had arrived. Arrival of charge at long times is attributed to ions from ionization or from attachment. Since the average energy of the electron swarm was less than 2 eV for most of the mixtures and conditions where ions were observed, electron attachment to form negative ions was assumed. The attachment coefficient then is determined by measuring the ratio of the collected ion charge to that of the electron charge. The equation below gives the relationship between the ion and electron signals, the attachment coefficient, η, and the drift distance, d.

$$\frac{V_{-ion} + V_{electron}}{V_{electron}} = \frac{\eta d}{[1 - \exp(-\eta d)]}$$

Electron attachment coefficients were measured as a function of E/N for gas mixtures from 2.5% to 10%. The electron collision cross section set determined above was used to convert the electron attachment coefficient from its dependence on E/N to a dependence on average electron energy. All measurements collapsed to the single curve shown in Figure 5.

The peak in the low energy electron attachment would generally indicate a stabilized attachment to the parent ion CHF_3. This kind of attachment-stabilization often has a pressure dependence which enhances the stabilization. However, we observed no pressure dependence from 200 torr to 1000 torr total pressure. The exact negative ion or ions cannot be directly determined with this technique, therefore a highly attaching impurity is a possibility. However, sensitive mass analysis of the sample gas used in this study could not identify a likely candidate responsible for this behavior[6].

The attachment coefficients determined here would indicate a moderately attaching gas at these low electron energies. Previous studies of attachment in CHF_3 at low energy or thermal energy have measurements that are two orders of magnitude higher[7] or lower[8] than the current measurements. The differences in the low energy attachment coefficient/rate and product channels of the attachment have yet to be resolved.

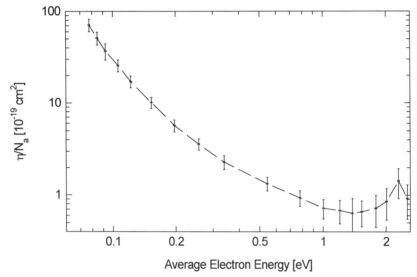

Figure 5. Density reduced attachment coefficient versus average electron energy. The error bars represent the statistical deviation between several data points at the same average energy.

CONCLUSIONS

Measurements of electron transport have been made for mixtures of technical interest employing CHF_3. A set of electron-molecule collision cross sections has been derived consistent with measured electron drift velocities to enable EEDFs to be calculated for other mixtures that use CHF_3. Three low energy collision cross sections were included in the set, total momentum transfer and two vibrational excitation cross sections. In addition, low energy electron attachment was observed with a peak in the attachment coefficient towards zero energy.

REFERENCES

1. L.G. Christophorou, J.K. Olthoff, and M.V.V.S. Roa, , J. Phys Chem. Ref. Data **26**, 1-15(1997).
2. J. E. Sanabia, G. D. Cooper, J. A. Tossell, and J. H. Moore, J. Chem. Phys. **108**, 389 (1998).
3. D. J. Mosteller, M.L. Andrews, J.D. Clark, and A. Garscadden, J. Applied Phys. 74,2247 (1993).
4. Y.Nakamura and M. Kurachi, J. Phys. D. **75**, 703 (1994)
5. T.D. Kolomiitsova, S.M. Melikova, and D.N. Shchepkin, Opt. Spectrosc.(USSR) **67**, 347 (1989).
6. C. Q. Jiao, R. Nagpal, and P.D. Haaland, Chem. Phys. Let. **269**, 117 (1997)
7. T. G. Lee, J. Phys. Chem. **67**, 360 (1963).
8. A. A. Christodoulides, R. Schumacher, and R. N. Schindler, Int. J. Chem. Kin. **X**, 1215 (1978).

DISCUSSION

G. R. G. RAJU: The peak observed in the electron drift velocity as a function of E/N is interesting. Is it due to the Ramsauer cross section of the parent gas or due to vibrational excitation?

J. D. CLARK: It is due to both the presence of inelastic vibrational excitation (modifying the electron energy distribution function) and the Ramsauer cross section of argon in this case.

M.-C. BORDAGE: You determined your cross section set by comparing with your drift velocity measurements. Are there other measurements available in the literature such as D_T/μ, for example?

J. D. CLARK: There are no other electron transport data to our knowledge. However, reported at this meeting is the work of Wang et al. on electron drift velocities in CHF_3-Ar mixtures also.

YICHENG WANG: We did similar experiments and we had serious problems with the "supposedly" ultra-high purity commercial gases, particularly Ar. We found that Ar had to be further purified. The results on the electron drift velocity in pure Ar before purification can differ from those after purification by a factor of 4 to 5 for low E/N near 0.1 Td. I wonder whether you had similar problems in your experiments.

J. D. CLARK: While we have seen effects due to impurities in Kr and Xe (pure), we do not observe these effects in pure Ar after pumping and baking the experimental apparatus. A base line measurement of pure Ar was made before each sample run to ensure that the system was clean.

ION-MOLECULE REACTIONS AND ION KINETICS IN DC TOWNSEND DISCHARGES IN DIELECTRIC GASES

M. V. V. S. Rao[1] and J. K. Olthoff

Electricity Division
Electronics and Electrical Engineering Laboratory
National Institute of Standards and Technology
Gaithersburg, Maryland, USA 20899

ABSTRACT

The transport of positive and negative ions plays an important role in the initiation and behavior of discharges in gaseous dielectrics. In many cases, the identities, intensities, and kinetic energies of ions are determined by ion-molecule collisions experienced by ions while they are accelerated through the discharge. To help delineate the role of various collisional processes, such as charge-exchange collisions, collisional detachment, and ion conversion, we have measured the kinetic-energy distributions (KEDs) of positive and negative ions formed in dc Townsend discharges generated in O_2 and SF_6 at high electric field-to-gas density ratios (E/N). The relative abundances and mean energies of the ions are obtained from these measurements, and the effective cross sections for the ion-molecule reactions, in some cases, may be calculated. The availability of ion-molecule collision cross sections for the negative ions formed in O_2 and SF_6 enable a reasonable understanding of the KED data. However, attempts to analyze the positive ion KED data for these two gases highlight the lack of fundamental ion-molecule cross section data for positive ions formed in these discharges.

INTRODUCTION

Understanding the details of ion production and transport in electrical discharges is of importance to both the semiconductor industry, which uses gas-phase discharges for microelectronic device production, and the electric equipment industry, which uses electronegative gases as high voltage insulation. Oxygen is a gas of interest in both of these areas because of its common usage in plasma discharges for etching and cleaning processes,

---

[1]Current Address: MS229-1, NASA/Ames Research Center, Moffett Field, CA 94035-1000

and because of its nearly universal presence in high voltage insulation systems as an impurity. Sulfur hexafluoride (SF_6) is of interest due to its overwhelming use as a dielectric gas in high voltage equipment, and its use by the semiconductor industry in a wide variety of etching processes.

In this paper, we utilize a dc Townsend discharge to measure the kinetic energy distributions (KEDs) of positive and negative ions striking the grounded electrode for values of E/N ranging from 2×10^{-18} Vm^2 to 40×10^{-18} Vm^2 (2 kTd to 40 kTd). These values of E/N are comparable to those observed in the sheath regions of glow discharges.

Analysis of the identities, energies, and intensities of the ions generated in a discharge enables a qualitative understanding of the ion-production processes, and subsequent ion-molecule collision processes that affect the flux of ions through the discharge. Additionally, the shapes of the energy distributions allow a determination of the range of E/N for which equilibrium conditions exist for the ions. Information determined from this simple dc parallel-plate discharge system is useful in modeling complex discharges, and in guiding the investigation of the significant ion-molecule reactions occurring in other types of electrical discharges.

EXPERIMENT

The apparatus used for these experiments is essentially identical to that used to investigate Townsend discharges in the rare gases.[1] The only significant change is the ability to reverse the polarity of the electrodes, and of the appropriate voltages in the energy analyzer-mass spectrometer, to allow detection of negative ions. Briefly, the discharge cell consists of two parallel, stainless steel plates that are 11 cm in diameter and spaced 2 cm apart. A 100 µm hole in the center of the lower electrode (the cathode when detecting positive ions and the anode when detecting negative ions) allows the sampling of ions by a quadrupole mass spectrometer with an electrostatic ion-energy analyzer. The kinetic energy distribution of the flux of ions striking the grounded electrode is measured by setting the quadrupole mass spectrometer to a specific mass, and then scanning the transmission energy of the ion-energy analyzer. The energy resolution of the energy analyzer is approximately 4 eV (full width at half maximum), and analysis of the transmission characteristics of the ion-energy analyzer indicates that ion energies below 10 eV are not reliable. Above 10 eV, both the ion-energy analyzer and the quadrupole mass spectrometer were tuned such that the ion transmission is independent of the mass and energy of the ion.

OXYGEN RESULTS AND DISCUSSION

Kinetic energy distributions for ions sampled from a Townsend discharge in oxygen were measured for values of E/N ranging from 2×10^{-18} Vm^2 to 40×10^{-18} Vm^2. Figure 1 shows representative measured KEDs for a Townsend discharge in oxygen at $E/N \simeq 15 \times 10^{-18}$ Vm^2 (15 kTd). As can be seen from the figure, O_2^+ is the dominant ion, followed by the only other ions detected, O^+, O_2^-, and O^-. This order of intensities was observed for all values of E/N investigated here. For values of E/N less than 15×10^{-18} Vm^2 the measured KEDs exhibit a Maxwellian shape that is reflected by the linear slope of the distribution when plotted on a semi-logarithmic graph (see line in Fig. 1a). This behavior is indicative of ions (*i*) whose motion takes place in a uniform electric field, (*ii*) whose primary collision process is symmetric charge exchange, and (*iii*) for whom equilibrium conditions apply (i.e., the ions have experienced many collisions).[2] The KEDs for O_2^+ exhibit increasing deviations from this Maxwellian behavior as the E/N increases above 15×10^{-18} Vm^2. This is just barely evident

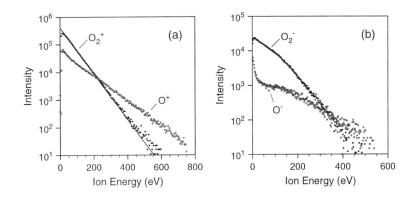

Fig. 1. Kinetic energy distributions of ions sampled from an oxygen Townsend discharge with $E/N \simeq 15 \times 10^{-18}$ Vm^2 (15 kTd).

in the O_2^+ KED shown in Fig. 1a, by the slight deviation at higher energies of the measured data from the linear fit. These deviations indicate the advent of additional energy-loss collision processes (other than symmetric charge transfer), or of the development of non-equilibrium conditions for the O_2^+ ions within the discharge.

The KEDs for O^+ ions exhibit higher mean energies than the KEDs for O_2^+, often with the maximum ion energy corresponding to the voltage across the discharge gap. This condition is illustrated in Fig. 1a where the maximum ion energy for O^+ equals 750 eV, corresponding to the discharge voltage of approximately 750 V, and is an indication of non-equilibrium conditions for the O^+ ions. The higher mean energies for O^+ ions are due to the fact that the cross section for asymmetric charge exchange is significantly less than for the symmetric process,[3] which does not exist for O^+ under these discharge conditions. These conditions are reflected in the observed non-Maxwellian shape of the KEDs for O^+ at all E/N.

The KEDs for O_2^- are determined primarily by symmetric charge-exchange collisions,[4] and by collisional detachment collisions[5] that destroy the ion. Below 5×10^{-18} Vm^2, the KEDs for O_2^- are Maxwellian in shape, indicating that symmetric charge exchange is the dominant collisional reaction experienced by the ion. At higher values of E/N, the effect of collisional detachment interactions increases due to the increasing cross section for this process with increasing ion energy.[5] The effect of collisional detachment is reflected in the KED for O_2^- shown in Fig. 1b by the drop-off in ion intensity at higher energies. The KEDs for O^- are highly non-Maxwellian at all E/N, indicating a complex dependence on several competing, energy-dependent collisional processes, including asymmetric charge exchange[6] and collisional detachment.[5]

The mean energies of the ions at each E/N may be calculated from the measured KEDs, and are observed to increase with increasing E/N for all ions. For O_2^+ and O_2^-, whose transport is primarily determined by symmetric charge-exchange interactions at low E/N, an effective charge-exchange cross section may be determined from the mean energies.[2] This effective cross section agrees well with beam-determined measurements of the charge-exchange cross sections for values of E/N below 15×10^{-18} Vm^2 for O_2^+ and 5×10^{-18} Vm^2 for O_2^-.[7]

SULFUR HEXAFLUORIDE RESULTS AND DISCUSSION

The investigation of ion transport in discharges in SF$_6$ is complicated by the increased number of ions generated in the discharge. Fig. 2 shows a comparison of a standard 70-eV electron-impact mass spectrum[8] for SF$_6$, and a mass spectrum of ions sampled from a

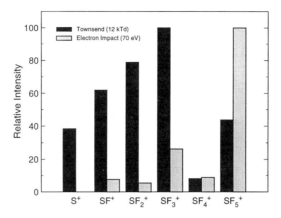

Fig. 2. Mass spectra of positive ions sampled from a Townsend discharge ($E/N \simeq 12 \times 10^{-18}$ Vm^2) in SF$_6$, compared with a standard 70-eV electron-impact mass spectrum for SF$_6$.[7]

Townsend discharge in SF$_6$ with $E/N \simeq 12 \times 10^{-18}$ Vm^2. In contrast to the electron-impact spectrum, the lower mass ions dominate the spectrum from the Townsend discharge, a clear indication of the role of ion-molecule reactions in determining the identity of the ions striking surfaces exposed to electrical discharges. The decreased SF$_5^+$ intensity and increased SF$_3^+$ intensity observed in the Townsend discharge, relative to the electron-impact mass spectrum, may be attributed to the previously investigated[8] ion-conversion process

$$SF_5^+ + SF_6 \rightarrow SF_3^+ + SF_6 + F_2. \qquad (1)$$

Other ion-conversion processes are responsible for the production of the SF$_2^+$, SF$^+$ and S$^+$ ions, but little is known about these processes.

Figure 3 shows representative KEDs for the positive ions detected from the SF$_6$ Townsend discharges at 12×10^{-18} Vm^2. The KEDs exhibit non-Maxwellian shapes for all the ions at all values of E/N investigated here (8 to 20 kTd), reflecting the impact of the complex ion-conversion processes taking place in the discharge. Interestingly, the mean energies of the ions are observed to increase with decreasing mass.

Negative ions in SF$_6$ are initially formed by electron attachment processes[10] that result primarily in the formation of SF$_6^-$ and SF$_5^-$. However, the mass spectrum of the negative ions detected from an SF$_6$ Townsend discharge ($E/N \simeq 12 \times 10^{-18}$ Vm^2), shown in Fig. 4, indicates that F$^-$ is the dominant ion striking the electrode. Since the yield of F$^-$ due to electron attachment processes is significantly lower than for SF$_6^-$ production,[10] this large F$^-$ signal is attributed to the ion-molecule reactions,

$$SF_6^- + SF_6 \rightarrow F^- + SF_5 + SF_6 \qquad (2)$$
$$SF_5^- + SF_6 \rightarrow F^- + SF_4 + SF_6, \qquad (3)$$

that have been shown to possess large cross sections.[11]

Figure 5 shows KEDs for the four negative ions detected in the SF$_6$ discharge. SF$_6^-$ and SF$_5^-$ exhibit significantly lower mean kinetic energies than the positive ions, due to reactions (2) and (3), large charge-exchange cross sections, and the manifestation of collisional detachment processes at higher energies.[11] In fact the maximum ion energies detected here for

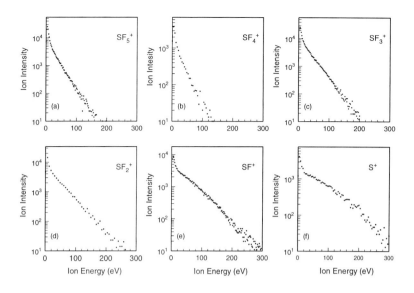

Fig. 3. Kinetic energy distributions for positive ions sampled from an SF_6 Townsend discharge with $E/N \cong 12 \times 10^{-18}$ Vm².

SF_6^- and SF_5^- ions nearly correspond with the 90-eV threshold for collisional detachment from SF_6^- and SF_5^-. F_2^- is also thought to be formed predominantly by collisional ion conversion of SF_6^- and SF_5^- ions, however, the ion-molecule collisions between F_2^- and SF_6 are unknown.

CONCLUSION

We have measured the mass and energies of the ions striking the grounded electrode in a dc Townsend discharge in pure oxygen and pure sulfur hexafluoride over a wide range of high E/N. For oxygen, O_2^+ is the dominant positive ion, while O_2^- is the dominant negative ion. Analysis of the ion kinetic energy distributions indicate that symmetric charge transfer is the dominant ion-molecule reaction for both ions, although the effects of collisional detachment

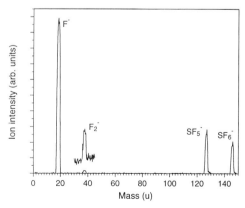

Fig. 4. Mass spectrum of negative ions sampled from an SF_6 Townsend discharge with $E/N = 12 \times 10^{-18}$ Vm².

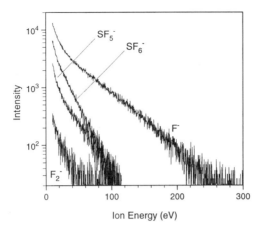

Fig. 5. Kinetic energy distributions of negative ions sampled from an SF_6 Townsend discharge with $E/N \simeq 12 \times 10^{-18}$ Vm^2. The ion signal below 10 eV is not shown (see Experiment section).

are evident in the measured KEDs for O_2^- for $E/N > 5 \times 10^{-18}$ Vm^2. Smaller, but significant, quantities of O^+ and O^- were detected with complex KEDs that are the results of multiple ion-collision processes.

Six positive ions are detected from Townsend discharges generated in SF_6. SF_3^+ is the dominant ion at all E/N, and is thought to be formed predominantly by an ion-collision process that converts SF_5^+ to SF_3^+. The processes responsible for the production of S^+, SF^+, and SF_2^+ are not fully understood. The dominant negative ion is F^- which is produced by ion-conversion processes involving collisions of SF_6^- and SF_5^- with SF_6. The measured KEDs for all of the ions detected in the SF_6 discharge exhibited non-Maxwellian shapes, again resulting from the effects of multiple ion-collision processes.

REFERENCES

1. M. V. V. S. Rao, R. J. Van Brunt, and J. K. Olthoff, Phys. Rev. E 54:5641 (1996).
2. S. B. Radovanov, R. J. Van Brunt, J. K. Olthoff, and B. M. Jelenkovic, Phys. Rev. E 51:6036 (1995).
3. R. F. Stebbings, B. R. Turner, and A. C. H. Smith, J. Chem. Phys. 38:227 (1963).
4. A. E. Roche and C. C. Goodyear, J. Phys. D: Appl. Phys. 4:1513 (1971).
5. R. Ranjan and C. C. Goodyear, J. Phys. B: Atom. Molec. Phys. 6:1070 (1973).
6. S. L. Lin, J. N. Bardsley, I. Dotan, F. C. Fehsenfeld, and D. L. Albritton, Int. J. Mass Spectrom. Ion Phys. 34:113 (1980).
7. M. V. V. S. Rao, R. J. Van Brunt, and J. K. Olthoff, in preparation.
8. NIST Standard Reference Database 69 - March 1998 Release: NIST Chemistry WebBook
9. Z. A. Talib and M. Saporoschenko, Int. J. Mass Spectrom. Ion Phys. Proc. 116:1 (1992).
10. L. G. Christophorou and R. J. Van Brunt, IEEE Trans. Dielectrics and Electr. Insul. 2:952 (1995).
11. Y. Wang, R. L. Champion, L. D. Doverspike, J. K. Olthoff, and R. J. Van Brunt, J. Chem. Phys. 91:2254 (1989).

DISCUSSION

K. BECKER: Is it possible to attribute the high F^- yield in the SF_6 Townsend discharge to dissociative attachment to SF_x free radicals produced by neutral dissociation of SF_6?

M. V. V. S. RAO: It is unlikely. It should be noted that F^- production by dissociative electron attachment to SF_6 is orders of magnitude lower than SF_6^- formation. Even if F^- production from SF_x radicals exists, the rate should be as small as for F^- from SF_6. Therefore, the enhancement of F^- intensity in our experiment is mainly from $SF_5^- + SF_6 \rightarrow F^- +$ products and $SF_6^- + SF_6 \rightarrow F^- +$ products, and it is consistent with earlier beam-beam experiments by R. L. Champion and collaborators [J. Chem. Phys. **91**, 2254 (1989)].

W. BOECK: The ion-molecule reactions are important for the extremely low pressure used in your experiments. They have low influence on the process of ionization in the case of usual pressure of 0.1 MPa since the ion energy is much smaller. What is your opinion in this respect?

M. V. V. S. RAO: Obviously the E/N will be lower in discharges at such a high pressure, so the energy distribution of the ions will be completely different than those shown here. However, the ion-molecule reactions discussed here are important for discharges at nearly all pressures.

YICHENG WANG: I have a question about negative ions in O_2 discharges. As you said, negative ions in O_2 discharges are mainly formed via dissociative electron attachment producing O^-. Indeed, it has been observed that O^- ions are dominant in RF O_2 discharges. Yet, your results (Fig.1) show that O_2^- is dominant. Could you say a few words about possible reasons?

M. V. V. S. RAO: The primary process of O^- production in O_2 discharge is by dissociative electron attachment. The production of O_2^- is by the asymmetric charge transfer reaction $O^- + O_2 \rightarrow O + O_2^-$ which has a large cross section compared to the detachment cross section for O^-. There is also a possibility of O_2^- formed at the electrode surface by ion and/or neutral bombardment during the discharge.

A. GARSCADDEN: Could you please define the discharge instability that you alluded to? Is it temporal or spatial, current runaway or oscillation? The point is that there is information in the oscillation limits on the saturation of attachment and ionization.

M. V. V. S. RAO: It is oscillation. We did not investigate it as it was not the point of the present experiment.

ELECTRON DRIFT VELOCITIES AND ELECTRON ATTACHMENT COEFFICIENTS IN PURE CHF_3 AND ITS MIXTURES WITH ARGON

Yicheng Wang, L. G. Christophorou, J. K. Olthoff, and J. K. Verbrugge

Electricity Division
Electronics and Electrical Engineering Laboratory
National Institute of Standards and Technology
Gaithersburg, Maryland, USA 20899

ABSTRACT

Measurements are reported of (*i*) the electron drift velocity in pure trifluoromethane (CHF_3) gas and in its mixtures with argon, and (*ii*) electron attachment in pure CHF_3. The electric field-to-gas density ratio (E/N) dependence of the electron drift velocity in the mixtures exhibits regions of distinct negative differential conductivity. The values of the electron attachment coefficients in pure CHF_3 are small and decrease with E/N. The measurements were made at room temperature and cover the E/N range from 0.05×10^{-17} V cm^2 to 60×10^{-17} V cm^2 (0.05 Td to 60 Td, 1 Td = 10^{-17} V cm^2). The electron attachment rate constant is virtually independent of E/N below about 50×10^{-17} V cm^2 and equal to $\sim 13 \times 10^{-14}$ cm^3 s^{-1}. This small attachment rate constant may be due to impurities.

INTRODUCTION

The CHF_3 molecule is polar, and its electric dipole moment is 5.504×10^{-30} C m (1.65 debye).[1] It has been studied in the past as a polar buffer gas in dielectric gas mixtures with electronegative components.[2] It is also used as a plasma processing gas in place of CF_4 in view of its lower global warming potential and lifetime in the atmosphere.[3] Very limited electron collision cross section and electron transport coefficient data are available for this gas,[3] although recently there has been a new report on measurement of the total electron scattering cross section[4] for CHF_3 and also another report on the electron transport coefficients in mixtures of CHF_3 with argon.[5]

In an effort to fill the need for data on electron interactions with CHF_3, we measured electron drift velocities in pure CHF_3 and in mixtures of CHF_3 with Ar (percentages of CHF_3 in argon ranged from 0.1 % to 10 %) for electric field-to-gas density ratios (E/N) ranging from $\sim 0.05 \times 10^{-17}$ V cm^2 to $\sim 60 \times 10^{-17}$ V cm^2. Measurements were also made of the electron

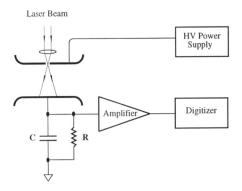

Fig. 1. Schematic diagram of the operating principle of the pulsed Townsend technique as employed in the present study.

attachment coefficient in pure CHF_3. The measurements with the gas mixtures can aid Boltzmann transport equation analyses aimed at calculating collision cross section sets for the CHF_3 molecule. They may also find use in particle radiation detectors.

EXPERIMENTAL METHOD

The experimental method employed is a pulsed Townsend technique with electron swarms photoelectrically produced using a 5 ns, frequency quadrupled (266 nm) Nd:YAG laser. A schematic diagram showing its operating principle is given in Fig. 1. The two parallel electrodes are circular stainless steel disks 6.2 cm in diameter. They are separated by a distance of 1.664 cm and are contained in a six-way cubic stainless steel chamber. The laser beam entering the chamber through a sapphire window is focused with a converging lens through a small hole (~0.6 mm diameter) at the center of the anode electrode before striking the cathode electrode. The induced current due to the motion of the electrons in the drift region is integrated by the RC (R = 100 GΩ and C ~50 pF) circuit in front of a high-impedance, unity-gain, buffer amplifier with a slew rate of 0.22 V/ns. The output voltage of the amplifier is then digitized by a digital oscilloscope with a resolution of 8 bits and a maximum sample rate of 10^8 samples/s. To minimize the influence of the AC line noise, laser pulses are synchronized with the zero-crossings of the AC line voltage. The synchronization scheme allows subtraction of the line noise and thus improves the overall performance of the pulsed Townsend method.

A typical voltage waveform is shown in Fig. 2(a). It was acquired in a 1% CHF_3 + 99% Ar mixture with a gap voltage of 200 V and a total pressure of 1.33 kPa (10 Torr). The intersection between a line fitted to the rising signal and a line fitted in the plateau region gives the electron transit time, $t = 0.522$ µs. The smooth transition is caused by the thermal diffusion of electrons. Figure 2(b) shows a waveform that was obtained at 100 V with the chamber under vacuum. It shows that the rise time of the electronic system is shorter than 20 ns. The estimated uncertainty for the measured electron transit time and the drift velocity is ±5%, due primarily to the uncertainty in determining the intersection points of the measured waveform.

All measurements were made at room temperature (about 298 K) at pressures ranging from 1.33 kPa to 66.7 kPa. Both the CHF_3 and the Ar gases were high purity (i.e., research grade). However, both gases had to be purified further by fractional distillation. This was found to be necessary as can be seen from the measurements on "pure" argon shown in Fig. 3.

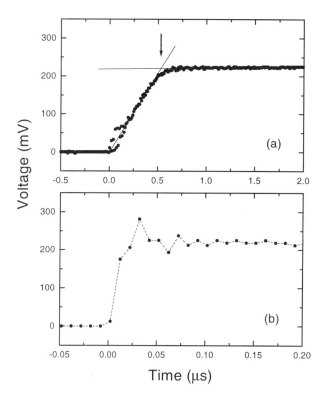

Fig. 2. (a) Typical voltage waveform ($E/N = 37 \times 10^{-17}$ V cm^2; $P_{total} = 1.33$ kPa; 1 % CHF$_3$ + 99 % Ar). (b) Waveform with the chamber under vacuum (applied voltage = 100 V).

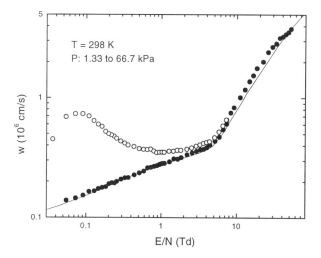

Fig. 3. Electron drift velocity, *w*, as a function of *E/N* in pure argon. ○, "ultra high purity argon" (quoted purity 99.999 %) used as received; •, the same measurements after gas purification by several freeze-pump-thaw cycles; (—), average of values as given in Ref. 6.

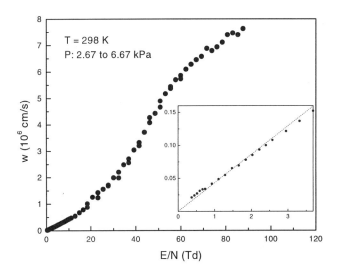

Fig. 4. Electron drift velocity, w, as a function of E/N for purified CHF_3. Inset: Comparison of the present measurements (●) with the w values reported earlier[2] for $E/N \leq 3.69 \times 10^{-17}$ V cm^2 (- - -).

The measurements on the ultra-high purity gas (quoted purity 99.999 %) used as provided by the manufacturer clearly show the effect of impurities. Upon purification by fractional distillation the measured electron drift velocities agreed with those reported in the literature.[6] The total pressures and the mixture ratios were measured using two temperature-controlled, high accuracy capacitance manometers. The estimated measurement uncertainty for the total pressure, the mixture ratio, and E/N is less than ±1%.

RESULTS AND DISCUSSION

Electron Drift Velocity, w, as a Function of E/N

In Fig. 4 are shown the present measurements of the electron drift velocity, w, in pure CHF_3 taken at pressures ranging from 2.67 kPa to 6.67 kPa. There are no other data to compare these measurements with besides an earlier value of the slope of the w versus E/N line measured[2] at low E/N ($<3.69 \times 10^{-17}$ V cm^2) where the electrons are in thermal equilibrium with the gas and w varies linearly with E/N. This is shown by the broken line in the inset of Fig. 4. It is in good agreement with the present measurements.

In Fig. 5 are shown the measured electron drift velocities in mixtures of CHF_3 with argon containing 0.1 %, 0.5 %, 1 %, 5 %, and 10 % of CHF_3. For comparison, the $w(E/N)$ for pure CHF_3 and for pure argon are also shown. The most distinct characteristic of the $w(E/N)$ data for the mixtures are the regions of pronounced negative differential conductivity and its dependence on mixture composition. The values, $(E/N)_{max}$, of E/N at which the w is maximum are plotted in Fig. 6 as a function of the percentage of the CHF_3 in the mixture. If one assumes that the drift velocity maxima are the result of electrons scattered by CHF_3 (principally through inelastic vibrational excitation of the CHF_3 molecules) into the energy region where the electron scattering cross section in argon has a minimum (~ 0.23 eV), the values of $(E/N)_{max}$ would represent the E/N value at which the average electron energy in the mixture is ~0.23 eV. Interestingly, this value increases linearly with the percentage of CHF_3 in Ar for the percentages used. If the linear dependence observed in Fig. 6 holds all the way to pure CHF_3,

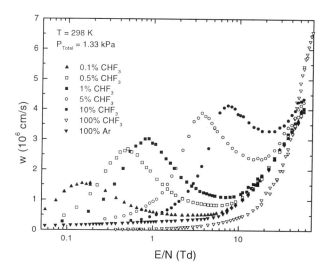

Fig. 5. Electron drift velocity, w, as a function of E/N for mixtures of CHF_3 and argon at the following compositions: 0.1 % CHF_3 +99.9 % Ar (▲), 0.5 % CHF_3 + 99.5 % Ar (□), 1 % CHF_3 + 99 % Ar (■), 5 % CHF_3 + 95 % Ar (○), and 10 % CHF_3 + 90 % Ar (●). For comparison the drift velocities in pure CHF_3 (▽) and pure argon (▼) are shown.

it would indicate that the maximum in the $w(E/N)$ for pure CHF_3 should be at an E/N value of about 75×10^{-17} V cm^2. This value is not inconsistent with the measurements in Fig. 5. However, the drift velocity maximum in pure CHF_3 may not be as distinct as those in the low-concentration mixtures because the data indicate that the minima in the w versus E/N dependence become shallower as the percentage of CHF_3 in Ar is increased. The data in Fig. 5 can be useful in Boltzmann code analyses, as studies of other gases have indicated.[7,8]

Electron Attachment in Pure CHF_3

In Fig. 7 are shown preliminary measurements of the electron attachment coefficient, η/N, in pure CHF_3 as a function of E/N for a number of CHF_3 pressures. These are the average of three runs. The $w(E/N)$ data for pure CHF_3 in Fig. 4 and the data on η/N (E/N) in pure CHF_3 in Fig. 7 have been used to determine the electron attachment rate constant

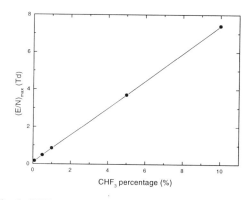

Fig. 6. $(E/N)_{max}$ versus CHF_3 volume fraction (in percent).

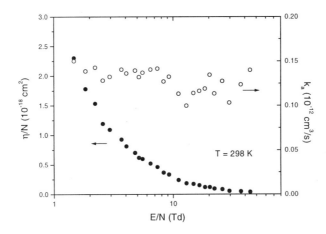

Fig. 7. Electron attachment coefficient, η/N, (●) and electron attachment rate constant, k_a, (○) as a function of E/N for pure CHF_3 measured at T = 298 K. Average values at three pressures (3.34 kPa, 6.67 kPa, and 13.3 kPa).

$$k_a(E/N) = \eta/N \, (E/N) \times w(E/N)$$

as a function of E/N. These are shown in Fig. 7 by the open circles. The values of k_a are virtually independent of E/N at about 13×10^{-14} cm^3 s^{-1}. Earlier electron swarm studies at only thermal energies showed that the thermal electron attachment rate constant for CHF_3 is less than about 6×10^{-14} cm^3 s^{-1} (see Ref. 3). Although efforts have been made to purify the gas by fractional distillation, it is not certain that the observed attachment is due to the CHF_3 molecule itself or due to possible residual traces of strongly electron attaching impurities. The CHF_3 molecule has not been reported to have a positive electron affinity. Its lowest negative ion state is at ~4.5 eV above the ground state of the neutral molecule, that is, the lowest vertical electron affinity of the CHF_3 molecule is about −4.5 eV. Earlier electron beam studies discussed in Ref. 3 showed that negative ions (mostly F$^-$) are produced by dissociative electron attachment and begin at about 2.2 eV. These are much higher energies than those expected in the present swarm study. It is highly unlikely that the observed attachment is due to dissociative electron attachment to the CHF_3 molecule. Clearly, electron attachment to CHF_3 is very weak or absent below about 50×10^{-17} V cm^2.

REFERENCES

1. A. L. McClellan, *Tables of Experimental Dipole Moments,* W. H. Freeman and Company, San Francisco, 1963, p. 38.
2. L. G. Christophorou, D. R. James, and R. A. Mathis, J. Phys. D **14**, 675, 1981.
3. L. G. Christophorou, J. K. Olthoff, and M. V. V. S. Rao, J. Phys. Chem. Ref. Data **26**, 1 (1997).
4. O. Sueoka, H. Takagi, A. Hamada, and M. Kimura, Private Communication (January, 1998).
5. J. D. Clark, B. W. Wright, J. D. Wrbanek, and A. Garscadden, Bulletin of the American Physical Society 42 (No. 8), 1718 (1997) (abstract).
6. L. G. Christophorou, *Atomic and Molecular Radiation Physics,* Wiley-Interscience, New York, 1971, p. 239.
7. M. C. Bordage, P. Ségur, and A. Chouki, J. Appl. Phys. **80**, 1325 (1996).
8. M. Hayashi and A. Niwa, in *Gaseous Dielectrics V,* edited by L. G. Christophorou and D. W. Bouldin (Pergamon, New York, 1987), p. 27.

CROSS SECTION MEASUREMENTS FOR VARIOUS REACTIONS OCCURRING IN CF_4 AND CHF_3 DISCHARGES

B. L. Peko, I. V. Dyakov, and R. L. Champion

Department of Physics
College of William and Mary
Williamsburg, VA 23187

ABSTRACT

In an attempt to expand the somewhat limited database for collisional processes relevant to plasma chemistry and its modeling, absolute cross sections have been measured for positive and negative ion-molecule reactions for species typically found in discharges which contain CF_4 and CHF_3. The reactions investigated include electron detachment, collision induced dissociation (CID), and dissociative electron transfer resulting from collisions of CF_3^+, F^+, and F^- with the targets CF_4 and CHF_3. Impact energies range from near threshold to ≥ 200 eV. In general, the cross sections for all inelastic processes studied here reach maximum values and are constant for $E \geq 40$ eV. The importance and implications of these measurements to discharge modeling will be discussed.

INTRODUCTION

Carbon tetraflouride has been widely used as the primary feed gas for plasma processing discharges used in semiconductor manufacturing and as a component of binary dielectric gas mixtures.[1] Carbon tetraflouride alone is not a good dielectric gas; it is very weakly electronegative (CF_4^- has not been observed in the gas phase) and the cross section for dissociative electron attachment is extremely small (≤ 0.1 Å^2) for $1 \leq E \leq 50$ eV.[2] When CF_4 is combined with a well known gaseous dielectric, such as SF_6 for example, the dielectric properties of the mixture are reduced, but the chance of liquefaction and hence electrical breakdown under extreme low temperature conditions is reduced, and the cost, corrosiveness, and toxicity of the insulator are lowered.[3,4]

In plasma processing applications, CF_4 is a popular source gas because the parent molecule is relatively non-reactive and dissociates readily in a discharge into chemically reactive ionic and neutral species, owing to the fact that excited states of CF_4 and ground and excited states of CF_4^+ are highly unstable. An example of the ions observed in a CF_4 discharge is given in Fig. 1 for an electric field-to-number density ratio (E/N) of 12 kTd.[5]

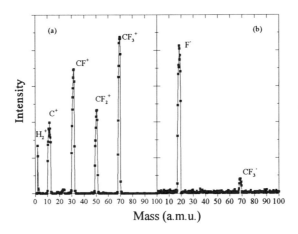

Figure 1. Relative intensities for positive (a) and negative (b) ions for a CF_4 discharge at E/N = 12 kTd.

While substantial progress has been made toward understanding microscopic phenomena within CF_4 and CHF_3 discharges through various electron impact studies,[2,6] scant data is available concerning ion-molecule reactions occurring in these plasmas. The purpose of this report is to provide cross sections for some of these reactions; such are essential for discharge modeling. The ultimate goal is to acquire a better understanding of macroscopic phenomena and enhance the ability to construct designer discharges specific to certain tasks. The cross sections for reaction channels for electron detachment, collision induced dissociation, and dissociative electron transfer will be referred to as $\sigma_1(E)$, $\sigma_2(E)$, and $\sigma_3(E)$ respectively, an energy level diagram for the ionic product reaction channels presented here is given in Fig. 2.

EXPERIMENTAL METHODS

Absolute cross sections presented here were determined using complimentary electrostatic trapping and crossed beam techniques, which have been describe in detail previously.[7,8] Briefly, absolute cross sections for electron detachment, collision induced dissociation, and the *sum* for all dissociative charge transfer reaction channels were measured directly with the trapping cell. In this arrangement, the primary ion current I_0, product ion current, I, target number density, n, and the path length are known. The cross sections are then determined for each collision energy by the equation,

$$\sigma = \frac{1}{nL} \ln\left(1 - \frac{I}{I_0}\right) \quad (1)$$

In order to ascertain the cross sections for the individual reaction channels associated with dissociative charge transfer, a crossed beam arrangement is implemented. In this scheme the relative probabilities for product ions of distinct mass are determined. These relative probabilities are then used in combination with the summed cross section determined via (1) to calculate the absolute cross sections for individual dissociative charge transfer reaction channels.

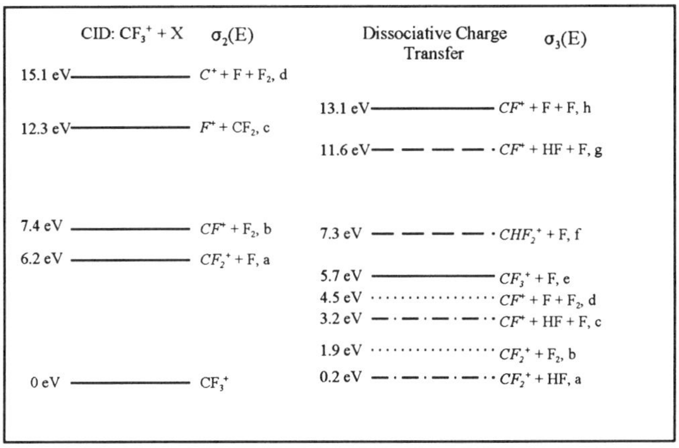

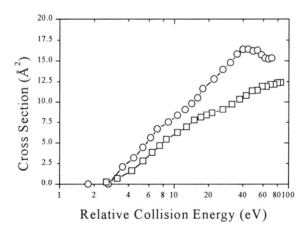

Figure 2. Energy levels for selected products in CID of CF_3^+ and dissociative charge transfer. For reactions 3, $CF_3^+ + CF_4$ (——), $CF_3^+ + CHF_3$ (- - - -), $F^+ + CF_4$ (••••), $F^+ + CHF_3$ (-•-•).

RESULTS

Electron Detachment

The cross sections for electron detachment for $F^- + CF_4$ and CHF_3 are given in Fig. 3. Electron detachment onsets at the electron affinity of F (3.4 eV) and increases rapidly with E for both targets. The cross sections are somewhat similar in magnitudes and behavior to those reported for F^- impacting rare gas targets[9] and show that electron detachment is a significant contributor to the population of chemically active fluorine, an important radical in plasma

Figure 3. Cross section for electron detachment for $F^- + CF_4$ - (O) and $F^- + CHF_3$ - (□) as a function of relative collision energy.

etching processes. Also from Fig. 3 it is observed that replacing one of the F atoms with a hydrogen atom has the effect of lowering the detachment cross section, suggesting that CHF_3 may be a better choice for a gaseous dielectric additive.

Collision Induced Dissociation

Cross sections for CID of CF_3^+ by CF_4 are presented in Fig. 4. The results for CF_3^+ + CHF_3 are similar in magnitude and structure and are not presented here. Cross sections

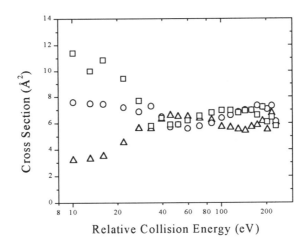

Figure 4. Total cross sections for collision induced dissociation from CF_3^+ + CF_4 producing CF_2^+ from reaction (2a) - (□), CF^+ from (2b) - (O) and the sum of C^+ and F^+ from (2c) and (2d) - (Δ).

for (2c) and (2d) could not be unambiguously assigned and are presented as a sum. The CID cross sections are largest for the least endothermic channel at low E and it is clear that the primary CF_3^+ ions are not completely void of some form of internal energy. Product ions C^+ and/or F^+ from (2c) and (2d) respectively are observed at lower collision energies than what would be predicted for ground state reactants. The trapping cell arrangement can only provide reliable cross sections for E ≥ 10 eV for these reactants.

Dissociative Charge Transfer

Total cross sections for the production of principal products from dissociative charge transfer are given in Figs. 5 and 6 for both targets for the CF_3^+ and F^+ projectiles respectively. Cross sections for producing C^+ and CF_2^+ from collisions of CF_3^+ with the CF_4 target and C^+, CH^+, CHF^+, CF_3^+ for the CHF_3 target (neither shown in Fig. 5) were relatively constant in magnitude, viz. ~1 Å^2, for all collision energies investigated. Also not shown in Fig. 6 but observed to be ≤ 3 Å^2 for all collision energies, are cross sections for F^+ + CF_4 producing C^+, F^+, CF_3^+ and F^+ + CHF_3 giving H^+, C^+, F^+, CH^+, C_2H^+, CHF^+, CHF_2^+ and CF_3^+ via reaction (3). The cross sections for reactions channels (3e)-(3h) displayed in Fig. 5 are most important for collision energies ≥ 40 eV and cross sections depend upon endothermicity in the same manner as observed for CID. The production of CF^+ from dissociative charge transfer of either target, (3g) or (3h), does not take place until the collision energy exceeds the endothermic threshold by ~ 10 eV, indicating that another energetic barrier exists for the formation of these products. What is most striking about the results presented in Fig. 5 is that production of CF_3^+, the least endothermic dissociative charge transfer reaction channel for CF_3^+ + CHF_3 system, is *not* the most probable channel.

In trying to interpret the results for the F^+ projectile, (presented in Fig. 6) it appears that no conventional wisdom holds. First, the cross sections for the least endothermic process for

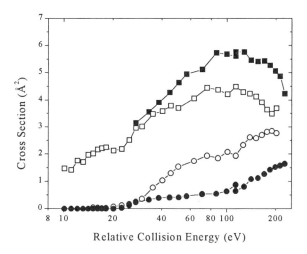

Figure 5. Total cross sections for dissociative charge transfer for $CF_3^+ + CF_4$ and CHF_3. $\sigma_{3e}(E)$ are presented as solid boxes, $\sigma_{3h}(E)$ - solid circles, $\sigma_{3f}(E)$ - open boxes, and $\sigma_{3g}(E)$ - open circles.

both targets, CF_3^+ production, are surprisingly small, ~3 Å^2 for the CF_4 target and ~ 0.25 Å^2 for the CHF_3 target (*exothermic* by 2.2 eV); neither are plotted in Fig. 6. In addition, the two product ions with the largest cross sections are identical for both targets, CF_2^+ and CF^+. However, their relative intensities are inverted: CF^+ production is *favored* over CF_2^+ production for the CHF_3 target, while the converse is true for CF_4. It is clear that much remains to be understood about the dissociative charge transfer process for these reactants.

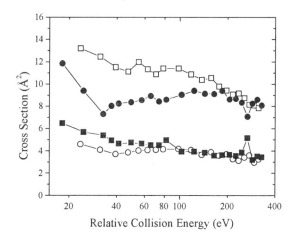

Figure 6. Total cross sections for dissociative charge transfer for $F^+ + CF_4$ and CHF_3. $\sigma_{3b}(E)$ - (●), $\sigma_{3d}(E)$ - (■), $\sigma_{3c}(E)$ - (□), and $\sigma_{3a}(E)$ - (○).

SUMMARY

Absolute cross sections have been measured for electron detachment, collision induced dissociation and dissociative charge transfer for collisions of F^-, F^+ and CF_3^+ with CF_4 and CHF_3. The cross sections for these inelastic reaction channels reach a maximum and are, for the most part constant, for relative collision energies greater than 40 eV. The cross section for producing a free electron via electron detachment increases sharply as the collision energy rises above the electron affinity of F. The cross sections for the collision induced dissociation of CF_3^+ are on the order of a few square angstroms and are independent of the targets studied here. From the dissociative charge transfer measurements, it is obvious that reaction barriers exist. The F atom, one of the most important radicals for semiconductor etching, is readily produced from the three ion-molecule reactions studied here. From the CID measurements, it is clear that the CF_3^+ beam used in these experiments contains internal energy, perhaps in a way similar to that in a discharge environment. It is difficult to quantify how much or how specifically it effects the measurements presented here other than to say product ions are observed at collision energies lower than what would be expected for ground state reactants. Clearly investigations concerning the internal energy and non-reactive de-excitation mechanisms of CF_3^+ are required.

Any credible model simulation of a plasma discharge which is designed to predict macroscopic phenomena based on particle interactions on a microscopic level requires the ion-molecule cross sections presented here. These measurements represent a small addition to the collisional database and more work needs to be done. A broader approach to determining ion-molecule and other particle interaction cross sections for reactive species is needed if a discharge is to be adequately characterized via modeling.

ACKNOWLEDGMENTS

This work was supported in part by the Division of Chemical Sciences, Office of Basic Energy Sciences of the U. S. Department of Energy.

REFERENCES

1. D. M. Manos and D. L. Flamm, *Plasma Etching*, (Academic, Boston, 1989).

2. L. G. Christophorou, J. K. Olthoff, and M. V. V. S. Rao, J. Chem. Phys. Ref. Data **25**, 1349 (1996), and references cited therein.

3. J. Berg, and E. Kuffel, IEEE 1995 Annual Report, 688 126 (1995) and The Ninth International Symposium on High Voltage Engineering, **2**, 9 (1995).

4. R. E. Wootton, S. J. Dale, and N. J. Zimmerman, *Gaseous Dielectrics II* (L. G. Chrisophorou (Ed.) Pergamon, New York, 1980) p. 137.

5. J. K. Olthoff and M. V. V. S. Rao, unpublished.

6. L. G. Christophorou, J. K. Oltoff, and M. V. V. S. Rao, J. Chem. Phys. Ref. Data **26**, 1 (1997) and references cited therein.

7. B. L. Peko, R. L. Champion, and Y. Wang, J. Chem. Phys. **104**, 6149 (1996).

8. B. L. Peko, and R. L. Champion, J. Chem. Phys. **107**, 1156 (1997).

9. M. S. Huq, L. D. Doverspike, R. L. Champion, and V. A. Esaulov, J. Phys. B. At. Mol. Phys. **15**, 951 (1982).

A STUDY OF ION-MOLECULE REACTIONS OF METHANE IONS IN PARENT GAS AT HIGH E/N

J. de Urquijo, E. Basurto[#], C. Cisneros and I. Alvarez
Instituto de Física, UNAM
P.O. Box 48-3, 62251, Cuernavaca, Mor.
México

INTRODUCTION

The wide number of applications of low-temperature plasmas in the semiconductor[1,2] and high voltage applications[3] demand more insight into the basic collisional processes taking place between electrons, ions and neutrals. For instance, methane and its mixtures with argon and hydrogen are used for etching GaAs wafers and diamond-like surface coatings. There is at present an increasing number of investigations dealing with the study of ion kinetics and mass spectrometry of fragment products as a means of providing more useful and comprehensive reaction schemes, as well as reliable quantitative data that may improve discharge simulation and modeling. Many of the above studies are now centred at high values of the density- reduced electric field intensity, E/N.

This paper deals with the study of the variation of ion fluxes of several positive ion species drifting and reacting in pure methane and in a 50-50% methane-argon mixture under swarm conditions at relatively high values of the reduced electric field strength, E/N, between 1 and 8 kTd (1 Townsend (Td) = 10 aV cm^2).

MEASUREMENT TECHNIQUE

The drift tube-mass spectrometer used for the ion abundance measurements has been described in detail previously[4,5]. Briefly, the drift tube consists of a hot filament ion source that can be moved along the drift space over a distance of 0-38 cm from the drift space end. The ion source, operated in the steady-state mode, emitted mostly CH_3^+ and CH_5^+ (pure methane), and additionally Ar^+ ions (mixture). All other ions were mostly formed in the drift space by secondary and tertiary reactions of the primary ions with the neutrals. A sample of the ions at the end of the drift space enters the mass analysis region through a small central orifice consisting of a mass spectrometer and channel electron multiplier. A

[#] Also at ESFM-IPN, México.

mass spectrum for each combination of E/N, N, and drift distance z was obtained by sweeping the mass spectrometer continuously.

CH_4 gas and the CH_4-Ar mixture with a quoted purity of 99.97% were admitted straight into the drift and ion source chambers without further purification. For the E/N range covered in this study, only traces of water vapour ions (H_2O^+, H_3O^+) with concentrations of less than 0.1 percent were detected.

Accuracies in the settings of E/N, drift distance and gas density were estimated to be 0.8, 0.5, and 0.3%, respectively. A mass resolution of $\Delta M/M < 1/100$ was normally obtained over the whole ion scan range from 1 to 60 amu.

RESULTS

Pure Methane

The range of measurement of E/N covered in this research was 1-8 kTd, and the gas pressure was varied between 0.8 and 7 Pa. Individual ion concentrations derived from each mass scan were normalised to the sum of all the ion peaks. Of particular interest was the observation of the variation of ion concentration with the Nz product. In all cases, a continuous, smooth curve was obtained, which was indicative of bimolecular reactions between a particular ion and the neutrals. This was also appreciated by the evolution of the ion concentration with drift distance. Thus, for each ion the variation of ion concentration [X] with the Nz product was fitted to the simple exponential functions

$$[X] = a \exp[-b(Nz)] \qquad (1)$$
$$[X] = a \{1 - \exp[-b(Nz)]\} \qquad (2)$$
$$[X] = a \exp[-b(Nz)] + c \qquad (3)$$

Values of these fitting parameters as functions of E/N are given in Figures 1 to 5. Equation (1) applies for CH_4^+, equation (2) for C^+, CH^+, CH_2^+, CH_3^+, $C_2H_2^+$, $C_2H_3^+$, and $C_3H_3^+$, and equation (3) for CH_5^+ and $C_2H_5^+$. Typical errors in the fitting process varied between 5-15%. The Nz range of validity is $4 \times 10^{14} - 4 \times 10^{16}$ cm^{-2}.

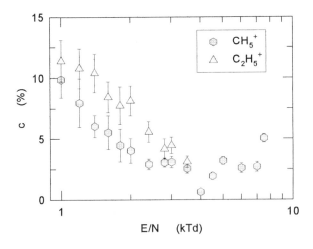

Figure 1. Values of the parameter c of equation 3.

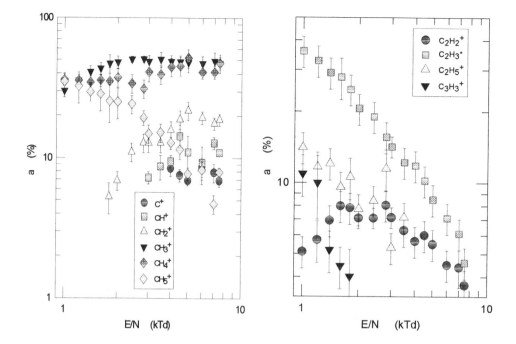

Figure 2. a for [C$^+$] to [CH$_5^+$]

Figure 3. a for [C$_2$H$_2^+$] to [C$_3$H$_3^+$]

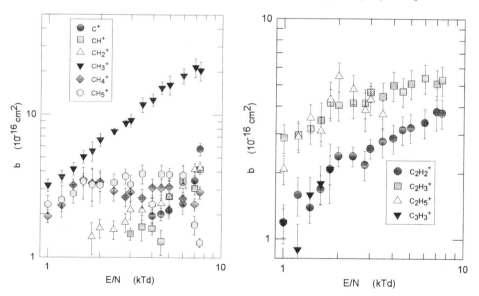

Figure 4. b for [C$^+$] to [CH$_5^+$]

Figure 5. b for [C$_2$H$_2^+$] to [C$_3$H$_3^+$]

Representative plots of the variation of the ion abundances as a function of E/N for Nz=10^{16} cm^{-2} are given in figures 6 to 9, which depict the remarkable increase in the concentrations of the lighter ions C$^+$, CH$^+$ and CH$_2^+$ and, to a lesser extent of C$_2$H$_2^+$ and C$_2$H$_3^+$.

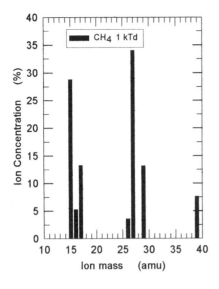
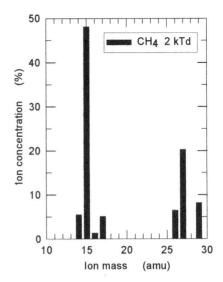

Figure 6. Ion abundance at 1 kTd

Figure 7. Ion abundance at 2 kTd

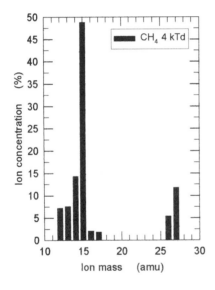
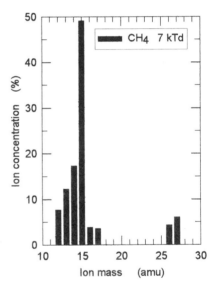

Figure 8. Ion abundance at 4 kTd

Figure 9. Ion abundance at 7 kTd

The above equations were not derived from a formal reaction scheme; however they can provide good qualitative insight into the processes responsible for the creation and destruction of the various ion species. In particular, it appears that the C^+, CH^+ and CH_2^+ species may result from the destruction of CH_4^+ and CH_5^+. A good acount of ion-molecule reactions has been given by Anicich[6] for thermal and near thermal reactions, and some of the above reactions were studied by Kline et al[7]. $C_2H_2^+$ and $C_2H_3^+$ may, in a similar way arise from the destruction of $C_2H_5^+$. It is also noticed that no $C_3H_3^+$ appears for E/N>2 kTd, since it is collisionally destroyed to enhance the lighter ion population.

Methane-Argon Mixture

The ion concentration ratios for the 50-50% CH_4-Ar mixture are shown plotted in figures 10 and 11 as a function of ion mass, for two extreme values of E/N, and $Nz=10^{15}cm^{-2}$. As for the case of pure methane, the same daughter ions were detected and, additionally, the presence of Ar^+ in substantial amounts at the highest E/N, while at E/N=1 kTd, the Ar^+ concentration is small and only low concentrations of Ar^{++} and ArH^+ could be measured. It has been pointed out[8] that the dissociative charge transfer process

$$Ar^+ + CH_4 \longrightarrow CH_3^+ + Ar + H$$

is important at E/N=2 kTd. On the other hand, $[Ar^+]$ increases drastically for E/N>4 kTd, and it reaches a value of about 15% at 11 kTd. Also, the dissociative collisional process

$$CH_3^+ + CH_4 \text{ (or Ar)} \longrightarrow CH^+ + H_2 + CH_4 \text{ (or Ar)}$$

has been suggested[8].

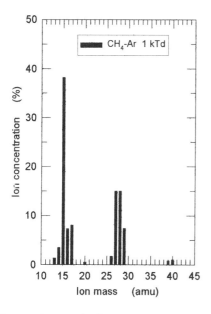
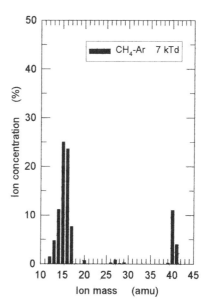

Figure 10. Ion abudance at 1 kTd   **Figure 11.** Ion abundance at 7 kTd

We have chosen to compare the case of pure methane with that of its mixture with Ar at different values of Nz where the yield of light hydrocarbon ions is about the same, as it is seen from figure 12. Furthermore, the second series of hydrocarbons ($C_2H_2^+$ and $C_2H_3^+$) is reduced. Therefore, it appears that approximately the same low-mass hydrocarbon ion yield is obtained with only one tenth the total gas pressure of the mixture, as compared with that of pure methane.

CONCLUSIONS

Thus far, most of the studies on the ion-molecule reactions of methane and its

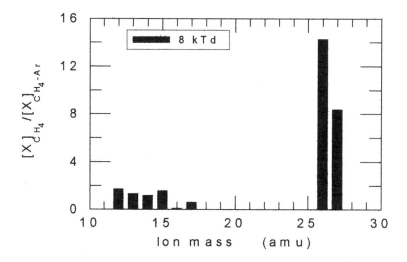

Figure 12. Comparison of the ion yield between pure methane and its mixture with Ar.

mixtures with other gases have concentrated mostly on the thermal region; however, there are many discharge conditions in which ions are far from thermal equilibrium. This study represents an approximation to the complex ion chemistry of methane and its mixtures.

ACKNOWLEDGEMENTS

This work was supported by DGAPA, Project IN104795. E. Basurto acknowledges CONACYT for a grant. The technical assistance of A. Bustos is appreciated.

REFERENCES

1. P.F. Williams, *Plasma Processing of Semiconductors*, Kluwer Academic Publishers, Dordrecht (1997)
2. Y. Mitsuda, Y. Kojima, T. Yoshida, and K. Akashi, "The growth of diamond in microwave plasma under low pressure", *J. Mat. Sci., 22:1557, 1987*
3. S.R. Hunter, J. G. Carter, and L.G. Christophorou, "Electron transport studies of gas mixtures for use in e-beam controlled diffuse discharge switches", *J. Appl. Phys. 58:3001-15, 1985*
4. J. de Urquijo, I. Alvarez, C. Cisneros and H. Martínez, "Mobility and longitudinal diffusion of SF_3^+ and SF_5^+ in SF_6", *J. Phys. D: Appl. Phys. 23:778-83, 1990*
5. J. de Urquijo, I. Domínguez, I. Alvarez and C. Cisneros, "Drift and proton transfer reactions of positive ions in methane parent gas", *J. Phys. B: At. Mol. Opt. Phys. 30:4395-4404, 1997*
6. V.G. Anicich, "Evaluated bimolecular ion-molecule gas phase kinetics of positive ions for use in modeling planetary atmospheres, cometary comae, and interstellar clouds", *J. Phys. Chem. Ref. Data, 22:1469-569, 1993*
7. L.E. Kline, W.D. Partlow, and W.E. Bies, "Electron and chemical kinetics in methane rf glow-discharge deposition plasmas", *J. Appl. Phys., 65: 70-8, 1989*
8 A.V. Phelps, Private communication

IONIZATION AND ION KINETICS IN c-C_4F_8

C.Q. Jiao, A. Garscadden, and P.D. Haaland

Air Force Research Laboratory
Wright-Patterson AFB, OH 45433

ABSTRACT

Cross sections of electron impact ionization of octafluorocyclobutane (c-C_4F_8) in the electron energy range of 16 to 200 eV have been measured by Fourier transform mass spectrometry (FTMS). $C_2F_4^+$ and $C_3F_5^+$ are the major fragment ions (total cross section of $1.5\pm0.2\times10^{-15}$ cm^2 at 70 eV). Among the fragment ions, only CF_2^+ and $C_2F_3^+$ react with neutral c-C_4F_8, yielding $C_3F_5^+$ as the major product ion with the rate coefficients of 4.7 ± 0.5 and $3.3\pm0.5\times10^{-10}$ cm^{3}s^{-1}, respectively. $C_4F_8^-$ and F^- are formed by low energy electron attachment to c-C_4F_8, and are inert to their parent gas molecule.

INTRODUCTION

Octafluorocyclobutane (c-C_4F_8) is a good high voltage gaseous insulator[1,2] due to its strong electron-attachment which has been well studied.[1-18] c-C_4F_8 is also used in the semiconductor industry as a substitute for CF_4 to obtain high etch selectivity of the SiO_2 film over silicon.[19] A recent study showed that positive ions are intimately involved in the surface deposition and etching processes.[20] Limited kinetic data measurements of the ionization cross sections at 35 eV[4] and 70 eV[21] were performed in the 1960's, and in 1972 the ionization cross sections in arbitrary units at energies near threshold were obtained.[9] We present a complete measurement of the ionization cross sections of c-C_4F_8 as a function of the electron energy in the range of 10 - 200 eV. Rate coefficients for the reactions of the fragment ions with c-C_4F_8 also are reported.

EXPERIMENTAL

Octafluorocyclobutane (99+%, TCI America) is mixed with Argon (99.999% Matheson Research Grade) with a ratio (c-C_4F_8 : Ar) of 1:2, and admitted through a precision leak valve into a modified Extrel Fourier-transform mass spectrometry system.[22] Ions are formed by electron impact in a cubic ion cyclotron resonance trap cell at pressures in the 10^{-7} Torr range. Ions of all mass to charge ratios are simultaneously and coherently excited into cyclotron orbits using a stored waveform[23,24] applied to two of the trap plates. Following excitation, the image currents induced on two of the remaining trap plates are amplified, digitized, and Fourier analyzed to yield a mass spectrum.

Calculation of cross sections from the mass spectra requires knowledge of the gas pressure, the electron beam current, and the number or ions produced. These calibration issues have been described previously.[22,25] The intensity ratios of the ions from c-C_4F_8 to Ar^+ give cross sections relative to those for argon ionization[26] since the c-C_4F_8:Ar pressure ratio is quantified by capacitance manometry of the gas mixture. The absolute c-C_4F_8 pressure at the ICR trap that is needed for ion-molecule kinetic analysis is inferred from the rate coefficient of the known reaction of O_2^+ with c-C_4F_8.[27]

RESULTS AND DISCUSSION

Electron impact ionization of c-C_4F_8 yields $C_2F_4^+$ and $C_3F_5^+$ as the two predominant ions over the energy range from the threshold to 200 eV. Other fragment ions include CF_x^+ and $C_2F_x^+$ with x=1-3, $C_3F_y^+$ with y=1-4, and F^+. Although the photoionization of c-C_4F_8 producing $C_4F_8^+$ has been observed with a threshold of 11.6±0.2 eV,[28] no detectable $C_4F_8^+$ has been found from the electron impact ionization in our study. However, as stated later, $C_4F_8^+$ is produced among other products from the reaction of Xe^+ with c-C_4F_8. The total ionization cross section reaches a maximum value of 1.6±0.2 x 10^{-15} cm^2 at 80 eV, and levels off thereafter. The cross sections for each dissociative ionization process as function of the electron energy are shown in figure 1.

Gas-phase reactions of the positive ions with neutral c-C_4F_8 are studied by experiments in which each of the ions is selected, followed by a reaction time. Only CF_2^+ and $C_2F_3^+$ are found to react, yielding reactions 1 - 10 and $C_3F_5^+$

CF_2^+ + c-C_4F_8 -----> CF_3^+ + (C_4F_7) 14% (1)
-----> $C_2F_4^+$ + (C_3F_6) 5% (2)
-----> $C_3F_3^+$ + (C_2F_7) 2% (3)
-----> $C_3F_5^+$ + (C_2F_5) 53% (4)
-----> $C_4F_6^+$ + (CF_4) 6% (5)
-----> $C_4F_7^+$ + (CF) 20% (6)

$C_2F_3^+$ + c-C_4F_8 -----> CF^+ + (C_5F_{10}) 3% (7)
-----> $C_2F_4^+$ + (C_4F_7) 7% (8)
-----> $C_3F_5^+$ + (C_3F_6) 81% (9)
-----> $C_4F_7^+$ + (C_2F_4) 9% (10)

as the major product ion. Formulae in the parentheses of the equations do not necessarily imply the actual neutral product composition. These reactions do not significantly change the overall ion composition of a c-C_4F_8 plasma, since

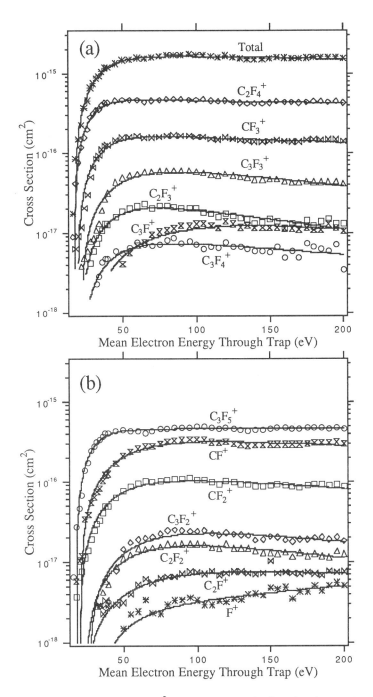

Figure 1. (a) and (b): Cross sections (cm^2) for ionization of c-C$_4$F$_8$ by electron impact. Points represent experimental data.

the relative abundances of the reactant ions, CF$_2^+$ and C$_2$F$_3^+$, are very small. Therefore, one can expect that under many conditions C$_2$F$_4^+$ and C$_3$F$_5^+$ are the major ionic species reaching surfaces from the plasma. Reactions of several selected rare-gas ions with c-C$_4$F$_8$ are also studied, with major product ions shown in equations 11-16. Minor product ions (with ~1% branching ratios)

$$Ar^+ + c\text{-}C_4F_8 \longrightarrow C_2F_4^+ + (C_2F_4) + Ar \quad\quad 50\% \quad\quad (11)$$
$$\longrightarrow C_3F_5^+ + (CF_3) + Ar \quad\quad 46\% \quad\quad (12)$$

$$Kr^+ + c\text{-}C_4F_8 \longrightarrow C_2F_4^+ + (C_2F_4) + Kr \quad\quad 81\% \quad\quad (13)$$
$$\longrightarrow C_3F_5^+ + (CF_3) + Kr \quad\quad 16\% \quad\quad (14)$$

$$Xe^+ + c\text{-}C_4F_8 \longrightarrow C_2F_4^+ + (C_2F_4) + Xe \quad\quad 46\% \quad\quad (15)$$
$$\longrightarrow C_3F_5^+ + (CF_3) + Xe \quad\quad 52\% \quad\quad (16)$$

from Ar^+ reaction include F^+, CF^+, CF_3^+, and C_2F^+, from Kr^+ reaction, F^+, CF_2^+, and C_2F^+, and from Xe^+ reaction, C_2F^+ and $C_4F_8^+$. It is interesting to note that the reaction with Xe^+ yields a small amount of the molecular ion, probably due to its "soft" chemical ionization nature. The above results are compared to those of Smith and Kevan[29] who studied in 1970 the reactions of rare-gas ions with 50 eV kinetic energy. Because of the difference in the kinetic energy of the ions, we saw different minor product ions, but the patterns of the branching ratios for the major product ions are similar in the two studies except for the case of Xe^+ where Smith and Kevan observed no $C_3F_5^+$ at all.

Reaction rate coefficients of the various fragment ions with the parent gas are obtained from the plots of the ion intensities as functions of the reaction time, as shown in figure 2 as an example for the case when 50 eV electrons are

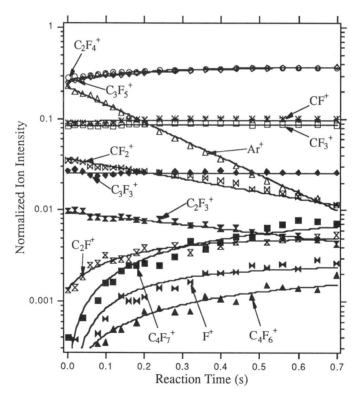

Figure 2. Time evolution of positive ion species produced by 50 eV electron impact in a mixture of c-C_4F_8 and Ar (1:2) with a total pressure of 3.3×10^{-7} Torr. Data are shown only for the ions whose intensities change with time. Points represent the experimental data.

used to generate the reactant ions. Table 1 lists the reaction rate coefficients obtained in experiments under conditions of different impinging electron energies, namely, 20, 35 and 50 eV. No significant variations of the rate coefficients with different electron energy are found in the energy range studied.

Table 1. Ion-molecule reaction rate coefficients in units of 10^{-9} cm^{3}s^{-1}, measured in three separate experiments, 1-3, in which the primary ions are formed by electron impact at 20, 35 and 50 eV, respectively.

Reaction	Experiment 1	Experiment 2	Experiment 3
$CF_2^+ + C_4F_8$	----	0.47±0.05	0.47±0.05
$C_2F_3^+ + C_4F_8$	----	0.33±0.05	0.30±0.05
$Ar^+ + C_4F_8$	1.5±0.1	1.3±0.1	1.3±0.1

Electron attachment of c-C_4F_8 is not probed by the actual beam electrons in our FTMS experiments because space-charge limits propagation of low energy electron beams along the 1 meter path from the electron gun to the ion trap. Instead, we use the secondary electrons that are produced by ionization of the reagent gases and are trapped in the ICR cell along with negative ions. The electrons produced at different locations along the trapping axis of the cell are born with different electrostatic potential energies. These electrons undergo damped harmonic motion in the trapping potential, where the damping arises from electron-neutral collisions with reagent gas. As the electrons cascade down in energy they have an opportunity to attach as well as be cooled by momentum transfer. Although this approach lacks the energy resolution of crossed beam measurements, it has the advantage that further reactions between anionic products of attachment and the parent gas may be conveniently probed. F$^-$ and $C_4F_8^-$ were observed in our experiments. The production of F$^-$ is a classic illustration of dissociative electron attachment, and has been extensively reported.[3,4,9,14] The observation of $C_4F_8^-$ at the very low pressures (10^{-6} Torr) and long observation times (100 ms) of FTMS is more unusual, since its lifetime was only 10-15 μs [2,7,8,12] without the third-body-collision condition. Ion-molecule reaction experiments showed that both F$^-$ and $C_4F_8^-$ are inert to their parent gas molecule, ruling out the possibility that $C_4F_8^-$ is the product of F$^-$ reaction with c-C_4F_8. We propose that one of the ring bonds in c-$C_4F_8^-$ has been broken to satisfy energy conservation, while retaining the stoichiometry of the parent molecule.

ACKNOWLEDGMENTS

The authors wish to acknowledge the Air Force Office of Scientific Research for supporting this research.

REFERENCE

1. A. A. Christodoulides, L. G. Christophorou, R. Y. Pai, and C. M. Tung, *J. Chem. Phys.* 70:1156 (1979).
2. I. Sauers, L. G. Christophorou, and J. G. Carter, *J. Chem. Phys.* 71:3016 (1979).
3. R. M. Reese, V. H. Dibeler, and F. L. Mohler, *J. Research NBS* 57:367 (1956).
4. M. M. Bibby and G. Carter, *Trans. Faraday Soc.* 59:2455 (1963).
5. M. V. Kurepa, 3rd Cz. Conference on Electronics and Vacuum Physics Transactions, 107 (1965).
6. R. Grajower and C. Lifshitz, *Israel J. Chem.* 6:847 (1968).
7. W. T. Naff and C. D. Cooper, *J. Chem. Phys.* 49:2784 (1968.
8. P. W. Harland and J. C. J. Thynne, *Int. J. Mass Spectrom. Ion Phys.* 10:11 (1972/73).
9. C. Lifshitz and R. Grajower, *Int. J. Mass Spectrom. Ion Phys.* 10:25 (1972/73).
10. K. M. Bansal and R. W. Fessenden, *J. Chem. Phys.* 59:1760 (1973).
11. F. J. Davis, R. N. Compton, and D. R. Nelson, *J. Chem. Phys.* 59:2324 (1973).
12. L. G. Christophorou, D. L. McCorkle, and D. Pittman, *J. Chem. Phys.* 60:1183 (1974).
13. P. W. Harland and J. L. Franklin, *J. Chem. Phys.* 61:1621 (1974).
14. R. L. Woodin, M. S. Foster, and J. L. Beauchamp, *J. Chem. Phys.* 72:4223 (1980).
15. L. G. Christophorou, R. A. Mathis, D. R. James, and D. L. McCorkle, *J. Phys. D: Appl. Phys.* 14:1889 (1981).
16. S. M. Spyrou, S. R. Hunter, and L. G. Christophorou, *J. Chem. Phys.* 83:641 (1985).
17. A. A. Christodoulides, L. G. Christophorou, and D. L. McCorkle, *Chem. Phys. Lett.* 139:350 (1987).
18. T. M. Miller, R. A. Morris, A. E. S. Miller, A. A. Viggiano, and J. F. Paulson, *Int. J. Mass Spectrom. Ion Processes* 135:195 (1994).
19. H. Kazumi and K. Tago, *Jpn. J. Appl. Phys.* 34:2125 (1995).
20. Y. Gotoh and T. Kure, *Jpn. J. Appl. Phys.* 34:2132 (1995).
21. J. A. Beran and L. Kevan *J. Phys. Chem.* 73:3866 (1969).
22. K. Riehl, *Collisional Detachment of Negative Ions Using FTMS*, Ph. D. Thesis, Air Force Institute of Technology (1992).
23. A. G. Marshall, T. L. Wang, and T. L. Ricca, *J. Am. Chem. Soc.* 107:7893 (1985).
24. S. Guan, *J. Chem. Phys.* 91:775 (1989).
25. P. D. Haaland, *Chem. Phys. Lett.* 170:146 (1990).
26. R. C. Wetzel, F. A. Baioochi, T. R. Hayes, and R. S. Freund, *Phys. Rev.* 35:559 (1987).
27. R. A. Morris, T. M. Miller, A. A. Viggiano, and J. F. Paulson, *J. Geophys. Res.* 100:1287 (1995).
28. G. K. Jarvis, K. J. Boyle, C. A. Mayhew, and R. P. Tuckett, submitted to *J. Chem. Phys.*
29. D. L. Smith and L. Kevan, *J. Chem. Phys.* 55:2290 (1971).

THE DEPENDENCE OF ELECTRON CAPTURE RATE CONSTANTS ON ELECTRONIC POLARIZABILITY

Iwona Szamrej

Chemistry Department
Agricultural & Teachers University
08-110 Siedlce, Poland

INTRODUCTION

The second order rate constants for the thermal electron capture have been determined for a long time in many laboratories with a wide spectrum of different experimental techniques. However, there are not too many compounds investigated. Moreover, in most cases a large scattering of the results from different laboratories is observed. This could be caused either by not achieving the truly thermal distribution of electrons energy and/or not including the contribution from higher order electron attachment processes.

Nevertheless these exists quite a big number of the kinetic data and the temptation to find some general approach to them arises.

There is also a lot of data on the higher than second order electron attachment processes, both collisionaly stabilized attachment and attachment by van der Waals complexes[1].

In seeking the link between the rate and the mechanism of the electron attachment in the gas phase and a structure of an accepting molecule a number of analyses have been done[1,2]. What seems to be obvious is to look for the dependence of the electron capture rate constant on the dipole moment, overall and electronic polarizability of the molecule as well as those of the accepting center.

RESULTS AND DISCUSSION

Only in the case of electron attachment by van der Waals (vdW) complexes consisted of inorganic hydrides there is a straightforward dependence of the mechanism and the kinetics of the process on the molecules dipole moments[3]. When two-body electron capture reaction with individual molecule is analyzed there is no such a dependence. The zero dipole moment molecules can attach thermal electrons extremely fast (SF_6, CCl_4) or extremely slow (CF_4). From the other side, the rate of the process can depend on total molecular polarizability or its component.

Molecular polarizability reflects an ability of core (mostly) electron orbitals of a molecule to deform under the influence of the electric field. It can be calculated from molar refraction:

$$R = \frac{4}{3}\pi N_a \alpha \quad (1)$$

where α is a polarizability, N_a - Avogadro's number and R - molar refraction. The last can be found using experimental index of refraction from the Lorenz-Lorenz equation:

$$R = (n^2 - 1)/(n^2 + 2)(M/d) \quad (2)$$

It is evident from the first look into the rate constant data that the overall molecular polarizability can not be used for the analysis, e.g. there are a big differences between the rate constants for isomers. In opposite, the polarizability of the accepting center shows the visible influence on the rate constants for two-body process. The considered polarizability of the accepting center is a sum for atoms which are able to accept or influence accepting electrons (fluorine, chlorine, bromine, iodine) as well as double or triple bond and aliphatic or aromatic rings.

In figures 1 and 2 the plots of the rate constants in the logarithmic scale for a two sets of saturated hydrocarbons containing halogen atoms vs polarizability of the accepting center are presented.

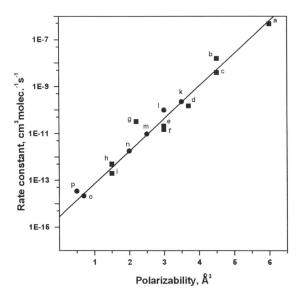

Figure 1. The dependence of the thermal electron capture rate constant on the calculated electronic polarizability of the accepting center for: ■ - Cl, and ● - F-containing aliphatic compounds. (a - CCl_4, b - 1,1,1-$C_2H_3Cl_3$, c - $CHCl_3$, d - 1,1,2-$C_2H_3Cl_3$, e - 1,1-$C_2H_4Cl_2$, f - CH_2Cl_2, g - 1,2-$C_2H_4Cl_2$, h - C2H5Cl, i - CH_3Cl)[4], (k - C_6H_{14}, l - C_5F_{12}, m - C_4F_{10}, n - C_3F_8)[5], (o - CHF_3, s - CH_2F_2)[6].

As it is seen in figure 1, the common, linear dependence, for different molecules containing chlorine or fluorine is obtained which does not depend on the number of halogen atoms and it also fits for different isomers.

The same linear dependence, but with a different slope and intercept, is shown in figure 2 for both chlorine and fluorine containing compounds.

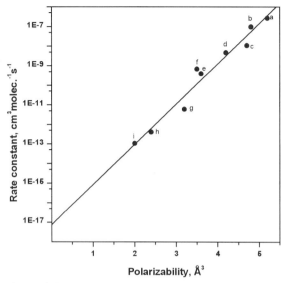

Figure 2. The dependence of the thermal electron capture rate constant on the calculated electronic polarizability of the accepting center for F and Cl containing compounds. (a - 1,1,1-C_2F_3Cl, c - 1,1,2-C_2F_3Cl, d - 1,1-$C_2F_4Cl_2$, f - 1,2-$C_2F_4Cl_2$)[7], (b - $CFCl_3$)[4], (e - CF_2Cl_2, g - $CHFCl_2$)[8], (h - CF_3Cl)[6], (i - CHF_2Cl)[9].

It is well known that the linear dependence of log(k) on some parameters, as e.g. 1/T, indicates the activation energy as a coefficient of proportionality. The same can be also accepted in this case. So, the activation energy of the thermal electron capture process should diminish linearly with the polarizability of the accepting center.

Activation energies for the thermal electron attachment processes are measured rather seldom, usually for individual molecules and they differ even more than one hundred percent when comparing results from different laboratories and methods. However, there is at least one set of data measured with the same method for several halogen substituted methanes, both with the same and different halogen atoms in one molecule [8]. The room temperature rate coefficients which have been measured differ six orders of magnitude between the lowest one (CF_3Cl) and the highest (CCl_4) one. The authors measured temperature dependence of the rate constants but did not calculate the activation energies. Still it is easy to find them from their data. Figure 3 shows the dependence of the activation energy on the corresponding polarizabilities of the accepting centers.

Moreover, the same authors, using semiempirical quantum mechanical method, calculated the geometry changes associated with negative ion formation. They found that the only significant change in geometry connected with addition of an extra electron to neutral molecule is an increase in the carbon-halogen bond length, R_e, that corresponds to the reaction coordinate for dissociative electron attachment. Changes in other bond lengths and in bond angles were relatively minor. Comparing the relative bond stretching, $\Delta R_e/R_e$, with corresponding room temperature rate constants they found the relationship inconclusive because of the large scattering.

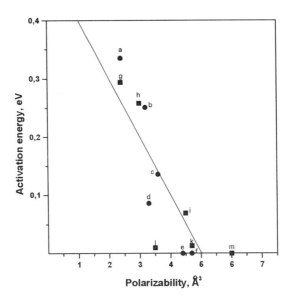

Figure 3. An activation energy of electron capture[10] as a function of electronic polarizability of the accepting center for: ■ - one and ● - two halogen containing halocarbons. a - CF_3Cl, b - $CHFCl_2$, c - CF_2Cl_2, d - CF_3Br, e - CF_3I, f - $CFCl_3$, g - CH_3Br, h - CH_2Cl_2, i - $CHCl_3$, k - CH_2Br_2, l - CH_3I, m - CCl_4.

This is now understood in view of the present work finding that different groups of components have different slopes in figures 1 and 2. However, if one plots their $\Delta R_e/R_e$ vs. reciprocal of polarizability of the accepting center a good straight line presented in figure 4 is obtained.

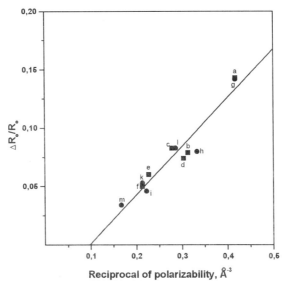

Figure 4. The dependence of the relative equilibrium bond changes on reciprocal of electronic polarizability of the accepting center. The legend as in figure 3.

As it was mentioned in the first section more complicated than just two-body electron attachment processes also occur quite abundantly in the gas phase. It is worth to add that also the last statement derived from the Burns et al.[10] work can be applied to the better understanding their kinetics. The rate constant of the electron capture by vdW complex (k_{vdW}) as compared with k_{II} for simple two-body process is much higher, usually few orders of magnitude. But the product of the attachment process in both cases is the same, halogen negative ion[11]. So, the supposition has been made that electrons are captured by filling the same orbital both in the isolated molecule and in the van der Waals complex[1]. The increase in k_{vdW} as compared with k_{II} can be ascribed to lowering the energy of this state by intermolecular forces in the vdW complex, thus decreasing the activation energy and increasing correspondingly the Boltzman factor - $\exp(-E_o/kT)$. If k_{vdW} and k_{II} are correspondingly equal to:

$$k_{vdW} = A_{vdW} \exp\left(-E_a^{vdW}/kT\right) \tag{3}$$

and

$$k_{II} = A_{II} \exp\left(-E_a^{II}/kT\right) \tag{4}$$

one can get the change in rate constants depending on the diminishing of the activation energy, assuming that the change in preexponential factor can be neglected

$$\Delta E_a = kT \ln(k_{vdW}/k_{II}) \tag{5}$$

These changes as calculated from the experimental data[6,8,9] are equal to a few kT which can be compared with the energy of van der Waals interaction. They can be even slightly lower if one assume that the preexponential factor increases in the vdW complex

CONCLUSIONS

The main observed features for the thermal electron capture dependence on the molecular parameters are following:
- The effectiveness of the electron capture does not depend on the dipole moment of the molecule for two-body process.
- It also does not depend on its orientational or overall electronic polarizabilities.
- The rate constant for halogenated hydrocarbons increases both when going from fluorine to iodine as well as with the number of given atoms in the molecule.
- The presence of the double and triple bonds, aliphatic and aromatic rings strongly increases the rate constant.
- Linear dependence of log(k) on α_{el} is observed, where k is the thermal electron capture rate constant and α_{el} is the electronic polarizability of the accepting center.
- The dependence of log(k) on α_{el} is different (but still linear) if hydrocarbon molecule is substituted with the same or different halogen atoms.
- $1/\alpha_{el}$ can be considered as a measure of an activation energy of the process.
- In terms of resonant attachment α_{el} determines the cross-section and energy position of the lowest resonance peak.

Thus the final conclusion is that the electronic polarizability of the accepting center determines the change in equilibrium bond length caused by the formation of the negative ion. This, in turn, determines the position of the crossing point of Morse curves for the molecule and the negative ion. The energy value of this point should correspond to the experimental activation energy.

The other, more practical conclusion, is that the expected rate constants for many compounds can be easily estimated or the measured ones verified on the basis of electronic polarizability of the accepting center.

REFERENCES

1. I.Szamrej and M.Foryś, The role of the van der Waals complexes in the thermal electron attachment processes in the gas phase, *Progress in Reaction Kinetics*, in press (1998).
2. M.Foryś and I.Szamrej, The dependence of electron capture rate constants on some molecular parameters, *J.Radioanal.Nucl.Chem.*, **232** (1998).
3. I.Szamrej, H.Janicka, I.Chrząścik and M.Foryś, Thermal electron attachment processes in gas mixtures containing H_2S, HCl and HBr, *Radiat.Phys.Chem.*, **33**, 387 (1989).
4. L.G.Christophorou, D.L.McCorkle and A.A.Christodoulides, Electron attachment processes, in *Electron-Molecule Interactions and their Applications*, L.G.Christophorou, ed., Academic Press, London (1984).
5. S.R.Hunter and L.G.Christophorou, Electron attachment to the perfluoroalkanes n-C_NF_{2N+2} (N=1-6) using high pressure swarm technique, *J.Chem.Phys.* **80**, 6150 (1974).
6. I.Szamrej, H.Kość and M.Foryś, Thermal electron attachment to halomethanes - part 2. CH_2F_2, CHF_3 and $CClF_3$, *Radiat.Phys.Chem.*, **48**, 69 (1996).
7. D.L.McCorkle, I.Szamrej and L.G.Christophorou, Electron attachment to halocarbons, *J.Chem.Phys.*, **77**, 5542 (1982).
8. I.Szamrej, W.Tchórzewska, H.Kość and M.Foryś, Thermal electron attachment to halomethanes - part 1. CH_2Cl_2, $CHFCl_2$ and CF_2Cl_2, *Radiat.Phys.Chem.*, **47**, 269 (1996).
9. I.Szamrej, J.Jówko and M.Foryś, Thermal electron attachment to CHF_2Cl in mixtures with CO_2 an N_2, *Radiat.Phys.Chem.*, **48**, 65 (1996).
10. S.J.Burns, J.M.Matthews and D.L.McFadden, Rate coefficients for dissociative electron attachment by halomethane compounds between 300 and 800 K, *J.Phys.Chem.*, **100**, 19436 (1996).
11. E.Illenberger, Electron attachment reactions in molecular clusters, *Chem.Rev.*, **92**, 1589 (1992).

NEW APPROACH TO ANALYSIS OF THE DATA OF ELECTRON SWARM EXPERIMENT IN IONIZATION CHAMBER.

Andrzej Rosa, Mieczysław Foryś and Iwona Szamrej

Chemistry Department
Agricultural & Teachers University
08-110 Siedlce, Poland

INTRODUCTION

The electron swarm method for studying electron attachment processes usually consists of monitoring the rate of electron disappearance from the swarm as a function of density-reduced electric field, E/N. The swarm of electrons produced by an α-particle drifts under the influence of a uniform electric field to a positive collector plate. The electrode is connected to a linear pulse amplifier. The amplified output pulse is registered in a multichannel analyzer. The pulse height is a function both of drift velocity, W, and electron capture rate constant, k. From its dependence on E/N the k value can be determined[1-3]. This method, while effective, uses rather complicated equations to fit simultaneously the electron drift velocity and the electron capture rate constant. Moreover, it is rather time-consuming as one needs a good statistics to determine accurately the pik height for each E/N.

Here we present a new approach to this problem. Namely, instead of multichannel analyzer at the output of linear amplifier we have put an oscilloscope with digital memory. It enables us to register time evolution of the electric pulse. The details of the swarm experiment are fully described elsewhere[2]. The response of our (Camac) linear amplifier to the step function is described by eq.1:

$$G(t) = \exp\left(\frac{t}{\sqrt{3} \cdot t_1}\right) \cdot \sin\left(\frac{t}{t_1}\right) \qquad (1)$$

where t_1 is a shaping time of the linear amplifier.

The change in electrode potential due to the drift of the electron swarm is described by eq.2:

$$W(t) = \frac{A}{k \cdot N_a \cdot t_0} \times \left[1 - \exp(-k \cdot t)\right] \qquad (2)$$

Here $t_0 = d/W$ is the drift time, d is the distance traversed by electrons on their way to the collecting electrode and N_a is a number density of the attaching gas. As $W(t)$ is not a step function the output pulse of the amplifier is an integral:

$$V(\tau) = \int \frac{dW(t)}{dt} \cdot G(\tau - t) dt \qquad (3)$$

with the integral limits $(0,\tau)$ if $\tau < t_0$ and $(0,t_0)$ if $\tau > t_0$. The integration gives following expressions for $V(\tau)$:

$$V(\tau) = F \cdot \left\{ \sqrt{3} \exp(-k \cdot \tau) - \left[\sqrt{3} \cdot \cos\left(\frac{\tau}{t_1}\right) + E \cdot \sin\left(\frac{\tau}{t_1}\right) \right] \cdot \exp\left(-\frac{\tau}{\sqrt{3} \cdot t_1}\right) \right\} \qquad (4)$$

if $\tau < t_0$ and

$$V(\tau) = F \cdot \exp\left(-\frac{\tau}{\sqrt{3} \cdot t_1}\right) \cdot$$
$$\left\{ H \cdot \left[\sqrt{3} \cdot \cos\left(\frac{\tau \cdot W - d}{W \cdot t_1}\right) + G \cdot \sin\left(\frac{\tau \cdot W - d}{W \cdot t_1}\right) \right] - \sqrt{3} \cdot \cos\left(\frac{\tau}{t_1}\right) - G \cdot \sin\left(\frac{\tau}{t_1}\right) \right\} \qquad (5)$$

if $\tau > t_0$, where A is

$$E = 1 - \sqrt{3} \cdot t_1 \cdot k \qquad (6)$$

$$F = \frac{\sqrt{3} \cdot A \cdot t_1 \cdot W}{d \cdot (3 + E^2)} \qquad (7)$$

$$H = \exp\left(\frac{d \cdot E}{\sqrt{3} \cdot t_1 \cdot W}\right) \qquad (8)$$

Eqs (4) and (5) together describe the time evolution of the output potential which is observed by the oscilloscope. Computer fitting the experimental peak with these equations allows to determine both the drift velocity and electron capture rate constant. Because of background noise it is necessary to average some number of registered pulses. Our experience shows that it is around 100-200. An example peak for the C_2H_5Br-CO_2 mixture is shown in figure 1 together with the fitted theoretical curve. Figure 2a shows the electron capture rate constant for C_2H_5Br at different CO_2 pressures. The k value corresponds to the rate of the electron disappearance divided by the acceptor concentration. Figure 2b shows the density normalized electron mobility, $\mu_d = W/(E/N)$, at different concentrations of CO_2 and constant $[C_2H_5Br]/[CO_2]$ ratio measured at different E/N. The pictures show good reproducibility of the results. Both values are in the range of expected ones. As the paper is aimed to show just the ability of the proposed method we no comment farther on the electron capture process itself.

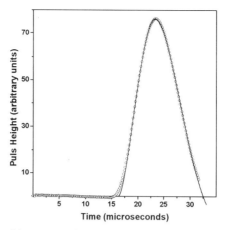

Figure 1. The time evolution of the output pulse at $t_1 = 4.4$ µs. The experimental data are the average of 100 pulses. The line is Simplex computer fit of the $V(\tau)$ function.

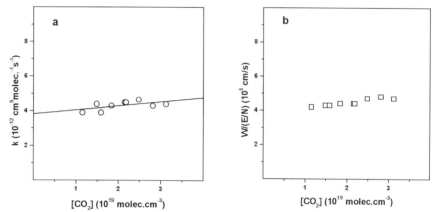

Figure 2. Thermal electron capture rate constant for C_2H_5Br (a) and thermal electron mobility (b) as a function of CO_2 concentration in the mixture C_2H_5Br - CO_2.

Concluding, we propose the new method for analysis of the electron swarm data consisting in fitting full output signal of the electron pulse. This approach is more straightforward and hence seems more reliable than the analysis of the pulse height. It also shortens greatly the time of experiment from many hours to just minutes.

REFERENCES

1. L.G. Christophorou, *Atomic and Molecular Radiation Physics*, Wiley Interscience (1971).
2. O.W. Dmitriev, W. Tchórzewska, I. Szamrej and M. Foryś, Thermal electron mobilities in low density gaseous mixtures, *Radiat.Phys.Chem.*, **40**, 547 (1992).
3. I. Szamrej, H. Kość and M. Foryś, Thermal electron attachment to halomethanes - part 2. CH_2F_2, CHF_3 and $CClF_3$, *Radiat.Phys.Chem.*, **48**, 69 (1996).

ELECTRON ATTACHMENT TO O_2 IN DENSE He AND Ar GASES

A. F. Borghesani and M. Santini

Istituto Nazionale per la Fisica della Materia
Department of Physics "G.Galilei"
University of Padua, Italy

INTRODUCTION

Electron attachment to molecular impurities in a gas sheds light on the energetics and density of states of excess electrons in disordered systems. At low densities, the attachment of an electron to O_2 is a resonant process described by the so-called Bloch-Bradbury two-stage mechanism[1]: firstly, an unstable O_2^- ion is formed into a vibrationally excited state. Then, collision of the ion with an atom of the host gas carries away the excess energy and stabilizes the ion.

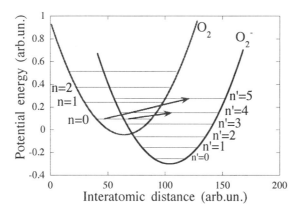

Fig. 1 Two-stage Bloch-Bradbury attachment process $O_2+e \to O_2^{-*}$; $O_2^{-*}+B \to O_2^- +B$ where B is a third body. The two lines show two possible transitions from the molecule ground state $n=0$ to the two first accessible vibrational levels of the ion $n'=4$ and $n'=5$.

The energy diagram of the process is shown in Fig. 1. The resonant energies of the transitions $n = 0 \to n' = 4$ and $n' = 5$, averaged over the spin-orbit doublets, are $E_R^{(4)} = 91\,meV$ and $E_R^{(5)} = 207\,meV$. Experiments carried out in dense He gas in the range $50 \leq T \leq 100\,K$ [2,3] have shown the presence of a sharp peak in the density dependence of the reduced attachment frequency ν_A/N at a density $N_M \approx 3.0\,atoms\,nm^{-3}$, and, at $T = 77\,K$, the first shoulder of a second peak at much higher N. Owing to the low T, it was argued that the most significant contribution to the resonant energy is due to the self-energy V_0 of excess electrons in dense gases, that, for He, can be approximated by the Fermi shift

$$E_F(N) = (2\pi\hbar^2/m)Na \equiv bN \qquad (1)$$

where a is the electron–He scattering length, m the electron mass, and N the gas density. Since $a = 0.063\,nm$, $E_F(N_M) \approx E_R^{(4)}$. Measurements in dense Ne gas at $45 \leq T \leq 100\,K$ confirmed the existence of both a first and second peak[4]. However, some observations remained unexplained, namely, the anomalous T dependence of N_M and of the width of the ν_A peaks. We therefore carried out measurements of ν_A in gaseous He and Ar, where $V_0 < 0$ in extended N and T ranges[5].

EXPERIMENTAL DETAILS AND RESULTS

We used the pulsed photoemission apparatus exploited for electrons and ions mobility measurements in high–density noble gases[6,7]. The cell can be filled with gas up to $10\,MPa$ and is thermoregulated within $0.01\,K$ in the range $25 \leq T \leq 330\,K$. Impurities like H_2O and hydrocarbons are removed by circulating the gas trough a LN_2–cooled activated charcoal trap. To lower the O_2 content to the desired level, an Oxisorb trap is inserted in the recirculation loop. The final O_2 impurity content is estimated to be $0.1 - 1\,p.p.m.$ The electrons photoinjected into the gap between two parallel-plate electrodes drift under the influence of an electric field. During the flight they are captured by O_2 and the measured current shows an exponential decay in time. By analyzing the signal waveform[8] the time constant τ and the attachment frequency $\nu_A = 1/\tau$ are determined.

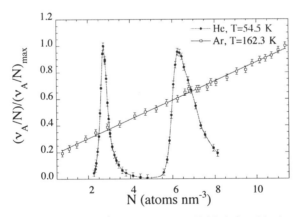

Fig. 2 Reduced attachment frequency ν^* in He at $T=54.5\,K$ (circles) and in Ar at $T=162.3\,K$ (squares) as a function of the gas density. The solid lines are a guide for the eye only.

In Fig. 2 we show the reduced attachment frequency $\nu^* = (\nu_A/N)/(\nu_A/N)_{max}$ as a function of N for He at $T = 54.5\,K$ and for Ar at $T = 162.3\,K$. In He we observe two complete peaks, while no structure is observed in Ar. An explanation will be proposed next. In Fig. 3 we show the first peak of ν^* as a function of N for He at $T = 77.2, 120.1, 152.3\,K$. One can observe that the width of the first peak (shown in

Fig.4) as well as its position N_M (shown in Fig. 5) increase with T.

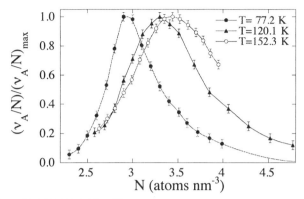

Fig. 3 $\nu^\star$ in He at $T=77.2, 120.1, 152.3\, K$.

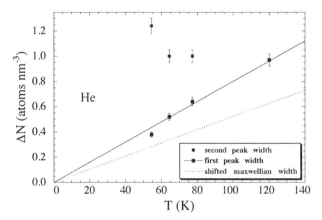

Fig. 4 Widths of the $\nu^\star$ peaks in He as a function of T. The solid line is only a guide for the eye; the dotted one is the width of a Maxwellian distribution function.

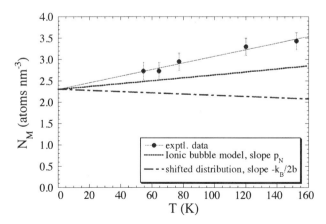

Fig. 5 Temperature dependence of the position N_M of the first attachment peak in He. Solid line: linear fit to the data; dash–dotted line: shifted Maxwellian distribution function model ; dotted line: ionic bubble model.

DISCUSSION

To interpret the data, we assume that the vibrational levels of O_2 and of O_2^- in a dense gas are the same as *in vacuo*. The resonant condition for the electron energy E is

$$E = (E_n - E_0) - E_A(N,T) \qquad (2)$$

where $(E_n - E_0)$ is the vibrational energy of O_2^- with respect to the O_2 ground state. $E_A(N,T)$ is the electron affinity of O_2 in a gas at given N and T. Thus, $E_R^{(n)} = (E_n - E_0) - E_A(N=0)$. The electron affinity can be written as $E_A(N,T) = E_A(N=0) + \Delta(N,T)$, where $\Delta(N,T) \approx (\alpha e^2 N/2) \int g(r) S(r) r^{-4} d^3r$ is due to polarization contributions[9]. α is the atomic polarizability of the gas, $g(r)$ is the ion–atom pair correlation function and $S(r) \approx (1 + 8\pi\alpha N/3)^{-1}$ is the Lorentz screening factor. An estimate of $\Delta(N_M = 3\, atoms\, nm^{-3}, T)$ is $\approx 10\, meV$ for He and $\approx 80\, meV$ for Ar. The attachment frequency can be written as[3]

$$\nu_A = p_s \sigma_c N_2 v(E_R) F(E_R, N) \qquad (3)$$

where p_s is the stabilization coefficient, σ_c is the integrated capture cross section, $N_2 = CN$ is the O_2 number density, $v(E_R)$ is the electron velocity at the resonance energy and $F(E_R, N)$ is the Maxwellian electron energy distribution function. Assuming[3] that the product $p_s \sigma_c$ does not depend much on N, and working at constant (even though unknown) concentration C of impurities, we obtain

$$\frac{\nu_A}{N} \propto F(E_R, N) \qquad (4)$$

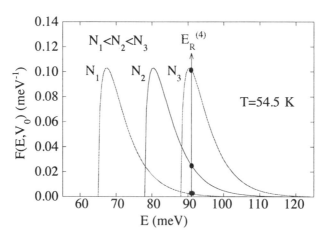

Fig. 6 Sampling of the shifted Maxwellian energy distribution function by means of the resonant attachment process at a fixed energy window of 91 meV.

In a dense gas, the electron ground–state energy is V_0 and, neglecting fluctuations, the electron energy distribution function is shifted by this amount with respect to vacuum. This shift of F has been put into evidence in excess electron mobility measurements [6,10,11]. The existence of a ν_A peak in He can be, therefore, understood by looking at Fig. 6, where the distribution function is sampled at the constant energy of 91 meV. By looking at Fig. 7, where the values of V_0 for He and Ar [12] are compared, and recalling that polarization lowers the ion energy, $\Delta(N,T) < 0$, it is easy to understand that in Ar both the distribution function and the ion vibrational levels are negatively shifted and the resonance conditions might not be met.

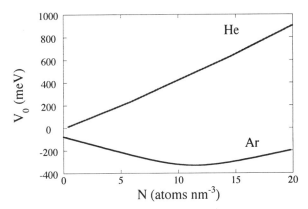

Fig. 7 Comparison between the self–energy V_0 of excess electrons in He and Ar as a function of N.

At a temperature T and at a density $N = N_M$, the most probable electron energy is $E_M = V_0(N_M) + k_B T/2$ and ν_A/N has the maximum when

$$E_M = V_0(N_M) + \frac{k_B T}{2} = E_R^{(4)} - \Delta(N_M, T) \tag{5}$$

The widtht at half height of a Maxwellian distribution is $\Delta E = 1.8 k_B T$. This width can be converted into the density width of the observed attachment peak by using the relation (1), thus obtaining $\Delta N = \Delta E/b = 1.8 k_B T/b \equiv aT$. In Fig. 4 we compare the experimental slope $a_{exp} = 8.0 \times 10^{-3}\ atoms\,nm^{-3}\,K^{-1}$ (solid line) with the value predicted by the Fermi shift (1) (dotted line) $a = 5.14 \times 10^{-3}\ atoms\,nm^{-3}\,K^{-1}$. On the contrary, the width of the second peak in He decreases with increasing T. This fact is probably due to the presence of a large amount, at lower T, of localized electrons self-trapped in bubble states [13] at energies $E_B \ll V_0$. These electrons can still get attached to O_2 even though N is so large that $V_0 \gg E_R$, because in this case $E_B \approx E_R$. Therefore, the attachment frequency is still large and the second attachment peak is broader than the first one, but the amount of trapped electrons decreases with increasing T and so does the width of the peak.

Within the simple model of the shifted Maxwellian distribution function, in He, where Δ can be neglected, the density of the first attachment peak N_M is related to T by the relation

$$N_M = b^{-1}\left(E_R^{(4)} - \frac{1}{2}k_B T\right) \tag{6}$$

and decreases with T as shown in Fig. 5 by the dash–dotted line, while experimentally N_M increases with T. The discrepancy between experiment and model can be partially reconciled by adopting the *ionic bubble model* developed by Khrapak et al. [14]. This model takes into account the fact that the extraelecton in the ion is pretty delocalized in space and its interaction with the atoms of the host gas in treated within the optical model. Because of the repulsive interaction of the outer electron and the gas atoms, leading to a positive V_0, an empty void of radius R surrounds the ion and the free energy of creating this bubble has to be taken into account in the energy balance, because the attachment process [15] and the creation of the bubble have comparable durations [16]. The free energy variation ΔA due to the formation of a bubble of radius R, neglecting surface tension contributions, is given by

$$\Delta A(R) = -\epsilon(R) + \frac{4}{3}\pi R^3 P \qquad (7)$$

where P is the gas pressure and $\epsilon(R)$ is the electron energy eigenvalue obtained by solving the Schrödinger equation for the electron immersed in the potential

$$V(r) = \begin{cases} V_0 - \alpha e^2/2(4\pi\epsilon_0)r^4 & \text{for } r \geq R \\ -\alpha e^2/2(4\pi\epsilon_0)r^4 & \text{for } R_0 < r \leq R \\ \infty & \text{for } r \leq R_0 \end{cases}$$

where R_0 is the hard-core effective radius of the ion. Eq. (7) is minimized with respect to R, thus obtaining the optimum radius R_B as a function of N and T. Since the ground state of the outer electron in the ion is shifted with respect to the case of the isolated ion, we assume that the vibrational levels are also shifted by the same amount. This means that the resonance energy is shifted to $E_R^{(n)} + \Delta E$ with $\Delta E = -E_A(N=0) + \Delta A(R_B(N,T))$. Owing to the large volume work $\propto PR_B^3$, the resonance energy increases with T and this explains why N_M also increases with T. We found that $\Delta E = \Delta E_0 + pT$. By recalling Eq.(1) and Eq. (5), we finally obtain

$$N_M = N_M(0) + \frac{p - k_B}{b}T = N_M(0) + p_N T \qquad (8)$$

The dotted line in Fig. 4 is Eq. (8). The slope of the fit to the data is $p_{exp} = 7.7 \times 10^{-3}\, atoms\, nm^{-3}\, K^{-1}$, to be compared to the calculated value $p_N = 3.4 \times 10^{-3}\, atoms\, nm^{-3}\, K^{-1}$. We observe a significative improvement over the simple model of the shifted Maxwell distibution. This model, in fact, predicts a positive slope of N_M as a function of T, as experimentally observed. The inclusion of surface tension contributions in the calculation of the free energy should even improve the results.

REFERENCES

1. Y. Hatano and H. Shimamori, in *Electron and Ion swarms*, edited by L. G. Christophorou (Pergamon, New York, N.Y., 1981)
2. A. K. Bartels, *Phys. Lett.*, **45 A**: 491 (1973)
3. L. Bruschi, M. Santini, and G. Torzo, *J. Phys.*, **B 17**: 1137 (1984)
4. A. F. Borghesani and M. Santini, in *Gaseous Dielectrics VI*, edited by L. G. Christophorou and I. Sauers (Plenum, New York, N.Y, 1991)
5. Dino Neri, A. F. Borghesani, and M. Santini, *Phys. Rev. E* **56**: 2137 (1997)
6. A. F. Borghesani, L. Bruschi, M. Santini, and G.Torzo, *Phys. Rev. A* **37**: 4828 (1988)
7. A. F. Borghesani, D. Neri, and M. Santini, *Phys. Rev. E* **48**: 1379 (1993)
8. A. F. Borghesani and M. Santini, *Meas. Sci. Technol.*, **1**: 939 (1990)
9. J. P. Hernandez and L. W. Martin, *Phys. Rev. A* **43**: 4568 (1991)
10. A. F. Borghesani and M. Santini, *Phys. Rev. A* **42**: 7377 (1990)
11. A. F. Borghesani, M. Santini, and P. Lamp, *Phys. Rev. A* **46**: 7902 (1992)
12. R. Reininger, U. Asaf, I. T. Steinberger, and S. Basak, *Phys. Rev. B* **28**: 4426 (1983)
13. J. P. Hernandez, *Rev. Mod. Phys.*, **B 63**: 675 (1991)
14. A. G. Khrapak, K. F. Volykhin, and W. F. Schmidt, *Phys. Rev. E* **51**: 4426 (1995)
15. L. G. Christophorou, *Adv. Electron. Phys.*, **B 46**: 55 (1978)
16. J. P. Hernandez and M. Silver, *Phys. Rev. A* **2**: 1949 (1970) and *Phys. Rev. A* **3**: 2152 (1971)

ELECTRON MOBILITY AND LOCALIZATION IN HIGH–DENSITY HELIUM GAS IN AN EXTENDED TEMPERATURE RANGE

A. F. Borghesani and M. Santini

Istituto Nazionale per la Fisica della Materia
Department of Physics "G.Galilei"
University of Padua, Italy

INTRODUCTION

Information on the electronic states of disordered systems can be gathered from the analysis of the transport properties of excess electrons in a dense gas. The coupling of a quantum light particle to its environment may result in different equilibrium states, depending on the coupling constant between them and on the response function of the environment. A very important problem is that of an electron interacting with a gas of hard–spheres. One experimental realization of such a system is an excess electron immersed in dense He gas. At low gas density the equilibrium state of excess electrons is a propagating one. The electron wavefunction is pretty much delocalized and the interaction of the excess electron with the atoms of the host gas is treated witihin the formalism of kinetic theory by introducing the electron–atom scattering cross section [1]. The observed excess electron mobility is relatively high. By increasing the gas density, and thus changing the response function of the environment, it happens that the particle non–perturbatively alters its environment so as to give origin to a new equilibrium state which is completely different from the state in the decoupled system of particle plus environment. In this case the electron deforms self–consistently the gas density around itself, and, owing to the electron–He atoms repulsive interaction, its wavefunction becomes localized in a narrow region of space where the local gas density is lower than average. As a result electrons become self–trapped in a partially filled cavity dug out in the gas and the electron state is non propagating. In this case, however, the complex electron plus cavity can still have a diffusive motion because the gas is compliant to deformations. Experimentally, the process of self-trapping of excess electrons can be observed as a strong drop of the electron mobility [2] as a function of density at constant temperature and it is a general behavior shown by gases, whose interaction with electrons is essentially repulsive. In fact, this self–trapping transition as a function of the gas density has been observed also in Ne [3]. Detailed reviews can be found in literature [4].

Measurements, and interpretations, have been aimed at the study of the transition between the quasi–free and localized electron regimes. However, they usually focused on the two phenomenologically distinct aspects of the problem: the low–density behavior of the mobility and the high–density one. On one hand, the high–temperature measurements, concerned with the low–density behavior, were not extended to high enough densities to observe the low–mobility regime typical of the bubble state. On the other hand, the low–T measurements did not produce data accurate enough at low–density to allow comparison with the high–T data. Also theoretically, the full problem has been treated in two separate halves. The goal of the low–density data analysis was to improve and extend the classical kinetic theory of free–electron transport taking into account the fact that even at high–T and low–N the electron thermal wavelength may be larger than the average interatomic distance and comparable to its mean free path. On the contrary, the analysis of the low–T, high–N data focused on the problem of how to calculate the most probable bubble state by suitably minimizing the free energy of the system. Until now, electron mobility measurements in dense He gas, aimed at the study of the full self–trapping transition, were limited to pretty low temperatures[2] between $T = 2.7\,K$ and $T = 20.3\,K$. We have therefore decided to extend this temperature range, in order to see the effect of temperature on the onset of self–trapping[5].

EXPERIMENTAL RESULTS AND DISCUSSION

We carried out electron mobility measurements using the same apparatus used for Ne[6]. We refer to literature for details. The temperature range explored is $26 \leq T \leq 300\,K$. Since the cell can be filled with gas up to $10\,MPa$ only, we were able to observe the self–trapping transition for $T \leq 77\,K$.

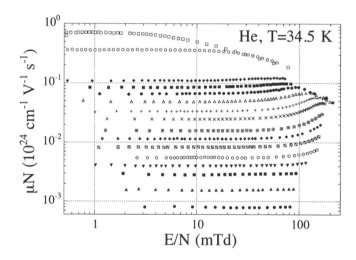

Fig. 1 μN as a function of the reduced electric field E/N ($1mTd=10^{-20}\,V cm^2$) for several $N=$ $(2.59, 17.0, 32.1, 34.5, 36.8, 39.2, 41.7, 43.8, 46.6, 48.3, 50, 52.1, 53.9, 55.6, 58.8, 63.5) \times 10^{20}\,cm^{-3}$ (from top).

In Fig. 1 we report a typical result for the electric field dependence of the electron density–normalized mobility μN for several densities N. At low N μN shows the usual behavior for a gas of hard–sphere scatterers. At small E/N electrons are in thermal equilibrium with the atoms of the gas and μN is independent of E/N. For higher E/N electrons get epithermal and μN decreases with increasing E/N, approximately as $(E/N)^{-1/2}$. As N increases above a threshold value around $N \approx (30-35) \times 10^{20}\,cm^{-3}$ which depends on T, the behavior of μN as a function of E/N radically changes. For small E/N μN is again independent of E/N. However, for larger N μN increases with E/N, until it meets the classically expected $(E/N)^{-1/2}$ behavior. The threshold value of E/N where μN starts increasing over its zero–field value shifts to larger E/N as N increases. This behavior was observed also in Ne[3]. Above the density threshold there is coexistence of self–trapped and quasi–free electrons and the resulting mobility is a weighted sum of the contributions of the two kind of charge carriers. As E/N is increased, the fraction of quasi–free states increases, either because the energy gained by the electrons from the electric field is large enough for them to escape from the cavity they are sitting in or because the electric field inhibits the formation of the self–trapped state.

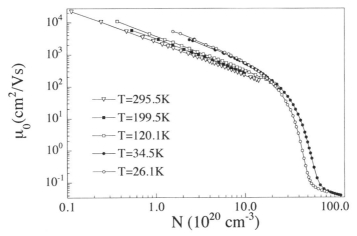

Fig. 2 Zero–field mobility μ_0 as a function of the gas density N for several temperatures $26.1 \leq T \leq 295.5\,K$. The lines are only q guide for the eye.

The mobility value extrapolated to zero–field μ_0 is reported for several T in Fig. 2. For the highest T we have been able to investigate only the low–N region, while for the lowest T we have been able to measure the complete self–trapping transition. In Fig. 3 μ_0 is shown in a restricted N–range for $26.1 \leq T \leq 77.2\,K$. The precipituous drop of μ_0 as a function of N clearly signals the self–trapping transition.

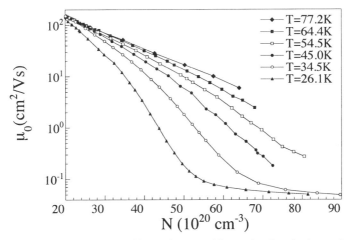

Fig. 3 μ_0 as a function of N in the self-trapping transition region for the lowest investigated $T = (26.1, 34.5, 45.0, 54.5, 64.4, 77.2)\,K$.

It can be seen that the transition region shifts to higher N as T increases. This behavior can be explained in terms of a simple quantum mechanical model of an electron in a well[3]. The variation of free energy E_B of an electron trapped into a partially empty cavity, whose walls height depends on N, with respect to the energy V_0 of a quasi–free propagating electron can be calculated as a function of N and T. Upon minimizing $E_B - V_0$, one obtains the optimum cavity radius and filling fraction describing the most probable self–trapped state. Self–trapped states form as soon as $(E_B - V_0) = 0$, although they are not absolutely stable against fluctuation until $-(E_B - V_0) \gg k_B T$. Within this model it turns out that the density for which $(E_B - V_0) = 0$ increases with T. This is essentially the result of the balance between the increase of V_0 with N and the increased volume work needed to expand the cavity. All these simple models based on the solution of the Schrödinger equation for an electron inside a cavity and on the minimization of the free energy of the system, however, are too crude to obtain agreeement with the experimental data. At best, they are able to reproduce semiquantitatively the density values for which the self–trapping transition occurs[7]. Moreover, they strongly rely on the model used for the description of the mobility of the quasi–free electron state.

Instead of pursuing here one such model, we prefer to illustrate how the suitable application of the heuristic Padua model developed for the description of the mobility of the quasi–free electron state in dense noble gases gives a reasonably good agreeement with the experimental data over the complete density range[8]. In this model, three multiple scattering effects concur to dress the electron–atom momentum–transfer scattering cross section: correlations among scatterers taken into account by means of the static structure factor $S(0)$, a density dependent quantum shift of the ground state energy of an excess electron in a dense gas (and, in particular, a shift E_k of its ground state kinetic energy), and, eventually, an increase of the scattering rate due to self–interference of the electron wavepacket multiply scattered off different atoms. All these effects depend on the ratio λ_T/ℓ, where λ_T is the thermal de Broglie wavelength of the electron and ℓ is its mean free path. When ℓ gets comparable with λ_T the quantum nature of the electron and the disordered character of the gas strongly modify the electron transport properties. Of the above mentioned effects, the first effect is not very relevant in He at the temperatures of this experiment since the gas is very far away from criticality and $S(0) \approx 1$. Also the second effect is not as important as in Ne[6] and Ar[8], because the electron–He atom scattering cross section does not depend strongly on the electron energy. Therefore, a shift of the ground state kinetic energy has little effect on the evaluation of the cross section. On the contrary, the third effect is of great importance here because the cross section is large and hence ℓ is relatively small. In this situation, when $\lambda_T/\ell \approx 1$ the Ioffe–Regel localization criterion is easily satisfied[9]. The onset of a mobility edge, an energy threshold below which electrons are Anderson–localized, can be explained by summing to all orders of perturbation the first order correction to the cross section due to Atrazhev and Iakubov[10], thus obtaining a result similar to that of Polischuk[11]. The effective scattering rate is given by $\nu = \nu_0/(1 - 2\lambda/\ell_0)$, where $\nu_0 = (2\epsilon/m)^{1/2} N \sigma_m(\epsilon)$ is the classical scattering rate with ϵ the electron energy and σ_m the energy–dependent momentum–transfer scattering cross section. $\ell_0 = 1/N\sigma_m$ is the classical mean free path and $\lambda = \hbar/(2m\epsilon)^{1/2}$ is the de Broglie wavelength of the electron. The mobility edge ϵ_c is naturally defined as the threshold energy at which the effective scattering rate diverge. We thus obtain $\epsilon_c = (2/m)[\hbar \sigma_m(\epsilon_c)N]^2$ corresponding to the realization $\lambda/\ell_0 = 1/2$ of the Ioffe–regel criterion for localization. Taking into account this mobility edge into the heuristic Padua model[8] we obtain for the zero–field density normalized mobilty $\mu_0 N$

$$\mu_0 N = \frac{4e}{3(2\pi m)^{1/2}(k_B T)^{5/2}} \int_{\epsilon_c}^{\infty} \epsilon\, e^{\epsilon/k_B T} \frac{\left[1 - f\frac{\hbar S(0) N \sigma_m(\epsilon+E_k)}{(2m\epsilon)^{1/2}}\right]}{S(0)\sigma_m(\epsilon+E_k)} d\epsilon \qquad (1)$$

and the mobility edge, corrected for multiple scattering effects, turns out to be given by $\epsilon_c = (1/2m)\left[f\hbar S(0)\sigma_m(\epsilon_c + E_k)N\right]^2$. In these formulas m is the free electron mass, E_k is the density dependent shift of the ground state kinetic energy of electrons, and f is a numerical factor of order unity. In Eq.(1) the contribution to the mobility of the localized states with energy $\epsilon < \epsilon_c$ is neglected.

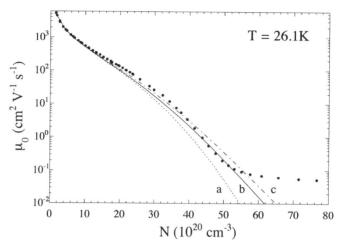

Fig. 4 Prediction of the heuristic Padua model with inclusion of the mobility edge. Curve a: $f=2\pi/3$ and $S(0)=1$. Curve b: $f=2\pi/3$ and the true $S(0)$. Curve c: $f=2$ and the true $S(0)$.

The predictions of this model are shown in Fig. 4 for $T = 26.1\,K$. Different curves have been obtained by using different f values suggested in literature. The agreement is very good over 5 decades. Similar results are obtained for all other temperatures. The low−N μ_0 is well described, since at low N $\epsilon_c \approx 0$ and the Padua model[8] is fully recovered, giving an agreement with the experimental data comparable to that obtained for Ne[6] and Ar[8]. The agreement with the data at the highest densities is not good because this model does not allow for the localized electron to move.

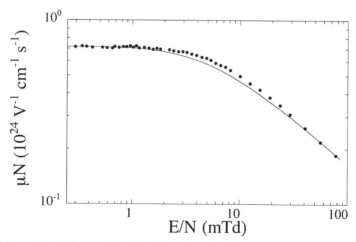

Fig. 5 Prediction of the Padua model with mobilty edge for μN as a function of E/N at $T=34.5\,K$ and $N=2.59\times 10^{20}\,cm^{-3}$.

This model can also be used for the description of the electric field dependence of μN. In Fig. 5 and Fig.6 we show the result of the model for low−density (Fig. 5) and high−density (Fig. 6). The field dependence is well described at low N. At higher N, although the agreement is not equally good (see Fig. 4), the model describes relatively well the data. By increasing E/N the electron energy distribution function gets broader and the fraction of states with energy $\epsilon < \epsilon_c$ decreases with increasing E/N. The resulting μ, a weighted sum over quasi–free and localized states, increases as well. We finally point out that the relationship between states below ϵ_c and Anderson–localized states, non–percolating states, or self–trapped one is not yet clear. Since the gas is compliant to deformations, it may happen that an electron below the mobility edge has time enough to deform the fluid getting self–trapped. It is not clear also if there is a dynamic exchange between states below and above ϵ_c. In other words, the dynamics of localization is not yet fully explained.

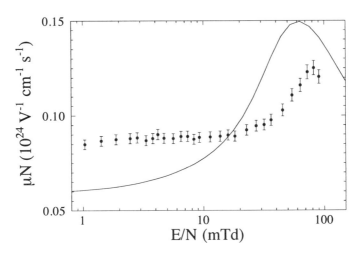

Fig. 6 Prediction of the Padua model with mobility edge for $\mu N(E/N)$ $\left(T=26.1\,K,\ N=29\times10^{20}\,cm^{-3}\right)$.

REFERENCES

1. L. G. H. Huxley and R. W. Crompton, *The diffusion and drift of electrons in gases* (Wiley, New York, N.Y., 1974)
2. J. L. Levine and T. M. Sanders, *Phys. Rev. Lett.* **8**: 159 (1962); H. R. Harrison and B. E. Springett, *Chem. Phys. Lett.*, **10**: 418 (1971) and *Phys. Lett.*, **A 35**: 73 (1971); J. A. Jahnke, M. Silver, and J. P Hernandez, *Phys. Rev.* **B12**: 3420 (1975); K. W. Schwarz, *Phys. Rev.* **B21**: 5125 (1980)
3. A. F. Borghesani and M. Santini, *Phys. Rev.* **A42**: 7377 (1990)
4. J. P. Hernandez, *Rev. Mod. Phys.*, **B63**: 675 (1991); A. F. Borghesani and M. Santini, in *Linking the Gaseous and Condensed Phases of Matter. The Behavior of Slow Electrons.*, L. G. Christophorou, E. Illenberger, and W. F. Schmidt Editors, NATO ASI Series vol. **326**, p.281 (Plenum, New York, N.Y. 1994)
5. A. M. De Riva, Ph.D. Thesis (University of Padua, 1994) unpublished; D. Neri, Ph.D. Thesis (University of Padua, 1997) unpublished
6. A. F. Borghesani, L. Bruschi, M. Santini, and G.Torzo, *Phys. Rev.* **A37**: 4828 (1988)
7. L. W. Martin, Ph.D. Thesis (University of North Carolina at Chapel Hill, 1991) unpublished
8. A. F. Borghesani, M. Santini, and P. Lamp, *Phys. Rev.* **A46**: 7902 (1992)
9. N. Mott, *Conduction in non–crystalline Materials* (Oxford University Press, Oxford, 1993)
10. V. M. Atrazhev and I. T. Iakubov, *J. Phys.*, **D10**: 2155 (1977)
11. A. Ya. Polischuk, *Physica*, **124 C**, 91 (1984)

O_2^- ION MOBILITY IN DENSE SUPERCRITICAL AND NEAR CRITICAL ARGON GAS

A. F. Borghesani and M. Santini

Istituto Nazionale per la Fisica della Materia
Department of Physics "G.Galilei"
University of Padua, Italy

INTRODUCTION

Understanding the properties of the motion of ions in gases and liquids is of great interest both from a fundamental as well as a practical point of view. Information on the scattering processes between a massive charged particle and the atoms of a host medium and on its microscopic structure can be deduced from the analysis of the ion motion. At very low density ($N \leq 10^{-5}\, atoms\, nm^{-3}$) the scattering cross sections for momentum transfer relative to the interaction between an ion and an atom or a molecule are extracted from ion drift mobility (μ) data within the Boltzmann formalism of kinetic theory [1]. On the other hand, hydrodynamic interactions explain the ion motion in high density liquids [2]. There is, however, a lack of information in the intermediate density region, important because of the transition from the kinetic to the hydrodynamic regime. Moreover, understanding the ionic conduction can explain the prebreakdown and breakdown phenomena in dielectric fluids.

In the recent past, we have carried out mobility measurements of O_2^- ions in noble gases [3,4,5]. This species is easily produced by resonant electron attachment to O_2 molecular impurities [6], always present in trace even in the best purified commercially available gas, in the pulsed photoemission experiment aimed at the measurement of electron mobility [7]. This technique has several advantages: the same ionic species is used as a probe for every gas, so that gas specific features may be put into evidence; the simple and reliable pulsed photoemission technique can be used; and the ionic concentration is low enough to neglect ion–ion interactions.

We report here mobility measurements of O_2^- impurity ions in dense supercritical and near critical Argon gas. The measurements were carried out in the temperature range $151.5 \leq T \leq 300\, K$, spanning a range of gas densities from dilute gas up to values comparable to those of the liquid, $0.1 \leq N \leq 14.0\, atoms\, nm^{-3}$ (Ar critical temperature $T_c = 150.86\, K$, critical density $N_c = 8.08\, atoms\, nm^{-3}$). The goal is to explore the transition from the kinetic to hydrodynamic transport regimes. At low density the experimental results show a strong temperature dependence in disagreement with the prediction of kinetic theory. At higher N the ionic mobility can be explained in terms of a hydrodynamic model including electrostriction effects and clustering of Ar atoms around the ion.

EXPERIMENTAL DETAILS AND RESULTS

We used the same pulsed photoinjection technique developed for the drift mobility measurements of excess electrons in dense nonpolar gases [7]. Details can be found in literature. We only recall that O_2^- ions are produced by electron attachment to O_2 impurities present in a concentration of $\approx 10^{-5}$. The ions move in the drift gap between two parallel–plate electrodes under the influence of an externally applied electric field and the drift time τ is measured by analyzing the signal waveform [8]. The mobility is calculated from τ as $\mu = d^2/\tau V$, where d is the electrode spacing and V is the applied voltage. High densities can be obtained because the experimental cell can be filled with gas up to $\approx 10\,MPa$ and can be thermoregulated in the neighborhood of the critical point of Ar.

ELECTRIC FIELD DEPENDENCE OF THE MOBILITY

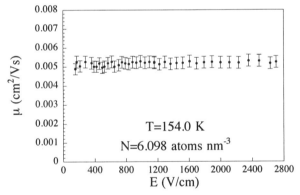

Fig. 1 Typical electric field dependence of the ion mobility. In this case $T=154.0\,K$ and $N=6.098\,atoms\,nm^{-3}$.

In Fig. 1 we show the mobility μ of O_2^- in Ar gas at $T = 154.0\,K$ and $N \approx 6.098\,atoms\,nm^{-3}$ as a function of the electric field E. μ shows a very small positive slope. This behavior, although not yet understood, is shown at every T and N and was detected previously also in Ne and He [3,4]. Anyway, it indicates that ions are practically in thermal equilibrium with the atoms of the gas even at the highest electric field strengths of this experiment.

LOW–DENSITY RESULTS

In order to compare different data sets with the same T but different N, we plot the density–normalized mobility extrapolated to zero–field $\mu_0 N$. In Fig. 2 we show $\mu_0 N$ for $T = 280\,K$ and $T = 180\,K$. $\mu_0 N$ depends linearly on N, as already observed in He [4,9] and Ne [3,4]. This behavior, as well as the E–dependence of μ, are in contrast with the kinetic theory prediction of μ independent of E and $\mu_0 N$ independent of N.

At the lowest densities of the experiment, $N \approx 0.1\,atoms\,nm^{-3}$, corresponding to an average interatomic distance $\bar{d} \approx 2.2\,nm$, the ion mean free path ℓ can be estimated from μ using the Drude result as $\ell = (3k_B TM)^{1/2}(\mu/e)$, where M is the reduced mass. We obtain $0.4 \leq \ell \leq 0.6\,nm$, a value comparable to the atom and ion radii σ. Although kinetic theory is fully applicable only if $\ell/\sigma \geq 200$, nonetheless we apply it to our case owing to the lack of a theory valid for this intermediate N range. In order to compare different isotherms we extrapolate $\mu_0 N$ to $N = 0$, yielding $(\mu_0 N)_0$, and plot in Fig. 3, as a function of T, $\mu_r = (\mu_0 N)_0 / N_{ig}$, where $N_{ig} = 2.687 \times 10^{-2}\,atoms\,nm^{-3}$ is the density of an ideal gas at S.T.P.

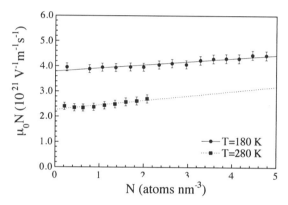

Fig. 2 Density dependence of $\mu_0 N$ for $T=180\,K$ and $T=280\,K$. The lines are only a guide for the eye.

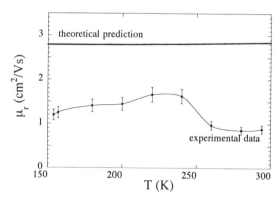

Fig. 3 Zero–field zero–density density–normalized mobility μ_r as a function of T in Ar gas. The thick solid line is the prediction of the kinetic theory, while the line through the data points is an eyeguide.

The thick solid line is practically the so-called *polarization limit* of the reduced mobility, calculated according to the kinetic theory as $(\mu_r)_P = 13.853(\alpha M)^{-1/2}$, where α is the atomic polarizability of the gas (in $Å^3$) and M is the reduced mass in $g/mole$. As already observed in He and Ne [3,4], also in Ar the predictions of the kinetic theory are in strong disagreement with the experiment. We can therefore conclude that even at the lowest densities of the experiment the conditions for the applicability of the kinetic theory are not fulfilled.

HIGH–DENSITY RESULTS

In Fig. 4 we plot $\mu_0 N$ as a function of N for $T = 154.0\,K$ and $T = 151.5\,K$. Its behavior is completely different than for the higher isotherms. In particular, at $T = 151.5\,K = T_c + 0.64\,K$, $\mu_0 N$ shows a deep minimum for $N = 6.2\,atoms\,nm^{-3} < N_c = 8.08\,atoms\,nm^{-3}$.

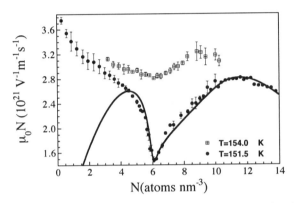

Fig. 4 Density–normalized zero–field mobility of O_2^- in Ar gas at $T=151.5\,K$ and $T=154.0\,K$. The solid line is the result of the present model.

Since the mobility data in Ne [3,10] were reproduced fairly well by the hydrodynamic Stokes formula for the mobility

$$\mu = \frac{e}{6\pi\eta R_h} \qquad (1)$$

where η is the gas viscosity and R_h is an effective hydrodynamic radius, we assume that it works well also in Ar at high N. In Fig. 5 we plot the effective hydrodynamic radius R_h for $T = 151.5\,K$ obtained by inverting the Stokes formula. We used literature viscosity data [11]. The behavior of R_h mirrors that of $\mu_0 N$. In order to explain the fact that the maximum of R_h and the minimum of $\mu_0 N$ occur for $N < N_c$, we assume that a layer of correlated fluid, proportional to the fluid correlation length and, hence, dependent on the gas static structure factor $S(0)$, enhances the effective ion radius. This effect is particularly strong near the critical point where the gas compressibility is very large. Owing to the electrostriction induced by the ion, there is an augmentation of the local density N_r of the gas around the ion in the cybotactic sphere. A density profile sets in around the ion and at a given distance from it the local density takes on the critical value, if at large distance from the ion the unperturbed gas density is $N < N_c$.

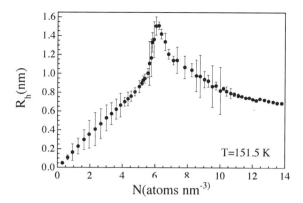

Fig. 5 Hydrodynamic Stokes radius R_h for $T=151.5\,K$.

The density profiles are calculated according to the electrostriction model [12]

$$\frac{e^2}{2(4\pi\epsilon_0 K)^2 r^4} = \int_N^{N_r} \left(\frac{\partial P}{\partial N'}\right)_T d\ln N' \quad (2)$$

where P is the gas pressure. We used a very recent equation of state of Argon [13]. Some of the density profiles are shown in Fig. 6.

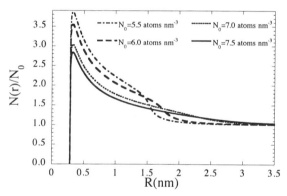

Fig. 6 Local density profiles of Ar around the ion. $T=151.5\,K$, unperturbed fluid densities $N_0 = (5.5, 6.0, 7.0, 7.5)\, atoms\, nm^{-3}$.

For a given N, the local static structure factor $S_r = S(0, N_r)$ differs significantly from the unperturbed $S(0)$. In particular, for $N < N_c$, S_r has a maximum at a given distance $r^\star$ from the ion. This fact suggests that the gas properties determining the transport properties of the ion should be evaluated at $r^\star$. We therefore use a modified version of the Stokes equation, where the local value η_r of the viscosity is used

$$\mu = \frac{e}{6\pi\eta_r R_h} \quad (3)$$

and assume that the effective radius R_h can be written as

$$R_h = a + bN + c\sqrt{S_r} \tag{4}$$

where a, b, and c are fitting parameters and S_r must be evaluated at $r^\star$. $r^\star$ is obtained by imposing that $S_r(r = r^\star)$ is maximum at the density where $\mu_0 N$ is minimum. This determines $r^\star = 1.95\,nm$. Finally, η_r must evaluated at the local density calculated for $r^\star$. The optimum values of the fitting parameters are $a = 0.58\,nm$, $b = 4.4 \times 10^{-4}\,nm^4$, and $c = 5.1 \times 10^{-2}\,nm$. We finally obtain

$$\mu_0 N = \frac{eN}{6\pi\eta_r \left[a + bN + c\sqrt{S_r}\right]} \tag{5}$$

The agreement with the experimental data is pretty good from medium $N \approx 4\,atoms\,nm^{-3}$ up to high densities, as shown in Fig. 4. However, we stress the fact that $R_h < r^\star$ where the thermodynamic properties of the fluid must be evaluated. We must therefore conclude that the simple picture of a solid-like sphere of radius R_h, where no–slip boundary conditions apply, is no longer valid because η_r and S_r must be evaluated at $r^\star$.

Finally, we would like to point out that the hydrodynamic description fails for $N \leq 4.0\,atoms\,nm^{-3}$, in the transition towards the kinetic region.

REFERENCES

1. E. A. Mason and E. W. McDaniel, *Transport Properties of Ions in Gases*, (Wiley, New York, 1988)
2. G. R. Freeman and D. A. Armstrong, in *Advances in Atomic and Molecular Physics*, vol. **20**, D. Bates and B. Bederson Editors, (Academic, New York, 1985) p.267
3. A. F. Borghesani, D. Neri, and M. Santini, *Phys. Rev. E* **48**: 1379 (1993)
4. A. F. Borghesani, F. Chiminello, D. Neri, and M. Santini, *Int. J. Thermophys.*, **16**: 1235 (1995)
5. A. F. Borghesani, D. Neri, and A. Barbarotto, *Chem. Phys. Lett.*, **267**: 116 (1997)
6. Dino Neri, A. F. Borghesani, and M. Santini, *Phys. Rev. E* **56**: 2137 (1997)
7. A. F. Borghesani, L. Bruschi, M. Santini, and G.Torzo, *Phys. Rev. A* **37**: 4828 (1988)
8. A. F. Borghesani and M. Santini, *Meas. Sci. Technol.*, **1**: 939 (1990)
9. A. K. Bartels, Ph.D. Thesis (University of Hamburg, RFG, 1971), unpublished
10. A. G. Khrapak, K. F. Volykhin, and W. F. Schmidt, *Structure and Mobility of Negative Ions in Simple Fluids*, in *Conference Record 12th International Conference on Conduction and Breakdown in Dielectrics Liquids*, Rome, Italy, July 15–19, 1996, C. Mazzetti and M. Pompili Editors (ELLEPI, Milan, Italy, 1996)
11. N. J. Trappeniers, P. S. van der Gulik, and H. van den Hooff, *Chem. Phys. Lett.*, **70**: 438 (1980)
12. K. R. Atkins, *Phys. Rev.* **116**, 1339 (1959)
13. C. Tegeler, R. Span, and W. Wagner: *Eine neue Fundamentalgleichung für das fluide Zustandsgebiet von Argon für Temperaturen von der Schmelzlinie bis 700 K und Drücke 1000 MPa*, VDI Fortschritt-Berichte, Reihe 3, Nr. 480, VDI-Verlag, Düsseldorf (1997).

ELECTRON MOBILITY MAXIMUM IN HIGH–DENSITY AR GAS

A. F. Borghesani and M. Santini

Istituto Nazionale per la Fisica della Materia
Department of Physics "G.Galilei"
University of Padua, Italy

INTRODUCTION

Transport properties of excess electrons in non polar gases or liquids give detailed information on the electronic states in a disordered medium and on the relationship among electron–atom interactions and fluid properties. In dense gases the electron zero–field density–normalized mobility $\mu_0 N$ is not independent of the gas density N, as predicted by kinetic theory. It rather shows anomalous density effects. In He and Ne, where the electron–atom interaction is essentially repulsive, $\mu_0 N$ decreases with increasing N at constant temperature T. In Ar, where polarization forces are most relevant, $\mu_0 N$ increases with N. A brief review can be found in literature[1].

From careful experiments[2,3,4,5] at high N and from their analysis an unified description of the electron–atom scattering process in dense gases has emerged[1,5]. For N not so large for qualitatively different phenomena to set in, such as electron self–trapping in very dense He and Ne[3,4], a model, henceforth the Padua (PD) model, has been developed[5], which heuristically incorporates into the kinetic theory formalism three multiple scattering effects arising from the fact that, at moderate N, the mean free path ℓ of the electron becomes comparable with its thermal de Broglie wavelength λ_T. The first effect is a density dependent quantum shift $V_0(N)$ of the mean electron energy. The electron kinetic energy is shifted by the kinetic energy part E_k of V_0, $\epsilon' = \epsilon + E_k$. The electron group velocity, that contributes to the energy equipartition value arising from the gas temperature, is obtained by subtracting this zero–point energy $v = (2/m)(\epsilon' - E_k)^{1/2}$. Consequently, also the electron velocity distribution function is shifted by E_k.

The second effect is an enhancement of the electron scattering rate due to quantum self–interference of the electron wavepacket multiply scattered by different scattering centers. To first order, it increases the momentum–transfer scattering cross section σ_m by the Atrazhev–Iakubov (AT) factor $(1 + a\lambda/\ell)$, where $\lambda = h/\sqrt{(2m\epsilon)}$ is the de Broglie wavelength of the electron at energy ϵ, $\ell = 1/N\sigma_m$, and $a = \mathcal{O}(1)$.

Finally, since the electron wavepacket encompasses many atoms and scatterers are correlated, σ_m must be increased by the Lekner factor, that takes into account scatterers correlations by means of the static structure factor $S(0)$. In any case, any dynamic properties of electrons, such as σ_m, have to be evaluated at the shifted energy $\epsilon + E_k$. The density–normalized mobility μN can be thus calculated by means of the following equations[6]

$$\mu N = -\frac{e}{3}\left(\frac{2}{m}\right)^{1/2}\int_0^\infty \frac{\epsilon}{\sigma_m^*}\frac{dg}{d\epsilon}d\epsilon \qquad (1)$$

g is the Davydov–Pidduck distribution function

$$g(\epsilon) = A\exp\left\{-\int_0^\epsilon dz\left[\frac{e^2 M}{6mz\sigma_m^{*2}(z)}\left(\frac{E}{N}\right)^2 + k_B T\right]^{-1}\right\} \qquad (2)$$

A is a normalization constant given by the condition $\int_0^\infty z^{1/2}g(z)dz = 1$. The effective momentum–transfer scattering cross section $\sigma_m^*(\epsilon)$ is related to the electron–isolated atom cross section σ_m by the relationship

$$\sigma_m^*(\epsilon) = \mathcal{F}(\epsilon + E_k)\sigma_m(\epsilon + E_k)\left[1 + 2\frac{N\hbar\mathcal{F}(\epsilon + E_k)\sigma_m(\epsilon + E_k)}{(2m\epsilon)^{1/2}}\right] \qquad (3)$$

where $\mathcal{F}$ is the Lekner factor $\mathcal{F}(k) = (1/4k^4)\int_0^{2k}q^3 S(q)\,dq$, with $\hbar^2 k^2/2m = \epsilon$. $S(q) = [S(0) + (qL)^2]/[1 + (qL)^2]$ is the gas static structure factor[7] and $S(0)$ is its long–wavelength limit. The correlation length is $L^2 = 0.1l^2[S(0) - 1]$ and $l \approx 1 nm$. The kinetic energy shift E_k is calculated according to the Wigner–Seitz (WS) model as $E_k = \hbar^2 k_0^2/2m$, where k_0 is obtained by a modified version[2] of the eigenvalue equation of the WS model using the total scattering cross section σ_T

$$\tan\left[k_0\left(r_s - \sqrt{\sigma_T(k_0)/4\pi}\right)\right] - k_0 r_s = 0; \qquad r_s = \left(\frac{3}{4\pi N}\right)^{1/3} \qquad (4)$$

The PD model agrees well with the experimental data in He and Ne up to densities where self–trapping produces a significant fraction of non quasifree electron propagating states. In Ar[5] the agreement of the model with the experimental data extends up to $N = N_0 \approx 70 \times 10^{20}\,cm^{-3}$.

Recent measurements[8] in Ar at $T = 152.5\,K$ up to $N = N_L \approx 125.1 \times 10^{20}\,cm^{-3}$, although they did not produce evidence for deviations from the positive density effect, have shown that the model agrees with the data up $N \approx 100 \times 10^{20}\,cm^{-3}$ if one realizes that the WS model is inappropriate at very high N. The obvious question is whether the PD model is pushed beyond its limits of applicability or whether different physical mechanisms determine the electron mobility rather than those described by kinetic theory. To this goal, it is worth recalling that in non polar liquids, such as Ar, Xe, CH_4, the observed μ maximum at a density not too close to the critical one has been interpreted in terms of electrons scattering off long–wavelength collective modes of the fluid, which modulate the bottom of the conduction band V_0[9]. Owing to these considerations, we have extended mobility measurements in Ar gas[10] up to $N \approx 140 \times 10^{20}\,cm^{-3}$. The main result is the observation of a mobility peak at a density close to that of the mobility maximum in liquid Ar.

EXPERIMENTAL DETAILS AND RESULTS

We used the pulsed photoemission apparatus for electron mobility measurements in high–density noble gases[2]. The cell can be filled with gas up to $10\,MPa$ and is thermoregulated within $0.01\,K$. Impurities, mainly O_2, are reduced to a final concentration of a fraction of p.p.b. by circulating the gas through an Oxisorb trap. Electrons photoinjected into the gap between two parallel-plate electrodes drift under the influence of an electric field and induce a current in the external circuit. The drift time τ is measured by analyzing the signal waveform [11]. The mobility is calculated from τ as $\mu = d^2/\tau V$, where d is the electrode spacing and V is the applied voltage.

EXPERIMENTAL RESULTS AND DISCUSSION

We have taken measurements at $T = 162.3\,K$ and $T = 152.15\,K$, close to the critical point of Ar ($T_c = 150.86\,K$ and $N_c = 80.8 \times 10^{20}\,cm^{-3}$). In Fig. 1 μN at $T = 152.15\,K$ is reported as a function of the reduced electric field E/N. For $N \leq N_L$ our measurements agree well with literature data [8]. At small N the maximum due to the Ramsauer-Townsend (RT) minimum of the momentum-transfer cross section [12] for $E/N \approx 4\,mTd$ is clearly visible. At higher N the RT μ maximum progressively shifts to smaller E/N and, finally, it disappears for $N > N_c$.

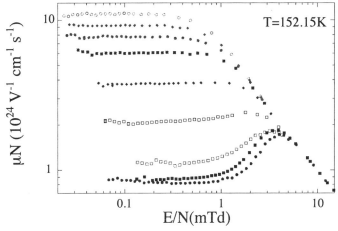

Fig. 1 Density-normalized mobility μN of excess electrons in Ar gas at $T=152.15\,K$ as a function of the reduced electric field E/N ($1\,mTd=10^{-20}\,Vcm^2$) for several $N=(104.0, 101.6, 99.35, 93.52, 80.48, 67.08, 61.46, 31.17, 10.97) \times 10^{20}\,cm^{-3}$ (from top).

In Fig. 2 we show the E/N value of the mobility maximum, $(E/N)_{max}$, as a function of N. Since at $(E/N)_{max}$ the mean electron energy $\langle\epsilon\rangle$ equals the energy of the RT minimum of the cross section, and since $\langle\epsilon\rangle$ increases with E/N, the reduction of $(E/N)_{max}$ with increasing N shows that $\langle\epsilon\rangle$ has a contribution increasing with N.

In Fig. 3 we show the zero-field density-normalized mobility $\mu_0 N$ as a function of N for $T = 162.3\,K$ and for $T = 152.15\,K$. Previous data [5] at $T = 162.7\,K$ are also plotted to show the excellent agreement with the present data at $T = 162.3\,K$. However, the most significant behavior is observed at higher N on the $T = 152.15\,K$ isotherm. Here, a sharp $\mu_0 N$ maximum occurs for $N \approx 124.7 \times 10^{20}\,cm^{-3}$, close to the value where it is observed in liquid Ar [9]. A similar behavior was previously observed [13] only in gaseous CH_4 for $N \approx 100 \times 10^{20}\,cm^{-3}$, once again at N close to the value observed in the liquid [14].

The presence of a mobility maximum in the dense gas at the same N as in the liquid raises the question whether there is a smooth, though definite, change in the physical mechanisms determining the electron transport properties as N increases beyond a certain threshold. At low and medium densities the modified single scatterer approximation is valid [5]: electrons are scattered off individual atoms, although the scattering properties must be modified to include multiple scattering effects. On the contrary, at higher N electrons might be scattered off density fluctuations [9,13].

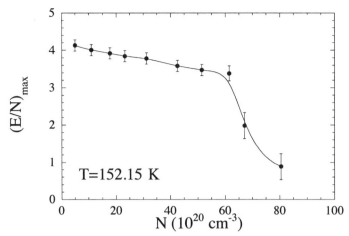

Fig. 2 Reduced electric field at mobility maximum $(E/N)_{max}$ as a function of N at $T=152.15\,K$. The line is only a guide for the eye.

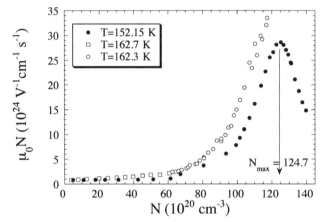

Fig. 3 $\mu_0 N$ as a function of the gas density N at $T=152.15\,K$ (closed symbols). The mobility maximum is for $N\approx 124.7\times 10^{20}\,cm^{-3}$. Open symbols represent $\mu_0 N$ measurements at $T=162.7\,K$[5] and at $T=162.3\,K$.

On one hand, this scenario is appealing. On the other hand, it is well known that the Basak and Cohen model[9] of electrons scattering off long-wavelenght modes of the fluid is in disagreement (by orders of magnitude) with the data, with the exception that it more or less correctly predicts the density of the maximum[15]. It is therefore interesting to push further the PD model and see if it can be improved to agree with the data, because it retains the simple single scattering picture of the kinetic theory. To this goal, we show in Fig. 4 the E_k values to be used to obtain agreement between the predictions of the model (Eqns. (1), (2), and (3)) and the experimental $\mu_0 N$ data. In the same figure we plot the WS approximation for E_k (Eq.(4)). Beside a small difference between the E_k values at different T, which might be attributed to the larger gas compressibility for the isotherm ($T = 152.15\,K$) closer to the critical one, we note that the WS approximation of E_k fairly well reproduces its experimental determination up to $N \approx 70 \times 10^{20}\,cm^{-3}$, as already observed[5]. For larger N the experimental determination of E_k increases faster with N than the WS model. This is not surprising because it is known that the WS model (Eq. (4)) is valid only when $r_s \gg \tilde{a}$, where $\tilde{a}$ is the hard-core radius of the short-range part of the effective electron-atom potential.

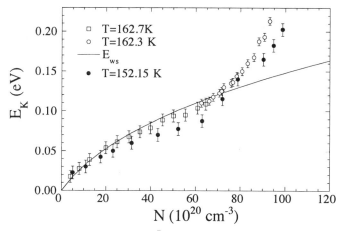

Fig. 4 Kinetic energy shift E_k for $T=162.7\,K$[5], $T=162.3\,K$ and $T=152.15\,K$. The line is the prediction of the WS model.

So, relaxing the condition that the WS model is valid up to very high N, the PD model reproduces accurately the experimental $\mu_0 N$ data up to $N \approx 100 \times 10^{20}\,cm^{-3}$. Moreover, in this density range it shows a good degree of consistency because the complete E/N dependence of μN is well reproduced by means of Eqns. (1)–(3) with E_k calculated from data at $E/N \to 0$, as shown in Fig.5 (dotted lines).

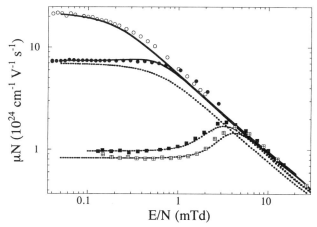

Fig. 5 μN as a function of N at $T=152.15\,K$. From the top the densities are $N=(133.3, 99.35, 51.44, 5.02) \times 10^{20}\,cm^{-3}$. The dotted lines are calculated according to the model described in the text; the solid lines are calculated introducing the effective scattering cross section σ_{eff}.

At low N the μ maximum due to the RT minimum of the Ar cross section is reproduced in position and strength and the disappearance of the maximum with increasing N is also described. In the frame of reference of this model the disappearance of the μ maximum as a function of E/N is essentially related to the density dependent shift of the energy distibution function. At large densities the shift is so relevant that the maximum of the distribution function has been shifted to energy larger than the energy of the RT minimum of the cross section. However, in the present form the PD model cannot account for the $\mu_0 N$ maximum at $N = 124.7 \times 10^{20}\,cm^{-3}$ and for the decrease of $\mu_0 N$ with increasing N, although we argue that this dependence of $\mu_0 N$ can be traced back to the quantum shift of the distribution function. Nonetheless, the E/N dependence of μN at the highest N is well reproduced if the cross section σ_m^* Eq.(3) is replaced by $\sigma_{eff} = c(N)\sigma_m^*$, with $c(N) = \mathcal{O}(1)$ is an adjustable parameter[8]. In this case E_k is assumed to be given by Eq. (4). $c(N)$ is then adjusted so as to reproduce $\mu_0 N$. In Fig. 5 the solid lines trough the data $\mu N(E/N)$ show the remarkable agreement of the improved PD model with the data at the highest N.

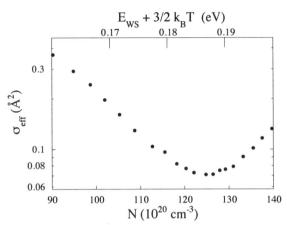

Fig. 6 Effective scattering cross section $\sigma_{eff} = c(N)\sigma_m^*$. The upper energy scale is obtained by converting density to energy by means of the WS relationship.

In Fig.6 we finally show the values of σ_{eff} at thermal energies as a function of N. In the upper scale density has been converted into energy by means of Eq.(4). We note a strong similarity of σ_{eff} with the electron–Ar one.

REFERENCES

1. A. F. Borghesani and M. Santini, in *Linking the Gaseous and Condensed Phases of Matter. The Behavior of Slow Electrons.*, L. G. Christophorou, E. Illenberger, and W. F. Schmidt Editors, NATO ASI Series vol. **326**, p.259 (Plenum, New York, N.Y. 1994)
2. A. F. Borghesani, L. Bruschi, M. Santini, and G.Torzo, *Phys. Rev.* **A37**: 4828 (1988)
3. A. F. Borghesani and M. Santini, *Phys. Rev.* **A42**: 7377 (1990)
4. K. W. Schwarz, *Phys. Rev.* **B21**: 5125 (1980); A. K. Bartels, *Phys. Lett.*, **44 A**: 403 (1973); A. K. Bartels, *Appl. Phys.*, **B8**: 59 (1975)
5. A. F. Borghesani, M. Santini, and P. Lamp, *Phys. Rev.* **A46**: 7902 (1992)
6. L. G. H. Huxley and R. W. Crompton, *The diffusion and drift of electrons in gases* (Wiley, New York, N.Y., 1974)
7. G. E. Thomas and P. W. Schmidt, *J. Chem. Phys.* **39**: 2506 (1960)
8. P. Lamp and G. Buschhorn, *Phys. Rev.* **B50**: 16824 (1994)
9. S. Basak and M. H. Cohen, *Phys. Rev.* **B20**: 3404 (1979)
10. D. Neri, Ph.D. Thesis (Padua, 1997, unpublished)
11. A. F. Borghesani and M. Santini, *Meas. Sci. Technol.*, **1**: 939 (1990)
12. G. N. Haddad and T. F. O'Malley, *Aust. J. Phys.*, **35**:35 (1982)
13. N. E. Cipollini and R. A. Holroyd, *J. Chem. Phys.* **67**: 4636 (1977)
14. J. M. L. Engels and A. J. M. Van Kimmenade, *Phys. Lett.*, **59A**: 43 (1976)
15. R. Reininger, U. Asaf, I. T. Steinberger, and S. Basak, *Phys. Rev.* **B28**: 4426 (1983)

LOW-ENERGY ELECTRON COLLISION BY EXCITED ATOMS

L. Vušković, C. H. Ying, Y. Wang, and S. Popović

Physics Department, Old Dominion University, Norfolk, VA 23529

INTRODUCTION

Low-energy (1 - 30 eV) electrons play very important role in the kinetics of gas discharges. Electrons in this energy range present the dominant majority in most low-temperature plasmas. Mechanisms underlying many phenomena in weakly ionized gas are based, in part, on the low-energy electron collisions with atoms and molecules. Ionization balance in partially ionized gas is maintained by a combination of excitation and ionization processes involving electrons. In addition, the momentum transfer from low-energy electrons to neutral particles is a significant mechanism in the heating of heavy particles. On the other hand, it is well known that excited-state atoms, especially metastable atoms, are the intermediate states in the multi-step ionization processes from ground-state atoms and molecules. Further, collisions between electrons and excited states are the dominant mechanism for ionization at higher gas pressures, usually competing with the energy pooling processes involving two excited-state atomic particles.

The fact that large quantity of external energy is stored in the excited states was extensively used in the past to develop a vast variety of gas lasers.[1] Recently, another important class of phenomena related to weakly ionized gas has been related to the excited states. Shock waves propagating through weakly ionized gas[2] demonstrated many anomalies that have no explanation in the framework of traditional aerodynamic models. At present, a comprehensive interpretation has not yet been established which could describe the underlying mechanisms. Among numerous observed facts about anomalous dispersion and propagation of shock waves through weakly ionized gas, in our opinion, three effects deserve special attention: first, the anomalies continue to be present for a relatively long time in the afterglow; second, anomalies increase with the ultraviolet radiation; third, dispersion and propagation velocity decrease with transversal magnetic field. Although the complete interpretation is still missing, especially in relation to the anomalous increase of shock speed, these three effects suggest the excited states as possible intermediaries offering the coupling link between the charged and neutral particles even at very low degree of ionization. In addition, it has been observed[3] that the anomalies in shock wave propagation and dispersion in ionized gas were closely linked to the increased presence of excited (especially metastable) states. The appearance of local maxima of their concentration coincides with the apparent increase of shock propagation velocity in the two generic experiments. Detailed spectroscopic studies[4] of ionizing shock waves in neon have also revealed an unusually abundant presence of metastable neon states in the precursor and in the translational shock regions.

However, data necessary for detailed analysis of the contribution of excited states to the anomalous dispersion and propagation of shock waves in weakly ionized gas are still insufficient. Experimental work on the low-energy electron scattering is particularly rare, because of many practical difficulties, ranging from problems of controllable production of excited states to the problems with interpretation of measurements, since in most cases many different scattering channels are involved. Moreover, experimental database on absolute differential cross-sections in electron scattering with excited states is virtually nonexistent,[5] apart from few reported experiments in barium, sodium, and rare gases.

Our ongoing work on electron collisions with excited-state sodium is partly motivated by the fact that a well-controlled experiment can serve as a test for various approximations used in the calculation of the cross sections and other parameters of collisional dynamics. In addition, sodium electronic configuration has some similarities with the configuration of metastable states of neon, including very close ionization potentials and almost equal transition energies. Therefore, sodium electronic system may be considered as a representative configuration for the electronic structure of excited neon, and the differential cross sections and collisional dynamics in sodium may be considered, in a rough approximation, as equivalent to the collisional dynamics of metastable neon. Therefore, in search of a plausible interpretation of anomalous concentration of excited states in the experiments with ionizing shock waves in neon, one may use the equivalent data taken from measurements in sodium. Similar approach[6] was taken in the extrapolation of collisional dynamics database to the excited systems.

EXPERIMENTAL TECHNIQUE

In our scattering experiments, we combined a crossed-beam configuration with recoil atom detection technique. Recoil atom experiments in a crossed-beams configuration are characterized by the fact that observations of the scattering events are made on the recoiled atoms rather than the scattered electrons. By measuring the corresponding recoiled atoms the momentum of scattered electrons can be found through a kinematic analysis. The scheme of electron and atom scattering angles of the experiment is shown in Fig. 1. Coordinates in the interaction region are denoted by x',y',z'. The electron (momentum magnitude mv) and atom (momentum magnitude MV) beams propagate along the +z' and +y' axes, respectively. In our analysis we assume the condition mv << MV and define electron scattering angles, θ and φ, in the laboratory frame. The difference between laboratory-frame angles and center-of-mass frame angles is negligible except for very small polar angles which cannot be resolved anyhow in the experiment.

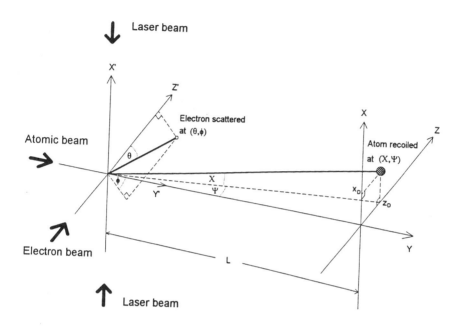

Figure 1. Schematic diagram of the crossed-beam configuration. Propagation of photon, atom, and electron beams are along x', y', and z' directions, respectively.

In our experiments, the recoiled-atom currents were measured along the z-axis. The relationship[7] between the differential cross-section (denoted as σ(θ)) and the atomic current recoiled into detector $I_{in}(z_D)$ for a single scattering channel at the detector position z_D is

$$I_{in}(z_D) = \frac{i_0 I_0(0)}{2h\Delta x \Delta z} \int_0^\pi \sigma(\theta)\sin\theta d\theta \int_0^\infty \mathcal{E}(E)dE \int_{z_1}^{z_2} Z(z)dz$$

$$\times \int_{V_1}^{V_2} \frac{\mathcal{V}(V)}{V}dV \int_{x_1}^{x_2} X(x)dx \int_{\phi_1}^{\phi_2} d\phi$$

where $I_o(0)$ is the atomic beam current at $z_D=0$, h is the height of the interaction region perpendicular to the scattering plane, $\varepsilon(E)$ is the electron energy distribution, $Z(z)$ and $X(z)$ are the horizontal and vertical atomic dc profiles, respectively, $\nu(V)$ is the velocity distribution of the atomic beam. V_1, V_2, ϕ_1, and ϕ_2, are the implicitly dependent on other variables due to the conservation of momentum and energy. After a set of $I_{in}(z_D)$'s for different z_D, $X(x)$, $Z(z)$, and $\nu(V)$ are measured, the $\sigma(\theta)$ can be deconvoluted.

By employing circularly polarized laser light propagating perpendicularly to the scattering plane, sodium atoms were prepared in $3^2P_{3/2}$, F=3 (M_F=+3 or M_F=-3) polarized states. To prepare[7] the initially excited polarized target a ribbon-shaped laser beam was used, with the cross-section overlapping the scattering region. The average fraction of sodium excited states achieved in the experiments[7] was about 25%.

RESULTS

Absolute differential cross section measurements of condensable targets are restricted to methods where the knowledge of the number of target-particles, often measured through the pressure, is not necessary. The atomic recoil technique is unique approach of this kind of experimental method. The recoil atom analysis[8] is based on the approximation that the change of the longitudinal velocity in the collision is negligibly small as compared to the incident atomic velocity. No calibration or normalization procedures were involved. One particular advantage of this experimental method, the monitoring of atoms after the collision, allows determination of the absolute cross sections in the domain where large impact parameters play an important role.

In the low energy range used in these experiments, 1 to 30 eV, many collision channels (elastic, inelastic, ionization) can be simultaneously open. This energy range is particularly interesting because it represents a regime where many aspects of the collision process, including target distortion, correlations, post-collision interaction, and exchange effects, simultaneously play important roles.

We have measured the absolute differential cross sections for elastic, inelastic, and superelastic scattering. The final results can be related to an unpolarized initial excited state. In the case of superelastic scattering, detailed balance can be directly invoked to compare the inverse reactions 3S =>3P. Thus we obtain additional insight into this collision process. Specifically, the polarized recoil-atom scattering results can be related[9] to the corresponding time-reversal experiment with unpolarized atoms, as long as the final state is an S state of even parity and the initial P state has odd parity. This is based on the assumptions[10] that spin dependent forces, such as the spin-orbit interaction as well as any effect of the nuclear spin can be neglected in the collision process.

The total ionization cross section of excited sodium has been determined in the energy range from threshold to 30 eV by measurements of collisionally produced ions as a function of electron energy. The relative ionization total cross sections as a function of energy for the ground- σ_g, excited-state σ_e, as well as the ratio between them were determined experimentally. Finally, σ_g was normalized to theory[11] and σ_e calibrated with respect to ground-state data. The results are given elsewhere.[12]

In Fig. 2 we present some experimental results of 3P=>3P elastic electron-sodium scattering. Large discrepancies (up to a factor of 2) between calculations and measurement are observed. From the theoretical point of view, if the quasi-three body approximation is not appropriate, then it should have a similar effect on both 3S=>3S and 3P=>3P results. However, our 3S=>3S measurements did not show big discrepancies. The only thing that both the theory, and the experimental analysis ignored is the coherence between the 3S and 3P initial states. No strong evidence or argument support this assumption in 3P=>3P elastic scattering.

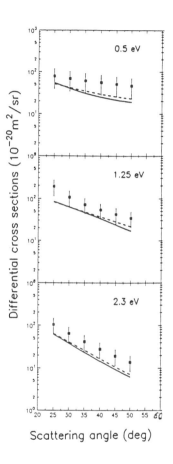

Figure 2. Differential cross-sections for the 3P=>3P electron sodium scattering plotted as a function of polar scattering angle θ. Incident electron energies are indicated. Present results are indicated by solid squares. Full line is the CC calculation by Zhou et al.[12] and dashed line is the CCC calculation by Bray et al.[13]

REFERENCES

1. J.-L. Delcroix, C. M. Ferreira and A. Ricard, Ch. 5 in "Principles of Laser Plasmas" (G. Bekefi, ed.) Wiley&Sons, New York (1976).
2. B. N. Ganguly, P. Bletzinger and A. Garscadden, Phys. Lett. A **230**, 218 (1997);
   I. V. Basargin and G. I. Mishin, Sov. Tech. Phys. Lett. **11**, 535 (1985);

G. I. Mishin, Yu. L. Serov, and I. P. Yavor, Sov.Tech.Phys.Lett. **17**, 311 (1991).
3. L. Vušković and S. Popović, in Proc. of 2nd Workshop on Weakly Ionized Gases, Lee Bain (Ed.), Norfolk, April 1998.
4. T. J. McIntyre, A. F. P. Houwing, R. J. Sandeman, and H.-A. Bachor, J. Fluid Mech. **227**, 617 (1991).
5. S. Trajmar and J. C. Nickel, Adv. Atom. Mol. Phys. **30**, 45 (1992).
6. J. Weiner, Private communication.
7. T. Y. Jiang, C. H. Ying, L. Vušković, and B. Bederson, Phys. Rev. A **42**, 3852 (1992).
8. L. Vušković, M. Zuo, G. F. Shen, and B. Bederson, Phys. Rev. A **40**, 133 (1989).
9. K. Bartschat and D. Madison, Phys. Rev. A **48**, 836 (1993).
10. N. Andersen and K. Bartschat, Comments At. Mol. Phys. **29**, 157 (1993).
11. I. Bray, Phys. Rev. Lett. **73** , 1088 (1994).
12. H. L. Zhou, D. W. Norcross, and B. L. Whitten, in "Correlation and polarization in electronic and atomic collisions and (e,2e) reactions" (Institute of Physics, Adelaide, 1991) p. 39.
13. I. Bray, D. V. Fursa, and I. E. McCarthy, Phys. Rev. A **50**, 4400, (1994).

SECTION 2: BASIC MECHANISMS

A STUDY OF PREBREAKDOWN IN SF_6

Henk H.R. Gaxiola, Joseph M. Wetzer

High Voltage and EMC Group, Eindhoven University of Technology
PO Box 513, 5600 MB Eindhoven, The Netherlands
phone +31 40 2473993   fax +31 40 2450735
e-mail: e.h.r.gaxiola@ele.tue.nl   j.m.wetzer@ele.tue.nl

INTRODUCTION

This work is a continuation of work presented earlier for nitrogen and dry air[1], and deals with a time-resolved study of the prebreakdown phase in SF_6. The goal of this work is to develop and verify discharge models for practical insulating geometries on the basis of physical discharge processes.

In this paper we present time-resolved optical and electrical measurements of streamer formation leading to breakdown in atmospheric SF_6. Through these diagnostics we obtain a better understanding of the processes fundamental to gas-breakdown. Here we will discuss results obtained for SF_6, from experiments in a uniform and non-uniform field. An experimental setup is used to measure the discharge current and the optical discharge activity. An ICCD camera is used with a minimum shutter time of *5 ns*.

EXPERIMENTS

An experimental setup with a *0.45 GHz* bandwidth is used to measure the discharge current, an ICCD camera and two photomultipliers with U.V./I.R. filters record the optical discharge activity. A more detailed description of the experimental setup is given by Gaxiola[2]. The type of gas under study is SF_6; pressure = $1,02*10^5$ *Pa*; voltage range 4 to *90 kV*; electrode gap range *5 mm* to *1 cm*.

RESULTS AND DISCUSSION

We present results for a uniform applied field, without and with space charge field distortion, and for non-uniform fields.

Figure 1 shows an example of measured prebreakdown currents for SF_6 in a uniform field. From earlier data[1] we found that increasing the initial electron density (by

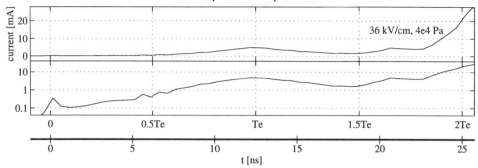

Fig.1 Measured current waveform (on a linear- and log-scale) for prebreakdown in a uniform field.

decreasing the area of photo-emission on the cathode) results in a shorter time to first current maximum. A slight increase in applied field (just beyond the static breakdown fieldstrength) results in a significant reduction of the time-to-breakdown for laser-induced breakdown in SF_6.

In Fig.2 the measured prebreakdown currents for *-37 kV* ($\leq 1.5\mu s$ and $\leq 1.6\mu s$) and *-38 kV* ($\leq 843ns$), the photomultiplier output with "filter 89B (*>700nm*)" and "filter 03FCG163 (*300nm< <700nm*)" or "filter 2B (*>400nm*)" and "filter 03FCG121 (*300nm< <400nm*)", and ICCD camera images during prebreakdown are shown for an inhomogeneous electrode geometry. On the horizontal axis the time is normalized to the electron transit time in a uniform field T_e. For the halfsphere plane electrode geometry the maximum field is calculated to be $3*E_0$, with E_0 the uniform laplacian field value.

The photomultiplier output signal (PM2) shows the start of the laser-induced discharge activity. Filters 89B and 2B filter this dominant N_2-laserpulsline (*337nm*) out completely. At breakdown and overexponential current growth also optical activity in the I.R. (*>700nm*) is generated. No difference was observed in photomultiplier signal (PM2) with or without filter. Use of photomultipliers helps to depict the laser-induced start of activity, and the time occurence of actual breakdown. The wavelength range for optical activity before breakdown is found to be below *400 nm*. The ICCD camera used is however far more sensitive to optical activity.

The gap partially breaks down and recovers. After five subsequent Trichel-like-pulse generations in which the complete gap was bridged we observe complete breakdown. The current shows the laser-induced avalanche, the ion drift phase and subsequent Trichel-pulse. The negative SF_6^- ions, as proposed by Gravendeel[3], do not play a role in the corona stabilization, since the attachment time for electrons to SF_6 molecules is too large to be able to contribute to the cut-off of the discharge. It is the electron avalanche cloud itself which creates the growth quenching space charge field. In its "tail" at the cathode halfsphere tip the next avalanche is started, initiated by secondary electrons from positive ion impact or cathode photo-emission (Trichel-pulses also occur for electrical fieldstrength values below which there is no field emission at the cathode tip, so field emission is of less importance). At later moments in time (after *100ns* and until the time moment of the Trichel-like-pulse peak) no optical discharge path can be distinguished in the drift phase, in agreement with the observation of negative ion clouds slowly drifting towards the plane anode electrode. For voltages below *-36 kV* we observe only small corona activity as shown in image *a* (illuminated area *0.002 cm²*). Images *b* through *e* show corona activity around the sphere surface at different time instances in which we see the formation of a discharge path of optical activity between halfsphere and plane electrode. At this voltage level we obtain laser-induced gap-breakdown.

Increase in voltage up to *-37 kV*, as shown in images *f* and *g*, causes typical "feathers",

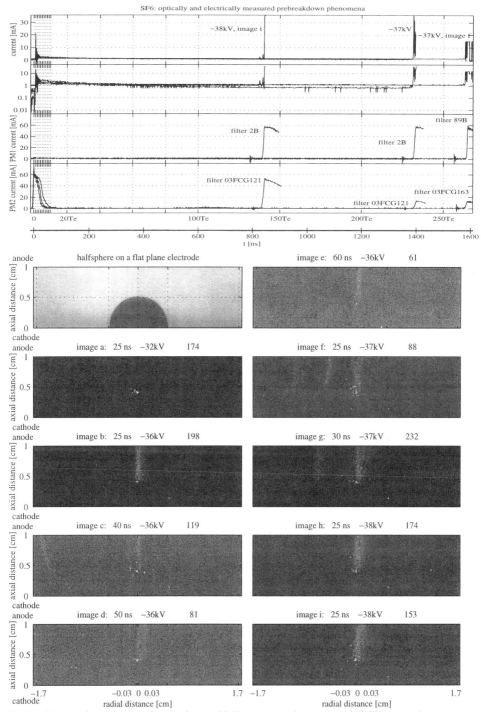

Fig.2 Measured current waveforms, photomultiplier output and sequence of ICCD camera images describing laser-induced discharge activity in an inhomogeneous electrode geometry. The corresponding times on the measured current waveforms are indicated. The parallel plane cathode (bottom) and anode (top) surfaces are located at 0 and 1 cm respectively on the vertical axis. The horizontal scaling in centimetres is also indicated. The initial electrons are released from the halfsphere cathode centre at t=0s indicated by the region around x=0 (-0.03<x<0.03). The maximum photon count per pixel is denoted in the upper right corner of each image. 5 ns ICCD camera gate-shutter time.

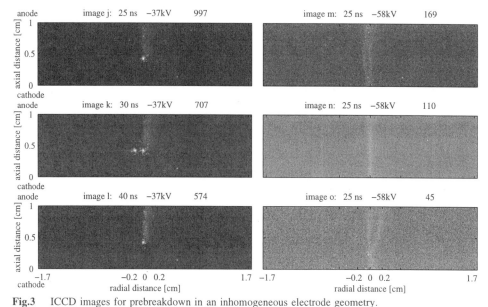

Fig.3 ICCD images for prebreakdown in an inhomogeneous electrode geometry.
5 ns ICCD camera gate-shutter time. Halfsphere plane electrode geometry.
left: Optical activity for a N_2-laserspot area of 0.1 cm^2. Top of sphere located at 0.5 cm on the vertical axis.
right: Optical activity for an increased gapwidth (gap=1cm). Top of sphere located at 0 cm on the vertical axis.

which (for wide enough electrode gaps) are superimposed on the corona activity present around the electrode, as discussed by Gravendeel[3] for negative corona discharges in air. Increasing the applied voltage even further leads to more optical activity and complete gap-breakdown as shown in images *h* and *i*.

Figure 3 (left) shows the optical activity for a N_2-laserspot area of *0.1 cm^2*, resulting in more active corona regions and much more optical activity. The number of subsequent Trichel-like-pulses is reduced to three (or four) after which gap-breakdown occurs. A smaller initial electron density results in a slower build-up of the space charge field and therefore longer formation-time of the Trichel-like current pulse peak.

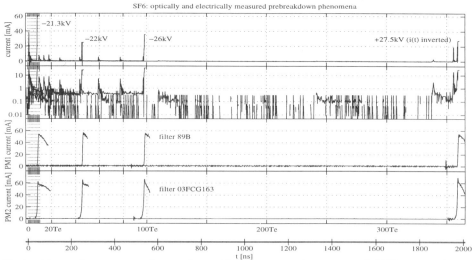

Fig.4 Measured current waveforms, photomultiplier output and sequence of ICCD camera images describing laser-induced discharge activity in an inhomogeneous electrode geometry.

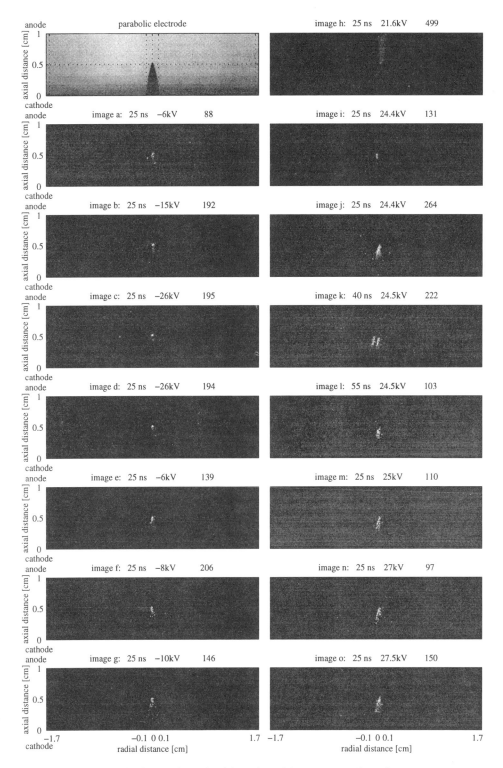

Fig.4 (continued) ICCD images for prebreakdown in an inhomogeneous electrode geometry. 5 ns ICCD camera gate-shutter time. Paraboloid plane electrode configuration.
left: Negative voltage　　　　　　　right: Positive voltage.

On the right hand side in Fig.3 the activity is shown for an increased gapwidth (gap=*1cm*). We observe no or only one subsequent Trichel-like pulse. The time between laser-induced activity and first Trichel-like pulse is larger than *10 μs*. The first negative ion cloud arriving at the anode results in immediate gap-breakdown. Images *m* and *o* are examples of the multiple discharge activity paths in the gap region (the top surface of the halfsphere is located at the bottom of these images) and the applied voltage at which breakdown occurred is *-58 kV*.

In Fig.4 a paraboloid-plane geometry (tipradius *0.2 mm*) is shown. The images *a* through *g* (images *a* through *d*: illuminated area *0.002 cm^2*, images *e* through *g*: illuminated area *0.1 cm^2*) show the corona activity around the tip for a negative (tip) voltage and images *h* through *o* for a positive voltage respectively. The maximum field is calculated to be *23*$*E_0$*, with E_0 the uniform laplacian field value. The activity around the tip is more intense compared to the phenomena seen in Fig.2 but we do not observe any "feather"-like phenomena anymore. For negative applied voltages we enter a Trichel-like-pulse regime, in which we do not find regular (self-sustained) Trichel-pulses, as for the case of air, but only irregular Trichel-pulses as shown in Fig.4 for *-21.3 kV* (≤*42ns*), *-22 kV* (≤*248ns*) and *-26 kV* (≤*490ns*). For even higher voltages we found the gap to break down for negative voltages larger than *-22 kV*. For a positive applied voltage discharge activity starts at a higher voltage (also because we are not able to induce sufficient laser-induced initial photo-electron numbers), but the gap breaks down around the same voltage level (*22 kV*). This behaviour is also described for SF_6 by Kuffel[4]. We observe several electron avalanche generations arriving at the anode tip electrode. In the positive space charge cloud left behind a cathode directed streamer forms leading to gap-breakdown. Initial electrons have to originate from atmospheric background irradiation since in the low field around the plane cathode no significant growth takes place, created electrons are attached to the strongly electro-negative SF_6 molecules forming negative ions preventing growth in discharge activity in these regions at low applied positive voltages.

CONCLUSIONS

* We have succeeded in observing prebreakdown phenomena in SF_6 at pressures up to atmospheric pressure.
* Applying our 2-D model, which correctly describes the externally measured current and the streamer formation leading to breakdown in N_2 and dry air, to SF_6 would help to make progress in understanding prebreakdown phenomena in SF_6.
* For non-uniform fields we have been able to make measurements of the optical activity related to prebreakdown phenomena in atmospheric SF_6.
* We observe corona-like discharges at voltages well below the d.c. breakdown fieldstrenghts and multiple sites of optical activity for larger laser-illuminated areas.
* For large electrode radii we observe "feathers": different simultaneous discharge paths.
* For sharp electrodes we observe one path of optical activity.

REFERENCES
1. E.H.R. Gaxiola, J.M. Wetzer, "Optical and electrical study of the avalanch and streamer formation in sulphurhexafluoride". 10th Int. Symp. on High Volt. Eng., Montréal, Canada, Vol.2, pp.293-296 (1997).
2. E.H.R. Gaxiola, J.M. Wetzer, "A study of streamer formation in nitrogen". 12th Int. Conf. on Gas Discharges and Their Applications, Greifswald, Germany, Vol.1, pp.232-235 (1997).
3. B. Gravendeel, "Negative Corona Discharges; a Fundamental Study". Ph.D. thesis, Eindhoven Univ. of Techn., The Netherlands, pp.9-20, 143-145 (1987).
4. E. Kuffel, W.S. Zaengl, "High Voltage Engineering", Pergamon Press, pp.377-383 (1984).

DISCUSSION

W. PFEIFFER: This is an excellent piece of work, and it is of special importance that the investigations have now been extended to the non-homogeneous field which is of higher practical relevance. The work has been done for a gap width of 0.5 cm to 1.0 cm and a pressure of 1 bar. This is rather far from the practical application in GIS with gaps of some centimeters and pressures of some bars. Do you think that the same mechanisms being investigated can be expected under this practical situation, or how can the results be transferred to this situation?

E. H. R. GAXIOLA: There is still a long way to go. The strategy is to first tackle the simple situation. It is similar to our earlier work in which we first started experiments and modeling of nitrogen as a simple modeling gas and air as a practical application. Although, as indicated by the work done by Professor Pfeiffer himself, it will be quite some work. I think that due to practical limitations/implications we have to start translating and understanding the mechanisms to higher realistic pressures and distances in, for example, SF_6-filled GIS, through modeling of prebreakdown in SF_6. From that point on we can start working on realistic practical experiments at such conditions.

A. BULINSKI: Were the optical signals associated always with partial discharges (PD) or did they indicate some pre-PD phenomena?

E. H. R. GAXIOLA: As pointed out in earlier work, for example, in a uniform field far below the electrical measurement threshold in our setup, we observe optical activity related to subsequent avalanche generations crossing the gap. The optical activity seen here was associated with either the corona around the high-field electrode or the optical activity of bridging of the gap prior to breakdown.

Y. QIU: To my understanding, there should not be any space charge field distortion in a uniform-field gap before breakdown. So what do you mean by "a uniform applied field with space charge field distortion?"

E. H. R. GAXIOLA: We can have a uniform Laplacian field and, for example, initiate space charge effects upon the creation of a large initial starting electron density at the cathode by the pulsed N_2 laser through cathode photoemission. This we would designate by "uniform applied field with space charge field distortion."

M. GOLDMAN: Have you seen some streamer propagation in SF_6? If yes, at which speed? If no, do you think that there is a propagation in SF_6?

E. H. R. GAXIOLA: We hoped and expected to, but did not. We do think there is streamer propagation, though at very high speed. We hope to present data on such propagation in the near future.

FROM CORONA STABILIZATION TO SPARK BREAKDOWN IN POINT-TO-PLANE SF$_6$ GAPS

Hervé Champain [1], Alice Goldman, Micail Lalmas [2]

Laboratoire de Physique des Gaz et des Plasmas (Univ. Paris Sud/CNRS)
Equipe Décharges Electriques et Environnement à SUPELEC
Plateau de Moulon
F 91192 Gif-sur-Yvette Cedex, France

[1] now with Thomson CSF, 92231 Gennevilliers, France
[2] now with Motorola, 31100 Toulouse, France

INTRODUCTION

Spark breakdown in gaseous media represents a fugitive phenomenon often studied with high voltage impulses long enough in duration to allow its formation and development. In SF$_6$, threshold voltages for spark breakdown under such conditions correspond to mean applied electric fields around 90 kV/cm/bar and people using such impulses to get high currents in gaseous gaps but without the risk of spark breakdown work with pulses generally of higher amplitude but short enough to avoid sparking.

On the other hand in corona stabilised gaps operated under DC or AC conditions, the threshold fields fall to values around or less than 40 kV/cm/bar, i.e. more than two times lower. This means that the corona discharges bring, in the properties of the gaseous medium prior to spark breakdown, changes able to create a privileged path where the effective electric field will reach the necessary 90 kV/cm/bar. The aim of our paper is to improve the knowledge on the corona effects which can explain the phenomena observed and allow a better evaluation of the risk of an insulating gap to go to spark breakdown under established voltage conditions. For this purpose, we shall use two different approaches :
  - a phenomenological approach with experimental results essentially got from investigations under DC applied voltages,
  - a temporal approach with results gained from investigations performed under AC applied voltages.

EXPERIMENTAL SET-UP AND WORKING CONDITIONS

All experiments were performed with point-to-plane gaps in 2 bars SF_6 of technical quality (99.9 %) in a 9 litres stainless steel vessel refilled with new gas for every new series of measurements after a turbomolecular pumping till a residual pressure $p < 10^{-2}$ Pa. This vessel was fitted with quartz windows for optical spectroscopy studies.

Point electrodes were generally made of aluminium or copper rods of 3 mm diameter with hyperboloidal tips of 5 to 300 µm radius of curvature and plane electrodes also made of aluminium or copper but generally used with point electrodes of the opposite material. In general, the gap length was fixed at 10 mm, but series of measurements were made as a function of gap length, then over a 2 or 3 mm to 10 mm range.

Main measurements of interest for this paper were current measurements and temperature measurements. Current measurements were carried out with a simple amperemeter in the case of discharges under DC applied voltage and with an oscillograph under AC applied voltage. Temperature measurements[1] were performed spectroscopically in the ionisation region of the discharge by means of the spectral emission of nitrogen used for this purpose as an additive with a proportion of 1%.

Finally, it is important to note here that all measurements concerned with this paper were carried out with electrodes every time renewed for each new series of measurements and that all experiments were conducted by increasing the applied voltage gradually but rapidly, so that all ageing phenomena able to affect the conditions of occurrence of spark breakdown can be here neglected, except in one specific case for which it will be specified.

RESULTS AND DISCUSSION

A. Phenomenological approach of the gaseous gap properties gained just prior to spark breakdown

1. Structural shape of the discharge till spark breakdown. It must be emphasised that in all our operating conditions (i.e. at 2 bars SF_6 with point electrodes of relatively small radius of curvature and gap lengths of 10 mm maximum), till spark breakdown and apart a spike phenomenon, sometimes observed on axis, all luminous phenomena, i.e. all ionising processes, remained concentrated at the point electrode, in a ball region (of maximum 2 to 3 mm diameter), as pointed out by other authors[2]. This is not surprising. Effectively, as pointed out for instance by Gallimberti and Wiegart[3], streamers propagation in SF_6 needs electric fields of the same order of magnitude as to ionise (~ 90 kV/cm at 1 bar). However, it may be added that even if the luminous region macroscopically appears as a glow region, the numerous erratic current pulses present in the discharge current do not allow to consider it as a pure glow region.

2. Electrical measurements. The most significant results we got on the electrical behaviour and properties of the gaseous gap at spark breakdown were obtained by studying the variations with the gap length d of the values reached, just before spark breakdown, by the applied voltage V_{br}, by the mean total current I_{br}, totalizing the continuous current and the current pulses, and by the power consumption P_{br} determined as $P_{br} = V_{br}.I_{br}$. Figures 1 and 2 present results obtained in positive discharge polarity, i.e. with the point electrode working as an anode. Their curves respectively show the variations of V_{br} and P_{br} as a function of d. Worthy of note are first their linearity and secondly their quasi insensitivity, in the limits of our investigations, to the point electrode parameters, especially to its radius of curvature r. These results are in accordance with the common idea, for SF_6 as for air, that close to the corona-to-spark transition the electric field tends to become uniformly distributed along the gap[4]. Obviously, this is not valid for too small gaps (d ≤ 3 mm) which give direct spark breakdown, without going through any kind of predisruptive discharges.

The linear increases with gap length of the breakdown voltage V_{br} and of the power consumption P_{br} provide us with, at 2 bars, a value of about 70 kV/cm (i.e. 35 kV/cm/bar) and about 1.7 mW/mm respectively for the mean electric field and for the power needed to bring the gap ready for spark breakdown in our operating conditions. Added to the fact that streamers are unable, even close to spark breakdown as already evoked, to propagate far away from the point electrode, the linear increases of V_{br} and P_{br} are consistent with the idea of the formation of a prebreakdown channel preparing the way across the gap for spark breakdown.

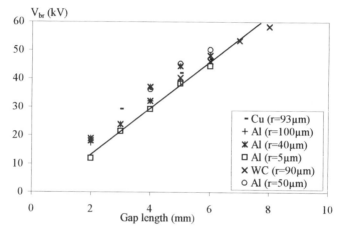

Figure 1. Breakdown voltage vs. gap length for various anodic point electrodes

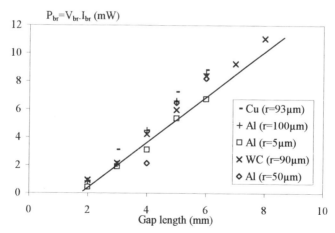

Figure 2. Power consumption just prior to breakdown vs. gap length for the same anodic point electrodes as in Fig.1.

Two phenomena will play an essential role in the formation of this channel, as predicted by R.S. Sigmond[5] :

- the aerodynamic forces created on the neutral gas by the ions motion in the vicinity of the point electrode: forces which are at the origin of the gas movement known as the « electric wind »[6] and from which is derived the Sigmond's narrow hot jet model[5].

- the energy of the drifting ions, providing a power density j.E which heats the gas more specifically on the gap axis where the current density j and the electric field E are maximum, as it was already previously stipulated to explain the transition from the Trichel regime to the pulseless regime in the case of negative air coronas[7,8]. Subsequent effects are a correlated decrease of the gas density n, inversely proportional to the temperature increase (n=p/kT), and a correlated increase of local fields E/n, privileging excitation and ionisation processes in the point electrode vicinity as well as along the so-formed prebreakdown channel.

3. A spark breakdown threshold criterion expressed in terms of a dynamical conductance.

In the introduction was introduced the idea of such a privileged path bringing at 1 bar to an effective mean electric field of 90 kV/cm, while the apparent mean applied field is only of ~35 kV/cm in our experiments. Such an increase of E/n in this path must be accompanied by a conductance increase coherent with the increase of temperature which will be discussed in next section. Effectively, when the corona-to-spark transition is approached, the slope of the I(V) discharge characteristics is seen to undergo a rapid increase. This slope can be expressed in terms of a dynamical conductance S, and its variations controlled by $\Delta I/\Delta V$ measurements.

We made such measurements under different working conditions and in all cases of steady state space-charge monitored discharges, it was found a critical threshold value S_{crit} only above which spark breakdown should be observed. Figure 3 shows that this value for positive corona discharges lies around 8 µA/kV. This value appears to be not sensitive to the point electrode characteristics (material and radius of curvature r) and spark breakdown is seen to occur generally for S values close to S_{crit}, except in the cases of very sharp point electrodes (r=5µm), privileging corona stabilisation.

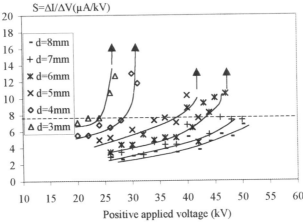

Figure 3. Dynamical conductance S as a function of applied voltage and gap length for anodic point electrodes. Spark breakdown is visualized by arrows.

With negative corona discharges, the S_{crit} value tends to be a little higher ($\approx$ 10 µA/kV) but the general shape of the S(V) characteristics is very similar to that one in the positive case.

We did not extend this type of investigations to AC coronas under corona onset conditions (t=0). However we made many S measurements with DC[9] and AC[10] coronas submitted during long duration to voltages clearly insufficient to provoke breakdown at time t=0, but able to lead to delayed breakdown because of ageing modifications brought meanwhile to the gap, especially on the electrodes surfaces. During this time, the dynamical conductance was seen to increase, from values which should be first as small as 2 µA/kV to values around 8 µA/kV for which again spark breakdown should occur, even if sometimes one sees the mean current decreasing meanwhile in amplitude.

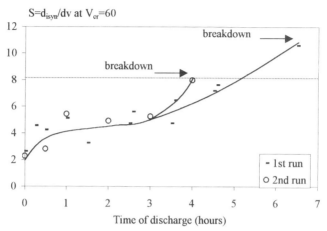

Figure 4. Temporal evolution of the dynamical conductance S, measured under AC conditions in the positive half-cycles at a crest voltage V_{cr} of 60 kV.

To illustrate this aspect, we present the Fig.4 which shows how evolves the dynamical conductance S with time, till a delayed spark breakdown, in a AC corona case (Al point, 60 µm radius of curvature, 10 mm gap length, 50 Hz applied voltage, 60 kV crest value). Here, S measurements refer to the slope $di_{syn}/dv = f(v)$ of the variation, as a function of the voltage v in instantaneous values, for of a series of positive half-cycles, of the current component i_{syn} synchronous with the voltage v which corresponds to the continuous current in a DC corona case. The numerical method used to perform these measurements with the current oscillographic signals will not be described here. But we shall mention that these measurements allowed to see that the i_{syn} current obeys, as in DC conditions, to the Townsend law : $i_{syn}=kv(v-v_0)$ with v_0 representing the corona threshold in instantaneous values.

Two last remarks concerning the AC corona case are that (i) S increases more rapidly in the positive half-cycles than in the negative ones, thus leading the delayed spark breakdowns to always occur in these positive half-cycles and that (ii) after breakdown, S comes back to its initial values while the discharge again stabilises, which means that the gap has then to a large extent recovered its dielectric strength.

Thus, the S_{crit} value appears as a general criterion for the risk of spark breakdown in SF_6 corona gaps under established voltages, AC as well as DC voltages.

4. Temperature measurements. The question of how a temperature increase prior to breakdown should explain the apparent reduction of the reduced field critical for breakdown $(E/n)_{crit}$ was already evoked above. Temperature measurements were important for the validation of this hypothesis. Rotational temperature measurements should be used for this purpose because, in gases at pressures as high as 2 bars, the relaxation times of rotationally excited states are so short that the transfer of rotational energy of molecules into translational energy can be considered as immediate. Indeed, such measurements are only possible where light is emitted. This means that the applicability of the method is limited to the ionisation region and, moreover, to make up for the lack of emissivity of SF_6 a small quantity of a better emitting gas had to be added to it[11,12]. In practice, the addition of 1% of N_2 to our 2 bars SF_6 was sufficient to enable rotational temperature measurements on the light emission of N_2 excited states ($C^3\Pi_u$) without appreciable changes in the discharge behaviour.

We made such measurements with positive and negative DC voltages applied to the point electrode and also with AC voltages. Typical temperature profiles, as those presented in Fig.5 for positive and negative coronas for a same mean current, were so obtained and from series of such profiles were deduced T(I) characteristics describing the temperature evolution in the region close to the point electrode as a function of the mean discharge current. The most significant results for our purpose here were obtained under AC conditions and are presented in Fig.6. The results show at first that the temperature increases with current quite linearly, which is also valid for the DC discharges.

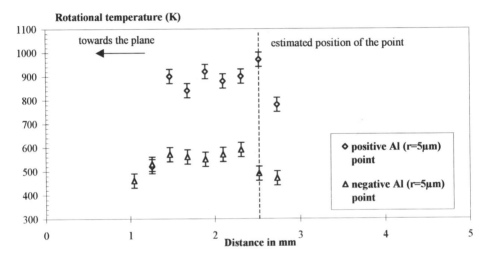

Figure 5. Temperature profiles along the gap for two 5 μm aluminium points under positive and negative polarity for a 100 μA discharge (d=10 mm).

But the mean current appears not to be by itself the most significant parameter for the temperature evolution, since the measurements with the two different radius of curvature for the point electrode ($\approx$ 10 μm and 300 μm) give very different temperatures for equal mean discharge currents ; this is evidenced by the slopes of the curves, respectively equal to ~ 5.5 kV/(μA)$_{eff}$ for the 5-20 μm points and to ~ 11 kV/(μA)$_{eff}$ for the 300 μm points. However the most relevant result for our purpose here is that spark breakdowns appears to occur for a same quite high temperature (between 850 and 900 K) independently of the radius of curvature of the point electrode. This tends to prove that the gas heating is the determining factor in the mechanisms which yield the gap ready for breakdown. The result also brings a validation to the temperature predicted by Sigmond[5], in relation with his narrow jet model evoked above, to show that a gas heating phenomenon should explain the spark breakdown mysteries. Effectively, his calculation ended with a core temperature at the start of the jet evaluated to 770 K, under conditions clearly different from ours (positive coronas in SF_6 at 48 kPa) but however implying a similar local gas density reduction (by a factor of 2.56) to allow breakdown achievement.

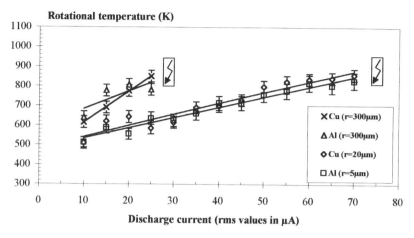

Figure 6. Rotational temperature as a function of the mean discharge current in rms values for new point electrodes under AC voltages (d=10 mm).

Now, since our temperature measurements were only possible to be made in the light emitting ionisation region, two questions arise : what about the temperature increase in the rest of the gap and how can the prebreakdown channel build up across the totality of this gap ? Should the gas flow associated with the electric wind, already evoked above, and directed from the high-field electrode (the point) towards the low-field electrode (the plane), be able to ensure a movement of heat propagation across the whole gap ? This will be discussed in next section in terms of time arguments. But supposed to be laminar, the hot gas jet can survive some centimetres of travel through atmospheric gas density without implying a mixing with the cold gas added on by friction to it as it was put forward by Sigmond[5]. Moreover, if, in the case of our investigations, we compare the gas density reduction attached to the temperature close to the point electrode ($n/n_0 = T/T_0 \approx 3$) and that one needed to achieve breakdown according to our electrical measurements [$(E/n)_{crit}$ / $(E/n)_{apparent}$ = 90/35 = 2.57], we see that we have 100 to 150 K in excess for the compensation of the thermal losses along the prebreakdown channel.

B. Temporal approach of the phenomena just prior to spark breakdown ; spark breakdown time-lags.

Let us, for this temporal approach, again refer to our AC corona discharges, the most convenient ones for our purpose because they enable, with the current oscillograms recorded, to see the recurrent kinetic evolution of the phenomena from corona reignition inside individual half-cycles. Unhappily, it was generally avoided to go till spark breakdown with the oscilloscope in operation, because of the risks of damage for it. But in the train of pulses exhibited by every half-cycle of the oscillographic current signals, it appears some emerging pulses which we assume to constitute preferential triggering pulses for the achievement of spark breakdown or at least for the formation of leaders, as it was also assumed by Kato et al.[13]. However, to be able to ensure this function, the pulses must satisfy threshold requirements which can be expressed through parameters attached to the first big current pulse observed in the positive individual half-cycles :

(i) in Kato's et al. experiments, the parameter chosen for this purpose was the charge q_{max} transported by such pulses approximated to be equal, close to breakdown in the working conditions of these authors (4 bars SF_6), to ~ 2.10^{-10} C.

(ii) in our experiments, performed at 2 bars SF_6, the parameter chosen was the height i_{max} of these pulses of maximal amplitude, with a threshold value for the triggering of breakdown around 1.2 mA.

But here, our interest is not so much for the big pulse charge or height, but more for the time parameters attached to its arrival with reference to the corona ignition parameters in the half-cycles ; let say, in terms of time-lags, interesting parameters will be (see Fig.7) the time-lags t_0 and t_{max} respectively attached to corona ignition and to the apparition of the current pulse of maximal amplitude with reference to the zero of the applied voltage waveform.

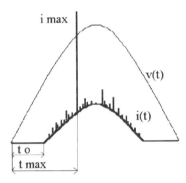

Figure 7. Schematic representation of v(t) and i(t) oscillograms showing the main parameters attached, in AC coronas, to corona ignition and to the current pulse of maximal height in the individual half-cycles.

From statistical analyses on the t_0 and t_{max} values, it came out the main following results :

- Corona onset time-lags t_0 tend to decrease, but only very slightly when the applied voltage is increased, let say for instance from 1.6 ms at V_{crest}=55 kV to 1.4 ms at 70 kV.

- As for the time-lags t_{max} of the pulses of maximal amplitude, they appear, over the same range of applied voltages (55 < V_{crest} < 70 kV), quite insensitive to the applied voltage, with a mean value between 3.3 and 3.5 ms for the maximal pulses considered in the positive half-cycles, the only ones directly concerned with breakdown ; effectively, the maximal pulses have higher amplitudes in the positive half-cycles than in the negative ones and breakdown always occur in these positive half-cycles.

So, if we now consider the time interval (t_{max} - t_0) elapsing between corona ignition and the apparition of the maximal height pulse, we get a value around 2 ms for all our operating conditions. Let us now suppose that this duration corresponds to the propagation time of a thermic wave which should start at the high-field electrode and cross the whole gap. This should imply a mean propagation speed of 3.5 to 5 ms according whether the propagation is supposed to start from the border of the ionisation region or from the point electrode itself. Now, this speed is very similar to that one of the electric wind. May we conclude that the prebreakdown channel should be formed by such a thermic wave ? This will have to be checked. But, though they do not yet bring a definitive proof about the way taken by the hot front to propagate, these results are also coherent with Sigmond's previous hypotheses[5] about the positive corona breakdown based on his narrow jet model.

CONCLUSION

Our aim in this paper was to show, through a selection of experimental results, gained with DC and AC voltages close to the spark breakdown voltages, that in such conditions of established voltages spark breakdown can occur in the track of a prebreakdown channel built across the gap by the propagation of a thermic wave initiated at the high-field electrode. Three steps characterised our approach :

1. Temperature measurements showing that, in the ionisation region, the temperature can reach high enough values (typically 850 to 900 K under AC voltages) to locally reduce gas density by an important factor (of about 3) that should explain a sufficient increase of the apparent reduced field E/n to get spark breakdown for applied voltages providing apparent reduced fields largely insufficient (35 kV/cm/bar instead of 90 kV/cm/bar).

2. Time measurements which were performed under AC voltages between corona ignition and the apparition, in the same positive half-cycles, of maximal height current pulses, showing that the gap can be the seat of a propagative movement of hot gas throughout its length.

3. Electrical measurements showing finally that the gap electrical conductance withstand variations in agreement with this scheme.

REFERENCES

1. H. Champain, G. Hartmann, M. Lalmas and A. Goldman, *Spectroscopic study of high pressure dc corona discharges in SF_6*, Proc. 11th Int. Conf. on Gas Discharges and their Applications, Vol. 1, pp. 152-155, Tokyo (1995).
2. R.S. Sigmond, R. Hegerberg and V.V. Baranov, *Positive glow pulses and sparks in 1 cm point-to-plane gaps in SF_6 up to one atmosphere : no streamers ?*, Proc. 7th Int. Conf. on Gas Discharges and their Applications, pp. 227-230, London (1982).
3. I. Gallimberti and N.J. Wiegart, *An investigation of the streamer to leader transition and related phenomena in SF_6 and SF_6 mixtures under positive impulse conditions*, Proc. 8th Int. Conf. on Gas Discharges and their Applications, pp. 219-222, Oxford (1985).
4. O. Farish, O.E.Ibrahim and A. Kurimoto, *Prebreakdown corona processes in SF_6 and SF_6/N_2 mixtures*, Proc. 3rd Int. Symp. on High Voltage Engineering, paper 31.15, Milan (1979).
5. R.S. Sigmond, *A narrow-jet model of dc corona breakdown*, Proc. 7th Int. Conf. on Gas Discharges and their Applications, pp.140-142, London (1982).
6. R.S. Sigmond, *Mass transfer in corona discharges*, Rev. Int. Hautes Tempér. Réfract., Fr. 25, pp. 201-206 (1989).

7. A. Goldman, M. Goldman and J.E. Jones, *On the behaviour of the planar current distribution in the pulseless regime of negative DC point-plane coronas in air*, Proc. 10th Int. Conf. on Gas Discharges and their Applications, pp. 270-273, Swansea, U.K. (1992).
8. J.E. Jones, A. Goldman and M. Goldman, *Dimensional methods and the Trichel pulse transition for negative DC point-plane coronas in air*, Proc. 12th Int. Conf. on Gas Discharges and their Applications, pp. 149-152, Greifswald, Germany (1997).
9. M.Lalmas, K. Hadidi, H. Champain and A. Goldman, *Corona discharges long-term behaviour and delayed spark breakdown in SF_6 under dc voltages*, Proc. 4th Int. Symp. on High Pressure Low Temperature Plasma Chemistry, pp. 189-194, Bratislava (1993).
10. M.Lalmas, H. Champain, A. Goldman and E. Fernandez, *Long-term evolution of point-to-plane SF_6 discharges under alternating voltage*, Gaseous Dielectrics VII, L.G. Christophorou and I. Sauers Eds, Pergamon Press, New York, pp. 617-623 (1994).
11. V. Zengin, S. Suzer, A. Gokmen, A. Rumeli and M.S. Dincer, *Analysis of SF_6 discharge by optical spectroscopy*, Gaseous Dielectrics VI, L.G. Christophorou and I. Sauers Eds, Plenum Press, New York, pp. 595-599 (1990).
12. A.M. Casanovas, J. Casanovas, V. Dubroca, F. Lagarde and A. Belarbi, *Optical detection of corona discharges in SF_6, CF_4 and SO_2 under dc and 50 Hz ac voltages*, J. Appl. Phys, 70, 3, pp. 1220-1226 (1991).
13. T. Kato, N. Hayakawa, M. Hikita and H. Okubo, *Breakdown prediction in SF_6 gas viewed from transition of partial discharge characteristics*, Proc. 11th Int. Conf. on Gas Discharges and their Applications, Vol. 1, pp. 280-283, Tokyo (1995).

DISCUSSION

A. BULINSKI: Have you tried to make IR measurements to see the presence of the hot channels that you assumed are taking place?

A. GOLDMAN: No IR measurements were performed. Such measurements should be interesting to attempt, but perhaps with a risk of too small light emission and too short duration because of the small diameter of the channel (< 1 mm) and since, in our steady-state working conditions, minimal energy expenses are implied for spark breakdown.

GENERATION MECHANISM OF PARTIAL DISCHARGE IN DIFFERENT KIND OF PURE GASES AND GAS MIXTURES WITH SF6

T. Yamada,[1] T. Takahashi,[1] T. Toda[2], and H. Okubo[1]

[1]Department of Electrical Engineering
Nagoya University, Nagoya, 464-8603, Japan
[2]Electric Power Research & Development Center
Chubu Electric Power Co., Inc., Nagoya, 459-8522, Japan

INTRODUCTION

SF6 is widely used in electric power apparatus such as gas insulated switchgears (GIS), not only because of its superior dielectric properties, but also its chemically inert and nontoxic characteristics.[1,2] However, it has several weak points; firstly, it has high liquified temperature. Secondly, its dielectric strength is drastically reduced by concentration of the electric field on insulation defects like metallic particles. Moreover, SF6 is the gas with high greenhouse effect. From these reasons, the development of new gas or gas mixture for the electrical insulation alternative to SF6 are strongly required and recently studied.[2-4] Much studies on gas mixtures are done for breakdown characteristics under lightning impulse voltage application,[5,6] while few study on partial discharge (PD) characteristics in gas mixtures are reported.

Our aim is to investigate application possibility of gas mixtures for GIS from the aspect of PD generation mechanism. So far, we have considered the process of PD characteristics from inception to breakdown (BD) in SF6. As a result, we have indicated the availability of PD measurement method based on its mechanism.[7,8]

We examined PD generation mechanism based on analysis of PD current pulses in gas mixtures. This paper describes the experimental results on SF6/N2, SF6/CO2 and SF6/CF4 mixtures, each gas used in this experiment mixed with SF6 has lower liquified temperature than SF6, nontoxicity and chemical inertness. The experimental results show current pulses, light intensity pulses and light images of PD simultaneously measured in pure gases and gas mixtures with SF6 for the purpose of clarifying the PD generation mechanism. As a result, we made it clear that PD in different pure gases had their own different characteristics based on the physical properties of gases themselves. The PD characteristics of gas mixtures differed with that of pure gases, but by mixing small amount of SF6, they approached to those of pure SF6 drastically. Moreover, from the measurement of PD extension length, it was shown that the PD in gas mixtures extended within the limited space that determined by their own critical electric fields.

As a result, we concluded that breakdown characteristics in gas mixtures resulted from the change in PD mechanism by mixing of SF6.

EXPERIMENTAL

Figure 1 shows a 66kV model GIS and the measuring circuit for simultaneous measurement of current and light emission of a single PD pulse. A needle with tip radius 500μm and length 20mm was fixed on the high voltage conductor to simulate a metallic particle condition in GIS. The gap

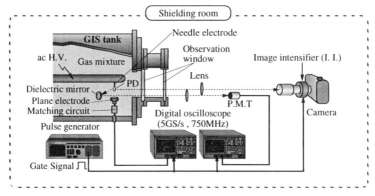

Figure 1. Experimental setup.

length of needle-plane electrodes was 10mm. PD occurred at the tip of the needle by applying ac 60Hz voltage to the conductor in GIS filled with different gas mixtures. A signal current pulse of PD was fed into a digital oscilloscope (5GS/s, 750MHz) through a plane electrode and a matching circuit. The matching circuit has high frequency response more than 1GHz.[8] On the other hand, PD light intensity was observed using a dielectric mirror, optical lenses and a photo multiplier tube (P.M.T.). Moreover, PD image was expanded by lenses, multiplied by an image intensifier (I. I.), and observed by a still camera. Controlling the gate pulse from a pulse generator, we constructed a measuring system which allowed simultaneous observation of current pulse, light intensity pulse and light image of PD occurring at a designated phase angle of ac applied voltage.[9]

In this experiment, we applied SF_6, N_2, CO_2 and CF_4 under room temperature. Mixture rate of gas mixture was determined by partial pressure of each gas.

RESULTS AND DISCUSSION

Breakdown Voltage of Pure Gases

Figure 2 shows breakdown voltage (BDV) as a function of gas pressure in N_2, CO_2, CF_4 and SF_6. BDV in SF_6 is higher than that in the other gases and shows nonlinear property for gas pressure. This nonlinear property results from corona stabilization effect caused by electron attachment of electronegative gas. CF_4 is also electronegative gas but its electronegativity is much smaller and BDV characteristics in CF_4 have smaller peak than that in SF_6. Moreover, BDV characteristics in CO_2 also shows nonlinearity. This property of CO_2 depends on dissociative attachment ($CO_2 + e^- \rightarrow CO + O^-$).[10,11] On the other hand, BDV in N_2 is lower than that in the other gases and shows linear property for gas pressure. This is because N_2 is not electronegative gas.[2]

Figure 3 shows partial discharge inception voltage (PDIV) and BDV as a function of mixture rate for SF_6 in different gas mixtures at 0.1MPa. It is clear that BDV of the gas mixtures exhibits a significant nonlinearity with increasing mixture rate of SF_6 than PDIV. BDV in SF_6/N_2 and SF_6/CO_2 containing 0-5% SF_6 and SF_6/CF_4 containing 0-1% SF_6 drastically increase. This drastic increase of BDV by mixing a small amount of SF_6 is seemed to be affected by the behavior of PD followed by breakdown.

PD Measurement in Pure Gases

Figure 4 shows current and light intensity waveforms and light emission image of PD occurring at 90° of applied ac voltage phase in CO_2, CF_4 and SF_6 (applied voltage: slightly above positive PDIV of each pure gas, gas pressure: 0.1MPa). PD light emission images were taken as the 60 cycles integrated images of PD occurring only between 87.5°-92.5° (for 0.23ms) of applied ac voltage phase. The data of the PD in N_2, however, express PD occurring at around 60° because PD in N_2 only occur at this phase. The light images reveal that PD in SF_6 and CF_4 expand within a restricted area near the needle electrode tip, while PD in N_2 and CO_2 expand longer close to the plane electrode.

Table 1 shows the amplitude and risetime of PD current pulse in different pure gases at slightly above the PDIV under 0.1MPa (average of 30 samples). It is obvious that the risetime of PD current pulse in each gas have a relation, $N_2 \fallingdotseq CO_2 > CF_4 > SF_6$. Thus, it can be said that in the gas with bigger molecular weight and the higher electronegativity, the PD current pulse has steeper risetime. Table 1 also shows the current increase rate di/dt of PD current pulse. As seen in Table 1, di/dt in SF_6 is much larger than that in the other pure gases. This is because the effective ionization coefficient of SF_6 depends highly on the electric field, compared to that of other pure gases.

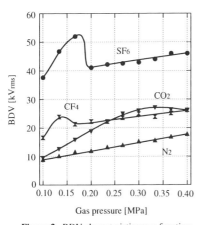

Figure 2. BDV characteristics as a function of gas pressure in different gases.

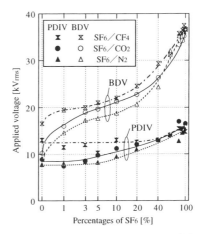

Figure 3. Positive PDIV and BDV characteristics as a function of mixture rate of SF_6 in gas mixtures at 0.1MPa.

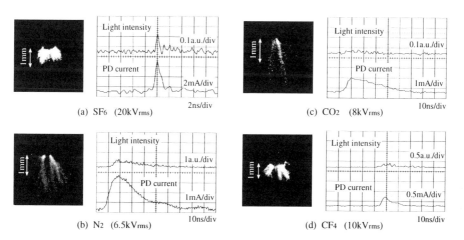

Figure 4. Light emission images, current and light intensity waveforms of PD pulses in different pure gases at 0.1MPa.

Table 1. Peak value and rise time of PD current pulses at positive PDIV in different pure gases at 0.1MPa. (in parenthese: applied voltage)

	SF_6 (20kVrms)	N_2 (6.5kVrms)	CO_2 (8kVrms)	CF_4 (10kVrms)
Amplitude	5.78mA	3.31mA	1.54mA	0.50mA
Rise time	0.58ns	6.64ns	6.84ns	4.45ns
$\frac{di}{dt}$	10.0×10^6A/s	0.50×10^6A/s	0.24×10^6A/s	0.11×10^6A/s

PD Measurement in Gas Mixtures

Figures 5 and 6 show current pulses, light intensity pulses and light emission images of PD occurring at 90° of applied ac voltage phase in SF_6/N_2, SF_6/CO_2 and SF_6/CF_4 mixtures containing 1% and 5% SF_6 (applied voltage: slightly above positive PDIV of each gas mixture, gas pressure: 0.1MPa). Containing 1% and 5% SF_6 in another gas intensify PD light emission of gas mixtures and shorten the risetime of PD current and light intensity waveforms drastically. The result of PD light images from Fig. 6 reveal that the PD in gas mixtures containing 5% SF_6 extend within a restricted area near the needle electrode tip like that in SF_6.

Table 2 shows the risetime of PD current pulses in different gas mixtures and comparison with PD in different pure gases at 0.1MPa (average of 30 samples). The risetime of PD current pulse was shorten drastically with increasing mixture ratio of SF_6 for each gas. Especially, by containing 1% SF_6 in another gas, SF_6/CO_2 shows the greatest reduction ratio of the risetime. The inquiry into this

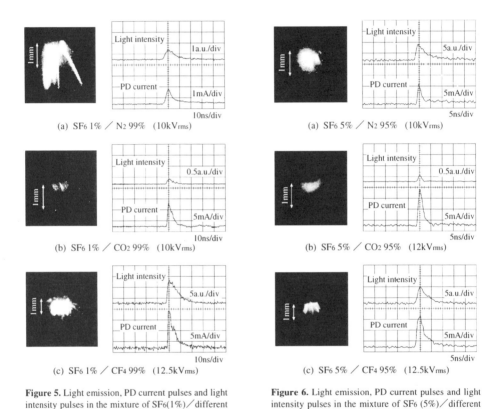

Figure 5. Light emission, PD current pulses and light intensity pulses in the mixture of $SF_6(1\%)$/different gases at 0.1MPa.

Figure 6. Light emission, PD current pulses and light intensity pulses in the mixture of $SF_6(5\%)$/different gases at 0.1MPa.

Table 2. Rise time of PD current pulses at positive PDIV in pure gases and gas mixtures at 0.1MPa. (in parenthese: applied voltage)

	N_2	CO_2	CF_4
Pure gas: A	6.6ns (6.5kVrms)	6.8ns (8kVrms)	4.5ns (10kVrms)
+1%SF_6: B Ratio: B/A	4.0ns (10kVrms) 60%	2.0ns (10kVrms) 29%	1.5ns (12.5kVrms) 33%
+3%SF_6: C Ratio: C/A	0.84ns (10kVrms) 13%	1.0ns (10kVrms) 15%	1.3ns (12.5kVrms) 29%
+5%SF_6: D Ratio: D/A	0.82ns (10kVrms) 12%	0.86ns (12kVrms) 13%	1.3ns (12.5kVrms) 29%

result needs the consideration of electron attachment process. In the case of N_2, non-attaching gas, the electron attachment effect results from only 1% SF_6 in SF_6/N_2 mixture. On the other hand, in the case of pure CO_2, weakly-attaching gas only by dissociative attachment,[11] electron attachment does not act effectively, but mixing 1% SF_6 in pure CO_2, electron attachment works in whole area between electrodes. Therefore, electron attachment acts effectively in SF_6/CO_2 containing 1% SF_6.

Investigation of PD Extension

Figure 7 shows the extension length of PD in gas mixtures as a function of applied voltage for different SF_6 content and applied voltage. The PD lengths in SF_6/N_2 and SF_6/CF_4 are shortened as increasing mixture rate of SF_6 and grow similarly as increasing applied voltage, while the PD lengths of SF_6/CO_2 appear to be shortened drastically and are shorter than that of SF_6.

Figure 8 shows the electric field at the PD tip calculated from the measured length of PD extension in gas mixtures shown in Fig.7. Broken lines represent the critical electric field E_{cr} of

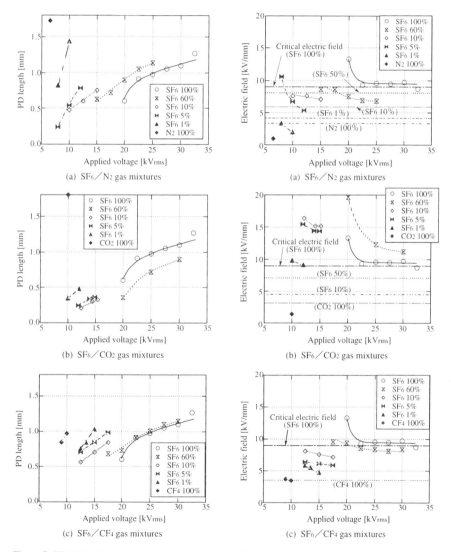

Figure 7. PD light emission length as a function of applied voltage in different gas mixture at 0.1MPa.

Figure 8. Electric field calculated from light emission length of PD in different gas mixture at 0.1MPa.

each gas mixture.[11-13] The data in SF$_6$ shown in Fig. 8 reveal good agreement of the calculated electric field at PD tip with the critical electric field in SF$_6$, E$_{cr}$=8.9kV/(mm·atm).[9,14] Moreover, it is obvious that the PD in gas mixtures occur within the electric field region over the E$_{cr}$ of their own gas mixtures. In the case of N$_2$, CO$_2$ and SF$_6$(1%)/N$_2$(99%), however, PD extend to the region whose electric field is below the E$_{cr}$ of their own gas mixtures, because PD types in N$_2$ and CO$_2$ are long brush (Figs. 5(b) and (c)), PD types in SF$_6$(1%)/N$_2$(99%) is short brush and filament type (Fig. 6(a)). Therefore, we consider this result being the fact that the electric field at PD tip occurring in SF$_6$/air containing 1-5% SF$_6$ changes E$_s$ into E$_{cr}$ (>E$_s$), where E$_s$ corresponds to a balance between the injected energy and the loss in the avalanches at the ionization front.[15]

CONCLUSIONS

To investigate PD generation mechanism in gas mixtures, we measured the current pulse, light intensity pulse and light emission image of PD occurring in pure gases and gas mixtures. In this paper, it was revealed that PD characteristics in pure gases change drastically into that in SF$_6$ by mixing a small amount of SF$_6$. We clarified the mechanism of this phenomenon through the measurement of PD lengths and the analysis of electric field strength at the PD tip in gas mixtures. As a consequence, in gas mixtures with SF$_6$, the drastic approaching of PD generation mechanism into SF$_6$ influences sensitively the sharp increase of BDV. In addition, PD and BDV characteristics of gas mixtures containing more than 5% SF$_6$ were nearly similar even though the original gas for SF$_6$ gas mixture is non-attaching gas like N$_2$ or weekly-attaching gas like CO$_2$ and CF$_4$. From this point of view, SF$_6$/N$_2$ mixture has efficient performance as the gas applicable to electrical power apparatus like GIS.

REFERENCES

1. S. Okabe, et al., Recent Development in Diagnostic Techniques for Substation Equipment, *CIGRE, Session Paper* 15/21/33-08, (1996).
2. L.G.Christophorou, et al., SF$_6$/N$_2$ Mixtures, *IEEE Trans. on Elect. Insul.*, Vol. 2, No. 5, pp. 952-991, (1995).
3. K. Mardikyan, et al., AC Breakdown Strength of N$_2$ SF$_6$, and a Mixture of N$_2$+SF$_6$ Containing a Small Amount of SF$_6$, *1996 IEEE Int. Symp. on Elect. Insul.*, June 16-19 Vol. 2, pp. 763-765, (1996).
4. Y. Qiu, et al., Investigation of SF$_6$-N$_2$, SF$_6$-CO$_2$ and SF$_6$-Air as Substitutes for SF$_6$ Insulation, *1996 IEEE Int. Symp. on Elect. Insul.*, Vol. 2, pp. 766-769, (1996).
5. H. L. Marsden, et al., High Voltage Performance of a Gas Insulated Cable with N$_2$ and N$_2$/SF$_6$ Mixtures, *10th Int. Symp. on High Voltage Eng.*, Vol. 2, pp 9-12, (1997).
6. T. B. Diarra, et al., N$_2$-SF$_6$ Mixtures For High Voltage Gas Insulated Lines, *10th Int. Symp. on High Voltage Eng.*, Vol. 2, pp 105-108, (1997).
7. T. Kato, et al., Breakdown Prediction in SF$_6$ Gas Viewed from Transition of Partial Discharge Characteristics, *11th Int. Conf. on Gas Dis. and Their Appl.*, pp. I-280-283, (1995).
8. M. Hikita, et al., Phase Dependence of Partial Discharge Current Pulse Waveform and its Freqency Characteristics in SF$_6$ Gas, *IEEE Int. Symp. on Elect. Insul.*, pp. 103-106, (1996).
9. H. Okubo, et al., Discrimination of Partial Discharge Type in SF$_6$ Gas by Simultaneous Measurement of Current Waveform and Light Emission, *IEEE Int. Symp. on Elect. Insul.*, pp. 107-110, (1996).
10. M. S. Bhalla, et al., Measurement of Ionization and Attachment Coefficients in Carbon Dioxide in Uniform Fields, *Proc. Phys. Soc.*, Vol. 76, pp. 369-377, (1960).
11. L. G. Christophorou, et al., Basic Physics of Gaseous Discharges, *IEEE Trans. on Elect. Insul.*, Vol. 325, No. 1, pp. 55-74, (1990).
12. Th. Aschwanden, Swarm Parameters in SF$_6$/N$_2$ Mixtures Determined from a Time Resolved Discharge Study, *Gaseous Dielectrics IV*, pp. 24-33, (1984).
13. Y. Qiu, et al., Dielectric Strength of SF$_6$/CO$_2$, Gas Mixture in Different Electric Fields, *Proc. 9th Int. Symp. on High Voltage Engineering*, Vol. S2, paper 2255, (1995).
14. L. Niemeyer, et al., The Mechanism of Leader Breakdown in Electronegative Gases, *IEEE Trans. on Dielect. and Elect. Insul.*, Vol. 24, No. 2, pp. 309-324, (1989).
15. I. Gallimberti, Breakdown Mechanisms in Electronegative Gases, *Gaseous Dielectrics V*, pp. 61-80, (1987).

DISCUSSION

A. M. MUFTI: The experimental work was carried out at a pressure of 0.1 MPa. Can you explain the experimental results at pressures of 0.3 MPa and 0.5 MPa considering the V/P characteristics?

T. YAMADA: We examined the pressure dependence of the PDIV and BDV characteristics at 0.1 MPa to ~ 0.4 MPa. The BDV characteristics show a peak in V/P characteristics. This peak at which the streamer changed into a leader appeared at gas pressures of ~ 0.2 MPa to ~ 0.4 MPa depending on the gas-mixture content. Recently, we examined PD in SF_6/N_2 mixtures and found that a 10%SF_6/90% N_2 shows a streamer-type characteristic while pure SF_6 shows the leader type. In addition, we would like to point out that the PDIV characteristics are mainly determined by the N_2 properties, but the BDV characteristics are strongly influenced by the PD behavior, that is, by the SF_6 gas content in the gas mixture. This means that in the gas mixture with SF_6 the PDIV shows nearly linear characteristics, while the BDV shows a non-linear and complicated pressure dependence.

E. H. R. GAXIOLA: How do you determine the "PD light emission length?" Is this by a linear or by a log-scale representation of the observed optical activity?

T. YAMADA: We determined the PD light emission length from the analog picture image taken by a still camera with an image intensifier. For the identification of the PD length in particular, we detected a PD image of a single discharge with highly non-uniform tip radius geometry, and thus we could measure the PD length with high precision.

K. NAKANISHI: The figure concerning corona inception voltages indicates that the corona inception voltages increase with the percentage of SF_6 gas, rather independently of the kind of the admixed gases, such as N_2, CO_2, and CF_4. Could you give a mechanism for this phenomenon?

T. YAMADA: From the results of our experiments, we clarified that PDIV characteristics of gas mixtures were different depending on the kind of the admixed gases. Especially the PDIV characteristics of SF_6/N_2 mixtures were lower than those of SF_6/CO_2 and SF_6/CF_4 mixtures as is shown in Fig. 3. In other words, PDIV characteristics would depend on the non-electronegative gas in the gas mixture. However, as you pointed out, the PDIV characteristics at higher SF_6 gas content show characteristics independent of the kind of the admixed gas. This would be because of the small difference of the PDIV by N_2, CO_2, and CF_4.

DIELECTRIC CHARACTERISTICS OF SF_6-N_2-MIXTURES FOR VERY FAST TRANSIENT VOLTAGES

W. Pfeiffer, D. Schoen, C. Zender

Institut für Hochspannungs-und Meßtechnik
Darmstadt, University of Technology, Germany

ABSTRACT

This paper deals with prebreakdown and breakdown phenomena of SF_6-N_2-mixtures in non-uniform gaps, stressed by very fast transient voltages (VFT-voltages). In these investigations the shape of the applied voltage is similar to the voltages that can occur in gas insulated substations (GIS) due to switching operations. Because of the very fast voltage collapse travelling waves are generated within the bus duct and together with reflections at impedance discontinuities high frequency oscillations are generated.

For these investigations the test voltage is generated by superposition of a sinusoidal voltage with a frequency of 3 MHz to a voltage step. The maximum voltage amplitude to be generated was 400 kV peak. The parameters being investigated are the frequency of the superimposed sinusoidal voltage, the polarity, the gas pressure and the rate of the SF_6 admixture. Two different inhomogenous field configurations were used. The gas pressure was varied from 0.05 MPa to 0.8 MPa.

For diagnostics both, optical and electrical measuring techniques were used to analyse the prebreakdown and breakdown phenomena. Optical investigations were performed with ultraviolet- and infrared-sensitive photomultiplier tubes to analyse the temporal development of prebreakdown phenomena. For spatially resolved optical measurements of the prebreakdown development a high speed camera system with an exposure time of 1 ns and with an improved shutter ratio was used.

In general terms a positive VFT voltage provides a smaller dielectric strength than a negative polarity VFT voltage. The difference in dielectric strength for the different polarities is much more pronounced for the more inhomogenous field configuration, whereas for a configuration with a higher field utilisation factor the dielectric strength can coincide for both polarities at certain pressures.

For negative polarity the more inhomogenous field configuration provides a higher breakdown voltage than the less inhomogenous field configuration in spite of the smaller field utilisation. This is explained with corona stabilisation caused by the anode directed streamer corona. The corona stabilisation for positive polarity is not as pronounced as for negative polarity. For positive polarity of the test voltage the positive space charge produced in the back of the streamer corona homogenises the field distribution so that due to the electron attaching characteristics of the SF_6-N_2-mixture discharge activity is reduced.

INTRODUCTION

Steep fronted oscillating voltages can occur in gas insulated high voltage apparatus after opening and closing of busduct connectors. For that kind of stress local distortion of the electric field for instance caused by the presence of small metallic particles either moving free or being fixed to the conductor, can cause severe reduction of the dielectric strength.

Because of synergetic effects SF_6-N_2-mixtures can exhibit advantages in dielectric strength, especially for non-uniform field distribution. Such gas mixtures are of increasing interest for application in compressed gas insulated transmission lines (CGITL).

EXPERIMENTAL SET-UP

The test voltage is generated with the experimental set-up shown in figure 1. The test voltage shape is a voltage step superimposed with a damped oscillating voltage. By charging a low inductive load capacitor C_L with a HVDC source and by firing a pressurised switching gap (SG), oscillation of the test circuit is initiated. The resonance frequency was tuned by varying the capacitor C_x and the inductor L_x. Oscillations from 3 to 12 MHz were generated during these investigations. A test voltage amplitude U_{VFT} up to 400 kV was generated (Pfeiffer et al. 1997). Two different electrode configurations within the test vessel (TV) were used to evaluate the influence of the field distribution on the prebreakdown development. These were on the one hand a needle-plane and on the other hand a hemisphere-plane configuration (figure 2).

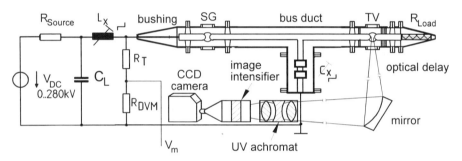

Figure 1. Experimental set-up

For optical investigations both, a high speed framing camera with an exposure time of 1 ns and a set of photomultiplier tubes for UV (180 nm - 330 nm) and IR (694 nm and 593 nm) were used. UV radiation is emitted by exited molecules, typically during streamer formation. 694 nm radiation is emitted when SF_6 is dissociated into fluorine atoms and 593 nm radiation is emitted when atomic nitrogen is produced. So radiation with a wavelength of 593 nm and 694 nm is an indication for the leader phase.

The camera is based on an image intensifier tube attached to a CCD camera (figure 1, lower part). The image intensifier tube basically consists of a UV photosensitive cathode (S20), a set of electron multiplying micro channel plates (MCP) and a phosphor screen. To provide a high shutter ratio, both, the cathode voltage and the MCP voltage were gated. By this means a maximum shutter ratio of $2.4 \cdot 10^{14}$ at 450 nm was achieved. Because the camera is gated by the current pulse of the discharge and due to the optical delay line, it is possible to display the final jump movement of the discharge in a picture series beginning at a gating time $t_{exp} = -22$ ns before the beginning of the voltage collapse. Furthermore each photograph is characterised by the gain factor G, which describes the relative luminous gain of the camera referring to the last picture of the series. To improve the visibility the photographs are displayed inversely.

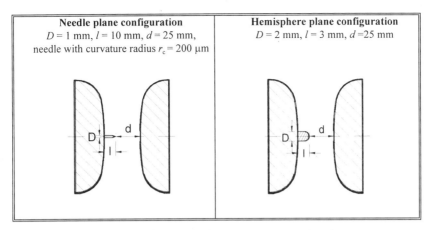

Figure 2: Electrode configurations being investigated

RESULTS

Pfeiffer et al. (1997) found that between an admixture of 0 % up to 10 % SF_6 to N_2 the gradient of dielectric strength versus the admixture has a maximum for a gas pressure of 0.1 MPa. Since within this range the 10 % SF_6 admixture exhibits the highest dielectric strength, research is focused on this mixture ratio. Figure 3 shows the dielectric strength of this mixture versus the pressure for different polarity of the test voltage and different field configurations.

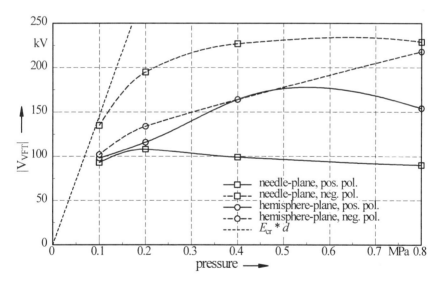

Figure 3. Dielectric strength of a SF_6-N_2-mixture with a SF_6 admixture of 10 % for two different inhomogenous electrode configurations and different polarity, f_{VFT} = 3 MHz, gap width d = 25 mm

The following aspects are of importance:
- The pressure reduced critical field strength of a SF_6-N_2-mixture with a SF_6 admixture of 10 % is $(E/p)_{cr}$= 58,3 kV/cm·bar (Aschwanden 1985). For small pressures (< 0.1 MPa) the dielectric strength coincides with the product of the critical strength E_{cr} and the gap width d (Farish 1983). Since E_{cr} is proportional to the pressure, the dielectric strength is reduced with decreasing pressure.

- For pressures higher than 0.1 MPa saturation of dielectric strength occurs and maxima of dielectric strength can be observed.
- In general terms a positive test voltage provides a smaller dielectric strength than a test voltage with negative polarity. This can be found here as well for both gap configurations, but for the hemisphere-plane configuration at a pressure of 0.4 MPa the dielectric strength coincides for both polarities. Since only four discrete pressures (0.1, 0.2, 0.4, and 0.8 MPa) were investigated, a configuration for positive polarity which provides even a higher dielectric strength can not be excluded. The difference in dielectric strength for both polarities is much more significant for the needle-plane configuration.
- In spite of the smaller field utilisation factor the needle-plane configuration provides for negative polarity a higher breakdown voltage than for the hemisphere-plane configuration. For the latter configuration a higher dielectric strength can only be expected for pressures beyond 0.8 MPa.

Corona stabilisation in SF_6-N_2 mixtures

Corona stabilisation effects are the main reason for the observed dielectric characteristics of the SF_6-N_2-mixture. In order to explain the mechanism of corona stabilisation the spatial and temporal prebreakdown development in a SF_6-N_2-mixture with a SF_6 admixture of 10 % for needle-plane configuration and positive polarity and a pressure of 0.1 MPa is shown in figure 4. The following conclusions can be drawn :

- The discharge inception occurs at the first positive slope of the test voltage with a streamer corona as indicated by the UV radiation. The positive space charge of the streamer corona and the corresponding field distribution is regarded to be the reason for the reduction of ionisation processes to a nearly constant level (UV radiation) despite increasing test voltage.
- The energetic condition for streamer leader transition (Buchner 1995) is fulfilled at $t = 105$ ns. Within a few nanoseconds one streamer within the corona is thermoionized and dissociated as indicated with the peaks in the 694 nm and 593 nm radiation.
- After the first leader inception the intensity of UV radiation is reduced to the plateau before the first streamer leader transition. The next streamer leader transition appears 20 ns before breakdown, and, hence, its spatial development can be seen in the series of pictures of figure 4. Up to $t_{exp} = -20$ ns the streamer corona with filamentary streamer channels directing to a luminous spot can be seen. The luminous spot is the head of the actual leader channel (parallel lines added in the figure). In between the filamentary streamers and the leader head a region of reduced radiation activity is detected as indicated by the added semicircle. Obviously the positive space charges left behind the streamers homogenise the field distribution and are the reason for the reduced discharge activity within the the area between the semicircle and the leader head. At $t_{exp} = -18$ ns the leader step is indicated by the channel between the two luminous spots. The lower luminous spot is the new corona being generated at the new leader head. The upper luminous spot is at the same place as the head of the previous leader and is, hence, the origin of the newly developed leader. The corona that is developed after this streamer leader transition reaches the cathode ($t_{exp} = -10$ ns) and within a few nanoseconds a highly conductive plasma channel is generated.

Negative polarity. Since the streamer heads of negative or anode directed streamers move towards a divergent field and are formed by electrons with high mobility, a larger radius is obtained compared with the radius of the head of the positive streamer. Therefore for negative polarity the minimum field strength which is required for streamer propagation is higher, leading to a higher breakdown strength. In order to understand the higher dielectric strength of the needle-plane configuration for negative polarity compared with the hemisphere-plane configuration photomultiplier signals were taken and compared for each pressure. Despite the different field configurations for negative polarity the amount of streamer leader transitions is the same for each pressure. For the needle-plane configuration a delay time between corona inception and the first streamer leader transition can be detected, whereas the hemisphere-plane configuration provides immediate streamer leader transition with corona inception. Since this delay time of the needle-plane configuration is decreased with increasing pressure a critical pressure can be detected beyond which corona inception coincides with streamer leader transition and subsequently with breakdown. The critical pressure for the needle plane configuration is ≈ 0.8 MPa. Beyond this pressure the dielectric strength can be calculated with the corona inception voltage using the streamer criterion (Farish, 1983).

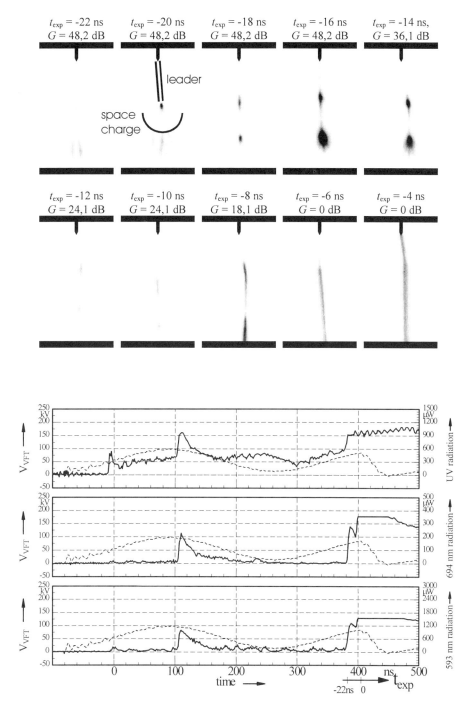

Figure 4. Spatial and temporal development of the prebreakdown phenomena in a SF_6-N_2-mixture with a SF_6 admixture of 10 %, f_{VFT} = 3 MHz, p = 0.1 MPa, d = 25 mm, positive polarity, needle-plane configuration

Positive polarity: For positive polarity also a delay time between corona inception an the first streamer leader transition for the needle-plane configuration, as well as immediate streamer leader transition with corona inception for the hemisphere-plane configuration can be detected. For small pressures

($p \leq 0.2$ MPa) corona stabilisation within the needle-plane configuration leads to a similar breakdown strength as for the hemisphere-plane configuration. The latter provides a higher dielectric strength for higher pressures. Obviously corona stabilisation is not so effective for positive polarity compared with negative polarity.

Coincidence of dielectric strength for both polarities: For the hemisphere-plane configuration at 0.4 MPa the dielectric strength for positive polarity coincides with the dielectric strength for negative polarity. For the negative VFT voltage breakdown occurs after a series of four streamer leader transitions whereas for positive polarity breakdown development is totally different. In the latter case the breakdown is initiated within 20 ns without streamer leader transition. The corresponding breakdown velocity of 1.25 mm/ns is very close to the breakdown velocity of the streamer as derived in (Pfeiffer et al. 1998).

Spatial development of the discharge channel

Positive polarity: For the needle plane configuration at pressures up to 0.2 MPa the transition from the streamer corona to a leader step usually takes place within one of the streamers close to the (vertical) axis so that the discharge path is more or less a straight line. For pressures beyond 0.4 MPa two or even more streamer channels within the streamer corona can be transferred into leaders and therefore the prebreakdown path is branched. In consequence the streamer corona becomes a set of streamer coronae, each one at the head of the actual leader branch, provided that the resulting field distribution allows streamer propagation for each branch. Therefore the shape of the discharge channel is usually not a straight but a crooked line.

Negative polarity: At negative polarity no branching could be observed.

CONCLUSIONS

Corona stabilisation plays the decisive role in prebreakdown development within SF_6-N_2 mixtures in inhomogenous fields for both polarities of VFT stress. For negative polarity it is the reason why a field distribution with a smaller field utilisation factor provides a significantly higher dielectric strength for a pressure of 0.1 MPa to 0.4 MPa. At 0.8 MPa no corona stabilisation was found.

Due to corona stabilisation the general assumption, that the positive VFT voltage provides a smaller dielectric strength than a negative polarity has to be reconsidered for SF_6-N_2 mixtures, since a configuration for positive polarity which provides a higher dielectric strength can not be excluded. Hence, in certain cases testing of gas insulated HV components with negative VFT voltages can be required.

REFERENCES

Aschwanden, Th. : "*Die Ermittlung physikalischer Entladungsparameter in Isoliergasen und Isoliergasgemischen mit einer verbesserten Swarm-Methode*", Dissertation ETH Zürich (1985)

Buchner, D. : "*Der Energie-Durchschlagmechanismus in SF_6 bei steilen transienten Überspannungen*", Dissertation TU München, 1995

Farish, O. : "Corona-Controlled Breakdown in SF_6 and SF_6 Mixtures", 16th Int. Conf on Phenomena in Ionized Gases, Düsseldorf S. 187-195, 1983

Pfeiffer, W.; Boeck, W. : "*Conduction and Breakdown in Gases*", Encyclopedia of Electrical and Electronic Engineering, John Wiley and Sons, to be published 1998

Pfeiffer, W., Schoen, D., Zender, C. : "*Dielectric Strength of SF_6/N_2 Mixtures for Non-uniform Field Distribution and Very Fast Transient Voltage Stress*",.VIII Internat. Symposium on Gas Discharges, Vol. I, pp. 286-289, 1997

DISCUSSION

A. H. MUFTI: Have you correlated your experimental work with a computer model?

W. PFEIFFER: Actually the development of appropriate discharge modeling is one of our further goals. We are grateful for any relevant contribution.

L. G. CHRISTOPHOROU: I would like to draw attention to your finding that the behavior of SF_6 mixtures with regard to transients does not differ much from that of pure SF_6 once the concentration of SF_6 in the mixture exceeds ~ 20%. This is significant for applications.

W. PFEIFFER: This is exactly a very important point, thank you for raising it. Actually, if the concentration of SF_6 in the mixtures with N_2 is equal to or greater than 10%, the dielectric characteristics for this kind of stress are very similar to those of pure SF_6.

THEORETICAL AND EXPERIMENTAL ANALYSIS OF A HIGH PRESSURE GLOW DISCHARGE CONTROLED BY A DIELECTRIC BARRIER

Pierre Ségur[1], Françoise Massines[2], Ahmed Rabehi[3], and Marie-Claude Bordage[1]

[1]Centre de Physique des Plasmas et de leurs Applications de Toulouse
[2]Laboratoire de Génie Electrique de Toulouse
Université Paul Sabatier
118, route de Narbonne, 31062 Toulouse Cedex 4, France
[3]SPIDELEC-MPC
9, Chemin d'EL PEY, 31770 Colomiers, France

INTRODUCTION

Recently many experimental works have been devoted to the study of an atmospheric pressure glow discharge controlled by a dielectric barrier[1,2]. There is a strong evidence that this type of discharge works well in pure helium. For other gases it seems to be necessary to introduce an additive (for example acetone in argon) in order to obtain a glow discharge or, in other situations (for example in air), small metallic mesh grids must be added inside the dielectric walls. Our experimental work[2], associated with our theoretical calculations[3] showed that, in helium, a glow discharge can only be obtained if there is a small amount of impurity inside the gas. The main role of these impurities (N_2 for example) is to allow, through the various metastable levels of helium, an increase of the ionization mainly in regions in which the electric field is so low that it cannot produce direct ionization of the molecules. It is clear then from these considerations that a glow

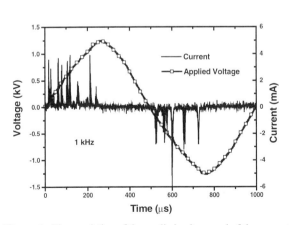

Figure 1 : Time variation of the applied voltage and of the current in helium for 1 kHz frequency, $P=10^5$ Pa, Vmax=1500 V.

discharge cannot occur in very pure helium in which the amount of impurities is too low.

In spite of these results, many phenomena remain unexplained in this type of discharge. The exact role played on the discharge by a change in the frequency of the oscillating applied voltage has to be clarified. Typically, experimental determinations (Figure 1) show that, at very low frequency (of the order of 1 kHz) or less, a glow discharge can no longer be obtained and the discharge becomes filamentary.

The objective of this paper is to try to explain this phenomenon from the information obtained from systematic numerical modeling of the discharge.

THE NUMERICAL AND PHYSICAL MODEL

Details of experiment and numerical modeling are given elsewhere[2-4]. As a glow discharge is usually homogeneous in the radial direction, for the moment, it is assumed that the discharge extends radially to infinity. It follows that the equations used to describe the space and time variation of the various physical quantities only depend upon the longitudinal position. Our numerical approach is then one-dimensional and cannot simulate the propagation of a filamentary discharge. Furthermore, the secondary emission of electrons at the cathode is assumed to be only due to the impact of ions. It is then clear that the only type of discharge that we can obtain from our calculations can only be a glow discharge.

Table 1. Reactions employed in the calculations together with the reaction rates

Direct processes	Collision frequency
$He + e \rightarrow He^+ + e + e$	Calculated by solving the Boltzmann equation
$N_2 + e \rightarrow N_2^+ + e$	Calculated by solving the Boltzmann equation
$He + e \rightarrow He(2^3S) + e$	Calculated by solving the Boltzmann equation
$He(2^3S) + e \rightarrow He + e$	Calculated by solving the Boltzmann equation
Three-body interaction	**Reaction rate**
$He^+ + 2He \rightarrow He_2^+ + He$	$6,3 \ 10^{-32} \ cm^6 s^{-1}$
$He_2^+ + e + He \rightarrow He_2^* + He$	$5 \ 10^{-27} \ cm^6 s^{-1}$
$He(2^3S) + 2He \rightarrow He_2^* + He$	$2,5 \ 10^{-34} \ cm^6 s^{-1}$
Recombination	**Reaction rate**
$He^+ + e \rightarrow He + h\nu$	$2 \ 10^{-12} \ cm^3 s^{-1}$
$He^+ + 2e \rightarrow He + e$	$7,1 \ 10^{-20} \ cm^6 s^{-1}$
$He^+ + e + He \rightarrow He^* + He$	$1 \ 10^{-27} \ cm^6 s^{-1}$
$He_2^+ + e \rightarrow He_2^* + h\nu$	$5 \ 10^{-10} \ cm^3 s^{-1}$
$He_2^+ + e \rightarrow He(2^3S) + He$	$5 \ 10^{-9} \ cm^3 s^{-1}$
$He_2^+ + 2e \rightarrow 2He + e$	$2 \ 10^{-20} \ cm^6 s^{-1}$
$N_2^+ + e \rightarrow N + N$	$4,8 \ 10^{-8} \ cm^3 s^{-1}$
$N_2^+ + 2e \rightarrow N_2 + e$	$1,4 \ 10^{-26} \ cm^6 s^{-1}$
Two-body interaction	**Reaction rate**
$He(2^3S) + He(2^3S) \rightarrow e + He^+ + He$	$2,9 \ 10^{-9} \ cm^3 s^{-1}$
Penning ionization	**Reaction rate**
$He(2^3S) + N_2 \rightarrow N_2^+ + He + e$	$8 \ 10^{-11} \ cm^3 s^{-1}$

The various reactions taken into account in this work are shown in table 1. As the discharge propagates at atmospheric pressure, molecular He_2^+ ions must be introduced together with atomic He^+ ions and molecular N_2^+ ions. For the moment, atomic N^+ ions are not considered. A simplified kinetic scheme is used for helium in which only the first metastable level $He(2^3S)$ is taken into account. According to the reactions given in table 1, this metastable level is created either by direct electron-atom excitation or by recombination between an ion He_2^+ and an electron. Population of the 2^3S level by cascade effects is not considered. Destruction of this state occurs through two-body and three-body interactions. Two-body and three-body recombination are introduced, three-body recombination being a very important process at atmospheric pressure.

Drift velocity of electrons, diffusion coefficient and ionization frequencies are calculated using a numerical solution of the equilibrium form of the Boltzmann equation. Direct and inverse excitation of 2^3S levels are obtained in the same way. All these quantities are functions of the E/N ratio in which E is the electric field and N the density of the neutral background gas.

The drift velocities and diffusion coefficients for ions are obtained from[5]. The secondary emission coefficient was assumed to be equal to 0.01 for helium ions and to 0.02 for nitrogen ions.

RESULTS

Calculations were made in standard experimental conditions i.e. for an applied voltage of amplitude 1500 volts, a frequency of 10 kHz, a gap distance of 0.5 cm. The thickness of the dielectric on the electrodes was 0.06 cm The radius of the electrodes was 2 cm. The study was made for a mixture of helium with 500 PPM of nitrogen.

Figure 2 shows the variation of the various electrical quantities (i.e. the discharge current, the applied voltage, the gas voltage and the memory voltage (this voltage is induced by the charge deposited on the dielectrics by the discharge)). This figure shows

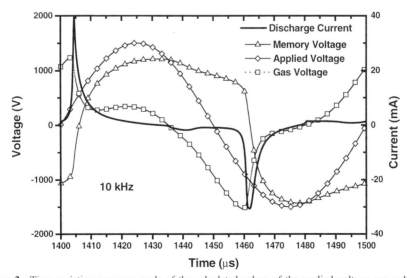

Figure 2 : Time variation over one cycle of the calculated values of the applied voltage, gas voltage, memory voltage and discharge current

that a stable state is reached. However, the two current pulses are not identical. In spite of that, their shape is very close to the experimental results reported in[4]. It was already assumed in our previous publications that the ionization of nitrogen molecules through Penning effect (see table 1) plays a very important role for the existence of a glow discharge. We can quantify this effect by introducing an effective ionization frequency ν_{eff} defined by the following relation

$$\nu_{eff} = \nu_{ion} + \frac{\nu_m \cdot n_m}{n_e}$$

in which ν_{ion} is the direct ionization frequency, ν_m the Penning ionization frequency, n_e the density of electrons and n_m the density of metastable atoms. This relation shows that the total ionization process is the sum of a direct plus an indirect process. Its interest is to allow the estimation of the relative importance of direct compared to indirect ionization. Figure 3 shows the space and time variation of these two quantities. It is clear that the effective ionization frequency is higher in the regions where the electric field is low (at the beginning of current growth for example) and that it allows the creation of electrons (and consequently of ions) in these regions. Furthermore, strong ionization appears in the decreasing part of the discharge which justifies the typical shape of the current after the current maximum. It is interesting now to plot the space and time variation of the density of metastable 2^3S levels together with the space and time variation of electron density. Obviously, these two quantities are strongly connected to the behavior of the effective ionization frequency. Figure 4 shows the space and time variation of electron and metastable 2^3S densities. We note that the density of 2^3S level varies in close agreement with the corresponding effective ionization frequency. The consequence is that, at each instant, when the density of metastable levels is sufficiently high, there is a continuous creation of electrons through the Penning reaction. Furthermore, as the voltage frequency is rather high, these electrons cannot disappear completely between two discharges and, since their effective ionization coefficient is large even when the electric field is small, electrons are able to initiate a new discharge for low values of the applied voltage.

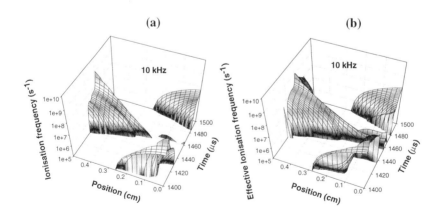

Figure 3: Evolution during one cycle of the direct ionisation frequency (a) and the effective ionisation frequency (b)

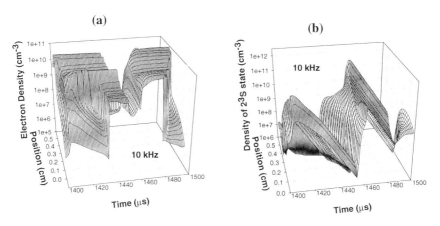

Figure 4: Evolution during one cycle of electron (a) and metastable densities (b), 10 kHz

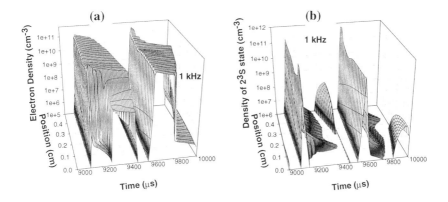

Figure 5: Evolution during one cycle of electron (a) and metastable densities (b), 1 kHz

We must note that, taking into account the various assumptions used, the results obtained from the present numerical simulation only correspond to a glow discharge. To simulate a filamentary discharge for example, it would be necessary to replace the Poisson equation by the well-known 'disk' method. In our conditions, as we assume that the discharge extends radially to infinity, the electric field obtained from the Poisson equation is much higher than the electric field given by the 'disk' method (since in this case the radius of the discharge would be assumed very small). In these conditions, if the voltage frequency is decreased to 1 kHz, the present current cannot be immediately compared with the measured currents shown in figure 1 for filamentary discharges. However, as we will see below important information can be obtained from these calculations.

Figure 5 shows, for a voltage frequency of 1 kHz, the space and time variation of electrons and metastable densities. We see now that unlike in the previous case (10 kHz), the number of electrons and of metastable levels between two discharges is very small. In these conditions, there is no possibility of strong ionization at low field and the discharge can only occur for higher values of the applied voltage.

COMMENTS AND CONCLUSION

The results given above clearly show the importance of the existence of secondary electrons for the production of a glow discharge at atmospheric pressure. Although a complete description of the transition from a filamentary to a glow discharge would need at least a two dimensional numerical model, it is possible, from the results above, to give a qualitative explanation of the occurrence of filamentary discharges for very low voltage frequencies. Whatever the frequency, the first discharge is initiated from a certain number of electrons created in the gas mainly by cosmic rays. At 10 kHz, many electron avalanches are produced which, if the applied voltage is high enough, are able to generate filamentary discharges. These discharges are produced during the first initial periods and create an increasing number of metastable levels. This 'cloud' of metastable levels quickly fills up the gap and, as the voltage frequency is high, they create conditions similar to those shown in figure 4 and a glow discharge appears.

At 1 kHz, even though the initial conditions are similar, the number of atoms excited on a metastable levels, during a filamentary discharge, is not so high than previously because, due to the long time between two cycles, most of them disappear in close agreement with the results reported in figure 5. It follows that, in this case, as a sufficiently dense 'cloud' of metastable levels cannot occur, the spatial distribution of electrons is sparse, most electron avalanches are isolated and can only generate a filamentary discharge when the gas voltage becomes sufficiently high.

REFERENCES

1. Y. Sawada, S. Ogawa and M. Kogoma, Synthesis of plasma-polymerized tetraethoxysilane and hexamethyldisiloxane films prepared by atmospheric pressure glow discharge, *J. Phys. D: Appl. Phys.* 28: 1661 (1995)
2. F. Massines, R. Ben Gadri, Ph. Descomps, A. Rabehi, P. Ségur, and C. Mayoux, Atmospheric pressure dielectric controled glow discharge: diagnostics and modeling, *XXII ICPIG*, Hoboken USA, Invited paper: 306 (1995)
3. A. Rabehi, P. Ségur, F. Massines, R. Ben Gadri, and M.C. Bordage, Investigation of the role of nitrogen impurities on the behavior of an atmospheric-pressure glow discharge in helium, *XXIII ICPIG*, Toulouse France, IV-44 (1997)
4. F. Massines, A. Rabehi, Ph. Descomps, R. Ben Gadri, P. Ségur, and C. Mayoux, Experimental and theoretical study of a glow discharge at atmospheric pressure controlled by dielectric barrier, *J. A. P.* 83-6: 2950 (1998)
5. H.W. Ellis, R.Y Rai, and E.W. McDaniel, E.A. Mason, L. A. Viehland, Transport properties of gaseous ions over a wide energy range I, *At. Dat. and Nucl. Data Tables* 17: 177 (1976); L. A. Viehland and E.A. Mason, Transport properties of gaseous ions over a wide energy range IV, *At. Data and Nucl. Data Tables* 60: 37 (1995)

ELECTRICAL AND PHYSICAL BEHAVIOUR OF A POINT-TO-PLANE BARRIER DISCHARGE IN AIR GOVERNED BY THE WATER VAPOUR CONTENT

L. Parissi, E. Odic, A. Goldman, M. Goldman, S. Koch
Laboratoire de Physique des Gaz et des Plasmas
(Univ. Paris Sud - CNRS)
Equipe Décharges Electriques et Environnement
Supélec, 91192 Gif-sur-Yvette Cedex, France.

INTRODUCTION

Increasing environmental remediation activities and global pollution control have motivated research into alternative methods to remove toxic wastes in gas phase. Non thermal plasmas are attractive in this regard due to their high efficiency for producing highly reactive species (i.e. secondary electrons, ions, excited species and radicals) and oxidising agents that will then destroy the pollutants. Volatile Organic Compounds (VOCs) are candidates for plasma remediation due to new restrictions on their combustion and disposal. For this purpose, methods using electron beams[1], microwaves[2], pulsed coronas[3], surface coronas[4], packed-bed[3] and dielectric barrier[5] discharges have already been reported.
Dielectric Barrier Discharges (DBD) seem to be particularly attractive for this purpose because of their ability to be operational at atmospheric pressure, of their low cost in energy and of their relative maturity in generating ozone, widely used. But this technology applied for VOCs conversion is in its early stage of physical and chemical development and characterisation. The products and intermediates of destruction are not well characterised, kinetics and mechanisms of destruction are not yet completely understood. Lots of destruction mechanisms suggested and used in modelling are mainly based on the chemistry of radicals such as oxygen atoms O (^{3}P) and O (^{1}D), and hydroxyl radicals (OH°)[6,7], but neglect the pure electrical effects of added water in the DBD, due to the utilisation of a dielectric layer covering at least one of the electrodes, and their resulting effects on the destruction of VOCs. This work brings a contribution to the understanding of phenomena involved in VOCs conversion by presenting an experimental study of the modifications brought on the discharge behaviour and effects in a point-to-plane barrier discharge system by the gas phase water content.

EXPERIMENTAL

Experimental Set-up

The experimental cell (~ 50 cm^3) is illustrated schematically in Figure 1. A point-to-plane gap is used for the discharge with a stainless steel needle as a high field electrode and, in front of it, a brass plane electrode covered with a dielectric barrier (alumina-silicate ceramics) of 5 mm thickness. The gap length between the point electrode and the dielectric barrier is fixed to 6 mm. The point electrode is connected to a high voltage generator supplying AC voltages up to 7 kV rms with a frequency of ~ 60 kHz.

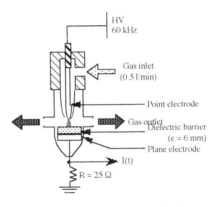

Figure 1: The experimental cell.

Figure 2 presents a block diagram of the complete experimental set-up including the measurement equipment. The experiments are carried out with synthetic air which is brought through a mixing chamber (where gas humidity and temperature are measured) to the reactor by a circuit comprising two branches. The main one, directly for dry air, is controlled by the mass flow controller MFC1. The flow controller MFC2 allows the passage of a fractional flow of dry air through a water bubbler to adjust the relative humidity as desired. The gas is introduced in the reactor in a direction parallel with the discharge axis, with a flow fixed at 0.5 l/min for all experiments concerned with the paper.

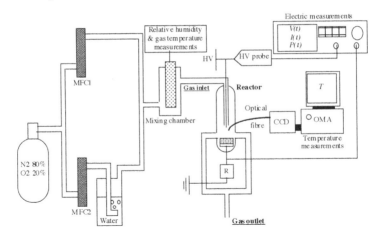

Figure 2: Block diagram of the complete experimental set-up.

Electrical Measurements

A high voltage probe (1:1000) is used for the voltage measurements V(t), and the discharge current I(t) is determined from the voltage drop through a 25 Ω resistance placed between the plane electrode and earth. Both signals are stored on a 200 MHz digital oscilloscope (input resistance for both V(t) and I(t) signals: 1MΩ). The power injected in the discharge is calculated by two different ways:
- by the product of the instantaneous values of the discharge current and applied voltage. The resulting signal is integrated over 10 cycles, providing the injected discharge energy during this time interval. Then multiplying the result by the generator frequency f, gives the power values as follows:

$$P(W) = f \cdot \frac{1}{10} \cdot \int_0^{10 \text{ periods}} V(t) \cdot I(t) \, dt$$

- using a Lissajous's display method. Integrated current measurements are achieved by means of a capacity C_o, much greater than the stray capacitance of the discharge gap, and put in series with it. Using the oscilloscope in a x-y display mode, the applied voltage and the integrated current give a Lissajous's loop and the value of the energy consumed is deduced from its area. Then, the power consumption of the discharge can be expressed as:

$$P(W) = \text{"area } U_x \cdot U_y \text{"} \cdot C_o \cdot f$$

The results obtained with the two methods agree within an error range of 5 %.
Systematic statistical analysis of the current pulses of amplitude I_p provided by the discharge was carried out and the charge Q_p transferred by each of the elementary discharges corresponding to the current pulses was calculated by integration of current over the pulse duration dt:

$$Q_p(nC) = \int I_p(mA) \cdot dt(\mu s)$$

Mean active discharge currents (I_m) were directly deduced from the power values by using the relation:

$$I_m(mA) = \frac{P(W)}{V_{rms}(kV)}$$

Temperature Measurements

Gas discharge temperatures are determined spectroscopically in the discharge volume close to the point electrode by the evaluation of the rotational temperatures measured on nitrogen N_2. The molecular spectrum of the second positive system of N_2 corresponding to the $C^3\Pi_u \rightarrow B^3\Pi_g$ transition is used for this purpose, assuming that there is an equilibrium established between the rotational temperature of the molecular excited N_2 and the kinetic temperature of the gas, that is a reasonable assumption at atmospheric pressure[8]. From a practical point of view, light is collected close to the point electrode by means of a quartz optical fibre which brings it, after passing through a Jobin Yvon HR 320 monochromator, to an Optical Multichannel Analyser (EG&G OMA), via a bidimensional charge coupled device (CCD) detector. The OMA system allows an analogic-digital conversion of the experimental spectra for the determination of the rotational temperature by comparison with simulated spectra according to a method previously developed in the laboratory for this purpose[9,10,11].

RESULTS AND DISCUSSION

Influence of the Water Vapour Content on the Discharge Current vs the Applied Voltage

Figure 3 shows the mean active current (I_m) as a function of the applied voltage for different water vapour contents in synthetic air at 0.5 l/min. In contrast with the monotonous rise of I_m in dry air, the presence of water vapour leads to a discontinuous behaviour. Now, the mean current progresses by steps with increasing applied voltage: the region 1 is followed by an abrupt rise of current (S1) leading to a plateau (2) until reaching a second step (S2) is reached; then the current increases abruptly(3), only mastered by the electrical and geometrical properties of the ceramics.

Now focusing on the pulses behaviour, it is noticeable that each domain of the I_m (V_{rms}) curve corresponds to a specific pulses distribution, as shown in Figure 4 where typical current and point voltage oscilloscope waveforms in a wet case are reproduced:
- step (S1) introduces the addition to the positive pulse I_{0+} of a new positive pulse I_1 and of a negative pulse I_{0-}, and this system remains constant all along the current plateau (2),
- the second step (S2) corresponds to the advent of a third positive pulse I_2, correlated to the disappearance of the negative streamer I_{0-} (3).

At this stage, some characteristic points have to be emphasised:
(i) A relevant feature is the regularity acquired by the discharge (current plateau 2) when water vapour is added, compared to its erratic behaviour in dry air.
(ii) The two positive pulses I_1 and I_2 advents, respectively about two and four times more energetic than I_{0+}, are demonstrative of changes in the discharge physics, and explain the gas temperature increase at step S2, which will be further discussed.
- As shown in Figure 4, I_{0+} and I_{0-} pulses occur when the voltage signal goes through zeros. This is well understandable, since the discharge is governed by the gaseous gap potential in fact given by the difference of the potentials varying with time, between the point electrode and the ceramics surface. The potential of which depends on charges accumulating on it[12,13].

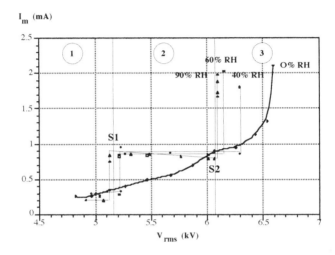

Figure 3: Mean discharge current vs applied voltage for different water contents.

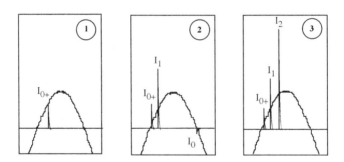

Figure 4: Typical current waveforms for the different discharge regions (1, 2 and 3).
Elementary discharges currents I_{0+}: 8-40 mA, I_1: 50-80 mA, I_2: 100-150 mA, I_{0-}: 10-20 mA. Pulse duration I_{0+}: 360µs, I_1: 360µs, I_2: 360µs, I_{0-}: 720µs. Charge transferred per elementary discharge I_{0+}: 2.2-2.4 nC, I_1: 4.5-5.5nC, I_2: 8.5-12nC, I_{0-}: 2.2-2.8nC. Total transferred charge per cycle changes from 8.9nC to 19nC.

Surface Effects of Water Vapour Content

The discharge can be described in two parts: its gaseous component and its dielectric surface component. Figure 5 illustrates the influence of the water vapour content on the surface discharge component, for a constant applied voltage (6.19 kV$_{rms}$): it clearly appears that the discharge extension on the dielectric surface is much larger in the wet case (30% RH), than in the dry one, while oscilloscope measurements show that this modification corresponds to what occurs at the transition step S2 of Figure 3.

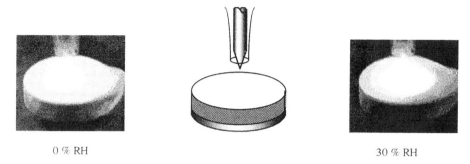

0 % RH   30 % RH

Figure 5: Photographs of the discharge before and after the RH transition value (30% RH at 6.19 kV$_{rms}$).

The change in the discharge behaviour, in correlation with the water vapour content, seems consequently due to a dielectric surface effect. Furthermore, this assumption is confirmed by the negative pulse I_{0-} disappearance occurring at this transition step S2. The current pulse I_{0-} comes from the electric field inversion (in advance with respect to the point voltage polarity change) due to charge accumulation on the ceramics surface. Since surface charge mobilities increase, this charge accumulation becomes weak and then I_{0-} has no chance to develop. This mechanism laying on the surface conductivity is governed on the one hand by the adsorption of water, and on the other hand by the ceramics surface heating due to the discharge.

Coming back to Figure 3, this combined effect on surface conductivity can explain the different transition voltage values observed for steps S1 and S2 with different RH: for the step S1, the transition voltage remains constant for 40 and 60% RH, and relative humidity has to reach 90% to obtain a lower transition voltage while for the second step, only 60% RH is required to observe a transition voltage decrease. This difference is attributed to the ceramics intrinsic surface conductivity magnified by temperature: when surface temperature increases, the transition potential is shifted towards lower relative humidities.

Correlation with the Discharge Temperature and Resulting Effects on VOCs Destruction

In order to investigate the RH threshold values more precisely, the average pulse amplitude of each type of elementary discharges (I_{0+}, I_1, I_2, I_{0-}), and the resulting mean discharge current (I_m) were studied at a constant applied voltage (6.19 kV$_{rms}$) for several water contents in the air flow (0.5 l/min). Results obtained were correlated with simultaneous in-situ measurements of the gas kinetic temperature (T_g) in the plasma bulk (Fig 6). First, the elementary discharges currents and thus I_m tend to decrease slowly due to a charge mobilities decrease, bred by the creation of heavy ions ($M^+[H_2O]_n$; $M^-[H_2O]_n$) in the discharge volume; for this point voltage value, the RH required to reach the second step S2 is 30 %, value for which the discharge regime changes as discussed above and remains constant afterwards.

Concerning the discharge temperature, the transition step S2 implies an abrupt rise, correlated to the sudden jump in the mean current. The advent of the I_2 pulse is responsible for both phenomena.

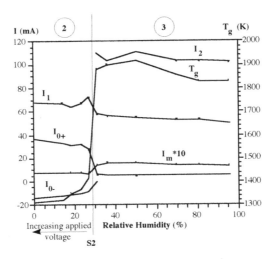

Figure 6: Current pulses amplitudes (I_{0+}, I_1, I_2, I_{0-}) and mean discharge current (I_m) measurements with regard to the plasma temperature (T_g) at constant applied voltage (6.19 kV_{rms}).

The efficiency of plasma depollution processing depends on three reliable parameters: conversion rate of the target pollutant, by-products control and power required. With regard to the conversion rate, experiments carried out on toluene showed that the destruction efficiency increased linearly with the discharge temperature, suggesting a temperature dominating mechanism[14]. Now, considering Figure 6, if sufficient water vapour is added to air in order to reach the transition step, a constant point voltage leads to a non negligible temperature increase (from 1300 to 1850 K). Since this phenomenon is correlated to a mean current rise, power injected in the discharge increases too. However, Figure 7 which represents the discharge temperature versus the injected power in wet (90% RH) and dry air exhibits a temperature gain of 10 to 16 % with wet air when the injected power increased from 5 and 10 W. Higher temperatures correspond to an increase of the surface component of the discharge; since the onset voltage required for surface discharges is lower than for discharges in gaseous volumes, it seems reasonable to observe a reduction of the energy cost for such a temperature increase[15].

So our measurements suggest an improvement in toluene destruction of almost 10 % by the single effect of water vapour on the discharge behaviour, apart from any chemical reaction with OH° radicals in the gas phase. Furthermore, from a processing point of view, the spreading of the discharge on the dielectric surface increases the interacting volume between plasma and gas flow and consequently the residence time, allowing a better conversion of the initial target molecule and of the by-products to CO_2 and H_2O.

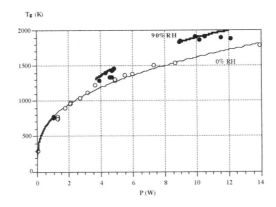

Figure 7: Plasma temperature evolution with increasing injected power in dry air and wet air (90% RH).

CONCLUSIONS

This paper demonstrated that, concurrently with the chemical effects produced by the addition of water vapour in VOCs remediation, widely discussed, another important effect of water is linked to the modification of the dielectric surface properties, which influences the discharge behaviour. In the case of plane-to-plane barrier discharges, it has been established that water vapour has a dramatic effect by decreasing the number of microdischarges so reducing the volume of the reactive plasma within the system and consequently its performance[16]. In the case of an inhomogeneous field configuration as that one our point-to-plane geometry, this problem is non-existent.

Water vapour introduces a discontinuity in the discharge behaviour in the form of mean current and temperature variations. The transition voltage values are themselves submitted to a same effect depending on relative humidity. The transition steps are characterised by a spreading of the surface discharge component due to the dielectric surface conductivity increase.

Taking advantage of these observations should be of great interest in terms of depollution efficiency and energy cost in VOCs remediation with DBD processing.

Acknowledgements

The authors acknowledge fruitful discussions with J.P. Borra and financial support from Electricité de France which helped them to progress with this work.

REFERENCES

1. B.M. Penetrante, M.C. Hsiao, J. N. Bardsley, B.T. Merritt, G.E. Vogtlin, A. Kuthi, C. P Burkhart and J.R. Bayless, Identification of mechanims for decomposition pf air pollutants by non-thermal plasma processing, *Plasma Sources Sci. Technol.*. 6, pp 251-59 (1997).
2. M. Chaker, M. Moisan and Z. Zakrzewski, microwave and RF surface wave sustained discharges as plasma sources for plasma chemistry and plasma processing, *Plasma Chem. Plasma Process.* Vol 6, no 1, pp 79-96 (1986).
3. T. Yamamoto, K. Ramanathan, P.A. Lawless, D.S. Ensor, J.R. Newsome, N. Plaks, G.H. Ramsey, C.A. Vogel and L. Hamel, Control of volatile organic compounds by an ac energized ferroelectric pellet reactor and a pulsed-corona reactor, *IEEE Transactions on Industry Applications.* 28, pp 528-34 (1992).
4. T. Oda, A. Kumada, K. Tanaka, T. Takahashi and S. Masuda, Low temperature atmospheric pressure discharge plasma processing for volatile organic compounds, *J. of Electrostatic.* 35, pp 93-101 (1995).
5. D. Evans, L.A. Rosocha, G.K. Anderson, J.J. Coogan and M.J. Kushner, Plasma remediation of Trichlorethylene in silent discharge plasmas, *J. Appl. Phys.*, 74 (9), pp 5378-86 (Nov 1993).
6. L.A. Rosocha and al, Treatment of hazardous organic wastes using silent discharges plasmas, *Non Thermal Plasma Techniques for Pollution Control.* Series G: Ecological Sciences, 34, part B, pp 281-308 (1993).
7. B. Eliasson and U. Kogelschaltz, Modeling and applications of silent discharge plasmas, *IEEE Trans. Plasma Sci..* Vol19, pp 309-22 (1991).
8. H. Champain, G. Hartmann, M. Lalmas and A. Goldman, 11th *Intern. Conf. on Gas Discharge and Their Applications* . Chuo Univ., Tokyo, (Sept. 1995).
9. A. Chelouah, E. Marode, G. Hartmann and S. Achat, *J. Phys. D: Appl. Phys.* 27, pp 940-45 (1994).
10. G. Hartmann and P.C. Johnson, *J. Phys. B: Atom. Molec. Phys.* 11, No 9, pp 1597-612 (1978).
11. G. Hartmann, E. Marode and A. Chelouah, ESCAMPIG 90 Orléans, pp 423-24 (1990).
12. S. Khabthani, A. Goldman, M. Goldman, È Odic, R. S. Sigmond, In-situ measurements of the electrical stresses induced by corona streamers on polymer foils under ac conditions, *5th Intern. Conf. on Conduction and Breakdown in solird Dielectrics*, Leicester (1995).
13. S. Khabthani, A. Goldman, M. Goldman, E. Odic, Influence of water layers on the electrical behaviour of polymer foils submitted to ac air coronas, *IEEE Intern. Conf. on Electrical Insulation*, Montreal (1996).
14. E. Odic, L. Parissi, A. Goldman, M. Goldman, S. Koch, Chemical by-products in air with admixtures in correlation with the gas temperature enhanced by non-equilibrium discharges, *2nd Intern. Symp. on Non Thermal Plasma Technology for Pollution Control.* pp 107-12 (1997).
15. M. Toepler, *Ann. Phys.* 21, 193 (1906).
16. Z. Falkenstein and J.J. Coogan, Microdischarge behavior in the silent discharge of nitrogen-oxygen and water-air mixtures, *J. Phys. D: Appl. Phys.* 30, pp 817-25 (1997).

EXPERIMENTAL AND THEORETICAL INVESTIGATIONS ON THE STATISTICS OF TIME LAGS TO CORONA INCEPTION AND BREAKDOWN OF SF$_6$ IN NON-UNIFORM ELECTRIC FIELDS

S K Venkatesh

Center for Air Borne Systems
Defence R&D Organisation
Bangalore 560 037, INDIA

M S Naidu

Department of High Voltage Engineering
Indian Institute of Science
Bangalore 560 012, INDIA

INTRODUCTION

Extensive experiments have been carried out earlier to study the discharge behaviour of SF$_6$ gas over a wide range of non-uniform fields using different types of voltages [1,2,3]. In non-uniform fields, under impulse voltages, the space charge effect and the time lags, namely statistical and formative time lags have a random dispersion and hence the impulse breakdown voltage of insulation is subject to random variation. Most of the time this random variation is often ignored and what tends to be concentrated on, instead are the average trends in order to interpret the impulse breakdown phenomena/processes. Also quite often it is not the average value, but an extreme value like the lower limit below which failure will not occur, that determines the performance of the system.

Extreme values are usually estimated by multiplying known mean values by some safety factors. This extrapolation to low failure rates has large uncertainties and hence a better approach is required for a more precise description of the breakdown behaviour and also for the economical design of the insulation. Therefore it will be useful to consider in greater detail the basic processes, essentially the statistical and formative time lags which introduces the statistical variations and to incorporate these into the evaluations of reliability.

The statistical time lag is calculated from the critical volume-time criterion, proposed by Boeck [4]. The statistical time lag refers to the time necessary for an initiatory electron to appear in a favourable position in order to start a critical avalanche. In the case of electronegative gases like SF$_6$, electron multiplication can take place only within a critical volume in which the ionization coefficient exceeds the attachment coefficient. Further, not all the electrons produced within the volume will in reality be available for avalanche production, since some of them are swept away as they become attached to form negative ions. Calculating the probability of an effective electron appearing in the critical volume within time t, gives the probability of statistical time lag.

Knorr [6] had demonstrated that, in uniform and quasi-uniform field conditions, a simple equal voltage-time area criterion can be used to estimate the formative time lags. This criterion cannot be applied in non-uniform field conditions, as corona precedes breakdown and the associated random dispersion of the magnitude of the space charge strongly influences the breakdown process and hence the formative time lag. In this study, this criterion is applied at pressures above the critical pressure ($>p_c$), where no corona precedes breakdown. By combining the formative time lag with the volume-time criterion, which gives the distribution of statistical time lag, the distribution of breakdown times is obtained.

EXPERIMENTAL

Before one can reach the stage of statistical formulation of the random features of the statistical and formative time lags, these features need to be "phenomenologically known". One therefore has to start with experiments. In this study, extensive experiments were conducted to estimate the complete impulse breakdown voltage probability distribution function and also the cumulative probability distribution of breakdown times under various conditions of field non-uniformity of a hemispherically capped point-plane gaps with field factors ($f = E_{max}/E_{avg}$) varying between 3.9 to 32.8, gas pressure varying between 0.1 to 0.5 MPa (1 to 5 bar), impulse front time varying between 1.2 μs to 350 μs, and gap distances of 25 mm and 15 mm. The distribution of corona inception delay was also measured for impulses with front time of 60 μs and above, which also gives the distribution of statistical time lags. Some typical results of cumulative distribution of corona inception times and breakdown times are given in figures 1 (a)-(b) and 2 (a)-(b) respectively along with the calculated cumulative distributions.

STATISTICAL TIME LAG

The volume-time criterion gives the probability of streamer discharge inception, i.e., the probability of appearance of an effective electron capable of forming a critical avalanche within a time t under an impulse voltage. In non-uniform fields, this criterion gives the distribution of corona inception times (statistical time lag), whereas in uniform and quasi-uniform fields (co-axial), it gives the distribution of breakdown times, neglecting the small formative time lag [7].

A concise derivation of the volume-time criterion is given below,
Let dn_e/dt be the local production rate of initial electrons per unit volume and unit time in a given gas volume V.
Not all the electrons produced in the gas volume will be effective in forming a critical avalanche, i.e., some of the electrons may get attached forming negative SF_6 ions and the electrons must be available within a critical volume V_c, which is explained later.
The probability that an electron will attach is given by η/α.
So the probability that an electron will remain active will then be given by $P_o = 1 - \eta/\alpha$.
Therefore for any given geometry and voltage, the probability dP that a first critical avalanche will be created during the time interval between t and $t+dt$ is,

$$dP = (1-P) \int_{V_c} \frac{dn_e}{dt} P_o \, dV \, dt \tag{1}$$

Where $1-P$ gives the probability that no critical avalanche will have occurred previously.
The solution of equation (1) results in the cumulative distribution function of statistical time lags or the corona inception times,

$$P(t_s) = 1 - \exp\left[-\int_{t_o}^{t_s} \left(\int_{V_c} \frac{dn_e}{dt} P_o \, dV\right) dt\right] \tag{2}$$

where t_s is the statistical time lag and t_o is the time at which the streamer inception level is exceeded, i.e., when the integrand of equation (2) becomes greater than zero.

The frequency distribution of statistical time lags is given by,

$$p(t_s) = \frac{dP(t_s)}{dt} = \int_{V_c} \frac{dn_e}{dt} P_o \, dV \exp\left[-\int_{t_o}^{t_s} \left(\int_{V_c} \frac{dn_e}{dt} P_o \, dV\right) dt\right] \tag{3}$$

As seen from the above equations, the quantities required in applying the volume-time criterion are the growth of the critical volume $V_c(t)$ with respect to the applied impulse voltage and the rate of production of electrons dn_e/dt. The next two sections deal with the critical volume and the electron production rate.

Critical volume

In electronegative gases such as SF_6 and in non-uniform electric fields, the initiating electrons must be available within a critical volume to be able to start a critical avalanche, so as to form a corona. Critical avalanche is when it reaches a critical size which ensures its transition to a streamer. The critical volume is practically bounded by two limiting surfaces;

i) The outer surface defines the zone in which the ionization coefficient (α) exceeds the attachment coefficient (η), so that an avalanche growth would be possible. In other words the field within this zone must exceed the limiting value of $E_{crit} = 8.9\,p$ kV/mm. p is the gas pressure in bar.

ii) The inner surface limits the size of the avalanche to exceed a critical value. i.e., the avalanche must grow over a certain minimum distance. According to streamer criterion a critical length x_c is necessary for the formation of a critical avalanche.

The critical volume is a function of the field distribution and thus varies with time when an impulse voltage is applied. The critical volume is evaluated from the distribution of the electric field in the point-plane gap, which is calculated numerically by Modified Charge Simulation Method (5).

Electron production rate as a function of electric field.

The fundamental quantity required for calculating the statistical time lags is the rate of production of initiatory electrons. It is well established [7] that, these electrons are produced mostly by field induced collisional detachment of SF_6^- ions, and hence a field dependent electron production rate must be considered. The electron production rate dn_e/dt being a function of the electric field, is also a function of space. Therefore the rate of production of electrons at any instant t is the volume integration of dn_e/dt over the critical volume V_c.

The rate of production of effective electrons is,

$$R_e(t) = \int_{V_c(t)} \frac{dn_e}{dt}\left(1 - \eta/\alpha\right) dV \quad (4)$$

where $1-\eta/\alpha$ gives the probability of electrons remaining active without getting attached.

The above equation can be rewritten as the product of the volume by some mean production rate $\bar{n}_e(t)$ based on its spatial distribution.

From equation (4), $\quad R_e(t) = \bar{n}_e(t) V_w(t) \quad (5)$

where the weighted critical volume,

$$V_w(t) = \int_{V_c(t)} \left(1 - \eta/\alpha\right) dV \quad (6)$$

The critical volume, $\quad V_c(t) = \int_{V_c(t)} dV \quad (7)$

The equation (2) can be rewritten as

$$P(t_s) = 1 - \exp\left[-\int_{t_0}^{t_s} \bar{n}_e(t) V_w(t) dt\right] \quad (8)$$

the frequency distribution of corona pulse delays is

$$p(t_s) = \frac{dP(t_s)}{dt} = \bar{n}_e(t) V_w(t) \exp\left[-\int_{t_0}^{t_s} \bar{n}_e(t) V_w(t) dt\right] \quad (9)$$

The measured frequency distribution of corona pulse delays $p(t_s)$ is used in equations (9) and (2) to estimate the value of the mean production rate of initiatory electrons,

$$\bar{n}_e(t) = \frac{p(t_s)}{V_w(t)\left[1 - \int_{t_0}^{t} p(t) dt\right]} \quad (10)$$

where the values of the weighted volume $V_w(t)$ are calculated numerically from the field spatial distribution.

For a given gas pressure the mean production rate of initiatory electrons $\bar{n}_e$ is only a function of electric field which is, inturn a function of time under impulse voltages. The relation of $\bar{n}_e$ to the electric field E is a result of the mechanism of collisional detachment of negative ions and is therefore believed to be independent of the gap configuration and the consequent field distribution. The aim of this approach is to find a functional relationship between the two from the experimentally determined distribution of corona inception time delays. Once this relationship is established, the volume-time criterion can be applied to other gap configurations to estimate the distribution of corona inception delays.

The relationship between the $\bar{n}_e$ and E, however, cannot be obtained directly from the test results since in non-uniform fields both are functions of space. A reasonable alternative would then be to relate the mean values of the two variables taken over the critical volume. At every given instant along the impulse voltage front, the mean value of production rate $\bar{n}_e(t)$ given by the equation (10) has a corresponding mean value of the electric field throughout the critical volume, which is given by,

$$\bar{E}(t) = \frac{1}{V_c(t)} \int_{V_c(t)} E(V) \, dV \tag{11}$$

The equation (11) is applied numerically, at various voltage levels, for the tested gaps under gas pressures from 0.1 to 0.5 MPa.

For each impulse waveform, the gap was tested under several voltage magnitudes at low probability breakdown levels and a number of discharge pulse delay distributions $p(t)$ were recorded. From each of these distributions, the mean electron production rate was calculated as a function of the mean field from the equation (10) and was found to be independent of the rate of rise of the voltage under a given impulse shape. A Normal distribution function could be fitted to the corona inception delay times. A chi-square test was used to verify the goodness of fit at levels of significance between 5 and 95 %. The corona inception time distribution was found experimentally at different point diameters at each gas pressure, in order to estimate $\bar{n}_e(t)$ as a function of $E(t)$ over a wide range.

Figure 1 (a) shows the calculated and measured cumulative probability distribution of corona inception delays for 1 mm diameter point, 25 mm gap distance at 0.3 MPa pressure under 86 kV and 80 kV peak, 250 μs impulse voltage. The electron production rate used in this calculation was obtained from the corona inception time measurements. From the figure 1 (a) it is observed that the agreement between the calculated and the experimental data is good.

Once the electron production rate is estimated as a function of the mean electric field from corona inception time delay measurements under a given gap configuration, and impulse front time, the volume-time criterion can be applied to other gap configurations using this electron production rate. The limitation of the method to be applied for other gap configurations is that the negative ion density distribution in the gap should be same as under the impulse and the gap configuration at which time delay measurements were made to find the functional relationship between the rate of production of electrons and the mean electric field. The electron production rate determined from the measurements made with 25 mm gap distance, was used in the volume-time criterion applied to a 15 mm gap distance and the results obtained are shown in the figure 1 (b). The agreement between the calculated and measured data was satisfactory, thus proving the accuracy of the volume-time criterion used in this study.

FORMATIVE TIME LAG

Knorr [6] has demonstrated the applicability of equal voltage-time area criterion for estimating the formative time lag to breakdown in SF_6 in slightly non-uniform fields such as co-axial cylinders. The voltage-time area criterion is;

$$\int_{t_o}^{t_b} (v(t) - V_o) \, dt = A \tag{12}$$

where, $v(t)$ is the impulse voltage, V_o is the minimum breakdown voltage which at high pressures ($> p_c$) is equal to the DC breakdown value (V_{dc}) and A is the area between the impulse curve and the minimum breakdown value V_o and is derived from the minimum breakdown time measured. The area A shown in the figure 2 (a) is independent of the shape of the voltage impulse but varies with gas pressure and field configuration. The time corresponding to the Area A gives the formative time lag.

TIME LAG TO BREAKDOWN

In nearly uniform fields, the formation of a streamer and the succeeding spark begins immediately as soon as the critical voltage V_o is exceeded at $t = t_o$, if an initiatory electron is available at a suitable place. In this case, there is no preceding statistical time lag ($t_s=0$) and the smallest delay and lowest breakdown voltage is then obtained. Since the statistical breakdown probability results from the statistical scatter of t_s, the v-t curve for 0 % breakdown probability can be found by adding area A at t_o as shown in figure 2 (a). By adding the

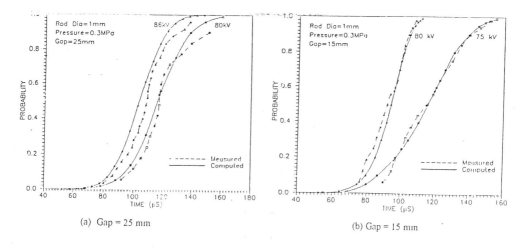

Figure 1 Computed and Measured Distribution of Corona Inception Times under 250 µs Front Impulse Voltage

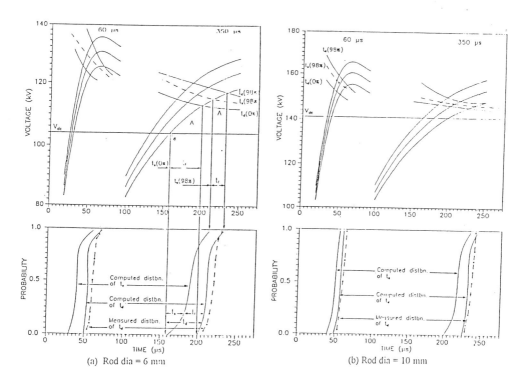

Figure 2 Voltage-Time characteristics and distribution of Statistical (t_s) & Breakdown (t_d) time lags under 60 µs and 350 µs front Impulse voltages. Gap = 25 mm, Pressure = 0.3 MPa

formative time lag corresponding to the area A to the distribution of statistical time lags, the distribution of breakdown times is obtained.

RESULTS

In this study, the equal voltage-time area criterion is applied to estimate the formative time lags, in the less non-uniform fields of 6 mm and 10 mm diameter rod electrodes at 0.3 MPa with 60 µs and SI voltages and the results are shown in the figures 2 (a) and 2 (b). Under each rod diameter, about fifty breakdowns were recorded and the minimum breakdown time was used to calculate the area A between the impulse waveshape and the DC breakdown value (V_{dc}). The area A is applied once at time corresponding to the V_{dc} value on the impulse to get the 0 % probability breakdown time (assuming statistical time lag is zero). Another application of the area A at the calculated 98 % cumulative probability statistical time lag gives the 98 % cumulative probability breakdown time. The criterion is applied at three over voltage levels under 60 µs and SI voltages and the limiting voltage-time curves are obtained as shown in the figures 2 (a) and 2 (b). At these three voltage levels, breakdown times were recorded and all the breakdown points were situated between the two limiting voltage-time curves, hence proving the applicability of equal voltage-time area criterion. Figures 2 (a) and 2 (b) also shows the cumulative probability distribution of breakdown times at one voltage level each under 60 µs and SI voltages, which are obtained by adding the time corresponding to the area A to the distribution of statistical time lags as calculated by volume-time criterion. Also shown in the figures are the measured distribution of breakdown times which agrees satisfactorily with the calculated distribution.

This criterion was also applied under the high non-uniform fields of 1 mm and 3 mm diameter rods and at pressures above p_c, but failed to give satisfactory results, because of large statistical variations in the formative time lag.

CONCLUSIONS

The Volume-Time criterion is successfully applied in highly non-uniform fields of point-plane gaps to predict the probability distribution of corona inception delays (statistical time lags) by considering a field dependent electron production rate.

The equal voltage-time area criterion for the estimation of formative time lags, is combined with the calculated distribution of statistical time lags for deriving the statistics of breakdown in fairly uniform fields and at pressures above p_c. It is shown that in highly non-uniform fields and at pressures less than p_c, where corona precedes breakdown, due to large statistical variations in the space charge dependent formative time lag, equal voltage-time area criterion cannot be used to estimate the formative time lags under different impulse voltage waveshapes.

REFERENCES

1. N H Malik, A H Qureshi. "Breakdown Mechanisms in Sulfur Hexaflouride", IEEE Trans. on Elect. Insul., Vol. EI-13, No.3 June 1978.

2. H Anis, K D Srivastava. "Prebreakdown Discharges in Highly Non-uniform Fields in relation to Gas Insulated systems", IEEE Trans. on Elect. Insul., Vol. EI-17, No. 2, 1982.

3. T Nitta, Y Shibuya, Y Fujiwara. "Voltage-Time characteristics of Electrical Breakdown in SF_6", IEEE Trans. on Power Appar. & Systems, Vol. PAS-94, No. 1, 1975.

4. W Boeck "SF_6 Insulation Breakdown behaviour under Impulse Stress", in *Surges in High Voltage Networks*, Edt. by K Ragaller, Plenum Press, New York, 1980.

5. N H Malick. "A Review of the Charge Simulation Method and its Applications" IEEE Trans. on Elect. Insul., Vol. EI-24, No. 1, February 1989

6. W Knorr "A Model to describe the Ignition Time lag of slightly Non-uniform arrangements in SF_6", Proc. of the 3rd Int. Symp. on High Voltage Engineering, Paper No. 31.11, Milan, Italy, 1979

7. I C Somerville, D J Tedford. "The Statistical time lag to Spark Breakdown in High Pressure SF_6", Proc. of 3rd Int. Symp. on Gaseous Dielectrics, Knoxville, Tennesse, USA, 1982.

PROPERTIES OF THE RF-DISCHARGE IN AR-TETRAETHOXYLSILANE

Rüdiger Foest[1], Ralf Basner[1], Martin Schmidt[1], Frank Hempel[1], and Kurt Becker[2]

[1]Institut für Niedertemperatur-Plasmaphysik e.V. an der Ernst-Moritz-Arndt Universität, R.- Blum-Str. 8-10, D 17489 Greifswald, Germany
[2]Department of Physics and Engineering Physics, Stevens Institute of Technology, Hoboken, NJ, USA

INTRODUCTION

Thin film deposition, surface treatment, plasma etching are expanding applications of plasma technology. Silicon compounds are important precursors for thin film deposition. Tetraethoxysilane (TEOS) $Si(OCH_2CH_3)_4$ is used for plasma polymerization or in mixtures with O_2 for SiO_x deposition[1]. Anisothermal rf-discharges are widely used in plasma technology. In this paper results of mass spectrometric investigations of the ions in an Ar-rf-discharge with admixtures of TEOS are presented. Plasma parameters of the Ar-TEOS-rf-discharge are estimated by means of the measured external voltages and the dark sheath thickness in front of the powered electrode[2]. The measured ion distribution is compared with the calculated ion production rate using the electron impact ionization cross sections[3] and the estimated electron temperature. The influence of the H_2 concentration on ion molecule reactions with proton transfer for the formation of H_3^+, COH^+ and ArH^+ ions is discussed. The calculated ion current to the grounded electrode is compared with the polymeric film deposition rate. The ion current is sufficiently high to be responsible for the thin film formation. In this way a contribution is given for the understanding of processes in such chemical active plasmas.

EXPERIMENTAL SET-UP

The asymmetric, capacitively coupled rf-discharge (13.56 MHz, 15-120 W) burns in a stainless steel vacuum chamber (⌀ 300 mm) with planar steel electrodes of 128 mm diameter and 40 mm spacing. The energy dispersive mass spectrometry is accomplished by a plasma monitor (SXP 300H / VG Instruments) with the particle effusion orifice (⌀ 0.1 mm) located at the cylindrical chamber side wall of the discharge vessel (Fig. 1). A more detailed description of the experimental equipment is given elsewhere[4]. The gas flow through the reactor was 10 sccm for an Ar-TEOS mixture of 50:1. The total pressure was varied between 1 and 10 Pa.

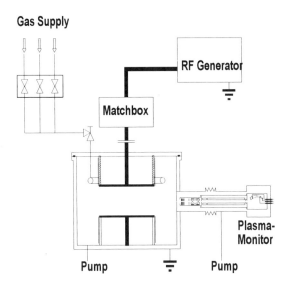

Figure 1. Set-up with rf-reactor and plasma monitor

RESULTS

The neutral gas composition in the plasma is marked by a decay of the TEOS concentration by one order of magnitude. Further the neutral gas mass spectrometry shows high relative concentrations (n_x/n_{Ar}) of H_2 (0.25) and CO (0.24). The CO_2 content is small (0.01). These values are obtained for a gas pressure of 10 Pa and a rf-power of 45 W. Lower hydrocarbons as CH_4 and C_2H_2 are observed too. The ion energy distribution (IED) at the grounded wall is measured with the plasma monitor. Typical results are presented in Fig. 2 (H_3^+, Ar^+, ArH^+, $SiO_3C_5H_{13}^+$) where the IEDs are normalized such that the area equals unity. The relative ion currents are obtained by integration over the IED. The observed plasma ions with m/z>60 (Fig. 3) in general are the same as the ions formed by electron impact ionization[3]. ArH^+, Ar^+, COH^+, H_3O^+, CH_3^+ and H_3^+ are the dominant ions in the lower mass region. The relative intensities of selected ions for different pressures are shown in Fig. 4.

DISCUSSION

The change in the neutral gas composition is related to the thin film deposition as a loss of the silicon containing particles. The H_2 and CO production may be mainly connected with

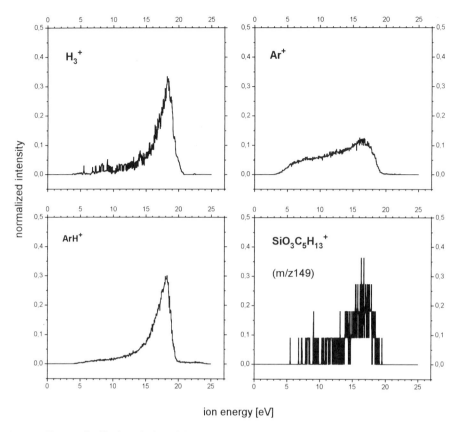

Figure 2. Energy distribution of selected ions at the wall of the Ar-TEOS-rf-discharge (60 W, 2 Pa, 2% TEOS)

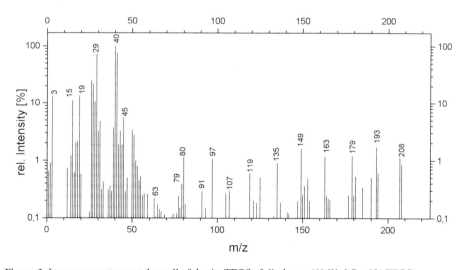

Figure 3. Ion mass spectrum at the wall of the Ar-TEOS-rf-discharge (60 W, 2 Pa, 2% TEOS)

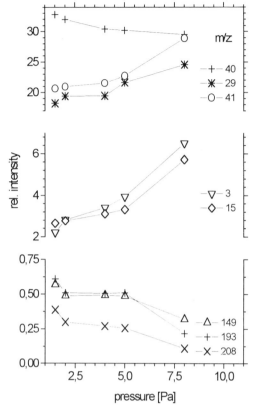

Figure 4. Relative ion currents (sum of all currents equals 100) of the Ar-TEOS-rf-discharge (45 W, 2% TEOS)

thin film formation and the plasma interaction with the deposited film. The formation of H or H_2 and CO as a product of electron impact dissociation or dissociative electron impact ionization of the TEOS molecule to a greater extent seems to be improbable, because the dissociative electron impact ionization[3] shows a very small production rate of H_2^+ and CO^+ (~10^{-2} of the base peak intensity). The neutral gas composition can be measured by neutral gas mass spectrometry with calibrated sensitivity of the mass spectrometer or by means of known ionization cross sections. An estimate of the H_2 and CO concentration with the currents of H_2^+, H_3^+, CO^+, COH^+ and Ar^+, ArH^+ ions assuming a formation of H_2^+, CO^+ and Ar^+ by electron impact ionization and H_3^+, COH^+ and ArH^+ ion by ion molecule reactions with the H_2 gas results in relative concentrations of CO and H_2 only near 1%. These low values in relation to the neutral gas mass spectrometry indicate that the reactions in the plasma are much more complex. Reactions of these protonated ions (H_3^+, COH^+, ArH^+) with other molecules connected with proton transfer must be taken into account.

Langmuir probe measurements for the determination of the plasma parameters are difficult in rf-plasmas for thin film deposition. A rough estimate of the plasma parameters is possible[2] for a sheath determined rf-discharge by measurement of the rf-voltage, the dc-bias[5], and the dark sheath thickness d_T in front of the powered electrode with the area A_T. The parameters of the sheaths near the grounded and the powered electrode can be calculated (thickness d_W, area A_W, voltages V_T, V_W). The charge carrier losses are determined by the Child

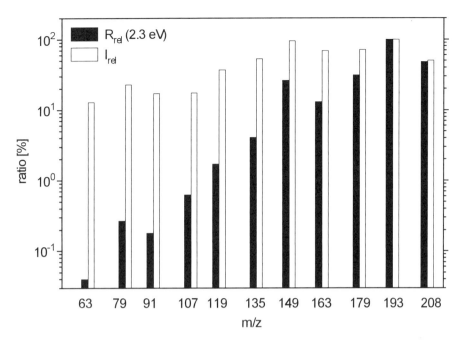

Figure 5. Relative rate coefficients for partial electron impact ionization (R_{rel}) and relative ion currents (I_{rel}) to the wall of the Ar-TEOS-rf-discharge (45 W, 5 Pa, 2% TEOS)

law[6]. A balance between the charge carrier production by electron impact ionization and the losses to the electrodes determines the electron temperature T_e:

$$\frac{k_{ie}}{\sqrt{T_e}} = \frac{1.22\sqrt{\frac{e_0}{m}}\left(\frac{V_T^{3/2}}{d_T^2}A_T + \frac{V_W^{3/2}}{d_W^2}A_W\right)}{n_g\left(A_T + \sqrt{A_T A_W} + A_W\right) \cdot \frac{h}{3}\left(\frac{V_T^{3/2}}{d_T^2} + \frac{V_W^{3/2}}{d_W^2}\right)}$$

Here k_{ie} is electron impact ionization rate, e_0 the electron charge and m the ion mass, n_g the neutral gas density, and h the distance between the electrodes. For the Ar TEOS rf discharge (45 W, 5 Pa) we obtained a value of 2.3 eV for T_e. Because of the dominance of Ar^+ ion in the observed spectrum only ionization of Ar-atoms is considered.

The IED indicates especially for Ar^+ the effect of charge transfer collisions in the sheath region. The energy distribution of the other ions is not disturbed by collisions according to the absence of a sufficient gas concentration of corresponding neutrals. The investigation of the ion population bears out the existence of all fragment ions observed also by electron impact ionization. Results of mass spectrometry of the ions in the Ar-TEOS plasmas together with calculated relative ion formation rates using the electron impact ionization cross-sections[3] and a Maxwellian distribution of the plasma electrons with a temperature of 2.3 eV are presented in Fig. 5. The smaller the ions the smaller is the formation rate in relation to the measured ion intensities. This behavior indicates the strong influence of secondary processes other than direct electron impact ionization of the TEOS molecule on the ion population. Secondary processes are e.g. dissociation into neutral particles and ionization of dissociation products.

A problem in PACVD is to decide whether the particles responsible for the mass transport from the plasma to the substrate surface are of neutral or ionic nature. For our conditions we can estimate the ion current density to the wall/grounded electrode using the Child law. With the evaluated parameters for the TEOS discharge $V_w = 11.7$ V, $d_w = 1.7 \cdot 10^{-3}$ m we receive a current density of $2 \cdot 10^{-5}$ A/cm^2 = $1.3 \cdot 10^{14}$ ions/cm^{2}s. The deposition rate of $7 \cdot 10^{-10}$ g/cm^{2}s was measured with a quartz crystal detector. A flux of $2 \cdot 10^{12}$ ions/cm^{2}s is necessary for the film formation by heavy ions (m/z~200) and of $2 \cdot 10^{13}$ ions/cm^{2}s by lighter ions (m/z~20). According to the measured mass spectrum the Si-containing heavy ions contribute ~1% to the ion current while about 20% of the total ion current is comprised of lighter ions. This leads to the conclusion that within the uncertainty of the used model a film formation governed by ions seems reasonable.

ACKNOWLEDGMENTS

The authors are grateful for the technical assistence provided by Ms. U. Haeder and Mr. K. H. Schmidt. One of us (KB) acknowledges partial support from the US National Science Foundation.

REFERENCES

1. A. Granier, F. Nicolazo, C. Vallée, A. Goullet, G. Turban, and B. Grolleau, Diagnostics in O$_2$ helicon plasmas for SiO$_2$ deposition, *Plasma Sources Sci. Technol.*, 6: 147 (1997)
2. M. Schmidt, F. Hempel, M. Hannemann, and R. Foest, Determination of electron density and temperature from external voltages and sheath thickness in a rf-discharge, *Bull. Am. Phys. Soc*, 42: 1733 (1997)
3. R. Basner, R. Foest, M. Schmidt, and K. Becker, Electron impact ionization of tetra-ethoxysilane and hexamethyldisiloxane, *Adv. Mass Spectr.* 14, in press
4. R. Foest, M. Schmidt, M. Hannemann, and R. Basner, On the ions in an argon-hexamethyl-disiloxane radio-frequency discharge, *Gaseous Dielectrics VII*, L. G. Christophorou, D. R. James, ed., Plenum Press, New York (1994)
5. K. Köhler, J.W.Coburn, D.E. Horne, and E. Kay, Plasma potentials of 13.56 -MHz rf-argon- glow discharges in a planar system, *J.Appl. Phys*, 57: 59 (1985)
6. M. A. Lieberman, A. J. Lichtenberg. *Principles of Plasma Discharges and Materials Processing*, John Wiley & Sons, New York (1994)

SECTION 3: SIMULATIONS / BREAKDOWN IN GAS MIXTURES

THERMODYNAMIC AND ELECTRICAL PROPERTIES OF SF_6-N_2 THERMAL PLASMAS

Alain GLEIZES

C. P. A. T., ESA 5002, Université Paul Sabatier
118 route de Narbonne, F31062 TOULOUSE Cedex 4, France

INTRODUCTION

High-voltage circuit breakers (HVCB) generally act in pure SF_6 under pressure of a few bars. The working of this kind of circuit-breaker consists first to create an electric arc due to the separation of the contacts and second to rapidly extinguish this arc by a high blowing during the current-zero phase of the alternative current. For a given gas, the rate of transition from the electrically conducting medium (the arc plasma) to a dielectric medium (the cold non-ionised gas) is called the interruption capability. SF_6 has a very high interruption capability because of its properties as a thermal plasma (thermal conductivity, radiation losses) and of its dielectric properties as an electro-negative gas. Nevertheless under severe weather conditions (external temperature of the order of -30°C) SF_6 may partially liquefy and thus may loss a part of its interrupting properties. For this reason SF_6-containing mixtures were considered for replacing pure SF_6 in HVCB. Among them SF_6-N_2 mixtures are attractive because nitrogen plasmas have a high thermal conductivity. Recently, environmental and safety considerations related to green house effect and to by-product toxicity tend also to decrease the use of SF_6 in gas insulated systems and in HVCB.

This paper deals with the determination of the interruption capability of SF_6-N_2 mixtures based on studies of the properties of SF_6-N_2 thermal plasmas and on the behaviour of stationary and decaying electric arcs burning in these mixtures. Although this simplified approach does not exactly correspond to HVCB conditions, we will see that it allows to have a rather good prediction of the interruption capability of the studied gases. The first part of this paper is devoted to the calculation of the equilibrium composition, thermodynamic properties, transport coefficients and radiative losses of the SF_6-N_2 thermal plasmas assumed to be in local thermodynamic equilibrium (LTE).

Then the influence of these material properties on stationary arc behaviour was studied by experiments and numerical modelling relative to a wall-stabilised arc. The interruption capability was determined by analysing the extinction of a wall-stabilised arc, i.e. by computing the evolution of arc plasma temperature and conductance ; we also made a comparison between SF_6-N_2 and SF_6-CF_4 mixtures. In the last part we will discuss about phenomena not taken into account in this study, but occurring in real conditions.

MATERIAL PROPERTIES

Equilibrium composition and thermodynamic properties

All the material properties were computed assuming LTE plasma, as functions of pressure (between 0.1 and 1 MPa), temperature (between 300 and 30 000 K) and SF_6

proportion in the SF$_6$-N$_2$ mixture. Note that this proportion was defined as a volume (or molar) proportion. The considered species are the following : SF$_6$, SF$_5$, SF$_4$, SF$_3$, SF$_2$, SF, S, S$_2$, F, F$_2$, N, N$_2$, F$^-$, S$^+$, S^{++}, F$^+$, F^{++}, N$^+$, N^{++} and the electrons.

Two methods have been used depending upon the temperature range considered : Gibbs' function minimisation at low temperature (T < 5 000 K) and resolution of equilibrium laws at high temperature (7 > 5 000 K). They are described in [1]. It is to be noted that the determination of the partition functions at high temperature depends on the lowering of ionisation potential ΔE defined as follows, according to Griem [2] :

$$\Delta E = \frac{1}{4\pi\varepsilon_0} \frac{e}{\rho_D} \quad (1)$$

where ρ_D is the Debye length. Taking the ionisation potential lowering into account may influence the results at high pressure (Debye-Hückel correction). The influence of pressure directly appears in the computation of Gibbs' function and through the perfect gas law. Using the composition of the plasma, we have determined the mass density ρ. The basic data needed for this kind of calculation are the internal partition functions derived from literature.[3,4] An example of equilibrium composition is given in figure 1.

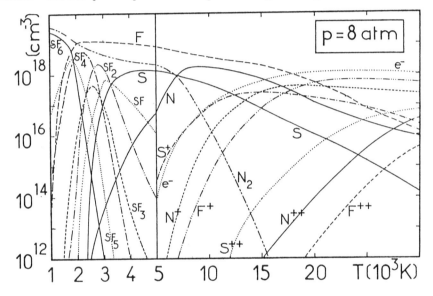

Figure 1. Equilibrium composition of a 40%SF$_6$ - 60%N$_2$ mixture, at a pressure of 0.8 MPa

Under the assumption of LTE, the thermodynamic functions can be derived from the composition of the plasma and from the knowledge of one of them. We have used the relations given by Capitelli and Molinari[5] for gas mixtures. At low temperature (T<5 000 K) the calculation of plasma composition requires the determination of Gibbs' potential G; Helmholtz's potential F, entropy S, enthalpy H, and internal energy U are then derived from it using the classical relations. For higher temperature values, composition determination requires the calculation of partition functions of all the species ; Gibbs' potential is given by :

$$G = -kT \sum_{j=1}^{v} N_j \ln\left(\frac{q_j}{N_j}\right) + \sum_{j=1}^{v} N_j \varepsilon_j \quad (2)$$

where q_j, N_j and ε_j respectively are the partition function, the number density, and the energy gap between the fundamental state and the common reference level of species j. The other thermodynamic functions are deduced from Gibbs' potential or directly calculated from partition functions.

Specific heat at constant pressure, C_p, is obtained by numerical derivation of the enthalpy. The variations of C_p with temperature, for pure SF$_6$, pure N$_2$ and 40 % SF$_6$ - 60 % N$_2$ mixture are shown in figure 2. The peaks of these curves correspond to chemical reactions : for SF$_6$ the peaks around 2 000 K are due to dissociation whereas the peak around 18 000 K is due to fluorine ionisation. In the case of nitrogen, we can observe the characteristic peak at 7 000 K corresponding to the dissociation of N$_2$ molecules.

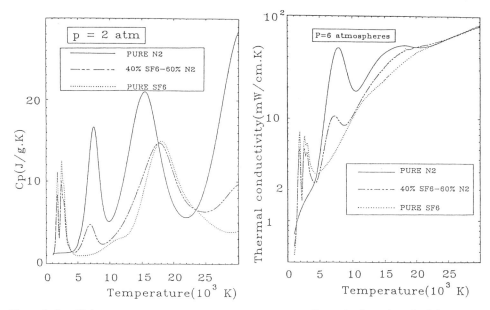

Figure 2. Specific heat at constant pressure

Figure 3. Thermal conductivity

Transport coefficients

The transport coefficients calculated were the electrical and thermal conductivities and the viscosity. The theoretical study of these coefficients was based on the resolution of Boltzmann's integro-differential equation using Chapman & Enskog's method.[6] The expression of the electrical conductivity used was that reported by Devoto.[7] The thermal conductivity was determined as the sum of four components: the heavy particle translation thermal conductivity, calculated to the 3rd approximation by the method of Chapman & Enskog[8], the electron translation thermal conductivity calculated to the 2nd approximation[9], the internal thermal conductivity[10] and the reaction conductivity which is determined by extension of the theory of Butler & Brokaw[11-12] to partially ionised gases.

The total thermal conductivity of a plasma can be written as the sum of several components

$$\kappa = \kappa_{tr} + \kappa_{int} + \kappa_r \qquad (3)$$

where κ_{tr} is the thermal conductivity due to the translation energies of the particles. It corresponds to a mixture of monoatomic gases which do not react with each other. κ_{int} is the thermal conductivity due to internal molecular energies. It represents the energy exchanges between the internal degrees of freedom of the particles such as vibration and rotation of the molecules. κ_r is the thermal conductivity due to various chemical reactions (dissociation, ionisation).

The total thermal conductivity κ for pure SF_6, pure N_2, and for one mixture is given in figure 3. Of the three components of κ the reactional thermal conductivity appears to be the one which is the most sensitive to the variation of the proportions of the gases in the mixtures[13]. Reactional thermal transfer is dependent on the number density of the elements which interact with each other. The thermal conductivity coefficient of SF_6 presents maxima at about 2000 K and that of nitrogen at 7000 K. Calculation clearly shows that for temperatures greater than 4000 K the reactional thermal conductivity of the mixture is higher than that of pure SF_6 but that it is lower at around 2 000 K.

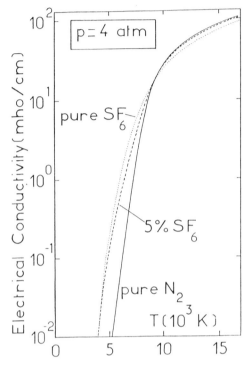

The variations of the electrical conductivity σ for different SF_6-N_2 mixtures are drawn in figure 4. At low temperature (T < 8 000 K), the conductivity of SF_6 and of SF_6-N_2 mixtures is greater than that of pure nitrogen. This is due to the lower ionisation potential of sulphur which leads, for a given value of T, to an electron number density which is greater in the presence of sulphur than in the presence of pure nitrogen. In figure 4, it can be seen that a small proportion of SF_6 has a strong influence on the electrical conductivity at low value of temperature. We will see that this effect has an important role in the decaying arc conductance and thus in the interruption capability of the mixture.

Figure 4. Electrical conductivity of SF_6-N_2 mixtures

Radiative losses

Treatment of the radiative transfer in thermal plasmas is very complicated. It has been shown that the total radiation emitted by the hottest regions of an arc plasma can be calculated by means of a net emission coefficient, computed with the assumption of an isothermal plasma. The bases of the method and the results for SF_6-N_2 mixtures were given in[14]. The net emission coefficient ε_N is the power radiated per unit volume and solid angle, on the axis of an isothermal column, that escapes radially from the plasma of radius R. Its calculation requires the knowledge of K'_v, the monochromatic absorption coefficient. Coefficient K'_v has two parts : the continuum and the lines.

For the continuum, all the mechanisms of emission and absorption were taken into account (radiative attachment and recombination ; bremsstrahlung). For the lines, the main difficulty was to determine the line profiles depending on several broadening phenomena. The results showed that it exists a strong self absorption of the radiation, in the first mm of plasma, mainly due to the resonance lines. The influence of the proportion of the gases in the mixture is indicated in figure 5. In an optically thin plasma (figure 5a) the radiation of the mixture is proportional to the relative concentration of the species present.

When self-absorption is taken into account (figure 5b) the relationship between ε_N and the proportion of the mixture is more complex. At R = 0.2 cm, ε_N of the mixture corresponds to ε_N of SF_6 at low temperature because of the ionisation of sulphur. In the range of average temperatures, ε_N is dependent on N_2 owing to the emissivity and the broadening of the NI lines. Finally, at high temperatures ε_N follows SF_6 because of the fluorine species.

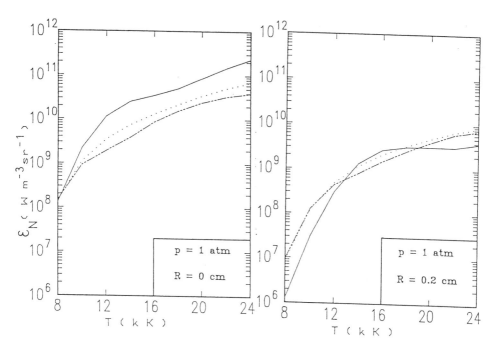

Figure 5. Net emission coefficient: ——— pure N_2; ········ 60%SF_6; —··— pure SF_6

STATIONARY STATE ARC

We have developed an experimental and numerical study of a wall-stabilised arc in SF_6-N_2 mixtures, in order to analyse the influence of the plasma properties on the temperature field. The arc chamber is constituted by a stack of 8 water cooled copper disks separated by bakelite rings. The discharge channel is 4 mm in diameter. The two tungsten electrodes at both ends of the channel are slightly shifted from the axis to allow end-on measurements (Figure 6). The gas mixture under study (SF_6-N_2) is injected in the central zone of the chamber and the electrodes are shielded by an argon flow. The pressure is atmospheric.

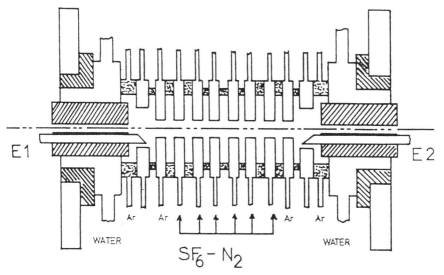

Figure 6. Arrangement of the cascade arc

In stationary state SF_6 plasmas the departures from equilibrium composition are mainly due to the demixing effect[15,16]. This arises from the combination of diffusion and ionisation mechanisms and results in an underpopulation in sulphur (atoms + ions) on the axis of an SF_6 arc : the real value of the concentration ratio M of fluorine and sulphur is greater than the stoichiometric value of 6. To take this effect into account, the temperature has been measured following the method proposed by Schulz-Gulde in the case of SF_6[15] : all the LTE laws, except the dissociation equilibrium, are assumed valid. In such conditions the measurement of the absolute intensities of one line of SI and one line of SII allows the determination, in an SF_6 plasma, of the temperature T, ratio M and all the other number densities deduced from these parameters using the laws of LTE.

In parallel with this experiment we have calculated the temperature profile by means of a 1D model well adapted to wall stabilised arc conditions, and based on the energy conservation equation

$$\sigma E^2 = -\frac{1}{r}\frac{d}{dr}\left(r\kappa\frac{dT}{dr}\right) + u \qquad (4)$$

where T is the temperature, u the radiation losses ($u = 4\pi\varepsilon_N$) and E the electric field, assumed to be uniform and constant, given by the Ohm's law :

$$I = 2\pi E \int_0^R r\sigma dr \qquad (5)$$

I being the current intensity and R the arc radius. We present in figure 7 (a and b) the experimental and calculated evolutions of the axis temperature T_0 versus current intensity. In both cases we observe a crossing of $T_0(I)$ curves obtained with different proportions. Although, the current value corresponding to the intersection is slightly different in both sets of results the conclusion is that the theoretical variations based on thermal equilibrium are analogous to the experimental variations. The explanation of the curves crossing may be explained if we consider the properties of pure SF_6 and pure N_2 :

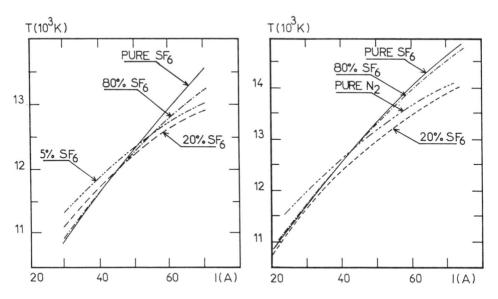

Figure 7. Experimental and calculated evolutions of the axis temperature versus current intensity

→ For high current value, the energy losses in the central zone are radiation dominated. The energy lost by radiation in a nitrogen plasma is higher than that lost in an SF_6 plasma. If follows (for fixed values of discharge current I and radius R) that the electric field of a stationary nitrogen arc is higher than that in SF_6. If we define the conductance G by $I = E.G$, we have from (Eq. 5):

$$G = 2\pi \int_0^R \sigma \, r dr \qquad (6)$$

Thus for a given current, the conductance in nitrogen is lower than that of SF_6 plasma. In wall stabilised arc, R is fixed, so that the electrical conductivity in nitrogen must be lower then that of SF_6 in the central region, which induced $(T_0)_{N_2} < (T_0)_{SF_6}$, which is observed in figure 7, for the high current.

→ For low current value (20 A ≤ T ≤ 50 A), the temperature profiles evolve in a different manner in nitrogen and SF_6 when the current intensity decreases. In figure 8 we have represented these computed profiles for a current of 38 A. Since the thermal conductivity of nitrogen exhibits a high maximum around 7 000 K the $T(r)$ profile presents a plateau around this temperature value. The conducting core of the arc is therefore stretched in comparison with that in SF_6 (the 10 000 K isotherm is at r = 0.76 mm for N_2 and r = 1.15 mm for SF_6); the effect of this is a higher axial temperature in N_2 than in SF_6 for 20 A < I < 50 A. This effect is accentuated by the higher radiation term in SF_6 than in nitrogen for T < 11 000 K.

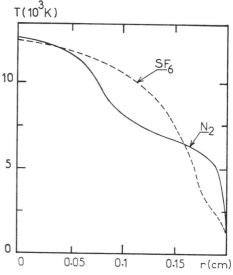

Figure 8. Calculated temperature profiles (I=38A).

DECAYING ARC

The purpose of this study is to analyse the extinction of a wall-stabilised arc by computing the evolution of arc plasma temperature and conductance. The transient evolution of conductance allows to evaluate the interrupting capability of the medium. Assuming an LTE plasma where the axial gas flow in negligible, the theory is based on a combination of the conservation equations for mass and energy[17]:

$$\frac{\partial \rho}{\partial t} + \frac{1}{r}\frac{\partial}{\partial r}(r\rho v) = 0$$

$$\rho C_p \frac{\partial T}{\partial t} + \rho C_p v \frac{\partial T}{\partial r} = \sigma E^2 - u + \frac{1}{r}\frac{\partial}{\partial r}\left(r\kappa \frac{\partial T}{\partial r}\right) \qquad (7)$$

where ρ is the mass density and v the radial velocity component, E is the externally applied electric field, C_p the constant pressure specific heat, σ the electrical conductivity, κ the thermal conductivity and u the net emission of radiation.

We start from a 100 A stationary arc and we assume a fast decrease of the electric field (exponential form with a time constant of 2 μs). Figure 9 gives the transient evolution of the axis temperature for various SF_6-N_2 mixtures. For pure nitrogen, a rapid cooling is observed at the beginning of the extinction (t < 50 μs) : this is due to the strong thermal conductivity of nitrogen for T ≥ 7 500 K (see fig. 4), thermal conduction being the main phenomenon of energy loss. After about 40 μs the axis temperature is below 7 500 K and exhibits a slower decrease. In the case of pure SF_6 the temperature decrease is more regular, with a slope intermediate to the two slopes observed with pure N_2. This effect is due to the thermal conductivity of SF_6 which has a pronounced peak around 2 000 K, corresponding to the dissociation of the molecule (see fig. 4).

What is remarkable is that the values of temperature computed for SF_6-N_2 mixtures are not systematically intermediate between the values computed in pure SF_6 and in pure N_2. For a mixture with 20 % SF_6, after 100 µs the axis temperature is markedly higher than that computed in all other cases. This is essentially due to the influence of the dissociation of SF_6 molecule on the thermal conductivity of this mixture. For a mixture of 20 % SF_6 - 80 % N_2 the computed equilibrium composition at atmospheric pressure shows that, at 7500 K, the partial pressure due to the species originating from SF_6 molecule (S, S^+, and F essentially) is 0.49 against 0.505 for the species originating from N_2 molecule (N, N_2, N^+, and N_2^+). This results in a thermal conductivity peak around 7 500 K markedly lower in a mixture than in pure nitrogen. In addition, the thermal conductivity peak around 2 000 K is markedly lower in the 20 % SF_6 mixture than in the case of pure SF_6. These characteristics determine the evolution of temperature profiles in the mixture and lead to the results in Fig. 9 concerning the axis temperature.

Figure 10 shows the evolution of decaying plasma conductance for different mixtures. For times beyond (or of the order of) 100 µs, the computed conductance for mixtures containing less than 50 % SF_6 is higher than for both pure SF_6 (which seems normal) and pure N_2. This phenomenon results from the combination of two effects. The first one, related to the thermal conductivity, has been described above : it is mainly responsible for the evoluton of temperature. The second one is due to the fact that the electrical conductivity at low temperature (5 000 K ≤ T ≤ 6 000 K) depends very closely on the ionisation potential of the constituent. The presence of sulphur, even in low proportion (see fig. 3), determines low temperature electrical conductivity values, which are much higher than in the case of pure nitrogen.

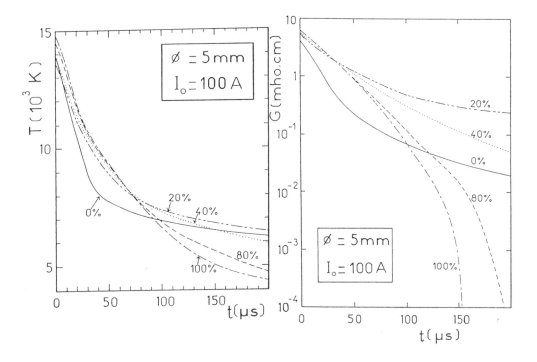

Figure 9. Transient evolution of axis temperature   **Figure 10.** Transient evolution of conductance

For mixtures with a proportion of SF_6 lower than 50 %, schematically, two effects are cumulative at low temperature (T < 7 500 K) : relatively low thermal conduction and relatively good electrical conductivity. Extinction properties of such mixtures, in the absence of blowing are, therefore, bad.

In a recent paper[18] we have performed a similar study for SF_6-CF_4 and SF_6-C_2F_6 mixtures. The main results are presented in figure 11 and show that the interruption capability of SF_6-CF_4 is much higher than that of SF_6-N_2, mainly when the SF_6 proportion is lower than 50 %. This is due to the fact that CF_4 plasma has similar properties of SF_6 plasma (same electrical conductivity ; high thermal conductivity at 2500 - 3000 K because of dissociation).

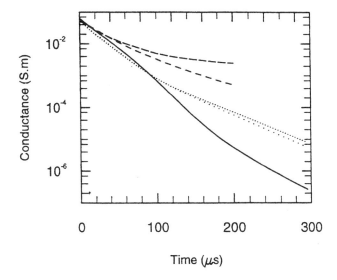

Figure 11. Variations of the conductance for various plasmas during the decay of 100-A arcs at 0.1 Mpa. SF_6: ; 50%SF_6+50%CF_4 ; 50%SF_6+50%C_2F_6 ; 40% SF_6+60%N_2: 20%SF_6 + 80% N2

DISCUSSION

This general study on the interruption capability of SF_6-N_2 mixtures was based on the behaviour of a wall-stabilised arc without convection, which corresponds to conditions quite different from those of a HVCB, where a strong blowing is imposed to the arc, leading to turbulence effects and very fast cooling. Let us examine the validity of an extrapolation of the previous results.

→ turbulence: the main effect of turbulence on a HVCB arc behaviour is the enhancement of the thermal conductivity. It can be shown (for example with a simple mixing-length model[19]) that the turbulent thermal conductivity is proportional to the terms (ρC_p), product of the mass density by the specific heat at constant pressure. The variations of this parameter are similar to those of the normal conductivity, in particular the characteristic peaks due to dissociation. Thus the superiority concerning the interruption capability of SF_6 on nitrogen was demonstrated for blast arc HVCB[19].

→ fast cooling and chemical kinetics : in circuit-breaker conditions the cooling of the plasma is very fast, due to turbulence. In our approach we have assumed that the plasma remains in LTE. But the interruption capability is depending on the rate of electron disappearance during the decay. With a rapid decrease of temperature, some quenching effect could happen leading to departures from the equilibrium conditions. A chemical kinetics study of SF_6 arc decay has been developed[20]; it has shown that the rate of electron disappearance is very high at intermediate temperature (4 000 < T < 8 000 K, which is the temperature range crucial for the interrupting capability) because of two mechanisms : dissociative attachment (with F_2 and SF molecules) and dissociative recombination (with S_2^+ and SF^+ ions).

So, departures from equilibrium in pure SF_6 remain low and decrease when the pressure increases[20]. In real circuit-breaker the pressure is in general higher than 4 bar, so that the LTE conditions are generally fulfilled. In presence of nitrogen we can expect more departure from equilibrium and thus an overpopulation of electrons which tends to diminish the interruption capability of nitrogen-containing mixtures.

CONCLUSION

From our study we can conclude that the use of SF_6-N_2 mixtures as a filling gas for circuit-breakers will be efficient (i.e. will partly preserve the interruption properties of SF_6) when the proportion of SF_6 is higher than 50 %. Furthermore, not all the aspects concerning the use of SF_6-N_2 mixtures in HVCB have been studied here. To do so, it would be necessary to investigate the dielectric rigidity of the cold mixture and the risk of the formation of toxic molecules among the decomposition products.

REFERENCES

1. A. Gleizes, M. Razafinimanana and S. Vacquié, Transport coefficients in arc plasma of SF_6-N_2 mixtures, *J. Appl. Phys.* 54:3777 (1983)
2. H.R. Griem, *Plasma Spectroscopy*, Mc Graw Hill, New York (1964)
3. H.W. Drawin and P. Felenbok, *Data for Plasmas in Local Thermodynamic Equilibrium*, Gauthiers-Villars, Paris, (1965)
4. M.W. Chase Jr., C.A. Davies, J.R. Downey Jr., D.R. Frurip, R.A. Mc Donald and A.N. Syverud, *Janaf Thermochemical Tables*, Third Edition (1985)
5. M. Capitelli and E. Molinari, Problems of determination of high temperature thermodynamic properties of rare gases with application to mixtures, *J. Plasma Phys.*, 4:335 (1970)
6. S. CHapman and T. Cowling, *The Mathematical Theory of Non Uniform Gases*, Cambridge University-Press, New York (1952)
7. R.S.Devoto, Simplified expressions for the transport properties of ionized monoatomic gases, *Phys. of Fluids* 10:2105 (1967)
8. C. Muckenfuss and C.F. Curtiss, Thermal conductivity of multicomponent gas mixtures, *J. Chem. Phys.* 29:1273 (1958)
9. R.S.Devoto, Transport coefficients of partially ionized argon, *Phys. of Fluids* 10:354 (1967)
10. J.T. Vanderslice, S. Weissman , E.A. Mason and R.J. Fallon, High temperature tranport properties of dissociating hydrog*en*, *Phys. of Fluids* 5:155 (1962)
11. K.S. Yun, S. Weissman and E.A. Mason, High temperature transport properties of dissociating nitrogen and dissociating oxygen, *Phys. of Fluids* 5:672, (1962)
12. J.N.Butler and R.S. Brokaw, Thermal conductivity of gas mixtures in chemical equilibrium, *J. Chem. Phys.*, 26:1636 (1957)
13. A. Gleizes, M. Razafinimanana and S. Vacquié, Calculation of thermodynamic properties and transport coefficients for SF_6-N_2 mixtures in the temperature range 1,000-30,000 K, *Plama Chem. Plasma Proc.* 6:67 (1986)
14. A. Gleizes, B. Rahmani, J.J. Gonzalez and B. Liani, Calculation of net emission coefficient in N_2, SF_6 and SF_6-N_2 plasmas, *J. Phys. D: Appl. Phys.* 24:1300 (1991)
15. E. Schulz-Gulde, Temperature determination for arcs in SF_6 accounting for demixing, *J. Phys D: Appl. Phys.* 13:793 (1980)
16. S. Vacquié, A. Gleizes and H. Kafrouni, Measurements of electron density in a SF_6 arc plasma, *J. Phys D: Appl. Phys.* 18:2193 (1985)
17. A. Gleizes, I. Sakalis, M. Razafinimanana and S. Vacquié, Decay of wall stabilized arcs in SF_6-N_2 mixtures, *J. Appl. Phys.* 61:510 (1987)
18. B. Chervy, J.J. Gonzalez and A. Gleizes, Calculation of the interruption capability of SF_6-CF_4 and SF_6-C_2F_6 mixtures- Part II: arc decay modeling, *IEEE Trans. Plasma Sc.*, 24:210 (1996)
19. A. Gleizes, A.M. Rahal, S. Papadopoulos and S. Vacquié, Study of a circuit-breaker arc with self-generated flow: II-The flow phase, *IEEE Trans. Plasma Sc.* 16:615 (1988)
20. A. Gleizes, F. Mbolidi and A.A.M. Habib, Kinetic model of a decaying SF_6 plasma over the temperature range 12000K to 3000K, *Plasma Sources Sci. Technol.* 2:173 (1993)

DISCUSSION

E. KUFFEL: You presented thermal conductivities for pure N_2, SF_6, and 40%SF_6/60%N_2. Do you have the corresponding values for CF_4 and where do they fall in the SF_6/N_2 picture?

A. GLEIZES: We performed the calculation of the thermal conductivity for CF_4 and for SF_6/CF_4 mixtures. I have not here the corresponding curve that was published in part I of a twin paper given as Ref. 18 in my paper. If we consider the evolution of the thermal conductivity versus the temperature (see Fig. 3), the variations for CF_4 are very close to those for SF_6, except the dissociation peaks (around T = 2000 K for SF_6) are shifted to ~3000 K to 4000 K for CF_4.

L. NIEMEYER: There is experimental evidence that the thermal arc interruption capability of SF_6/N_2 mixtures is limited in terms of composition, i.e., that there is no synergy between these gases with respect to thermal interruption. Will you be able to model this observation with your turbulent exchange model based on the ρC_p function? Is there really something like a limit concentration of N_2 up to which the thermal interruption performance is not degraded at all?

A. GLEIZES: First, I have two remarks: (1) We did not apply our turbulence model to SF_6/N_2 mixtures, thus I have not a direct answer to your question. (2) Our study deals mainly with the post-arc phase of H. V. circuit breakers and not in the dielectric phase, whereas experiments take implicitly into account these two phases. Now, as the turbulent thermal conductivity is proportional to ρC_p, we can expect that the cooling rate of the plasma will increase with the proportion of SF_6 because of the high mass density of the cold SF_6 gas. Thus, I think there does not exist a threshold in the nitrogen concentration up to which the thermal interruption performance is not degraded.

L. G. CHRISTOPHOROU: What do you mean with the statement that for the SF_6/N_2 mixture to be used in circuit breakers the SF_6 content has to be greater than or equal to 50%?

A. GLEIZES: From our study it is clear that all the SF_6/N_2 mixtures have an interruption capability lower than that of pure SF_6, whatever the nitrogen proportion. But our results show that, in the framework of the assumptions used in this study, the interruption capability of a mixture containing 20% of SF_6 (even with 40% but the difference is not so high) should be lower than that of pure SF_6 (what is expected) but also than that of pure nitrogen. For this reason we conclude that the SF_6 proportion should be greater than or equal to 50%. Of course it must be noticed that our work deals only with the thermal interruption and does not take into account the dielectric phase of H. V. circuit breakers. Thus, we did not include the fact that the dielectric strength of the mixture is substantially higher than that of pure nitrogen, even with low SF_6 proportions.

LEADER STEP TIME AND LOW PROBABILITY IMPULSE BREAKDOWN VOLTAGE MEASURED IN SF_6 GAS MIXTURES

Wenguo Gu, Qiaogen Zhang, and Yuchang Qiu

School of Electrical Engineering
Xi'an Jiaotong University
Xi'an, Shaanxi 710049, China

INTRODUCTION

SF_6 is used worldwide today owing to its high dielectric strength and good heat transfer properties. But SF_6 is extremely sensitive to non-uniformity of the electrical field. Therefore in some electrical apparatus, such as cubicle type GIS, SF_6 gas mixtures are used[1]. Experiments show that when SF_6 is mixed with low electrical strength gases, such as air or CO_2, the low probability impulse breakdown voltage U_i and dc breakdown voltage U_b of the mixtures are higher than that of SF_6[2,3]. The view of corona stabilization enhancement[4] is usually used to explain the increase of U_b: when low electrical strength gas is added to SF_6, the corona onset voltage of mixtures decreases, so more ions are produced leading to the enhancement of corona stabilization. But the increase of U_i can not be explained by this viewpoint because there is hardly any corona stabilization under impulse conditions. Moreover, it is reported[5] that when Freon 113 ($C_2Cl_3F_3$), whose electrical strength is higher than that of SF_6, is mixed with SF_6, not only will U_b and U_i increase, but the corona onset voltage will also increase. Obviously, this phenomenon conflicts with the view of corona stabilization enhancement.

It is well known that when mixtures contain more than 10% SF_6, the mechanism of low probability breakdown changes from streamer breakdown to stepped leader breakdown[6]. The leader step time t_s, which is the time interval between two leader steps, is therefore an important parameter to describe the leader propagation, a determinative factor influencing the breakdown process. In this paper t_s and U_i are measured for a point/plane gap in SF_6/N_2, SF_6/CO_2, SF_6/air and SF_6/CCl_2F_2(hereafter R12) gas mixtures with different mixing ratios and the effect of additives is discussed in the light of the mechanism of streamer to leader transition.

EXPERIMENTS

A set of point/plane brass electrodes is mounted in a pressure vessel fitted with quartz windows. Gas pressure in experiments is up to 0.35 MPa. The radius of the point is 0.5 mm; the plane electrode with a Rogowski profile has a diameter of 90 mm; the gap length is 15 mm. A photomultiplier tube (PMT), whose spectral response range is from 170 nm to 850 nm and the impulse risetime is about 15 ns, is used to detect the optical events in the discharge development. Both dc and lightning impulse voltages used in the experiments are of positive polarities.

Gas additives are divided into two groups. The first group includes air, CO_2 and N_2, whose electrical strengths are lower than that of SF_6; R12 belongs to the second group, whose electrical strength is slightly higher than that of SF_6. The pressure vessel is first evacuated, and then filled with the mixture to be investigated with the predetermined mixing ratio and total pressure according to Dalton's partial pressure law. In order to obtain a uniform mixture, the minor gas is first filled, then the major gas, and the experiments are performed 12 hours later than the gas filling process.

At each testing pressure, U_b is first measured; every recorded point in the figures represents the average of five readings and the scattering around the average value does not exceed 1 percent of the average breakdown except for the data in the negative-slope part of the U_b-P curve, where the scattering can reach 5 percent of the average breakdown. Then U_i is determined by progressively reducing the voltage until 25 applications of voltage do not result in one breakdown. After that, the PMT is mounted to observe the predischarge process so as not to damage the PMT by the high intensity light emitted when breakdown occurs.

RESULTS

The scaling law of t_s in mixtures is found to be of the same form as that in SF_6 reported in our early publication[7]:

$$t_s = K/(PU) \qquad (1)$$

where P is the gas pressure; U is the applied voltage; K is a constant. While the value of K is 1.68 for pure SF_6, K varies with the mixing ratio and is different for different gas mixtures. An example of the variation of t_s with the product of P and U in SF_6/N_2 mixtures of different mixing ratios is given in Fig. 1, where the values of K are 3.6 and 1.5 for mixing ratios of 0.8/0.2 and 0.6/0.4 respectively. It can be seen that the theoretical values of t_s are in agreement with the measured values. All the values of K measured in this paper are shown in Table 1. From the experimental results, it is found that at the mixing ratio when K reaches its maximum value, U_i also reaches its highest value. The variations of U_b and U_i with the mixing ratio and gas pressure are shown in Fig. 2 ~ Fig. 5.

SF_6/N_2 Mixtures

Figure 2 shows the variation of breakdown voltage with the gas pressure in SF_6/N_2 mixtures of different mixing ratios. It can be seen that in pure SF_6, U_b reaches the maximum value at the pressure of 0.15 MPa and gets to its minimum value at the pressure of 0.35 MPa. The pressures corresponding to the maximum and minimum values of U_b in mixtures increase with the concentrations of N_2. The U_i of #2 gas is the highest among the three gases, so is the value of K. It is interesting to note that the U_i of #3 gas is lower than that of pure

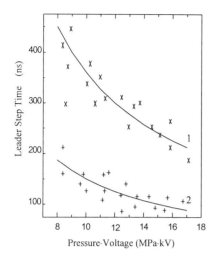
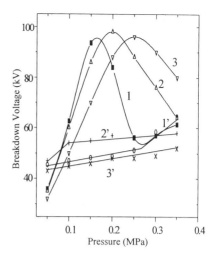

Figure 1. The variation of t_s with product of P and U in SF_6/N_2 mixtures of different mixing ratios

mixing ratio: 1- 0.8/0.2; 2- 0.6/0.4
value of K: 1 -3.6; 2- 1.5

Figure 2. The variation of breakdown voltage with gas pressure in SF_6/N_2 mixtures of different mixing ratios
mixing ratio: 1, 1'- 1.0/0.0; 2, 2'-0.8/0.2; 3,3' - 0.6/0.4   dc breakdown: 1, 2, 3;
low probability breakdown: 1', 2', 3'

SF_6, which indicates that the additive gas in excess of a certain concentration will have an adverse effect.

Table 1. The measured values of K in SF_6 mixtures

Mixture	Mixing ratio	Value of K
SF_6/N_2	0.8/0.2	3.6
	0.6/0.4	1.5
SF_6/Air	0.8/0.2	5.2
	0.5/0.5	3.1
SF_6/CO_2	0.8/0.2	5.6
	0.6/0.4	3.4
$SF_6/R12$	0.95/0.05	15.1
	0.8/0.2	7.6

SF_6/Air and SF_6/CO_2 Mixtures

Figures 3 and 4 show the experimental results in SF_6/air and SF_6/CO_2 mixtures respectively. The pressure where U_b of mixtures gets to its maximum value increases with the concentration of additive; U_i of all the mixtures is higher than that of SF_6. In both figures, U_i of #2 gas is the highest, so is the value of K.

$SF_6/R12$ Mixtures

The variation of breakdown voltage with the gas pressure in $SF_6/R12$ mixtures of

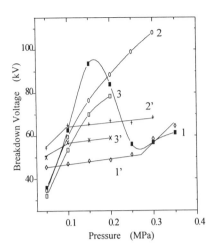
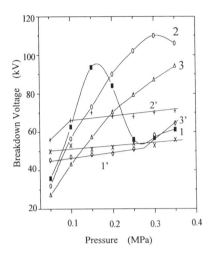

Figure 3. The variation of breakdown voltage with gas pressure in SF_6/air mixtures of different mixing ratios
mixing ratio: 1, 1'-1.0/0.0; 2, 2'-0.8/0.2; 3, 3'- 0.5/0.5   dc breakdown: 1, 2, 3
low probability breakdown: 1', 2', 3'

Figure 4. The variation of breakdown voltage with gas pressure in SF_6/CO_2 mixtures of different mixing ratios
mixing ratio: 1,1'-1.0/0.0; 2,2'-0.8/0.2; 3,3'- 0.6/0.4   dc breakdown: 1, 2, 3
low probability breakdown: 1', 2', 3'

different mixing ratios is shown in Fig. 5. The difference between SF_6/R12 mixtures and other mixtures is that U_b of SF_6/R12 mixtures is higher than that of SF_6 in the pressure range from 0.05 MPa to 0.2 MPa, but U_b of other mixtures is lower than that of SF_6 in the above pressure range. In Fig. 5, U_i of #2 gas is the highest, so is the value of K.

DISCUSSIONS

From the precursor mechanism, which describes the streamer to leader transition under positive impulse conditions, t_s can be obtained as follows[8]:

$$t_s = \frac{4\pi\varepsilon_0 R^2 \Delta E_{lim}}{V_i \lambda} \quad (2)$$

where R is the radius of the streamer; ΔE_{lim} is the precursor inception threshold value of space-charge-induced field enhancement; V_i is the ionic drift velocity; λ is the line charge density of the streamer. In corona periphery, the electrical field is the critical field E_{cr}, and then V_i is given by

$$V_i = E_{cr} \cdot \mu_i = (E/P)_{cr} P \mu_i \quad (3)$$

where $(E/P)_{cr} \approx 89$ kV/(mm·MPa); P is the gas pressure; μ_i is the ionic mobility. Inserting Equation (3) into Equation (2), one obtains

$$t_s = \frac{4\pi\varepsilon_0 R^2 \Delta E_{lim}}{(E/P)_{cr} P \mu_i \lambda} \quad (4)$$

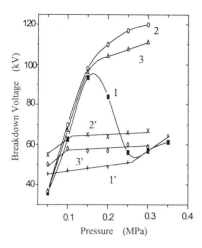 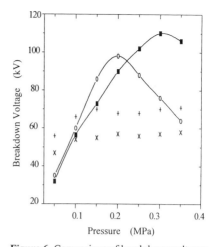

Figure 5. The variation of breakdown voltage with gas pressure in $SF_6/R12$ mixtures of different mixing ratios
mixing ratio: 1,1'-1.0/0.0; 2,2'-0.95/0.05; 3,3'-0.8/0.2   dc breakdown: 1, 2, 3
low probability breakdown: 1', 2', 3'

Figure 6. Comparison of breakdown voltages between SF_6/CO_2 and SF_6/N_2 mixtures at the optimum mixing ratio(0.8/0.2)
dc breakdown: ■ SF_6/CO_2 ○ SF_6/N_2
low probability breakdown: + SF_6/CO_2
╳ SF_6/N_2

Although $\Delta E_{\lim}$ and $(E/P)_{cr}$ vary with the category of mixture and mixing ratio, their variations tend to be in the same direction, i.e. when $\Delta E_{\lim}$ increases, $(E/P)_{cr}$ also increases, and *vice versa*. Therefore $\Delta E_{\lim}/(E/P)_{cr}$ can be considered as a constant, and the gas pressure does not change during the transition, so the proportionality is obtained as:

$$t_s \propto \frac{R^2}{\mu_i \lambda} \qquad (5)$$

When small concentrations of air (ionization energy of $O_2 \sim 12.2$ eV) or CO_2 (ionization energy ~ 13.7 eV) are added into SF_6 with their ionization energies being lower than that of SF_6 (ionization energy ~ 15.9 eV), these gases can be ionized by collision with low energy electrons or by absorption of low energy photons. This leads to an increase of the corona range, so the lengths of streamer increase, and thus λ decreases. On the other hand, the molecular weight and radius of these gases are much smaller than those of SF_6, so ions of additives can easily drift and diffuse, and therefore μ_i and R increase. The resultant effect of the variations of λ, μ_i and R lead to increases of t_s and U_i. When the concentration of the buffer gas further increases, ionized molecules increase, and the increase of λ will be the dominant factor, so t_s and U_i decrease. This explains why there exists an optimum mixing ratio which is 0.8/0.2 in our experiments.

When N_2 (ionization energy ~ 15.9 eV) is mixed with SF_6, the molecular weight and radius of N_2 are much smaller than those of SF_6, so ion drift and diffusion are easy, therefore μ_i and R increase, and this lead to increases of t_s and U_i. When the concentration of N_2 further increases, the ionized molecules increase because the electrical strength of N_2 is lower than that of SF_6, so increase of λ will be the dominant factor, and therefore t_s and U_i decrease. The optimum mixing ratios is 0.8/0.2.

Considering that the space-charge-induced field enhancement is produced by the competition between ion drift and ion recombination, the effect of R12 can be explained as follows: the molecules of R12 attach electrons and form negative ions which remain in the streamer channel, hence the chance of ion recombination increases, so λ decreases, therefore t_s and U_l increase. The phenomenon of the decrease of t_s and U_l with the further increase of the concentration of R12 can not be explained presently.

From the above discussions, one can guess that though N_2, CO_2 and air are all low electrical strength additives, the effect of N_2 on breakdown voltage is less than that of air or CO_2, because N_2 has only one effective factor, i.e. ion mobility, but air or CO_2 has two effective factors, i.e. ion mobility and ionization energy. From the comparison of breakdown voltages between SF_6/N_2 mixture and SF_6/CO_2 mixture at the optimum mixing ratio shown in Fig. 6, it can be seen that the speculation is reasonable.

CONCLUSIONS

The scaling law of leader step time in SF_6 gas mixtures has the same form as in pure SF_6: $t_s = K/(PU)$. In each mixture, the constant K varies with its mixing ratio; at the mixing ratio when K reaches its maximum value, U_l also reaches its highest value. Hence, K can be used as an indication in determining optimum mixing ratio.

The effect of gas additives on the low probability breakdown can be explained as follows: when additives are mixed with SF_6, not only the line charge density and the radius of the streamer but also ion mobility of mixture may vary, and the resultant effect leads to a delay of streamer to leader transition.

ACKNOWLEDGMENT

Financial support from the National Natural Science Foundation of China is highly appreciated.

REFERENCES

1. Y. Qiu and Y. P. Feng, Investigation of SF_6-N_2, SF_6-CO_2 and SF_6-air as substitutes for SF_6 insulation, In: *Conf. Record of the 1996 IEEE Int. Symp. on EI*: 766(1996)
2. N. H. Malik, DC voltage breakdown of SF_6-Air and SF_6-CO_2 mixtures in rod-plane gaps, *IEEE Trans. on Electrical Insulation*. 18: 629(1983)
3. L. M. Zhou, Y. Qiu and Y. P. Feng, The influence of O_2 and N_2 additives on impulse breakdown of SF_6 in a non-uniform field, In: *Conf. Record of 1994 Int. Symp. on Electrical Insulation*: 504(1994)
4. O. Farish and I. D. Chalmers, SF_6 mixtures with enhanced corona stabilization, In: *Proc 7th Int. Conf. On Gas Discharges and Their Applications*: 223(1982)
5. I. D. Chalmers, Y Qiu, A Sun and B Xu, Extended corona stabilisation in SF_6/Freon mixtures, *J. Phys. D: Appl. Phys.* 18: L107 (1985).
6. W. Voss, Discharge developments of non-uniform gaps in SF_6-N_2 mixtures, In: *Proc 3rd ISH*. Paper 31.13(1979)
7. W. Gu, Q. Zhang and Y. Qiu, Leader step time measured in a point/plane gap in SF_6 gas using a photomultiplier tube, In: *Proc 10th ISH*, Vol. 2: 181(1997)
8. L. Niemeyer, L. Ullrich and N. Wiegart, The mechanism of leader breakdown in electronegative gases, *IEEE Trans. on Electrical Insulation*. 24: 309(1989).

DISCUSSION

H. OKUBO: You measured the current pulse and the light intensity pulse. How did you identify the leader discharge physically? How did you distinguish the leader type discharge among the streamer type discharge?

Y. QIU: In an earlier publication (Ref. 7), we compared wave forms recorded by PMT for both streamer and leader discharges in SF_6 gas, which were apparently different from each other because of their different discharge mechanisms (for example, see Ref. 8).

THE STUDY OF SF_6 MIXTURES WITH BUFFER GASES

I.D.Chalmers, X.Q.Qiu and P.Coventry*

Centre for Electrical Power Engineering
University of Strathclyde
204 George Street
Glasgow G1 1XW
UNITED KINGDOM

INTRODUCTION

Because of its outstanding electrical and physical properties, SF_6 has gained wide acceptance as the dielectric of choice for high voltage applications. However, the relative high cost of SF_6 and other drawbacks, such as its sensitivity to particle effects and restricted working temperature range had limited its application. Moreover, the recent issue on SF_6 over its environmental concern[1] has made the wide use of the gas in the future much in doubt.

Alternative high strength gases and gas mixtures, notably those containing C and F (e.g., C_4F_6, c-C_5F_8) may have higher dielectric strength than SF_6. An extensive search, especially in the seventies and early eighties[2,3,4,5], showed that alternatives to SF_6 do exist with superior dielectric properties. For example, the breakdown strengths of some perfluorocarbons and their mixtures are as much as 2.5 times that of pure SF_6. Unfortunately, these are greenhouse gases and often have other undesirable properties such as toxicity, low vapour pressure, release of solid carbon during arcing etc. In summary, no single, pure gas has been found that is superior to SF_6 in all respects.

It is well known that the insulation performance of SF_6 is limited, not by its uniform field strength, but by the effects of local field enhancement. In a typical industrial application the non-uniform field breakdown predominates. It is clear that the requirement is not necessarily for a gas with superior uniform field strength, but rather for a gas or gas mixtures which offer a significant improvement over SF_6 under the non-uniform field conditions associated with particles or other stress raisers. The key factor is that there should be an increase in the voltage required for leader propagation, without a significant loss of uniform field strength. The propagation of the leader which cause breakdown in a non-uniform field can be inhibited by adding traces of compounds such as TEA[6], MEK[7] and some original refrigerants freon (e.g. R113, R12 etc)[8,9], which when added to SF_6 in small quantities (<1%), result in improved corona stabilisation and an increase in the leader

---

* The National Grid Company plc, Kelvin Avenue, Leatherhead, Surrey KT22 7ST

propagation field. Because of the small concentrations used, there is little change in the effective ionisation coefficient and hence in the uniform field strength relative to pure SF_6, but the particle triggered breakdown voltage may be increased by 50%. These additives, however, are unlikely to have any industrial prospects because TEA and MEK are highly toxic and the production of freon should be restricted from the point of view of protection of the ozone layer in stratosphere.

From a practical point of view, only SF_6 mixtures with those common gases or buffer gases show promise for industrial application. There are at least four ways in which SF_6 gas mixtures with buffer gases may be superior to SF_6 used alone in industrial practical high-voltage system. First of all, higher total gas pressure may be used without rating the minimum operating temperature. Second, the cost of mixtures for a given pressure can be reduced considerably depending on the cost of second component. Third, gas mixtures offer the possibility of some degree of immunity from the breakdown initiated by local (microscopic) field non-uniformities associated with electrode irregularities (e.g. roughness) or free conducting particles. Last but not the least, the use of SF_6 mixtures with buffer gases could be a short-term solution to the problem of eliminating the potential contribution of SF_6 to global warming.

The present study was undertaken using SF_6 mixtures with three buffer gases, N_2, air and CO_2. The dielectric characteristic of these mixtures were investigated. It has been found that SF_6 mixtures with buffer gases could be ideal substitutes for SF_6 in some applications in terms of their comparable high dielectric characteristics, some degree of immunity from the breakdown initiated by local field non-uniformities associated with electrode irregularities or free conducting particles, reduced gas cost and their environmental friendly property.

EXPERIMENTAL SETUP AND PROCEDURE

Positive lightning impulse voltage with a waveform of $1.2/50\mu s$ was applied to a point-plane gap (simulating the local field inhomogeneity). The pressure vessel housing the electrodes and the gas handling technology have been well documented[10]. Tests were carried out to determine the minimum impulse breakdown strength as this will correspond to the worst case in a practical system. In terms of physical mechanisms, that is the stress level at which the gap will just breakdown if all necessary statistically varying conditions are simultaneously satisfied which was discussed previously[10]. In each test sequence, the 50% breakdown level was first determined by the normal up-and-down method, and then the minimum (2.5% level) breakdown voltage was found by progressively reducing the voltage in steps of 0.5kV until 40 applications of voltage resulted in only 1 breakdown. The time interval between each impulse voltage application was 60 seconds to ensure complete deionisation of the gap.

THE DIELECTRIC CHARACTERISTICS AS A FUNCTION OF MIXING RATIO

A series of minimum impulse breakdown tests were carried out in SF_6/N_2, SF_6/CO_2 and SF_6/air at 0.1MPa, varying the SF_6 concentration at different gap length of 10, 15, 20 and 25mm. The optimum mixture ratio for a particular application, of course, will invariably depend on economic factor as well as design requirement, apart from the dielectric considerations.

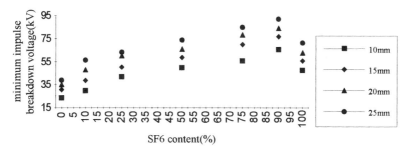

Figure 1. Minimum impulse breakdown voltage of SF_6/air mixtures, point-plane gap, P=0.1MPa

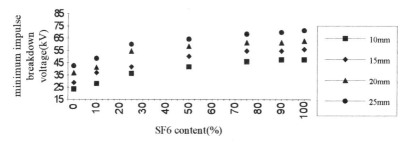

Figure 2. Minimum impulse breakdown voltage of SF_6/N_2 mixtures, point-plane gap, P=0.1MPa

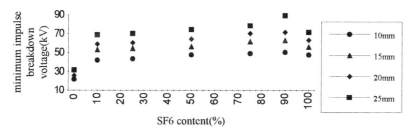

Figure 3. Minimum impulse breakdown voltage of SF_6/CO_2 mixtures, point-plane gap, P=0.1MPa

It can be noticed that SF_6/air and SF_6/CO_2 mixtures have higher impulse breakdown strength than pure SF_6 and exhibit a maximum value at the mixing ratio of 90/10. For example, the SF_6/air and SF_6/CO_2 mixtures at the mixing ratio of 50/50 increase the breakdown voltage by 5% and 0.7% (for 15mm gap). At the mixing ratio of 90/10, the increasing rates are 37.5% and 12.6%. The SF_6/N_2 mixtures, on the other hand, always has a breakdown voltage lower than that of pure SF_6 and its breakdown voltage versus SF_6 concentration being a monotonic rising curve. At the mixing ratio of 50/50, the breakdown voltage is 10% lower than that of pure SF_6 (for 15mm gap). At the ratio of 90/10, it is 2.5% lower. The results are consistent with those reported[11]. Based on the facts that the impulse breakdown level of SF_6/buffer gases mixtures are not too much lower or higher than that of pure SF_6, the use of these mixtures looks promising because it is the impulse withstand level that determines the insulation dimensions.

THE DIELECTRIC CHARACTERISTICS AS A FUNCTION OF GAS PRESSURE

It has been realised that though SF_6 has excellent insulating strength, in practice the electrical breakdown strength of compressed SF_6 is very often determined by local field enhancement due to protrusion, surface roughness and conducting particles left in the system. At working pressure above 0.2MPa, the benefit of higher SF_6 pressure could be largely eliminated. It has been shown that the gas breakdown voltage or insulator flashover voltage can be reduced by more than 50% under certain condition of local field enhancement[12]. The SF_6 mixtures with buffer gases are expected to exhibit less sensitivity than SF_6 to electric field nonuniformities associated with particles contamination and surface roughness based on the electrical breakdown strength of N_2 and some other buffer gases are less sensitive than that of SF_6. In non-uniform fields, these SF_6 mixtures may be comparable to pure SF_6 in terms of voltage withstand capability. Minimum impulse breakdown tests were carried out in the 10mm point-plane gap varying the gas pressure (up to 0.45MPa). The gas mixtures contained 50% of SF_6 with 50% of one of the common gases.

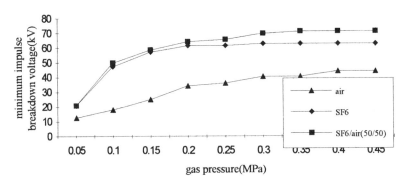

Figure 4 Minimum impulse breakdown voltage in the 10mm point-plane gap

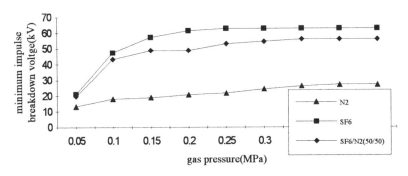

Figure 5 Minimum impulse breakdown voltage in the 10mm point-plane gap

The minimum impulse breakdown voltage of SF_6/air and SF_6/CO_2 mixtures is higher than that of pure SF_6 in the 10mm point-plane gap up to 0.45MPa. While on the other hand, the SF_6/N_2 always has a lower (by ~11% than pure SF_6 at 0.45MPa) minimum impulse breakdown. Also noticed was that above 0.2MPa, the minimum impulse breakdown voltage for pure SF_6 changes very little in the highly non-uniform field. This saturation effect is also observed in SF_6 mixtures with N_2, air and CO_2, but at a higher pressure (above 0.3MPa).

The results indicate that the mixtures have breakdown characteristics that are at least as desirable as those of pure SF_6.

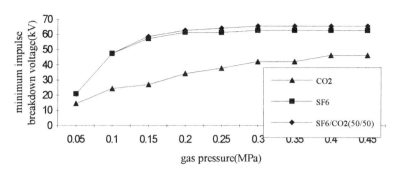

Figure 6 Minimum impulse breakdown voltage in the 10mm point-plane gap

DISCUSSION

Improved gaseous dielectrics have been studied and so designed on the basis of knowledge of fundamental electron-molecule interaction. The most important electron control mechanism in gas under electric stress is electron attachment to gas molecules. Breakdown experiments indicate that the best dielectrics are those that control electrons well at energies below the gas electronic excitation threshold energy, and the actual ionisation process details are apparently less important than the attachment process in preventing breakdown[13]. Mixing gases is then advantageous since no one gas attaches electrons well for all the energies of the electrons present. Attachment become more difficult for all attaching gases at high electron energies (typically above 2 eV), so that gas components are included which can remove energy from electrons escaping the efficient low-energy attachment range, returning them to the energy region where attachment is effective and away from the ionisation threshold. It is generally accepted that binary mixtures should team one gas that primarily de-energises free electrons and one gas that removes free electrons from the dielectric by electron attachment[14]. N_2, O_2, CO and CO_2 are some typical electron retarding gases in which fast electrons can be slowed and the electron energy can be reduced. The role of these electron retarding gases is to de-energise electrons reaching higher energies and return them to the lower energy range where attachment by the electronegative gas is most effective.

In highly non-uniform fields, why the SF_6/air and SF_6/CO_2 mixtures (when the content of SF_6 is more than 50%) have a higher minimum impulse breakdown voltage than that of pure SF_6 is not fully understood and should be further studied. Although there are suggestions that the faster space charge movement[15] and/or the production of photo-electrons[16] may be responsible. At higher pressure (above 0.3MPa), the minimum impulse breakdown voltage for SF_6/N_2(50/50), SF_6/CO_2(50/50) and SF_6/air(50/50) in highly non-uniform field is quite similar to those observed for SF_6, i.e., there seems to be a saturation effect, at least at the tested mixing ratio (50/50) as the breakdown level does not increase too much as the gas pressure increases. The degree of immunity from the breakdown in highly non-uniform field is not very pronounced when 50% of SF_6 is presented in the mixtures. The characteristic of sensitivity to non-uniformities of SF_6 seems to predominate in the 50/50 mixtures.

GENERAL CONCLUSIONS

The SF_6/air and SF_6/CO_2 have higher minimum impulse breakdown voltage than pure SF_6 in highly non-uniform field and exhibit a maximum value at the mixing ratio of 90/10. The SF_6/N_2, on the other hand, with its breakdown voltage versus SF_6 concentration being a monotonic rising curve, always has a breakdown level lower than pure SF_6. The expected insensitivity to electric field nonuniformities of the SF_6/buffer mixtures is not very pronounced at the current tested mixing ratio (50% SF_6 with one of the buffer gases). The results show that in non-uniform field which is inevitable and predominant in a practical system, SF_6/buffer gas mixtures are at least as good as SF_6 in terms of their impulse breakdown characteristics. There are indications that these mixtures are technically and economically attractive alternatives.

ACKNOWLEDGEMENTS

The financial support from the National Grid Company is gratefully acknowledged.

REFERENCES

1. L.G., Christophorou, and R.J. Van Brunt, "SF_6 insulation: possible greenhouse problems and solutions", CIGRE, paper 15.03.06 IWD4 (1996)
2. *Minutes of workshop on gaseous dielectric for use in future electric-power systems,* Edited by R.J.Van Brunt, Report NBSIR 80-1966, National Bureau of Standards (1980)
3. J.C.Devins, "Replacement gases for SF_6", IEEE Trans.Vol.EI-21(6):81-85 (1980)
4. L.G. Christophorou, et al, "Recent advances in gaseous dielectrics at Oak Ridge National Laboratory", IEEE Trans. Vol.EI-19(6):550-566 (1984)
5. R.E. Wooton, *Gases superior to SF_6 for insulation and interruption,* Report ERRI EL-2620,Electric Power Research Institute (1982)
6. O. Farish, et al, "SF_6 mixtures with enhanced corona stabilisation", 7th Int. Conf. on Gas Discharges and their Applications, 223-226 (1982)
7. A. Goldman, et al, "Influence of organic vapor traces on corona corrosion of electrodes in SF_6",4th International Symposium on High Voltage Engineering, paper 32.07 (1983)
8. Y.Qiu, et al, "Improved dielectric strength of SF_6 gas with a trichlorotrifluoroethane vapor additive", IEEE Trans. Vol.EI-22: 763-768 (1987)
9. I.D.Chalmers, et al, "Extended corona stabilisation in SF_6/freon mixtures", J.Phys.D:Appl.Phys.,18: L107-L112 (1985)
10. X.Q.Qiu and I.D.Chalmers, "On the effect of space charges on breakdown strength by using two different injection methods", Proceedings of 7th International Conference on Dielectric Materials Measurements & Applications, Bath:180-184 (1996)
11. Y.Qiu and Y.P Feng, "Investigation of SF_6-N_2, SF_6-CO_2 and SF_6-Air as substitutes for SF_6 insulation", Conf. Record of IEEE Int.Symp. on Electrical Insulation, 766-769 (1996)
12. Pedersen, A., "The effect of surface roughness on breakdown in SF_6", IEEE Trans.Vol.PAS-94:1749-1754 (1975)
13. L.G., Christophorou, "The use of basic physical data in the design of multicomponet gaseous insulators", Proceedings of the 5th International Conference on Gas Discharges,Liverpool:1-8 (1978)
14. Pace, O.M., et al, "Application of basic gas research to practical systems", Proceedings of the 7th IEEE/PES Transmission and Distribution Conference and Exposition, Atlanta:8-177 (1979)
15. Y.Qiu, et al, "Breakdown of SF_6/air mixtures in non-uniform field gaps under lightning impulse voltage", Proceedings of the 10th International Conference on Gas Discharges and Their Applications, Swansea:398-401 (1992)
16. I.Gallimberti and N.Wiegert, "Streamer and leader formation in SF_6 and SF_6 mixtures under positive impulse conditions", J.Phys.D:Appl.Phys., 19:2351-2373 (1986)

DISCUSSION

E. KUFFEL: How did you define the minimum impulse breakdown voltage? How many shots did you make to determine the minimum value?

X. Q. QIU: The 50% breakdown level was first determined using the normal up-and-down method. Then the minimum (2.5% level) breakdown voltage was found by progressively reducing the voltage in steps of 0.5 kV until 40 applications of voltage resulted in only one breakdown.

M. GOLDMAN: Are your results valid for any shot of a series of shots or does a conditioning exist? If yes, does the conditioning come from effects in the gas or on the electrodes?

X. Q. QIU: The conditioning effect existed in both SF_6 gas and the gas mixtures. It resulted from the inevitable surface defects of the electrode and had nothing to do with the gas composition.

A. H. MUFTI: Why do the SF_6/Air mixtures and the SF_6/CO_2 mixtures have the same characteristics at the ratio of 50%/50%?

X. Q. QIU: The effect is probably due to oxygen, which is the common element in both CO_2 and air.

W. BOECK: Your findings are valid for point-plane arrangements and they are not permitted in a sound GIS insulation. But your results point that such mixtures are less sensitive to defects than pure SF_6. That is an important point.

X. Q. QIU: Thanks for your comment.

L. G. CHRISTOPHOROU: Since SF_6/CO_2 mixtures work so well, do you care to comment as to possible problems of carbon deposits in a real system?

X. Q. QIU: Under our test conditions, no carbon deposits were observed. In a real system, this is a very important issue and further investigation is needed.

M. GOLDMAN: Electrode aging generally proceeds through the formation of more or less insulating layers on them [See M. Lalmas et al., *Gaseous Dielectrics VII*, L. G. Christophorou and D. J. James (Eds.), Plenum Press, New York, 1994, p. 617]. The introduction of small amounts of O_2 or CO_2 in the vessel should increase the conductivity of these layers and so protect the gap from the formation of micro-spark breakdowns across them.

X. Q. QIU: Thank you for your valuable comment. Although up to now we have not observed the influence of O_2 or CO_2 on electrode aging, we will try to do more work in this respect in the future.

BREAKDOWN STRENGTH OF SF_6/CF_4 MIXTURES IN HIGHLY NON-UNIFORM FIELDS

E. Kuffel and K. Toufani

Dept. of Electrical and Computer Engineering
University of Manitoba
Winnipeg, MB, CANADA R3T 5V6

INTRODUCTION

Dielectric properties of primary gas mixtures comprising of sulfurhexafluoride (SF_6) as the basic component have been studied extensively in the past. The underlying objective was to find optimal mixtures with dielectric properties superior to SF_6 alone. The added gas component can be a "buffer" gas which slows down high energy electrons via scattering process and allows these to be captured by SF_6, or an electronegative gas (halogen family) which captures free electrons by attachment and forms heavy negative ions. In cases when the basic physical properties of individual gases are known, gas mixtures can be tailored analytically to meet specific apparatus requirements[1, 2, 3], before performing experimental tests.

SF_6 gas because of its chemical stability and high dielectric strength (Ë 90 kV/cm bar), and its excellent arecquenching properties has been almost universally employed in high voltage gas insulated equipment and circuit breakers.

A major limitation is its relatively high liquefaction temperature (-33°C at 400 kPa) and apparatus insulated with pure SF_6 becomes unsuitable for use in cold climate regions. To overcome these difficulties earlier designs of gas insulated circuit breakers for northern applications used mixtures of SF_6/N_2. The introduction of N_2 reduced significantly the arc interruption capability below that of pure SF_6 and the circuit breaker required rating reduction.

In recent years gas mixtures comprising of sulfurhexafluoride (SF_6) and carbontetrafluoride (CF_4) have been successfully implemented in outdoor circuit breakers designed for operation in northern Manitoba[4]. A mixture of 50% SF_6/50% CF_4 exhibits excellent arc quenching properties and has a much reduced liquefaction temperature. There is little published information available on the dielectric strength of SF_6/CF_4 mixtures. Berg[5] reported data on a.c. and impulse breakdown voltages for guasi-uniform field gaps (sphere-sphere) for SF_6/CF_4 mixtures. He observed that with changing the proportions of SF_6 gas in CF_4 from 0 to 100% the breakdown increased linearly with the percentage of

SF_6, in accordance with Takuma's[6] empirical expression. For highly non-uniform field gaps Berg's studies showed that under direct and power frequency voltages addition of a few percent of SF_6 to CF_4 caused a large increase in the mixture's dielectric strength. The present paper presents experimental data on a.c., (+ve) and (-ve) d.c. and standard impulse breakdown characteristics in highly divergent fields, obtained for SF_6/CF_4 mixtures varying the percentage of SF_6 from zero to 100%.

Experimental Setup and Test Procedure

The electrodes consisted of a grounded copper sphere (ϕ 62.5 mm) and high voltage brass rod (ϕ 10 mm) tapered 30° to a tip of 0.5 mm radius. The gap arrangement was mounted coaxially inside a fiberglass reinforced resin high-pressure chamber of dimensions: ϕ 380 mm and 680 mm height, fitted with brass end plates. The cross section of the set up is shown in Figure 1.

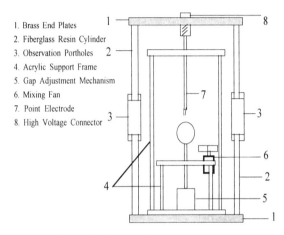

1. Brass End Plates
2. Fiberglass Resin Cylinder
3. Observation Portholes
4. Acrylic Support Frame
5. Gap Adjustment Mechanism
6. Mixing Fan
7. Point Electrode
8. High Voltage Connector

Figure 1. Cross-section of test chamber.

The rated pressure of the vessel was specified as 700 kPa. The experiments were made at a pressure of 300 kPa. A small d.c. motor geared to 3/4 rpm was used to rotate the lower electrode (sphere) for gap adjustment. The shaft of the earthed electrode had a thread pitch such that one revolution caused a 1mm linear motion. The shaft was graduated so that measurements of gap length could be made with 0.015 mm accuracy. The direction of rotation of the gap adjustment mechanism was controlled by means of two double pole – double throw switches which could be operated from outside the chamber.

Prior to filling the chamber with the desired gas mixture, the vessel was evacuated down to 0.0013 kPa ($\cong$ 0.01 mm Hg), flashed with purified N_2 (99.995%), then evacuated again. The residual and experimental pressures were measured by a Matheson pressure gauge ($\pm$ 0.25% accuracy).

To ensure uniform mixture of the gas components a small mixing fan, mounted in the lower part of the chamber was activated after introducing the gases for one hour prior to starting the experiments. During the filling procedure a predetermined amount of the gas constituting the smaller fraction was admitted first. The pressure was carefully measured and the second gas component was admitted.

The experiments were carried out under 60 Hz a.c. voltages, (+ve) and (-ve) direct voltages and (+ve) and (-ve) standard impulse voltages.

Experimental Results and Discussion

Figure 2 presents an example of experimentally determined a.c. breakdown voltages, plotted as function of the percentage of SF_6 gas in SF_6/CF_4 mixture, for 25 mm and 30 mm gaps. Included in the figure are the maximum, the minimum, and the average values. The figure also includes the corresponding corona inception voltages.

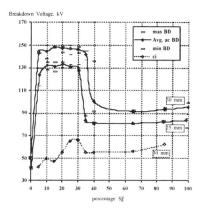 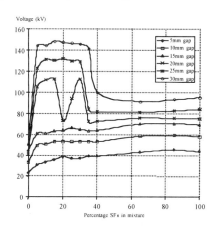

Figure 2. a.c. breakdown voltage and corona inception — % of SF_6 in CF_4/SF_6 mixture for 25 mm and 30 mm gaps.

Figure 3. a.c. breakdown voltage — % SF_6 in CF_4/SF_6 mixture for gap range 5 mm – 30 mm.

The results in Figure 2 show that addition of the first 5% of SF_6 to CF_4 increases the breakdown voltage approximately threefold above the corresponding value for pure CF_4. The breakdown values then level off and remain constant with adding a further 30% of SF_6. On further addition of SF_6, above $\cong 40\%$, the breakdown strength falls abruptly down to about half of its highest values. On further increase of SF_6, the breakdown strength remains unchanged at a value corresponding to pure SF_6.

Despite the rapid increase in the breakdown voltage with the addition of small amounts of SF_6, the measured high values remain very consistent with small scatter, except in the transition region 30% - 40% (SF_6).

Figure 3 summarizes a.c. breakdown voltages obtained for gaps ranging from 5 mm to 30 mm. The large initial increase is limited to gaps above 20 mm gaps. For gaps shorter than 20 mm, while there is an initial increase observed, the rate of rise is much smaller than for the larger gaps.

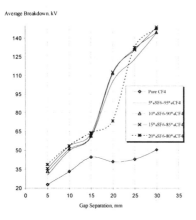

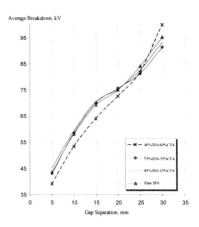

Figure 4. a.c. breakdown voltage – gap length; mixtures containing SF_6 0 – 20% by volume.

Figure 5. a.c. breakdown voltage – gap length; gas mixtures containing 50% - 100% SF_6.

The breakdown voltage – gap length relationship observed for mixtures obtaining small amounts of SF_6 ($\leq$ 20%) displayed in Figure 4, which compares the breakdown characteristics of CF_4 with CF_4/ SF_6 mixtures comprising of 5%, 10%, 15% and 20% of SF_6. Note the discontinuity in the curves at the 15 mm spacing with the exception of 20% SF_6/80% CF_4 mixture in which the discontinuity is shifted to 20 mm gap.

The breakdown voltage – gap length characteristics for mixtures with $\geq$ 40% SF_6 are shown in Figure 5. The relationship remains nearly linear with the breakdown voltage at values observed in pure SF_6 at pressures of 300 kPa.

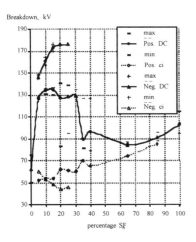

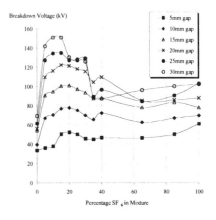

Figure 6. d.c. breakdown voltage and corona inception – % SF_6 mixture under (+ve) and (-ve – polarity voltages; gap 25 mm.

Figure 7. Positive direct breakdown voltage – % SF_6 for 5 mm – 30 mm.

Figure 6 compares the breakdown characteristics obtained for mixtures of CF_4/SF_6 under positive and negative direct voltages for a 25 mm gap. Included in the figure are the corona inception voltages. The positive breakdown voltage characteristics shown in Figure 6 resemble closely the data obtained under a.c. voltage (Figure 2). The results obtained under d.c. voltage are more scattered than the corresponding a.c. values especially in mixtures containing 20% of SF_6 or more. Under negative polarity the breakdown voltage is significantly higher, while the corona inception is observed at a lower voltage.

Figure 7 summarizes the breakdown date obtained for gaps ranging from 5 mm to 30 mm length in SF_6/CF_4 mixtures (% SF_6 from 1 – 100%) under positive polarity voltage. The characteristics follow closely the pattern observed under a.c. voltage (Figure 3).

Under standard impulse voltages of both polarities the 50% breakdown voltage - % SF_6 content characteristics did not display the positive "synergism" observed under a.c. and d.c. voltages for mixtures with small amounts of SF_6.

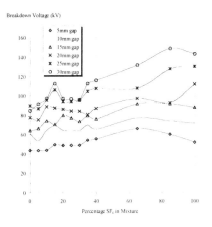

Figure 8. 50% positive standard impulse breakdown voltage – % SF_6 in mixture gaps 5 mm - 30 mm.

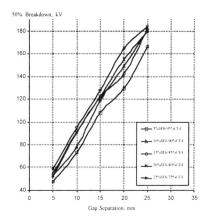

Figure 9. 50% negative standard impulse breakdown voltage – gap length (5 - 25% SF_6).

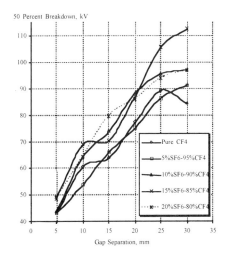

Figure 10. 50% positive standard impulse breakdown voltage – gap length (0 - 20% SF_6).

The 50% breakdown voltage characteristics measured under positive polarity for gaps 5 mm – 30 mm in mixtures with SF_6 content 0 - 100% are included in Figure 8. The results are somewhat erratic, but there is no clear indication of positive or negative synergism. When plotted in terms of 50% breakdown voltage – gap length relationship for mixtures containing 25% or less SF_6 the relationship remains linear as shown in Figures 9 and 10. For negative polarity (Figure 9) the relationship is clearly linear with the expected increments in the 50% breakdown values when SF_6 is added.

Under positive polarity the results are more erratic but the trend remains the same. The approximately linear behavior of the breakdown characteristics continued when the content of SF_6 was increased from 25% to 100%.

CONCLUSION

Breakdown voltage characteristics of SF_6/CF_4 mixtures in highly divergent fields have been studied under alternating, both polarity direct voltages and standard lightning impulse voltages. The content of SF_6 was varied from 0 - 100%. Under alternating and positive direct voltages the addition of the first few percent of SF_6 to CF_4 causes a large increase in the breakdown of the mixture voltage. This is more apparent for larger gaps when the field divergency increases and corona stabilization may occur, leading high breakdown values. Under impulse voltage with no corona stabilization the breakdown voltage increases nearly linearly with the % of SF_6 content.

ACKNOWLEDGMENT

The authors gratefully acknowledge the financial support provided by Manitoba Hydro and the National Science and Engineering Council of Canada, which made these studies possible.

REFERENCES

1. L.G. Christophorou, I. Sauers, D.R. James, H. Rodrigo, M.O. Pace, J.G. Carter, and R.M. Hunter, Recent advances in gaseous dielectrics, *Proceedings of IEEE International Symposium on Electrical Insulation*, Montréal, (1984).
2. P.J. Chantry and R.E. Wooton, A critique of methods for calculating the dielectric strength of gas mixtures, *Journal of Applied Physics*, 52:2731 (1981).
3. Y. Qiu, S.Y. Chen, Y.F. Liu and E. Kuffel, Comparison of SF_6/N_2 and SF_6/CO_2 mixtures based on figure of merit concept, *1988 Annual Report on Electrical Insulation and Dielectric Phenomena*, 299, Ottawa, (1988).
4. R. Middleton, V. Koshik, P. Hogg, P. Kulkarni and H. Heiermeirev, Development work for the application of 245 kV circuit breakers using SF_6 – CF_4 gas mixtures on the Manitoba Hydro system, *Proceedings of CEA Conference*, Toronto, (1994).
5. J. Berg and E. Kuffel, Breakdown voltage characteristics of SF_6/CF_4 mixtures in uniform and non-uniform field gaps, Proceedings of IEEE International Symposium on Electrical Insulation, Montréal, (1996).

DISCUSSION

M. S. NAIDU: Will the addition of CF_4 to SF_6 not increase the decomposition rates? Further, in tropical countries where the humidity content is very high, it is possible that HF formation will be very high leading to corrosion?

E. KUFFEL: The decomposition rate of CF_4 is unlikely to be higher than that of SF_6. However, it may be decomposed leading to carbon deposition, but that will be relatively small. As to HF formation due to very high humidity, water in SF_6 or SF_6/CF_4 insulated equipment has to be avoided at any cost.

THE EFFECT OF ELECTRODE GEOMETRY ON THE LIGHTNING IMPULSE BREAKDOWN IN SF_6 / N_2 MIXTURES

D. Raghavender

Dept. of Electrical Engg.
S.S.G.M. College of Engg.
Shegaon - 444203 - (M.S.)
INDIA

M.S. Naidu

Dept. of High Voltage Engg.
Indian Institute of Science
Bangalore - 560 012
INDIA

ABSTRACT

Investigations were undertaken to focus on the effect of methodical variation of the rod-diameter in a rod-plane geometry under lightning impulse voltages in SF_6 / N_2 mixtures (1 to 20 % of SF_6), over a pressure range of 0.1 MPa to 0.5 MPa for gaps of 5 mm to 80 mm with rod diameter of 0.1 (sharp needle), 0.8, 1.5, 3.0, 5.0 and 10.0 mm and a 230 mm dia. plane electrode of Rogowski profile. The experimental results indicate that the breakdown voltages are sensitive to small amounts of SF_6 gas impurities in N_2 under lightning impulse voltage conditions for different non-uniformities investigated. The highlight of this paper describes V_{50}-rod diameter characteristics, which shows maxima in their V_{50} at all gap spacings and gas pressures. These maxima (V_{50} min.) do not appear to occur at any fixed value of rod-diameter, but vary depending on the gap pressure and gap spacing.

The significance of V_{50} min. in breakdown characteristics imply for a given gap length a safe voltage level exists (generally referred to as "Forbiden-Zone") below which no breakdown can ocur irrespective of the rod-diameter and a gas pressure. The practical interpretation of the results suggested that the irregularities on the electrode surfaces such as roughness, scratches, etc., do not necessarily diminish the dielectric strength of either SF_6 or SF_6 gas mixtures, below certain voltage level.

INTRODUCTION

It is well known that the use of SF_6 / N_2 mixtures in SF_6 gas insulated equipment could solve the liquification problem, reduce the cost of the gas and to some extent lessen the sensitivity of the dielectric strength to the local field enhancement.

In recent years, concern has been expressed of the possible impact of SF_6 gas on global warming. Therefore, SF_6 in mixtures with N_2 has been investigated as a possible substitute to SF_6 gas. Most of the earlier studies in SF_6 / N_2 mixtures have been concentrated essentially to determine the breakdown strength by illustrating the breakdown characteristics using a limited number of concentrations of SF_6 gas in N_2, at one or two gap spacings.

Further, limited work has been reported on the effect of field divergence (use of different rod diameters in a rod-plane electrode geometry) on breakdown in SF_6 and SF_6 / N_2 mixtures under d.c, and a.c voltages, very little information is available on the effect of field divergence under lightning impulse voltages. While a variety of electrode configurations were considered, attention was not focussed on the effect that a methodical variation of the radius of curvature of the electrodes may have on the impulse breakdown characteristics of SF_6 or its mixtures.

In the present study, the investigations have been carried out to determine the breakdown characteristics of SF_6 / N_2 mixtures using standard lightning impulse (1.2 / 50 µs) voltages when the non-uniformity of the fields was varied over a wide range. Total gas pressures upto 0.5 MPa and gap spacings of 5 to 80 mm were investigated. The electrode arrangement consisted of hemispherically capped stainless steel rods of different diameters $D \cong 0.1$ (sharp needle), 0.8, 1.5, 3.0, 5.0 and 10.0 mm and a 230 mm dia. plane electrode of Rogowski profile. The mixtures investigated had a SF_6 concentration of upto 1%.

EXPERIMENTAL TECHNIQUE

The studies were carried out in a cylindrical steel chamber with an internal diameter of 450 mm and gross volume of 0.12 m^3 can withstand pressures upto 1.4 MPa. It could be evacuated to about 10^{-3} torr. The impulse voltages were applied from a Marx type generator of 500 kV rating. Suitable precautions were taken to see that the high pressure chamber was dry and the electrode arrangement facilitate accurate gap separations measured using an externally mounted dial gauge to an accuracy better than $\pm$ 0.1 %. All the electrodes used were made of stainless steel and were polished and cleaned. High purity (99.5 %) N_2 and SF_6 gases were used. For preparing the gas mixtures, the gas which formed the smaller percentage (SF_6) was first let in to partial pressure corresponding to the required total pressure. The total pressure was then slowly built up by admitting the other component (N_2). Sufficient time was allowed to the gas to mix thoroughly. The gas pressures were measured to an accuracy better than $\pm$ 0.5 %.

The lightning impulse 50 % breakdown voltages (V_{50}), of both positive and negative polarity were determined using the conventional step-by-step method as well as Bakken's method [4] w2ith the application of 20 repeated impulses of the same polarity at each voltage level. It is estimated that the absolute accuracy in V_{50} measurements is $\pm$ 5 %, while the relative accuracy is about $\pm$ 3 %.

RESULTS

Investigations were carried out to determine the breakdown characteristics of SF_6 / N_2 mixtures, with SF_6 content upto 20 %, using standard lightning impulse (1.2 / 50 μs) voltages, when the non-uniformity of the electric field was varied over a wide range. The electrode arrangement consisted of hemispherically capped rods of diameter $D \cong 0.1$ (sharp needle), 0.8, 1.5, 3.0, 5.0, 10.0 and 20 mm. The results of the study for both the positive and negative polarities are shown in figs. 1 and 2. Fig. 1 represents data for 1 % SF_6 / 99 % N_2 mixtures for positive polarity for different field non-uniformities (rod diameters from 0.1 mm to 10 mm). Each set of curves represents the breakdown voltage of a given electrode configuration and gap spacing as a function of gas pressure. Similarly, in fig. 2 data are shown for SF_6 concentrations varying from 0 % (pure N_2) to 20 % with a rod diameter of 5 mm for different gap spacings upto 80 mm.

It is seen from fig. 1 that as the field non-uniformity is decreased in increasing rod diameter from 0.1 to 10 mm, the width of the corona stabilization region gets progressively reduced and is hardly seen at rod diameters greater than 5 mm. Similar results were also observed when the applied voltage was of negative polarity. Also the critical pressures at which the maxima are seen reduce with increasing rod diameter. Further, the breakdown is corona free above the critical pressure, which depends on the rod diameter, gap spacing and gas pressure. The results clearly indicate that the breakdown of positive rod-plane gaps is corona stabilized in the low pressure region (0.1 to 0.3 MPa) for rod diameters upto 5.0 mm. At higher pressures (> 0.3 MPa) the breakdown phenomena appears to be corona free. Under negative polarity the corona stabilization is not very distinct and can be seen only over a limited range of pressures for gaps lower than 40 mm. The width of the corona stabilization region is much smaller for the positive and rod-plane gaps as compared to the negative ones, as was also observed by Malik and Qureshi [5]. Consequently the critical pressure for the negative rod-plane gaps is higher than that observed for the positive ones (see fig. 5.7). Furthermore, the breakdown voltage in 1 % SF_6 / 99 % N_2 mixtures studied using negative gaps is higher than those for the positive rod-plane gaps at the low pressure end of the corona stabilization region (i.e. at pressures lower than critical pressures) as well as higher pressures (> 0.3 MPa) where the breakdown is not corona stabilized and the negative V_{50} are substantially higher than the corresponding positive data. This is true over the entire range of gap distances and rod diameter investigated.

Fig. 2 shows the typical lightning breakdown voltages as a function of percentage concentration of SF_6. It is seen that under positive polarity, maxima in the characteristics occurs at 0.5 % SF_6 concentration at low pressures (< 0.3 MPa) while the same is seen at 20 % SF_6 at 0.5 MPa. On the other hand, under negative polarity the maxima occur at SF_6 concentrations greater than 10 % at all pressures. Similar observations were observed made at all gap spacings. Two effects were clearly seen. First, as the SF_6 content is reduced below 20 % the dielectric strength progressively reduces, decrease being more rapid at low pressures. Secondly, when SF_6 content is reduced below 1 % the breakdown strength of the mixtures increases upto 0.5 % of SF_6 content. Further, the negative V_{50} values of the mixtures are substantially higher than the corresponding V_{50} values.

PRACTICAL INTERPRETATION OF RESULTS

The data shown in fig. 1 imply that for a given gap length a safe voltage level exists below which no breakdown can occur irrespective of the value of the rod diameter and gas pressure. For example, under negative polarity at a pressure of 0.1 MPa the safe voltages at gaps of 10, 40 and 80 mm are 39, 74 and 130 kV respectively; corresponding to a critical rod diameter of 5 mm. Similar values can be obtained at other gap spacings. If the voltages higher than the safe voltages are chosen for operation, for example 66 kV for positive polarity at a gap of 40 mm, then a horizontal line drawn at these voltage levels in fig. 1 will give for each pressure two limiting points for the rod diameter. These limiting rod diameters are plotted against pressure as shown in fig. 3. These two curves drawn (A and B for positive polarity and A' and B' for negative polarity) through these points represent the upper and lower limits of a range of rod diameters with which breakdown will occur and is called the *forbidden zone"*. For values of the rod diameters outside the *forbidden zone* no breakdown will occur and these values of the rod diameter may be considered as safe values. It can be appreciated from fig. 3 that the use of pressures in the range of 0.4 - 0.5 MPa is advantageous since the *forbidden zones* are narrow over this pressure range than at over pressures. Similar observations were made by Azar and Comsa [5] under a.c. voltages.

CONCLUSIONS

The positive polarity impulse breakdown voltages in 1 % SF_6 / 99 % N_2 mixtures are higher than the corresponding pure N_2 values. This improvement is more pronounced under most divergent fields. The breakdown characteristics obtained as a function of varying field non-uniformity shown that for a given gap length a safe voltage level exists below which no breakdown can occur. There appear to exist a region called *forbidden zone* which gives the maximum and minimum limits to the rod diameter outside which breakdown will not occur.

REFERENCES

[1] Farish, O., (1978b), Proc. 1st Int. Symp. On Gaseous Dielectrics, Knoxville, Tenn., U.S.A., p. 60.
[2] Malik, N.H. and Qureshi, A.H., (1979a), IEEE Trans., EI-14, 1, p.1.
[3] Christophorou, L.H., Dale, S.J., (1987), Encyclopedia of Physical Science and Technology, Vol.4, p. 246.
[4] Bakken, H., (1967), IEEE Trans., PAS-86, p. 962.

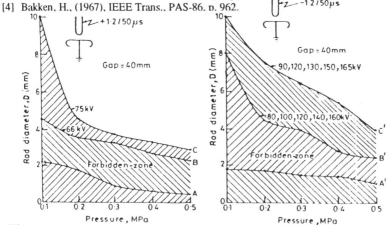

Figure 3. Forbidden- zone of the rod diameter against pressure for positive and negative rod plane gaps in SF_6/N_2 mixtures

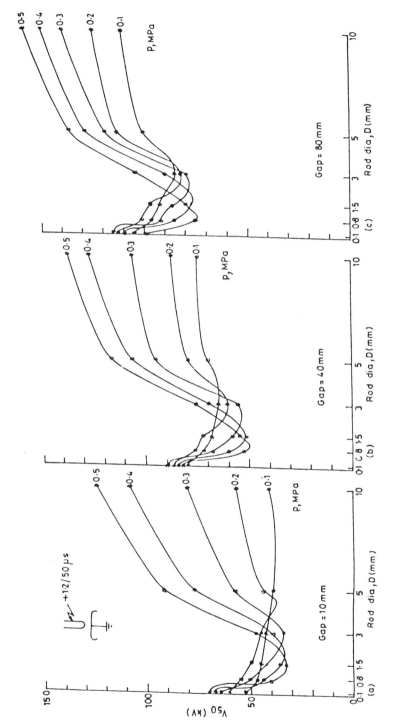

Figure 1. Impulse breakdown voltage V_{50} against rod diameter for a positive rod-plane gaps in SF_6/N_2 mixtures for different Pressures, p, at gaps of 10, 40, and 80 mm.

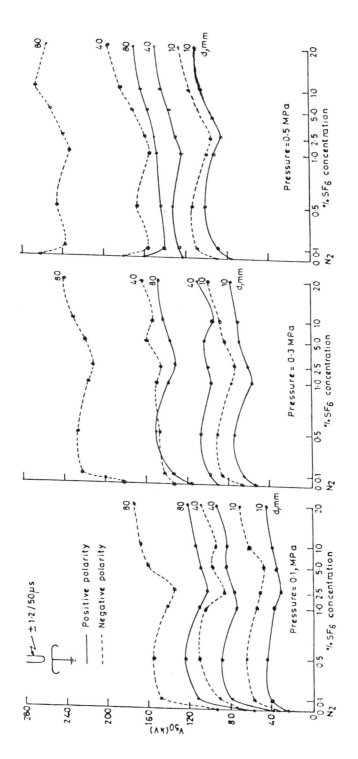

Figure 2. Impulse breakdown voltage V_{50} against % SF_6 concentration for positive and negative rod-plane gaps in SF_6/N_2 mixtures for different gaps, d, at different pressures, p.

DISCUSSION

E. KUFFEL: Very interesting observation of the "forbidden zone." Any suggestion for reasons of this phenomenon?

M. S. NAIDU: For very sharp points, the increased breakdown voltage may be due to low probability of availability of initiatory electron, attachment, and recombination processes. On the other hand, when the rod diameter is large, the field will be nearly uniform and the breakdown voltage increases with rod diameter. The forbidden zone occurs between the above two high and average field conditions.

H. HAMA: What is the optimum SF_6 content to improve the dielectric performance in the presence of metallic particles?

M. S. NAIDU: In the author's opinion it is $40\%SF_6/60\%N_2$ for insulation purposes.

B. HAMPTON: I wonder whether the apparently high breakdown voltage for a very sharp point could be due to the lower probability of finding an initiatory electron in the very small highly stressed region.

M. S. NAIDU: I agree that low probability of finding an initiatory electron can cause high breakdown voltages for a very sharp point. Further, negative ion-positive ion recombination as well as rapid removal of electrons due to attachment also contribute to the high breakdown voltage at a sharp point.

MACROSCOPIC MODELING OF INHOMOGENEOUS FIELD BREAKDOWN IN SF_6 UNDER LIGHTNING IMPULSE STRESS

H. N. Suresh[*], M. S. Naidu

Department of High Voltage Engineering
Indian Institute of Science
Bangalore - 560 012, INDIA

ABSTRACT

Gas Insulated Substations (GIS) have proven to be a viable solution for compact and environmentally clean substations. SF_6 gas has been widely used in GIS as insulating medium because of its excellent dielectric and arc quenching properties. During the course of their operation, GIS are subjected to Lightning Impulse (LI) and Very Fast Transient (VFT) over voltages which quite often are found to cause the SF_6 insulation failure. Experimental investigations in this regard have suggested that the breakdown of insulation is by the leader mechanism. The phenomena of leader inception and propagation leading to breakdown in SF_6 are currently being studied because of their relevance to the design of GIS. Attempts have been made in the recent past to macroscopically model the leader growth process for theoretical studies. The leader propagation is guided by the electric field at the leader tip, which inturn is dependent on the nature of voltage applied, electrode geometry and charge distributed in the leader. The charge build up process in the leader is governed by Resistive-Capacitive (R-C) transient fields. Modeling of breakdown in SF_6 based on field approach involves modeling the leader mechanism in SF_6 and computing the associated R-C transient field. In the present paper, an attempt made in that direction is discussed. Inhomogeneous field is realised by a hemisphereically tipped point-plane electrode system. A Laplace solver based on the principles of Charge Simulation Method (CSM) was coded and used in implementing the algorithm for the transient field computation. In order to reduce the computational burden without causing the numerical oscillations of the computed field quantities, a nonlinear numerical filter was employed. The final outcome of the work in terms of V-t characteristic for an inhomogeneous filed breakdown in SF_6 under LI stress is also presented.

---

[*]Senior Lecturer, Malnad College of Engineering, Hassan - 573 201, INDIA.
Currently, a Research student at Indian Institute of Science, Bangalore - 560 012, INDIA.

INTRODUCTION

SF$_6$ gas has very good insulating and arc quenching properties. In view of this, it is a popular dielectric in High Voltage apparatus like Gas Insulated Substation (GIS). Vast majority of the power components in a GIS are safely sealed in a clean environment of dry SF$_6$ gas. It is of interest to note that field non-uniformities would reduce the breakdown strength of SF$_6$ gas. This makes the GIS insulation very sensitive to field intensifications. At the same time, field non-uniformities are inevitable in practical systems due to surface roughness, floating metallic particles and fixed particles on electrodes and insulator surfaces. The presence of such defects may lead to SF$_6$ insulation failure due to overvoltage surges during the course of operation of GIS. Experimental investigations in this regard have suggested that the breakdown of SF$_6$ insulation is by the leader mechanism. Attempts have been made in the recent past to macroscopically model the leader growth process[1-7]. Modeling of breakdown in SF$_6$ under inhomogeneous fields involves modeling the leader mechanism in SF$_6$ and computing the associated R-C transient field.

Earlier works in this area have employed a lumped R-C circuit model for the purpose of transient field computation[3-6]. In these models streamer is represented by Resistance and Capacitance in parallel. Electrostatic coupling to ground is approximated by a single capacitance connecting streamer head to ground. With the lumped circuit model, it is not possible to account for the electrostatic coupling of leader and streamer with each other and with respect to the ground. A distributed circuit approach looks like an alternative. However, even with the distributed circuit approach, it is nearly impossible to compute the various capacitances represented in the equivalent ladder network. Therefore, it would be appropriate to employ field theoretical approach to such studies. In the present work a computational tool to asses the transient field quantities concerning such inhomogeneous field breakdown of SF$_6$ insulation is attempted. The concerned modeling technique, algorithm employed and results obtained in terms of V-t characteristic under LI stress are presented in this paper.

MODELING

Inhomogeneous field is realised by the electrode geometry chosen consisting of a high voltage hemispherically tipped needle electrode and a grounded plane electrode as shown in Fig.1. The needle length is 15 mm, hemispherical tip radius is 0.25 mm and electrode gap distance is 85mm. The dimensions employed are exactly same as the one mentioned in the earlier literature[6] for which experimental data in respect of insulation breakdown under LI stress are available. This would help in validating the simulation data against the measured data for a similar geometry.

The high voltage electrode is modeled using ring charges over major portion of the electrode and a point charge in the hemispherical tip portion. The electrostatic field in the electrode gap is governed by the Laplace equation

$$\nabla^2 \phi = 0 \qquad (1)$$

To solve the Laplace equation a computer code using 'C' language was written on the principles of CSM[8]. The field computations were initially carried out in a discharge free gap to asses the field non-uniformity factor. The calculation of non-uniformity factor for the given gap was cross checked with the one determined using empirical equations suggested in the literature[9]. The streamer/leader propagation is not only guided by the electric field at the streamer/leader tip, but also by the charge distributed in the streamer/leader. The charge build up process in the streamer/leader is governed by the resistance of the streamer/leader and the distributed capacitance of the gap.

The boundary conditions that are valid along the streamer/leader segments are

$$E_{1t} = E_{2t} \tag{2}$$

$$D_{n1} - D_{n2} = \sigma_s \tag{3}$$

$$J_{1n} - J_{2n} + \nabla_s \cdot J_s = -\partial \sigma_s / \partial t \tag{4}$$

The boundary condition (4) governs the time dependence of the field quantities. It suggests that at any instant, difference in normal components of conduction current densities and divergence of surface current density of the streamer/leader will be compensated by the displacement current densities[10]. Such class of fields are classified as *Resistive-Capacitive (R-C) fields*. For the present case, because of thin cross section of the streamer/leader, equation (4) can be modified as

$$J_{1n} - J_{2n} + \partial J_s / \partial z = -\partial \sigma_s / \partial t \tag{5}$$

Transient field computational studies associated with the streamer/leader propagation process involves modeling the streamer/leader and computation of associated R-C transient field. The leader is modeled using cylindrical charges placed axially. With the assumption of a vertical leader, the problem becomes axi-symmetric and hence the discretisation needs to be carried out along the axis of the high voltage electrode only. The leader is sectionalised into definite number of cylindrical charge segments each with a linearly varying charge density. In z-direction the variation in charge density for each segment is

$$\lambda = \lambda_2 (z-z_1)/(z_2-z_1) + \lambda_1 (z_2-z)/(z_2-z_1) \tag{6}$$

One would appreciate that it is necessary to cross check the authenticity of electric field calculation routines used to evaluate the potential and electric field at the tip of each of these segments. In order to achieve this, the electric field gradients of these leader segments were initially deliberately fixed. The tip potential/electric field gradients at the end of each of these segments was determined using the Laplace solver. The maximum error in computation of such tip gradients was 0.02%. The R-C transient algorithm mentioned in the next section is then applied to this leader to study the temporal variations of field quantities.

ALGORITHM

The R-C transient algorithm has been successfully developed and employed for field transient studies[11] and is proved to be very useful. Under the specific excitation and the given boundary conditions, the steps followed for the transient field computation are:

1. At zero time step t_0, the problem is solved as an electrostatic one i.e., by giving the permittivity and initial charge distribution (if any) as inputs to the Poisson's solver. The resulting current is assumed to be approximately constant over a short interval of time Δt. With this assumption, charge accumulated in that interval is computed using the relation

$$\lambda_z(t+\Delta t) = \lambda_z(t) - \Delta t \, [\, \partial J_z / \partial z \,] \tag{7}$$

The charge computed is used in the field calculation in the next time step

2. The charge densities computed for the time steps t_n from the time steps t_{n-1} are used for the field calculation (as explained in step **1**) at time steps t_n ($n = 0,1,2,3,4,5$ ......)

3. Using the field computed at time step t_n, the charge accumulated till time t_{n+1} is computed using the equation in step **1**

4. Steps **2** and **3** are repeated to cover the desired time duration.

RESULTS AND DISCUSSION

The model for the leader propagation in SF_6 under inhomogeneous field for the applied LI stress considers the discharge development to start with the formation of a streamer. When the electric field at the tip of the electrode or leader exceeds a critical value (E/p_{cr} = 89.6 kV/cm-bar) a streamer is initiated. The extent to which a streamer may propagate will be limited by the equation

$$V = -\int_0^d E.ds \tag{8}$$

where, V is the applied voltage and d is the gap distance.

Streamer propagation cannot be sustained in SF_6 with an electric field less than E/p_{cr}[7]. If the electric field in the streamer trail is E/p_{cr}, then the maximum streamer extension l is given by

$$V = E/p_{cr} * l \tag{9}$$

For the streamer to leader transition, a definite energy input W_{diso} is necessary to dissociate the SF_6 gas.

$$W_{diso} = \Delta h_{diso} \cdot (\rho/p_0) \cdot p_0 \cdot \pi \cdot r_0^2 \cdot l_{lead} \tag{10}$$

It is assumed that the leader can be described as a cylindrical segment with an initial channel radius r_0. The energy required to heat up the gas is provided by the discharge current.

Time to breakdown for different peak values of Standard Lightning Impulse voltage applied to the high voltage electrode were computed. V-t characteristic thereby obtained is shown in Fig. 2. The simulation data is validated against the measured data available in the literature[6] for a similar geometry. The results are in good agreement as can be observed from Fig. 2.

In the present work, computation of the transient field quantities associated with the streamer/leader were carried out using the algorithm described in the earlier section. For obtaining good accuracy and temporal resolution, it is necessary to have small time steps. This will be at the cost of larger computational time and storage. An approximate time step was initially chosen and a study was carried out with one small and one large time step. The signs of the interface charge densities computed are compared. Sign reversal takes place only if the second time step chosen is large. By trial and error, the second time step chosen was reduced to obtain no sign change, thereby getting an idea of the time step to be taken. In the present study, a time step of 10 pico seconds was employed.

Also, digital filtering technique[12] was incorporated to reduce the numerical oscillations of computed fields quantities. In the present study, time step chosen under the usage of filter is two order high (without causing instability) in comparison to that of without filter. Higher the time step chosen, lower is the computational time required. Thus, use of filter also brings down the computational time by a good order. Smooth build up of potential and field at the leader tip with or without the inclusion of the filter was possible, indicating the versatility of the algorithm.

CONCLUSION

In summary,

1. The leader development in SF_6 leading to inhomogeneous field breakdown under LI stress is macroscopically modeled. The important aspect of transient field computation involved in the process is made possible by the R-C transient algorithm. The algorithm employed efficiently traces the field quantities at the leader tip over space and time.

2. The incorporated technique is inherently complete in the sense that the streamer capacitance or streamer/leader head to ground capacitance is not considered as lumped equivalent for the streamer/leader sections. The suggested field approach doesnot give room to the omission of any of the associated distributed capacitances.

3. The filtering algorithm employed would help in the suppression of the numerical oscillations in the computation of the leader field quantities. Using the filter, a higher time step can be employed without causing numerical instability in the computation. This would naturally bring down the total computational time.

4. The results obtained in terms of V-t characteristic are in good agreement with the experimental data under LI stress available in the literature for a similar geometry.

REFERENCES

[1] I.D. Chalmers, I. Gallimberti, A. Gibert and O. Farish, "The development of electrical leader discharges in a Point-plane gap in SF_6", Proc. Royal Soc. London, **A 412**, pp 285 - 308, 1987.

[2] L. Niemeyer, L. Ullrich and N. Wiegart, "The mechanism of leader breakdown in electronegative gases", IEEE Transactions on EI, Vol. 24, No.2, pp 309 -324, April 1989.

[3] Heinrich Hiesinger, "The calculation of leader propagation in Point/plane gaps under very fast transient stress", Proc. VI International Symposium on Gaseous Dielectrics, Plenium Press, New York, pp 129 - 135, 1991.

[4] B. Heers, "Leader propagation in inhomogeneous SF_6 gaps under VFT stress with critical frequencies", Ninth International Symposium on High Voltage Engineering, Austria, pp 2267-1 to 2267-4, 1995.

[5] Dietmar Buchner, "Discharge development in SF_6 in case of composite voltage stress", Gaseous Dielectrics VII, Ed. by L.G. Christophorou and D.R. James, Plenium Press, New York, pp 291 - 297, 1994.

[6] Dietmar Buchner, "The calculation of Leader propagation in point/plane gaps under LI and FTO stress", Eigth International Symposium on High Voltage Engineering, Yokohama, Japan, paper 30.01, pp 255 - 258, 1993.

[7] R. Morrow, "Properties of streamers and streamer channels in ", Physical Review - A, Vol. 35, No. 4, pp 1778 - 1785, 1987.

[8] Nazar H. Malik, "A Review of the Charge Simulation Method and its applications", IEEE Transactions on EI, Vol. 24, No. 1, pp 3 - 20, Feb. 1989.

[9] Y. Qiu, "Simple expression of field Non-uniformity factor for Hemispherically capped Rod-plane gaps", IEEE Transactions on EI, Vol. 21, No. 4, pp 673 - 675, August 1986.

[10] H. H. Woodson and J. R. Melcher, "*Electromechanical dynamics*", part 1- Discrete systems, Wiley, pp 277 - 280, 1968.

[11] Udaya Kumar, G.R. Nagabhushana, "Solution of Capacitive - Resistive transient fields", Ninth International Symposium on High Voltage Engineering, Austria, pp 8351-1 to 8351-4, 1995.

[12] W. Shyy, M. H. chen, R. Mittal and H. S. Udaya Kumar, "On the suppression of Numerical oscillations using a Non-linear filter", Journal of Computational Physics, Vol. 102, pp 49 - 62, 1992.

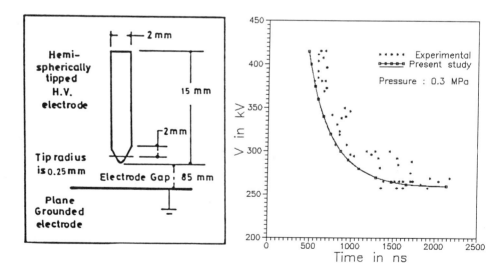

Fig. 1. Electrode geometry   Fig. 2. V-t characteristic

A COMPARATIVE EVALUATION OF COST / BENEFIT ANALYSIS WITH LIGHTNING IMPULSE BREAKDOWN STRENGTH OF BINARY AND TERNARY GAS MIXTURES CONTAINING SF_6 IN NON-UNIFORM FIELD GAPS

D. Raghavender

Dept. of Electrical Engg.
S.S.G.M. College of Engg.
SHEGAON-444203 (MS)
INDIA

M.S. Naidu

Dept. of High Voltage Engg.
Indian Institute of Science
Bangalore – 560 012.
INDIA

ABSTRACT

SF_6 is still expensive than other commonly used dielectrics, and is very sensitive to local field enhancements which are inevitable in engineering applications because of the electrode surface defects, sharp edges and presence of free conducting particles. In the most commonly used pressure range, SF_6 retains only one half of its potential dielectric strength for practical electrode finishes. A properly selected additive in ternary gas mixtures can effectively enhance the corona stabilization effect of the binary mixtures (SF_6 / N_2 and SF_6 / CO_2) in highly non uniform fields and thus increase its breakdown voltage within the corona stabilization region.

This paper describes the results of lightning impulse (1.2/ 50 µs) breakdown tests (up to 300 kV), carried out using a rod-plane geometry (rod dia: 5 mm and 230 mm dia plane electrode of Rogowski profile). The gap distances between electrodes and pressure range were varied from 5 to 80 mm and 0.1 to 0.5 MPa respectively. The binary gas mixtures (SF_6 / N_2 and SF_6 / CO_2) and ternary gas mixtures (SF_6 / N_2 / CCl_2F_2 and SF_6 / CO_2 / CCl_2F_2) are used for investigation.

The investigated ternary gas mixtures (1 % SF_6 / 39 % CCl_2F_2 / 60 % N_2) and (20 % SF_6 / 20 % CCl_2F_2 / 60 % N_2) contribute to the cost ratio by only about 0.12 and 0.27 respectively, at the same time giving a dielectric strength of about 0.83 to 0.86 relative to pure SF_6 (1.0). A 30 % SF_6 / 10 % CCl_2F_2 / 60 % CO_2 mixtures cost only 33 % of cost of pure SF_6 gas but gives typically 110 % to 170 % positive impulse breakdown strength, and 90 % to 110 % negative impulse breakdown strength as compared to that of pure SF_6 gas depending on the mixture pressure range (1 to 5 bar) investigated.

The ternary mixtures studied can possibly used in CGTL and GIS apparatus since they posses comparable or superior dielectric strength than pure SF_6 at reduced cost.

Gaseous Dielectrics VIII, Edited by Christophorou
and Olthoff, Plenum Press, New York, 1998

INTRODUCTION

SF_6 gas is widely used in High voltage technology, particularly in substation components insulated with compressed gas (GIS). However, SF_6 is also extremely sensitive to field non-uniformity within the GIS. In addition, in recent years, concern has been expressed over the possible impact of SF_6 gas on global warming. Many unitary, binary, ternary gas mixtures have been studied with a new to adopt for practical applications [1-3]. However, it is well known that SF_6 is compared to other gases and therefore the cost /benefit analysis of the various gas mixtures will be valuable for practicing engineers.

Over the years, extensive study has been carried out in various gas mixtures to measure 50 % breakdown voltages using Lightning voltages under non-uniform fields. Results of the some of the recent experimental studies are presented.

EXPERIMENTAL APPARATUS AND PROCEDURE

The experiments are performed using a Marx-type, 10 stage impulse voltage generator of 500 kV rating. A cylindrical mild steel pressure chamber having a volume of 0.12 m^3, and fitted with a high voltage bushing was used. The electrode arrangement consists of a 5 mm dia stainless steel rod of hemispherical termination and a 230 mm dia brass plane of Rogowski profile. The gap distance between the electrode was 20 mm, measured to an accuracy of better than ± 0.1 %. SF_6/CCl_2F_2/CO_2 gas mixture pressure used were in the range of 0.1 to 0.5 MPa (1 to 5 bar). The concentration of SF_6, CCl_2F_2 and CO_2 contents in the mixture was as follows:

(a) SF_6 (1 to 20%), CCl_2F_2 (20% fixed) and the rest CO_2 (79% to 60%) and

(b) SF_6 (1 to 30%), CCl_2F_2 (39% to 60%) and CO_2 (60% fixed)

The gases (SF_6, CCl_2F_2 and CO_2) of the cylinder grade purity (99.5%) were used in the study. Gas mixtures were prepared using the method of partial pressures to an accuracy of $\pm 0.2\%$. The 50% breakdown voltages (V_{50}) have been measured using a statistical methods, namely, the step-by-step method and the Bakken's method. The values obtained by both the methods were in very good agreement ($\pm 3\%$). These variations however do not indicate any uncertainty in V_{50}, during this study, the V_{50} could always be reproduced to within $\pm 2\%$. In evaluating cost analysis, the cost per volume of SF_6 (1.0) for CO_2 and CCl_2F_2 have been calculated as 0.007 and 0.21 respectively.

UNITARY GASES

Many gases have been investigated for their dielectric and other properties [1].

The following table shows the d.c. breakdown strength of such gases.

Table-1

Gas	Relative breakdown strength
N_2	1.0
CO_2	1.0
SF_6	2.4
C_3F_8	2.2
C_4F_8	3.5 – 4.0
C_4F_6	4.2 – 5.0

Although some of the above gases have superior dielectric strength than SF_6, they do not meet the other requirements such as chemical stability, high heat transfer capability and high arc quenching ability etc.. They also contribute to ozone depletion and global warming.

BINARY MIXTURES

Although extensive work has been carried out using SF_6 in mixtures with perfluorocarbons (see Table-1), due to their inherent disadvantages mixtures of SF_6 with N_2, Air or CO_2 are of interest. The following table shows the breakdown strength of some binary mixtures [4]

Table-2

Gas mixture	Dielectric strength* relative to SF_6
SF_6	1.0
0.1% SF_6 + 99.9% N_2	0.90
1% SF_6 + 99% N_2	0.90
5% SF_6 + 95% N_2	0.90
10% SF_6 + 90% N_2	0.88
20% SF_6 + 80% N_2	0.86
40% SF_6 + 60% N_2	0.85

*Note: At a pressure 2 bar, using impulse voltages

TERNARY GAS MIXTURES

Work has also been carried out with $SF_6/CO_2/N_2$ mixtures and also SF_6/Freon/ N_2 mixtures under various conditions of voltage for different mixture concentrations as shown below.

Table –3

Gas mixture	Dielectric strength relative to SF_6
SF_6	1.0
0.1% SF_6 + 20% freon + 79% N_2	0.74 - 0.84
10% SF_6 + 20% freon + 70% N_2	0.87 - 0.94
20% SF_6 + 20% freon + 60% N_2	1.0 - 1.09
10% SF_6 + 30% CO_2 + 60% N_2	0.85 - 0.88
20% SF_6 + 20% CO_2 + 60% N_2	0.85 - 0.86

BREAKDOWN IN SF_6/CCl_2F_2/CO_2 MIXTURES

The data of SF_6/CCl_2F_2/CO_2 mixtures (1/20/79, 5/20/75, 10/20/70 and 20/20/60) show that the positive breakdown voltages of 20% SF_6/ 20% CCl_2F_2/ 60% CO_2 mixture is higher than those of pure CO_2, pure SF_6 and 20% CCl_2F_2/ 80% CO_2 mixture over the complete pressure range studied. On the other hand, the negative breakdown voltage of pure SF_6 is higher than that of all the mixtures studied. A similar trend in voltage – pressure characteristics can also be seen in SF_6/CCl_2F_2/CO_2 mixtures (1/39/60, 5/35/60, 10/30/60, 20/20/60 and 30/10/60) with different compositions investigated. It was interesting to note that at higher pressures (> 0.4 MPa), the positive breakdown voltages of pure SF_6 are much more lower than those of binary mixtures (20 % CCl_2F_2 / 80 % CO_2) and all the ternary mixtures (SF_6/CCl_2F_2/CO_2) studied. It was seen that the dielectric strength of two ternary mixtures 20 % SF_6 / 20 % CCl_2F_2/ 60 % CO_2 and 30 % SF_6 / 10 % CCl_2F_2/ 60 % CO_2 were found superior to that of pure SF_6, especially under positive polarity in the pressure range of 0.1 to 0.5 MPa.

COST / BENEFIRT ANALYSIS OF SF_6/CCl_2F_2/CO_2 MIXTURES

In order to minimize the total cost of the gas in Compressed Gas Insulated System (GIS), CCl_2F_2 is also being tried as a third component gas with SF_6, CO_2, in binary, ternary mixtures as a promising candidate for practical use.

In view of the above consideration, a cost / benefit analysis was carried out from the various mixtures investigated in the present study. Fig. 1 and 2 present the positive and negative dielectric strengths, V_{50}^R (V_{50} of mixture / V_{50} of pure SF_6) with cost ratio C_R for different ternary mixtures studied at a gap of 20 mm. In these figures the results are shown in the form of histograms in which, for a given mixture the data shown separately at five pressures investigated.

It can be clearly seen from the fig. 1 (a to h) which show the positive relative dielectric strength ratios to lie between 0.73 to 0.1 in the pressure range 0.1 to 0.5 MPa, while at higher pressures (> 0.4 MPa) these ratios reach maximum levels even higher than that for SF_6 (1.0) to lie in the range 1.0 to 1.7. The negative relative breakdown strength $V_{50}{}^R$ of all the ternary mixture lie between 0.72 to 1.0 over all the pressure range investigated (fig. 2). The cost ratios of all the eight sets of mixtures vary between 0.06 to 0.33 relative to pure SF_6 (1.0).The increase in positive dielectric strength caused by adding CO_2 to SF_6 and CCl_2F_2 can be attributed to the effective slowing down of electrons via its strong low lying negative ions states [5]. Also CCl_2F_2 in combination with SF_6 and CO_2 in all the SF_6/ CCl_2F_2/ CO_2 mixtures was found very effective in suppressing formation of free carbon, which was examined by inspecting the pressure chamber, electrodes, etc., after breakdown.

REFERENCES

[1] D. R. James, L. G. Christophorou et al, Proc. Int.Symp. on Gaseous Dielectrics, Knoxvelle, Tennessee, U.S.A, March 6-8, 1978.
[2] T. V. Babu Rajendran, M. S. Naidu et al, Proc. IEE, Part A, Vol.130, p.134-138, 1983.
[3] R. Y. Pai and L. G. Christophorou et al (1980), Proc. VII Int. Conf. On Gas Discharges, London, p. 232.
[4] P. T.Medeiros, K. D.Srivastava et al, Proc. IVth Int.Symp. on High Voltage Engineering, Athens, Paper 33-03, 1983.
[5] L. G. Christophorou, et. al. (1982), "Applied Atomic Collision Physics", Academic Press, N.Y., p. 87.

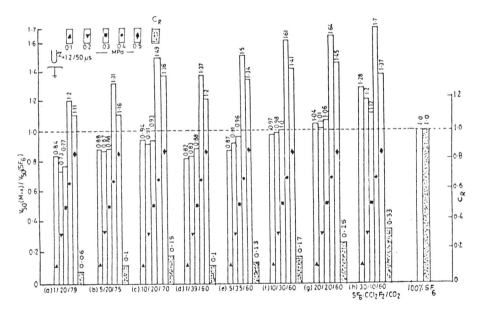

Figure 1. Histograms of relative Impulse breakdown strength $V_{50}(mix.)/V_{50}(SF_6)$ of positve rod-plane gaps with cost ratio C_R for different ternary gas mixtures ($SF_6/CCl_2F_2/CO_2$) for different pressures, p, at a gap of 20 mm

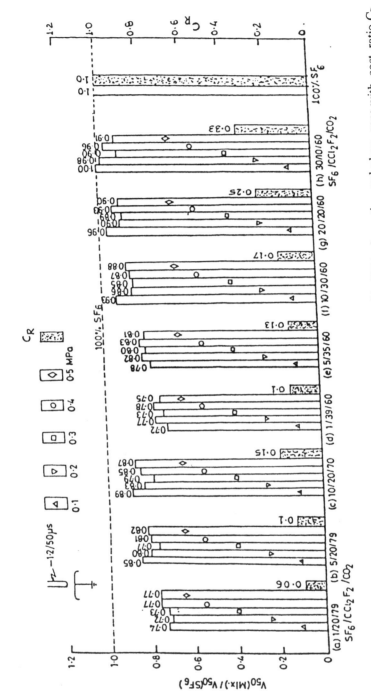

Figure 2. Histograms of relative Impulse breakdown strength $V_{50}(mix.)/V_{50}(SF_6)$ of negative rod-plane gaps with cost ratio C_R for different ternary gas mixtures ($SF_6/CCl_2F_2/CO_2$) for different pressures, p, at a gap of 20 mm

SIMULATION OF ELECTRICAL CIRCUIT OF PLASMA OPENING SWITCH USING PSPICE

Zhengzhong Zeng,[1,2] Yuchang Qiu,[1] and Aici Qiu[2]

[1]High Voltage Division, Xi'an Jiaotong University, Xi'an, 710049
[2]Northwest Institute of Nuclear Technology, P.O. Box 69-13, Xi'an, 710024
  People's Republic of China

INTRODUCTION

The plasma opening switch(POS) is a promising pulsed power circuit switching device which uses an injected plasma, embedded in a low density gaseous dielectric background, first to conduct high currents for a period ranging from tens of nanoseconds up to several microseconds and then to convert them into a load by means of some conduction and opening mechanisms.[1,2,3]

The POS is a strongly nonlinear device from the point of view of electrical circuit characteristics, and these characteristics depend tightly upon the electrical circuit containing the POS. Therefore, detailed electrical circuit simulations are necessary in designing POS-containing pulsed power systems for the purpose of optimizing the parameters and performances of the systems.

Conventional treatment for electrical simulations of POS-containing pulsed power systems is to solve the coupled differential equations consisting of the equations of state of all the electrical elements except the POS and the differential equations describing the POS physics in terms of the voltage across and the current through the switch. This method represents a long course that is complicated, easy to make mistakes and lack of flexibility.

In this paper, the simulation has been done on the basis of the combination of the POS model with the widely used electrical circuit simulating software package, PSPICE. A PSPICE input file module for the POS was built up for being easily combined with other arbitrary electrical circuit elements. Simulated results are generally in agreement with experimentally obtained data, with the erosion model overestimating the rate of rise and the value in opened phase of the switch impedance. Furthermore, the simulation indicates strong interac-tion between the switch and loads. The method proposed in this paper gives great simplicity, flexibility and efficiency to the preliminary design of pulsed power systems containing POS and to the analyses of nonlinear switch behaviors and interactions between POS and loads as well as the surrounding circuit system.

PHYSICAL EQUATIONS ON POS FOR ELECTRICAL SIMULATION

Erosion Model

The erosion model is considered to be suitable for explanation of the POS operation in the case of short conduction time(tens of nanoseconds).[1] In this model the POS operation is divided into four consequential phases, i.e., the conduction, erosion, enhanced erosion and magnetic insulation phases. From the point of view of electrical simulation, the physical equations for each of these phases can be arranged to build a module to be coupled with the equations of state of the surrounding circuit in a complete PSPICE input file.

The Conduction Phase. In this phase, the switch can be effectively taken as a short circuit until the current flowing through the switch reaches

$$I_s = e n_i v_d (1836 Z N_a)^{1/2} A \quad \quad (1)$$

where, e, n_i, v_d, Z, N_a is electron charge, plasma density, ion drift speed, ion charge state, ion mass, respectively, $A = 2\pi r_c \ell$ is the cathode area through which the switch current flows, with r_c being the cathode radius, ℓ the switch length. M.K.S. system of units is used all over the paper.

The Erosion Phase. A bipolar space-charge-limited current flows during this phase in the switch. The switch current is represented as

$$i_s = 4.34 \times 10^{-6} A u_s^{3/2} / D^2 \quad \quad (2)$$

where, $u_s = u_s(t)$ is the voltage across the switch, $D = D(t)$ is the thickness of plasma sheath at the cathode and satisfies

$$dD/dt = (1836 Z N_a)^{-1/2} / (e n_i A) \cdot i_s - v_d \quad \quad (3)$$

The Enhanced Erosion Phase. During this phase the switch current becomes

$$i_s = 1.36 \times 10^4 (\gamma^2 - 1)^{1/2} r_c / D \quad \quad (4)$$

where, $\gamma = 1 + e|u_s|/(m_e c^2) = 1 + |u_s|/5.11 \times 10^5$ is the ratio of electron mass to its rest mass, c is the speed of light in vacuum, m_e the mass of electron, and D is determined by

$$d\left(\frac{1}{2} \cdot D^2\right)/dt = ((\gamma + 1)/(918 Z N_a))^{1/2} \ell / (e n_i A) \cdot i_s \quad \quad (5)$$

The Magnetic Insulation Phase. A simple Child-Langmuir space-charge-limited switch current is given in this phase as

$$i_s = 5.45 \times 10^{-8} (Z/N_a)^{1/2} A u_s^{3/2} / D_o^2 \quad \quad (6)$$

where, D_o is the typical thickness of the plasma sheath at the end of the enhanced erosion phase.

Snowplow-Magnetic Insulation/Effective Gap Picture

According to the snowplow model,[2,3] the POS conduction phase sustains for a time specified by

$$T = 6^{1/4}\left(IG/\dot{I}\right)^{1/2} \rho^{1/4} \quad\quad\quad\quad\quad (7)$$

where, $I = \dot{I}T$ is the switch current at the end of the conduction phase, with $\dot{I}$ being the average rate of rise of the switch current, ρ the mass density of injected switch plasma taken to be constant, G a geometrical factor specified, for a coaxial switch geometry, by

$$G^2 = 10^7 \pi\left(r_a^2 - r_c^2\right)/\ln(r_a/r_c) \quad\quad\quad\quad\quad (8)$$

where r_a is the switch anode radius.

Then the switch opens and the opening is governed by[3]

$$i_s \propto \left(\gamma^2 - 1\right)^{1/2} r_c/D \approx u_s \cdot r_c \quad\quad\text{for}\quad u_s > 1\text{ MV} \quad\quad (9)$$

that in fact is the same as equation (4).

Time-dependent plasma sheath(gap) thickness, D, can be set to get good agreement between simulated and experimental data. This thickness is referred to "effective gap".

PSPICE MODULE FOR THE POS AS A NONLINEAR RESISTANCE[4,5]

Using the controlled voltage and current sources which can be arbitrarily expressed in terms of voltage, current and time provided by PSPICE, the POS can be taken as a nonlinear resistance whose value varies with time as indicated by the equations listed above.

Integration can be done by blocking R-C circuits with their output voltage taken across the capacitance. R-L circuits can also be used to do integration.

Although PSPICE(Evaluation Version 5.3) does not directly provide time-dependent switch function, this function can be fulfilled by the combination of pure time-dependent source and voltage/current controlled switch function.

Based on these exploitation of PSPICE, the PSPICE input file module of the POS can be built. An example of this module for the erosion model is as follows.

```
Vd2    4     5
Gs     5     0     value={V(701)*I(Vd2)+V(201)*V(501)*1e8+ V(301)*
+                  V(601)*1.36e4+V(801)*V(901)}
Vts1   201   0     pwl(0  0  60n  0  61n  1  70n  1  71n  0)
Rts1   201   0     1e6
Vts2   301   0     pwl(0  0  70n  0  71n  1  80n  1  81n  0)
Rts2   301   0     1e6
```

Vts3	701	0	pwl(0 1 60n 1 61n 0)
Rts3	701	0	1e6
Vts4	801	0	pwl(0 0 80n 0 81n 1)
Rts4	801	0	1e6
Es1	501	0	value={2.727e-5*r_c*ℓ*V(5)*sqrt(V(5)) /V(103)/V(103)}
Res1	501	0	1e6
Es2	601	0	value={sqrt((1+V(5)/E0)*(1+V(5)/E0)-1)*r_c/V(121)}
Res2	601	0	1e6
Es7	901	0	value={3.424e-7* r_c*ℓ*sqrt(Z/N_a)*V(5)*sqrt(V(5))/D_o/D_o}
Res7	901	0	1e6
Es3	101	0	value={2.319e-3/(r_c*ℓ*n_i)*sqrt(1/Z/N_a)*I(Vd2)-v_d}
S1	101	102	151 0 ms1
.model	ms1		Vswitch Voff=40 Von=50 Roff=1e9 Ron=1m
Res3	102	103	10k
Cs1	103	0	10n
Es4	111	0	value={3.279e-3/r_c/n_i*sqrt(1/Z/N_a)*sqrt(2+V(5)/E0)*I(Vd2)}
S2	111	112	151 0 ms2
.model	ms2		Vswitch Voff=50 Von=60 Roff=1e9 Ron=1m
Res4	112	113	10k
Cs2	113	0	10n
Es5	121	0	Value={sqrt(2*(V(113))*1e-4)}
Res5	121	0	1e6
Es6	151	0	value={time*1e9}
Res6	151	0	1e6

For the snowplow-magnetic insulation/effective gap picture, similar module can be formed based on this one with necessary modification.

RESULTS AND DISCUSSIONS

This module can be easily coupled with the equations of state of the surrounding circuit to form a complete PSPICE input file for simulation of electrical circuit of POS. Figure 1 is a lumped equivalent circuit of ACE 4 driver at Maxwell Technologies in San Diego, CA.[3]

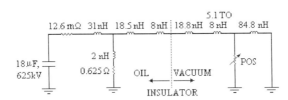

Figure 1. Lumped equivalent circuit of ACE 4 pulsed power driver,
© 1998 IEEE.

For this POS circuit under operational conditions specified in Table 1, if the snowplow-effective gap picture is used, simulated generator and load current will be as the curves shown in Figure 2, compared with experimental waveforms in Figure 3. Simulated effective gap is plotted in Figure 4.

Table 1. A group of operational parameters of ACE 4.

n_i (m^{-3})	ρ (kg/m^{-3})	$\dot{I}$ (A/s)	I (MA)	T (µs)	r_c (m)	r_a (m)	l (m)	G (m^3/H)
4×10^{21}	80×10^{-6}	6.2×10^{12}	5.3	0.85	0.2	0.23	0.12	1.7×10^3

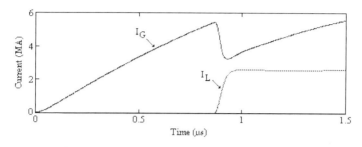

Figure 2. Simulated generator current I_G and load current I_L of ACE 4 power driver.

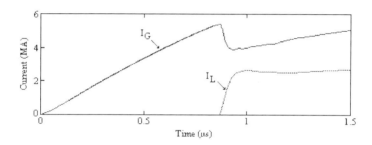

Figure 3. Experimental generator current I_G and load current I_L of ACE 4 power driver, © 1998 IEEE.

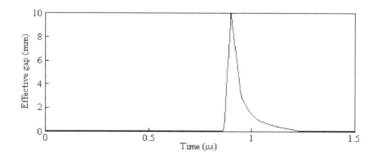

Figure 4. Simulated effective gap (plasma sheath thickness) of the POS of ACE 4 driver.

If the erosion model is used, one will obtain simulated generator and load current for Gamble II of NRL as shown in Figure 5. The results are in agreement with the simulated curves[6] and show much greater rate of rise and value in opened phase of the switch impedance than experimental measurements.[6]

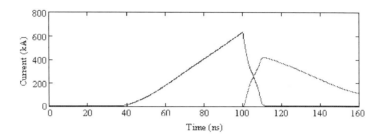

Figure 5. Simulated generator current I_G and load current I_L for Gamble II of NRL.

For Gamble I,[7] a constant resistance of 10Ω and an electron diode of steady impedance of 10Ω are considered to evaluate the interaction between the POS and diode. The results show that with the resistance the switch voltage and load current is 1.4 MV and 125 kA, and with the diode the voltage and current is 2.4 MV and 95 kA, respectively, indicating a strong interaction between the switch and load.

CONCLUSIONS

The combination of POS physical models with PSPICE is useful for giving simplicity, flexibility and efficiency to the preliminary design of pulsed power circuits containing the POS and to the analyses of nonlinear switch behaviors and interactions between the POS and load.

REFERENCES

1. P.F. Ottinger, S.A. Goldstein, and R.A. Meger, Theoretical modeling of the plasma erosion opening switch for inductive storage applications, *J. Appl. Phys.* 56:774(1984).
2. W. Rix, D. Parks, J. Shannon et al, Operation and empirical modeling of the plasma opening switch, *IEEE Trans. Plas. Sci.* 19:400(1991).
3. W. Rix, P. Coleman, John R. Thompson et al, Scaling microsecond-conduction-time plasma opening switch operation from 2 to 5 MA, *IEEE Trans. Plas. Sci.* 25:169(1997).
4. K.F. McDonald, Dependent source modeling for SPICE, in: *Digest of Technical Papers, Eighth IEEE International Pulsed Power Conference*, R. White and K. Prestwich, eds., IEEE Publishing Services, San Diego(1991).
5. T. Warren and L. Matheus, Modeling non-linear pulsed power component behaviors, in: *Digest of Technical Papers, Eighth IEEE International Pulsed Power Conference*, R. White and K. Prestwich, eds., IEEE Publishing Services, San Diego(1991).
6. P.F. Ottinger, S.A. Goldstein, and R.A.Meger, Theoretical Modeling of the Plasma Erosion Opening Switch for Inductive Storage Applications, *J. Appl. Phys.* 56:774(1984).
7. R.J. Comisso, G. Cooperstein, R.A. Meger et al, The plasma erosion opening switch, in: *Opening Switches*, A. Guenther et al, eds., Plenum Publishing Corporation, New York(1987).

POWER ABSORBED IN A RF DISCHARGE IN OXYGEN

G.R. Govinda Raju

Electrical Engineering Program
College of Engineering and Science
University of Windsor, Ont., N9B 3P4
Canada

INTRODUCTION:

An RF discharge in a low pressure gas has become an integral processing part of the semi conductor industry for fabrication of components. Processes for deposition of thin films and etching are well developed for regular application. One of the parameters which concerns these applications is the power absorbed and experiments have been carried out by a number of authors to determine the influence, among other factors, of electrode geometry, gas pressure and RF voltage magnitude.

Van Roosmalen et. al [1] have carried out experiments in oxygen in a commercial plasma reactor made of aluminum and operating at 13.56 MHz. A dark space shield was fitted above the powered electrode with an exposed area of 0.285 m^2 and the electrode spacing was either 0.05 m or 0.09 m. Measurements were carried out at 7-53 Pa and 50-800 W of power. For the purpose of comparison measurements were also carried out in nitrogen and air. The dc potential developed between the powered electrode and ground was measured with a probe and an oscilloscope. Dark-space thickness was measured visually by a travelling microscope.

Theoretical calculation of power absorbed in the discharge was carried out and compared with the measured values. They observed that the theoretical values calculated by using different models differed from those measured by as much as a factor of five, the latter being higher. For example a discharge at 7 Pa absorbed a power of 200W and an application of Bohm's criterion [2] yielded a value of 44W. After considering several other mechanisms they suggested that the disagreement between the measured and calculated power necessitates a consideration of power dissipation at the sheath-glow boundary. They have also referred to wave interaction or resonance processes as a possibility.

In this paper we recalculate the power absorbed in a RF discharge in oxygen at low gas pressures (< 100 Pa) by considering various fundamental parameters associated with the discharge. The procedure outlined by Lieberman and Lichtenberg [3] has been adopted with minor modifications.

BASIC DISCHARGE PROCESS:

Consider two electrodes separated by a distance in the range of 2-20 cm. in a low pressure discharge tube. With the application of dc voltage of sufficient magnitude a glow discharge is struck and a dark space develops near the cathode.

Application of a rf voltage renders each electrode cathode during alternating half cycle. A dark space is seen to develop at both the electrodes. An additional source of ionization in addition to secondary electrons comes into play, in the form of electrons in the glow oscillating between the two electrodes and acquiring energy from the applied field to augment the primary ionization process. The large difference in the mass and mobility of electrons and ions causes the glow to acquire a positive potential with respect to the electrodes and walls, to balance the loss of fluxes of positive and negative charge carriers. Both the electrodes now have a negative potential with respect to the glow.

In a symmetrical discharge the dc potential of both electrodes will be equal with respect to the glow and the rf voltage will be twice the dc bias. However in practice it is more common for the potential or the electric field at one electrode to be higher than that at the other. In this situation the ratio of rf voltage to dc bias has an intermediate value between 1.0 and 2.0.

POWER ABSORBED

The power absorbed in a rf discharge may be calculated in several different ways and we provide a summary:

(1) The density of power absorbed may be expressed by the simple relationship

$$P = \frac{1}{2}\sigma E_{dc}^{2} \tag{1}$$

In which E_{dc} is the dc electric field and σ is the conductivity of the glow. With an alternating applied electric field the power density becomes

$$P = \sigma E_{ac}^{2} \tag{2}$$

In which the electric field is the rms value. Electrons in the glow acquire energy and suffer both elastic and inelastic collisions. These processes give rise to Ohmic heating.

(2) The Bohm criterion [2] for the positive ion current in the dark space is given by

$$J_{i} = 0.4 e n_{e} W_{th} (m_{e}/m_{i})^{0.5} \tag{3}$$

where e is the electronic charge, n_e the electron density, m_e and m_I the electronic and ionic mass respectively, and W_{th} is the thermal velocity of electrons given by

$$W_{th} = \sqrt{(8kT/\pi m)} \tag{4}$$

and m is the mass and T the temperature of the particle. Following Cantin and Gagne [4] we substitute $T_e = 15000$ K and $n_e = 5 \times 10^{15}$ m^{-3} for electrons. The power absorbed is related to the positive ion current according to

$$P = J_i A E \tag{5}$$

where A is the area of the electrodes.

(3) The glow is viewed of a resistive medium and the power absorbed is given by

$$P = (V^2 A / L e n_e \mu) \tag{6}$$

where L is the length of the glow and μ the mobility.

(4) From a calculation of the potential distribution in the sheath the power absorbed is expressed as [1]

$$P = \frac{J_r}{eA}\left(2m_e \overline{W}^2\right) \tag{7}$$

in which

$$\overline{W} = \frac{\varepsilon \omega^2 V_0}{2en_e} \tag{8}$$

and

$$J_r = 0.25 e n_e W_{th} \tag{9}$$

Application of these models have not yielded absorbed power that agrees with that measured and van Roosemalen et. al. have suggested that additional processes such as wave interaction or resonance processes must be considered. In the following we show that calculated power agrees with that measured by application of a more rigorous theory.

POWER ABSORBED: PRESENT CALCULATION

To calculate the power absorbed we adopt the procedure developed by Liebermann and Lichtenberg [2] for capacitively coupled discharges at intermediate pressures for gap lengths in the range mentioned earlier. Though the theory is developed for electropositive discharges we explore whether it is also applicable to electronegative discharges with a modification for the fraction of the energy lost in collisions.

In the intermediate pressure range or intermediate mean free path range the density of charge carriers is relatively flat in the center and steep near the edges. At high pressures as defined above it tapers down sharply from the center toward the electrodes with a cosine distribution. An effective gap length d_{eff} may be defined in terms of the gap length and electron density.

The ratio of electron density from the center to edge is given by

$$n_{sc}/n_{se} = 0.86(3 + L/2\lambda_I)^{-0.5} \tag{10}$$

where λ_t is the mean free path of ions. With the help of equations (9) & (10) we can determine d_{eff} which is related to Bohm velocity W_B according to

$$n_g \, d_{eff} = W_B / K_{iz} \qquad (11)$$

in which K_{iz} is the rate constants for ionization. Bohm velocity is determined by the electron temperature T_e according to

$$W_B = \sqrt{\frac{e T_e}{m_i}} \qquad (12)$$

in which m_I is the mass of the ion. Note that the Bohm velocity involves the electron energy and ionic mass. For a given pressure and gap length the electron energy may now be determined.

As already stated the Ohmic component of the power absorbed results in bulk heating of the plasma. Following the analysis of Lieberman and Lichtenberg [2] the power density due to Ohmic component is expressed as

$$S_{ohm} = 1.73 \, (m/2e) \, (n_{es}/n_{ec}) \, \varepsilon_0 \omega^2 \, \nu_m \, T_e \, V_1^{0.5} \, d \qquad (13)$$

in which ε_0 is the permittivity of free space, ω the angular frequency, ν_m the momentum transfer collision frequency, T_e the electron temperature, V_1 the rf voltage across a single sheath, and, n_{es} and n_{ec} are the density of electrons at the sheath edge and center respectively. The relation between the rf voltage and dc voltage across a ion sheath is $V_{dc} \approx 0.83 \, V_1$.

The second component of the power absorbed is due to stochastic heating. Electrons from the plasma moving towards the electrode are reflected back at the edge of the sheath and from a consideration of the time and energy dependent distribution function the power absorbed is given by

$$S_{stoch} = 0.45 \, (m/2e)^{0.5} \, \varepsilon_0 \, \omega^2 \, T_e^{0.5} \, V_1 \qquad (14)$$

The power delivered to electrons is given by

$$S_e = 2 \, e \, n_{es} \, W_B \, (\varepsilon_c + 2T_e) \qquad (15)$$

in which ε_c represents the loss of energy per electron-ion pair generation. The total power absorbed is derived as

$$S_{tot} = (P_{ohm} + 2 \, P_{stoch}) \, [1 + 0.83 V_1 / (\varepsilon_c + 2Te)] \qquad (16)$$

A similar theory with important differences applies for high Pressure discharges.

RESULTS OF COMPUTATION

The experimental results of Van Roosemalen et al. (1) provide a convenient scope for the verification of the equations for power absorbed. The parameters required are gap length, the gas pressure and the pwer absorbed. From the gas pressure data the number of molecules are calculated by the simple relation $N_g = 3.28 \times 10^{17} \, p \, m^{-3}$ where p is expressed in units of Pa.

The ionization rate is expressed by the analytical function [2]:

$$K_{iz} = 9 \times 10^{-16} T_e^{0.5} \exp(-12.6/T_e) \tag{17}$$

An alternative function is given by

$$K_{iz} = 2.13 \times 10^{-14} \exp\left(-\frac{14.5}{T_e}\right) \tag{18}$$

Fig. 1 shows the ionization rates according to these representations.

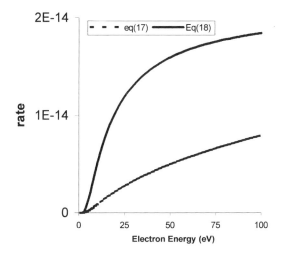

Fig. 1. Ionization rates in oxygen according to eq.(17) and (18).

We can also calculate the ionization rate as a function of the mean energy of electrons from the data of ionization cross section as a function of electron energy as measured in beam experiments. The procedure is rather involved and we have used Eq. (17) in our computations.

Making an initial assumption of sheath thickness of 0.01 m for a gap length of 0.05 m the parameters d/λ_I and $(\varepsilon_i/<\varepsilon>)d$ were calculated to determine the pressure range in which the discharge occurs according to the criteria set forth above. The calculations at 7 Pa give satisfactory result with the relationships given for the intermediate pressure and for the remaining pressures the high pressure considerations apply. The energy loss per ionization collision is adopted from the compilation of Vahedi quoted in reference [2]. The energy loss per ionization collision, ε_c, is expressed as a function of $<\varepsilon>$. A cubic spline interpolation was adopted to evaluate the energy loss at intermediate values of $<\varepsilon>$. Since ε_c increases with decreasing $<\varepsilon>$ a minimum threshold value for the latter was set at 1.8 eV to overcome difficulties with excessive (>1000V) energy loss at lower energies. Use of eq. (22) to determine V_1 for high pressures involves an iterative procedure because of the previously stated functional dependence of ε_c on $<\varepsilon>$. The calculated power absorbed is shown in Table 1 and good agreement is obtained.

Table 1. Calculated and Measured [1] Power in a RF discharge in Oxygen

Pressure (Pa)	Measured (W)	Calculated (W)
7	50.0	49.7
	100.0	99.6
	200	197
	400	409
	800	791
53	50.0	51.0
	100	85
	200	151
	400	379
	800	547

REFERENCES:

[1] A.J. van Roosemalen, W.G.M. van den Hoak and H. Kalter, "Electrical Properties of Planar RF Discharges For Dry Etching", J. Appl. Phys. Vol. 58, pp. 653-658, 1985
[2] M. A. Liebermann and A.J.Lichtenberg, "Principles of Plasma Discharges and Materials Processing, (Book), John Wiley & Sons, inc., New York, 1994.
[3] B. Chapman, "Glow Discharge Processes", John Wiley &sons, New York, 1980.
[4] A. Cantin and R.R.J. Gagne, "Pressure Dependence of Electron Temperature Using RF-Floated Electrostatic Probes in RF Plasmas", Appl. Phys. Lett. 30, 1977, pp. 316-319, 1977.

SECTION 4: PARTIAL DISCHARGES / DIAGNOSTICS*

* Dedicated to the memory of A. Pedersen

PARTIAL DISCHARGE TRANSIENTS: THE FIELD THEORETICAL APPROACH

I. W. McAllister and G. C. Crichton

Department of Electric Power Engineering
Technical University of Denmark
DK-2800 Lyngby, Denmark

INTRODUCTION

Up until the mid-1980s the theory of partial discharge transients was essentially static. This situation had arisen because of the fixation with the concept of void capacitance and the use of circuit theory to address what is in essence a field problem. Pedersen rejected this approach and instead began to apply field theory to the problem of partial discharge transients. This work was to occupy him for the last 10 years of his life. In that time Pedersen authored/coauthored 16 papers on this topic.[1-16] In the present manuscript the contributions of Pedersen using the field theoretical approach will be reviewed and discussed.

The *Status Quo*

For many years, an equivalent circuit approach was used when discussing partial discharges; see Figure 1. With respect to this circuit, C_a denotes the capacitance of the system; C_b is considered to be the capacitance of the solid dielectric in series with the void, while C_c is considered to be the capacitance of the void. The gap across C_c indicates that the gas-filled void breaks down when there is a sufficiently high field strength in the void.

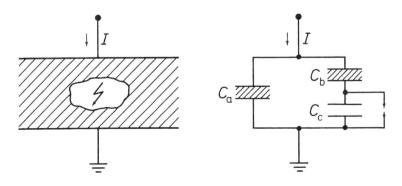

Figure 1. The equivalent circuit approach.

An equivalent circuit is *any* circuit which can generate, as faithfully as possible, the signals which are manifest at the terminals of the actual system. In this respect the simple equivalent circuit of Figure 1 has been very sucessful in promoting the development of methods for partial discharge detection.[17,18] Unfortunately, this success lead to the identification of the capacitive circuit with the actual dielectric system. However, it must be realized that the operation of an equivalent circuit need in no way correspond to the physical processes which occur in the actual system. With respect to partial discharge phenomenon, this limitation of equivalent circuits was apparently overlooked during the past 50 years.

With his background in electromagnetic field theory, Pedersen rejected the equivalent circuit approach on the basis that it was fundamentally incorrect. Even in a mathematical sense C_b and C_c cannot exist.[19] New thinking was required. The basic steps in Pedersen's approach will now be outlined.

INDUCED CHARGE: GENERAL ASPECTS

Concept and Definitions, Poissonian and Laplacian Components [7,9,12]

To understand the physical basis of partial discharge transients, Pedersen considered the fundamental topic of induced charge. Unfortunately, textbooks usually treat this subject in a superficial manner. So as was his habit, Pedersen began to study the original work of Faraday from the 1830/40s.[20] It was at that time that the famous ice-pail experiment was performed. However, with our topic in mind, the general aspects of induced charge are best illustrated by re-working one of the standard problems of electrostatics.

We consider an isolated electrode which is initially in a field-free environment; *i.e.* the electrode is at zero potential and carries zero net charge. The surface charge density σ is also zero. A charge source is then located at a known distance from this electrode. This situation is depicted in Figure 2a for a spherical conductor and a point charge.

For this condition, the field solution indicates that now $\sigma \neq 0$. On evaluating the charge q induced by the source charge, we find that

$$q = \int \sigma dA = 0 \tag{1}$$

There has however been a change in potential. This change ΔU is given by

$$\Delta U = \frac{Q}{4\pi\varepsilon b} \tag{2}$$

which may also be expressed as

$$\Delta U = \frac{(R/b)Q}{C} \tag{3}$$

where C is the capacitance of the sphere; *i.e.* the change in sphere potential is associated with a charge $(R/b)Q$. This charge has been named the *Laplacian induced charge*.

The sphere is now earthed, see Figure 2b, *i.e.* $\Delta U = 0$. On solving for the new field condition we still find that $\sigma \neq 0$, but in contrast to (1) we have

$$q = \int \sigma dA = -(R/b)Q \tag{4}$$

This finite value of q has been named as the *Poissonian induced charge*. The situation implies that the earthing of the sphere has resulted in the transfer of a charge to the sphere; *i.e.* the Laplacian induced charge can also be associated with a charge transfer.

To generalize the above concepts, the following definitions were

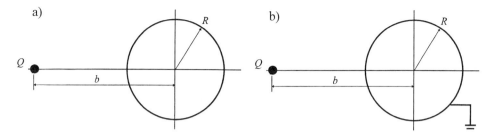

Figure 2. Characteristics of induced charge.

introduced.

The Poissonian induced charge is that component of the induced charge related to the Poissonian field established by the space charge.

The Laplacian induced charge is that component of the induced charge related to the Laplace field associated with the change in conductor potential.

The initial values of the Poissonian induced charge and the Laplacian induced charge are *numerically equal*, but of *opposite polarity*. The fact that the latter component of the induced charge can be associated with both a change in potential and a transfer of charge is the basis of partial discharge *detection*. This aspect will be discussed later.

It should be noted that there is no logical connection between the Poissonian induced charge and the apparent charge as defined in IEC 270 (1981), §3.2.2.[21] It is nevertheless evident from the actual usage of this latter term that apparent charge is in reality the Poissonian induced charge.

Evaluation of Induced Charge, λ and ϕ Functions [1,2,5,6,13]

To evaluate induced charge, Pedersen introduced two dimensionless functions: the λ function and the ϕ function. These functions are the proportionality factors between the source charge and the induced charge. The properties of these functions are as follows.

The λ function is a solution of the general Laplace equation

$$\text{div}(\varepsilon \text{grad}\lambda) = 0 \qquad (5)$$

where ε denotes permittivity. The boundary conditions are $\lambda = 1$ at the detecting electrode, and $\lambda = 0$ at all other electrodes. In addition, the following condition must be fulfilled at all dielectric interfaces, such as the walls of voids, viz.

$$\varepsilon_+ (\partial\lambda/\partial n)_+ = \varepsilon_- (\partial\lambda/\partial n)_- \qquad (6)$$

where λ is differentiated in the direction normal to the interface, and the signs + and - refer to each side of the interface, respectively.

The ϕ function is a solution of the reduced Laplace equation

$$\text{div}(\text{grad}\phi) = 0 \qquad (7)$$

The boundary conditions are $\phi = 1$ at the detecting electrode and $\phi = 0$ at all other electrodes.

The difference between the two functions is akin to the different representations of the $\vec{D}$-field, viz.

$$\vec{D} = \varepsilon_r \varepsilon_0 \vec{E} = \varepsilon_0 \vec{E} + \vec{P} \qquad (8)$$

where $\vec{P}$ denotes the dielectric polarization. The λ function is akin to the first relation, while the ϕ function relates to the second; *i.e.* the λ function takes account of dielectric polarization *implicitly* whereas ϕ does so *explicitly*.

In practice, the direct application of ϕ is limited, as, with respect to $\vec{P}$, the partial discharge source term is unknown. However it has great relevance for analytical studies with a known source term. In contrast the λ function can be readily applied to practical systems. Moreover it is a powerful analytical tool in other areas which involve space charge and solid dielectrics.[22,23]

INDUCED CHARGE FROM A PARTIAL DISCHARGE

Partial Discharge Source, Dipole Representation and λ_0 Function[2,5,6,12]

The question arises as to how an arbitrary distribution of charge may be represented. A partial discharge in a gaseous void in a bulk dielectric occurs in a closed volume, resulting in a deposition of charge on the void wall. In this confined space, the net charge produced by the discharge will be zero. Hence we can represent the charge distribution by a system of multipoles.

If however, the dimensions of the void are such that gradλ may be assumed constant within the void, then it is only necessary to consider the dipole moment of the charge distribution within the void. This gradλ assumption implies that the second and higher order derivatives of λ do not exist. Thus second and higher order multipoles will not contribute to the induced charge.

If gradλ is not constant within the void, then it becomes necessary to consider if the induced charge contributions from these multipoles are significant in comparison to the dipole contribution. A non-uniform gradλ within the void will occur if the spatial non-uniformity of λ external to the void is comparable with the dimensions of the void, or if the void is in the proximity of an interface between two media.

The dipole moment $\vec{\mu}$ of a distribution of charge is defined as

$$\vec{\mu} = \int \vec{r}\, dQ \qquad (9)$$

where $\vec{r}$ is a radius vector which locates the position of the charge element dQ. If the net charge associated with the distribution is zero, *i.e.* if

$$\int dQ = 0 \qquad (10)$$

then $\vec{\mu}$ is independent of the origin of $\vec{r}$.

We are principally interested in the final value of the induced charge, and thus the dipole integral reduces to one involving the surface charge density σ at the void wall; *i.e.* we have

$$dQ = \sigma dA \qquad (11)$$

where dA represents a surface element.

By considering the induced charge associated with two point charges of opposite polarity, separated by an infinitesimal distance, it can be readily shown that the induced charge q associated with a dipole $\vec{\mu}$ is given by

$$q = -\vec{\mu}\cdot\mathrm{grad}\lambda \qquad (12)$$

For small voids, gradλ may readily be taken as constant. This situation allows the introduction of another function, λ_0, which represents the λ function for the void-free system. By mathematical analogy with electrostatic fields, the relation between λ and λ_0 may be expressed as

$$\text{grad}\lambda = h\,\text{grad}\lambda_0 \tag{13}$$

where, for the void geometries under consideration, h is a constant dependent on the void geometry and the relative permittivity ε_r of the bulk dielectric, e.g. for a spherical void

$$h = \frac{3\varepsilon_r}{1 + 2\varepsilon_r} \tag{14}$$

With the introduction of λ_0, the induced charge from a dipole is given by

$$q = -h\vec{\mu}\cdot\text{grad}\lambda_0 \tag{15}$$

Influence of Void Parameters [3,4]

Having adopted a dipole representation, Pedersen proceeded to examine the influence of the void parameters upon the induced charge. This required the evaluation of the dipole moment. This moment is related to the void wall charge, which establishes a field within the void opposing the applied field. When the net field in the void is reduced sufficiently, the discharge development ceases. Thus with some assumptions about the discharge characteristics and using the streamer criterion, Pedersen derived the following expression for the dipole moment of a partial discharge in an ellipsoidal void

$$\vec{\mu} = \frac{K}{h}\varepsilon\Omega(\vec{E}_i - \vec{E}_\ell) \tag{16}$$

where K is a dimensionless parameter which is dependent upon the geometry of the ellipsoid. Ω is the ellipsoid volume and ε is the permittivity of the bulk dielectric. The field strengths E_i and E_ℓ represent the inception field strength for discharge development, and the limiting field strength for ionization growth, respectively. That is a partial discharge can develop when the void field attains a value of E_i and will be quenched when the field is reduced to E_ℓ. Insertion of (16) in (15) gives the induced charge q, viz.

$$q = -K\varepsilon\Omega(\vec{E}_i - \vec{E}_\ell)\cdot\text{grad}\lambda_0 \tag{17}$$

An examination of this expression indicates that the induced charge is dependent on
- void location
- void geometry
- void physical dimensions
- void gas
- void gas pressure
- permittivity of the (homogeneous) bulk dielectric.

More recently the induced charge has been shown to depend crucially upon the void orientation with respect to the applied field.[24]

NB: In using the streamer criterion, Pedersen assumed implicitly that the void dimensions and gas pressure were such that swarm conditions could develop. Such conditions are necessary to ensure the validity of the use of the macroscopic gaseous ionization coefficients.

Influence of Bulk Dielectric Polarization[1,13-16]

The occurrence of a partial discharge in gaseous voids leads not only to a charge being induced on the detecting electrode, but also to a change in the polarization of the bulk dielectric. Mathematically, the final value of the induced charge q can be resolved into two components:

$$q = q_\mu + q_P \qquad (18)$$

where q_μ is the induced charge directly associated with the space charge in the void. For a dipole $\vec{\mu}$, this component is given by

$$q_\mu = -\vec{\mu} \cdot \mathrm{grad}\phi \qquad (19)$$

The component q_P represents the induced charge related to the change in dielectric polarization ($\delta \vec{P}$) due to the presence of this space charge. By combining (12), (18) and (19) we arrive at an expression for q_P:

$$q_P = -\vec{\mu} \cdot (\mathrm{grad}\lambda - \mathrm{grad}\phi) \qquad (20)$$

It is seen that the q_P expression involves three *independent* vectors. Thus the general behaviour of q_P is difficult to envisage. Hence initial studies of q_P were concentrated on situations for which the three vectors were parallel.

For a *homogeneous* bulk dielectric, the magnitude of q_P is dependent on the geometry of the void, ε_r of the bulk dielectric and on the orientation of the void with respect to $\mathrm{grad}\lambda_0$.

For a *heterogeneous* bulk dielectric, the behaviour of q_P becomes more complex. For a two layer dielectric system, q_{Pn} increases when the void is located in the medium of lesser permittivity. In contrast a decrease occurs when located in the medium of higher permittivity. This reduction can be of such a degree that the *polarity* of q_{Pn} is *reversed*. This reversal occurs for permittivity ratios of about 0.5. In addition, when the void is in the dielectric of smaller physical extent, the greater is the relative increase/decrease in q_{Pn}. The influence of void geometry upon q_P for a heterogeneous dielectric system is presently under study.[25]

For a fixed void location, q_μ is dependent only on $\vec{\mu}$, which we assume constant. Hence variations in q_P will be reflected directly in the magnitude of the Poissonian induced charge q, see (18).

From the above results, it is evident that changes in dielectric polarization can significantly affect the magnitude of the induced charge. This is particularly so for heterogeneous bulk dielectric systems. The existence of the polarization component implies that the apparent charge approach to the assessment of the energy dissipated by a partial discharge *within* a void has no valid basis.

Partial Discharge Detection[6,7,9,12]

In the previous Sections, the application of field theory to partial discharges has been illustrated. When considering the question of partial discharge detection however, it is necessary to introduce circuits and circuit theory. The manner by which the two theories are combined will now be indicated. The essence of the situation is depicted in Figure 3.

The measuring system is composed of terminals and circuits. This sys-

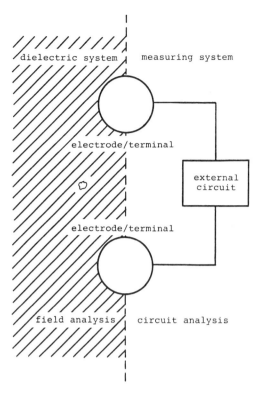

Figure 3. Partial discharge detection: Overall features.[12]

tem can be handled within the framework of circuit analysis. The dielectric system with its void and electrodes obeys the laws of fields, and is thus not a part of the measuring system. However it is through field theory that the terminal properties required by the circuit analysis are obtained.

With respect to the external circuitry, the primary source for the partial discharge transient is the Laplacian induced charge. It is this component of the induced charge which we can manipulate.

The circuital aspects of partial discharge detection can be illustrated through the basic circuit shown in Figure 4. We assume that a partial discharge occurs in a void external to the electrode. With reference to the discharge transient, the governing equations for this circuit are

$$q(t) = \Delta Q(t) + C\Delta U(t) \tag{21}$$

$$\Delta U(t) = Zi(t) \tag{22}$$

and

$$\Delta Q(t) = \int_0^t i(t')dt' \tag{23}$$

where t' is a dummy variable.

We have a single electrode and thus (21) represents the simplest form of induced charge expression. In (21) $\Delta Q(t)$ denotes the charge transferred to the detecting electrode from the external source, while $\Delta U(t)$ represents the change in potential of the detecting electrode, for which C is the capacitance to earth. In (22) Z is the impedance operator for the external circuit and $i(t)$ is the transient current associated with the circuit. On

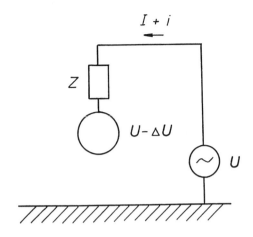

Figure 4. Basic detection system.[12]
I - power frequency current;
U - power frequency voltage.

selecting an appropriate $q(t)$ variation, the circuit was analysed. The results indicated that to achieve an accurate determination of the maximum value of the induced charge one should record the voltage rather than the current transient.

For a system of several electrodes, the expression for the induced charge becomes more complicated. The reason is that the partial discharge induces charge on all electrodes. Those electrodes not held at earth potential will undergo transient changes in potential. Owing to the partial capacitances between the electrodes, these changes in potential will be registered at the detecting electrode. Thus for a system of N electrodes, we have

$$q_i(t) = \Delta Q_i(t) + \sum_{j=0}^{N} C_{ij}[\Delta U_i(t) - \Delta U_j(t)] \qquad (24)$$

where $q_i(t)$ denotes the Poissonian induced charge on the ith electrode; $\Delta Q_i(t)$ denotes the charge transferred to the ith electrode from the external source; ΔU_i and ΔU_j represent the drops in potential of the ith and jth electrodes from U_i and U_j to $U_i - \Delta U_i$ and $U_j - \Delta U_j$, respectively, while C_{ij} is the partial capacitance between these electrodes. The subscript $j = 0$ refers to ground and the parameter t is time.

FIELD ENHANCEMENT IN BULK DIELECTRIC

Basic Aspects [8,9]

The fact that a partial discharge induces a charge on the detecting electrode is a direct consequence of the partial discharge producing a perturbation of the electric field in the bulk dielectric. This field perturbation also produces the change in the polarization of the bulk dielectric. External to the void, the effect of the void wall charges can be considered equivalent to that of an electric dipole located within the void. This allows the nature of the field perturbation in the bulk dielectric to be assessed through a simple model.

We consider a point dipole $\vec{\mu}$ located at the centre of a spherical void in an extended dielectric; see Figure 5. From the field solution, the maximum field strength within the bulk dielectric is found to occur at A and B. This field strength is given by

$$\delta\vec{E}_A = \delta\vec{E}_B = \frac{2h\vec{\mu}}{3\varepsilon\Omega} \qquad (25)$$

External to the void, the applied field and the dipole field are additive. For a spherical void $K = 3$, and thus from (16) the partial discharge dipole moment for a spherical void is given by

$$\vec{\mu} = \frac{3}{h}\varepsilon\Omega(\vec{E}_i - \vec{E}_\ell) \qquad (26)$$

Upon substituting in (25) we obtain

$$\delta\vec{E}_A = \delta\vec{E}_B = 2(\vec{E}_i - \vec{E}_\ell) \qquad (27)$$

To indicate the magnitude of this perturbation for a single discharge, reference is made to a spherical void of volume 1 mm³ filled with air at a pressure of 0.1 MPa. The streamer criterion yields the following value for E_i, viz.

$$E_i = 4.3 \text{ kV/mm, with } E_\ell = 2.4 \text{ kV/mm} \qquad (28)$$

Hence in the proximity of the void, (27) indicates that a *single* discharge could generate a field distortion *within the solid dielectric* of approximately 4 kV/mm. This value can quite easily exceed the applied field at A and B by a factor of 2.

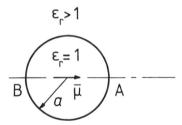

Figure 5. Model for field perturbation in solid dielectric.[8]

Practical Aspects [10,11]

In practice, the applied field strength in the void will be much greater than the inception field strength for a partial discharge. This situation will result in several partial discharges per cycle. If the surface conductivity of the void wall is taken as zero, then the partial discharge wall charge remains *in situ* and the field distortion remains at a constant level until the occurrence of the next discharge. If this discharge occurs in "the same direction", then the field distortion will be further increased. Consequently, a unidirectional sequence of partial discharges results in a cumulative field distortion within the solid dielectric in the vicinity of the void. For the spherical void under consideration, this distortion can be expressed as

$$\delta E_{max} = N(E_i - E_\ell) \qquad (29)$$

where N is the total number of either positive or negative discharges per cycle.

It may be noted that, compared with the very short time duration of each partial discharge *current* pulse, some nanoseconds, the maximum value of the induced charge associated with several partial discharges will persist essentially as a quasi-DC signal; *i.e.* with a duration of several milliseconds for a power frequency voltage.

The significance of the partial discharge field enhancement is that as electrostatic forces F are proportional to the square of the electric field strength E, *i.e.* $F \propto E^2$, then such field enhancement could lead to significant force levels within the solid dielectric. This electromechanical stressing may be the catalyst which promotes electrical treeing.

If E_v represents the field within the void, then the above analysis can be extended by considering the limiting condition of $E_v \rightarrow 0$ following several partial discharges. For this condition, the maximum possible field strength within the solid dielectric as a result of partial discharge activity in a void can be deduced. In turn this field value allows the corresponding maximum possible dipole moment to be estimated. Then, with respect to test procedures, this dipole moment value enables a meaningful, maximum permissible level of induced charge to be specified.

DISCUSSION

At the time of Pedersen's death in October 1995, a framework had evolved from the theoretical studies he had initiated 10 years earlier. This framework, which is shown in Figure 6, links in a logical manner the different topics touched upon in this manuscript. The result is a cohesive, self-consistent view of the theoretical and measurement aspects of partial discharge transients.

Beginning at the left-hand side, we have the <u>a</u>pplied voltage U generating an electric field <u>e</u>xternal to the void and also an <u>i</u>nternal field, *viz.* E_{ae} and E_{ai}, respectively. When the internal field is greater than the inception field strength for the void gas, a partial discharge can develop. As a consequence of this discharge, charge separation occurs within the void. Subjected to the applied field, these charges accumulate at the void <u>w</u>all, and by virtue of their polarities generate a field, E_{wi}, which opposes the applied field within the void. When the net field in the void ($E_{ai} - E_{wi}$) has been reduced sufficiently, *i.e.* less than the limiting field for ionization growth, the partial discharge will be quenched.

With respect to the detecting electrode, the effect of the void wall charges is equivalent to that of an electric dipole of moment $\vec{\mu}$ located within the void. Through the λ function, the Poissonian induced charge q_p arising from $\vec{\mu}$ can be evaluated. The associated Laplacian induced charge q_L serves as the source for the detecting and recording systems.

The surface conductivity Γ of the void wall affects the persistence of the wall charge.[26] If $\Gamma = 0$, then the wall charge will remain *in situ*. The occurrence of the next discharge will, depending on its polarity, either augment or reduce the magnitude of the wall charge.

External to the void, the wall charge generates a field, E_{we}, which adds to the applied field E_{ae}. As electrostatic forces F are proportional to E^2, the field enhancement arising from partial discharge activity could promote electromechanical fatigue of the solid dielectric.

E_{we} also perturbs the polarization of this dielectric. The induced charge associated with the change in polarization, $\delta \vec{P}$, can be evaluated by using the ϕ function.

Although the major components of the partial discharge framework are undoubtedly in place, there is still much detail to be uncovered; e.g. the response of different detecting systems, the influence of surface conductivity especially with reference to "DC voids", the field enhancement in the solid dielectric and the associated electromechanical effects.

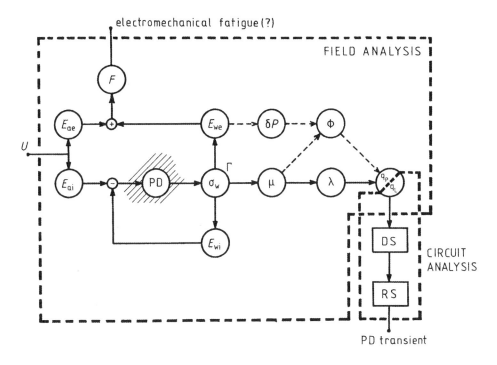

Figure 6. Partial discharge transients: Theoretical framework.
PD - partial discharge; DS - detecting system; RS - recording system; U - applied voltage; λ - the λ function; ϕ - the ϕ function; q_P - Poissonian induced charge; q_L - Laplacian induced charge;
With reference to the void:
σ_w - void wall charge; $\vec{\mu}$ - dipole moment of σ_w;
E_{ai} - internal applied field; E_{ae} - external applied field;
E_{wi} - internal wall-charge field; E_{we} - external wall-charge field;
F - electrical forces generated by $(E_{ae} + E_{we})$;
$\vec{\delta P}$ - change in dielectric polarization due to E_{we};
Γ - surface conductivity of void wall.

CONCLUSION

On comparing the contents of Figure 1 with those of Figure 6, only one conclusion is possible. Pedersen made a major contribution to our understanding of partial discharge transients. To paraphrase Heaviside, without question the concept of void capacitance has been committed to the flames!

REFERENCES

1. A. Pedersen, Current pulses generated by discharges in voids in solid dielectrics. A field theoretical approach, *Conference Record of the 1986 IEEE International Symposium on Electrical Insulation*, IEEE Publication 86CH2196-4-DEI, 112 (1986).
2. A. Pedersen, Partial discharges in voids in solid dielectrics. An alternative approach, *1987 Annual Report - Conference on Electrical Insulation and Dielectric Phenomena*, IEEE Publication 87CH2462-0, 58 (1987).
3. G.C. Crichton, P.W. Karlsson, and A. Pedersen, A theoretical derivation of the transients related to partial discharges in ellipsoidal voids, *Conference Record of the 1988 IEEE International Symposium on Electrical Insulation*, IEEE Publication 88CH2594-0-DEI, 238 (1988).
4. G.C. Crichton, P.W. Karlsson, and A. Pedersen, Partial discharges in ellipsoidal and spheriodal voids, *IEEE Trans. Elect. Insul.* 24:335 (1989).
5. A. Pedersen, On the electrodynamics of partial discharges in voids in solid dielectrics, *Proceedings of the 3rd International Conference on Conduction and Breakdown in Solid Dielectrics*. IEEE Publication 89CH2726-8, 107 (1989).
6. A. Pedersen, G.C. Crichton, and I.W. McAllister, The theory and measurement of partial discharge transients, *IEEE Trans. Elect. Insul.* 26:487 (1991).
7. A. Pedersen, G.C. Crichton, and I.W. McAllister, Partial discharge detection: Theoretical and practical aspects, *International Conference on Partial Discharge*, IEE Conference Proceedings No.378, 21 (1993).
8. A. Pedersen, G.C. Crichton, and I.W. McAllister, PD-related stresses in the bulk dielectric and their evaluation, *1993 Annual Report - Conference on Electrical Insulation and Dielectric Phenomena*, IEEE Publication 93CH3269-8, 474 (1993).
9. A. Pedersen, G.C. Crichton, and I.W. McAllister, Recent advances in the theory of PD-tansients, *NORD-IS 94, Nordic Insulation Symposium*, Vaasa Finland, 229 (1994).
10. A. Pedersen, G.C. Crichton, and I.W. McAllister, PD-related stresses in the bulk dielectric for ellipsoidal voids, *1994 Annual Report - Conference on Electrical Insulation and Dielectric Phenomena*, IEEE Publication 94CH3456-1, 79 (1994).
11. A. Pedersen, G.C. Crichton, and I.W. McAllister, PD related field enhancement in the bulk medium, in: *Gaseous Dielectrics VII*, L.G. Christophorou and D.R. James, eds., Plenum Press, New York, 223 (1994).
12. A. Pedersen, G.C. Crichton, and I.W. McAllister, Partial discharge detection: Theoretical and practical aspects, *IEE Proc. - Science, Measurement and Technology*, 142:29 (1995).
13. A. Pedersen, G.C. Crichton, and I.W. McAllister, The functional relation between partial discharges and induced charge, *IEEE Trans. Diel. & Elect. Insul.* 2:535 (1995).
14. A. Pedersen, G.C. Crichton, and I.W. McAllister, Influence of bulk dielectric polarization upon PD transients, *1995 Annual Report - Conference on Electrical Insulation and Dielectric Phenomena*, IEEE Publication 95CH35842, 323 (1995).
15. A. Pedersen, G.C. Crichton, and I.W. McAllister, Partial discharge transients and the influence of dielectric polarization, *NORD-IS 96, Nordic Insulation Symposium*, Bergen Norway, 71 (1996).
16. A. Pedersen, G.C. Crichton, and I.W. McAllister, Influence of system geometry and bulk dielectric polarization upon PD transients, *1996 Annual Report - Conference on Electrical Insulation and Dielectric Phenomena*, IEEE Publication 96CH35985, 545 (1996).
17. R. Bartnikas, A commentary on partial discharge measurement and detection, *IEEE Trans. Elect. Insul.* 22:629 (1987).

18. R. Bartnikas, Partial discharge diagnostics: A utility perspective, in: *Gaseous Dielectrics VII*, L.G. Christophorou and D.R. James, eds., Plenum Press, New York, 209 (1994).
19. I.W. McAllister, Electric field theory and the fallacy of void capacitance, *IEEE Trans. Elect. Insul.* 26:458 (1991).
20. M. Faraday, *Experimental Researches in Electricity*, vols.I & II, Bernard Quaritch, London (1839 & 1844).
21. IEC Publication 270, *Partial Discharge Measurements*, 2nd. edition, (1981).
22. I.W. McAllister, G.C. Crichton, and A. Pedersen, Space charge fields in DC cables, *Conference Record of the 1996 IEEE International Symposium on Electrical Insulation*, IEEE Publication 96CH3597-2, 661 (1996).
23. T.O. Rerup, G.C. Crichton, and I.W. McAllister, Using the λ function to evaluate probe measurements of charged dielctric surfaces, *IEEE Trans. Diel. & Elect. Insul.* 3:770 (1996).
24. I.W. McAllister, Partial discharges in spheroidal voids: Void orientation, *IEEE Trans. Diel. & Elect. Insul.* 4:456 (1997).
25. I.W. McAllister and G.C. Crichton, Influence of void geometry and bulk dielectric polarization upon PD transients, in: *Gaseous Dielectrics VIII*, L.G. Christophorou and J.K. Olthoff, eds., Plenum Press, New York, in press (1998).
26. I.W. McAllister, Decay of charge deposited on the wall of a gaseous void, *IEEE Trans. Elect. Insul.* 27:1202 (1992).

DISCUSSION

E. H. R. GAXIOLA: How many higher order terms do you estimate would be necessary to describe a cylindrical void by a term expanded λ function reducing π to a λ_o, k description, i.e., ε_r expression?

I. W. MCALLISTER: To estimate the number of terms required, it is necessary to have knowledge about the degree of non-uniformity of the λ function within the void. This non-uniformity will depend on the relative dimensions of the cylindrical void. As I have never undertaken numerical calculations on cylindrical voids, I lack the necessary background knowledge upon which to make an educated guess.

UHF DIAGNOSTICS AND MONITORING FOR GIS

Brian Hampton

Diagnostic Monitoring Systems Ltd
The Teacher Building
14 St. Enoch Square
Glasgow G1 4DB
UK

INTRODUCTION

The UHF technique of detecting partial discharges (PD) in gas insulated substations (GIS), which 10 years ago was seen as little more than an interesting laboratory investigation, is now recognised as an ideal means of giving an early warning of impending dielectric failures in a substation. Its rapid acceptance by both utilities and switchgear manufacturers worldwide has resulted in part because the technique is simple in concept and easy to install; is remarkably free from interference and does not require any screening; is suitable for use at the highest system and test voltages; gives unambiguous results which are interpreted fairly readily; and is ideally suited to on-line measurements.

However utilities worldwide are increasingly seeing UHF monitoring more as a cost effective means of avoiding unplanned outages of their GIS. This may be especially important where the supply industry has been privatised and a utility or transmission company is under contract to accept energy from, for example, a nuclear power station. The consequence of being unable to do so due to the failure of the station GIS, which in an extreme case could require the reactors to be shut down, is likely to be penalties against which the initial cost of a UHF monitoring system would be insignificant.

Utilities are also under mounting pressure to maximise their return on plant, which is a major investment and needs to be used to its limits for as long as possible. This has to be achieved with the minimum of maintenance, which is expensive and possibly disruptive when plant is taken out of service. The utility must also provide the highest possible quality of supply to customers, especially those electronics manufacturers using processes that are susceptible to dips in the supply.

All this points to the need for knowing the condition of plant through on-line monitoring, so that remedial actions may be taken in time, but only when shown to be necessary. Achieving this goal can mean interpreting a large amount of data, which is time consuming and requires the most important and scarce resource of all - experienced engineers. Clearly there is a need for routine data interpretation through some form of artificial intelligence, and it is reassuring to know of the many developments taking place in this field.

PRINCIPLE OF THE UHF TECHNIQUE

The principle of the UHF technique is now well known, but as a reminder the current pulse which forms the partial discharge has a very short risetime, which recent measurements have indicated can be less than 70 ps[1]. These pulses excite the GIS chambers into multiple resonances at frequencies of up to 1.5 GHz or more. Although the duration of the current pulse is less than one nanosecond, the microwave resonances persist for a relatively long time, typically a few microseconds. They may readily be picked up by UHF couplers fitted either inside the GIS chambers, or over dielectric apertures in the chamber wall. The latter are usually either the exposed edges of cast resin barriers, or glass windows, and in both cases couplers can be designed to give a perfectly acceptable output from the UHF signal which propagates through them.

Whether external or internal couplers are used, the UHF signals can be amplified and displayed in ways where their characteristic patterns reveal the nature of any defect that might be present in the GIS.

DATA INTERPRETATION

The features of the UHF discharge pulses that are most useful for interpretation purposes are their amplitude, point on wave, and the interval between pulses. These parameters enable typical defects such as fixed point corona, free metallic particles and floating electrodes to be identified. Other defects occur less commonly, but have their own distinctive features.

Most defects show PD on each cycle, or at least every 5-10 cycles, and we find it sufficient to look at the signals for only a second in order to identify the defect causing them. It is convenient to display all the pulses detected in 50 (60) consecutive cycles, in their correct phase relationships over the cycle. This provides the 3D display shown in the later figures. Having identified the defect, the probability of it causing failure can be estimated from the results of tests in laboratory sections of GIS.

Corona

Three distinct phases in which corona develops from a protrusion may be seen as the voltage is raised:

Inception, where discharges occur first on the half-cycle that makes the protrusion negative with respect to the other electrode. Inception is therefore on the negative half-cycle when the protrusion is on the busbar, and on the positive half-cycle when it is on the chamber wall. The initial discharges are of very low magnitude (less than 1 pC), and are centred on one of the voltage peaks.

Streamers start at a slightly higher voltage, and the negative discharges become larger and erratic. The positive discharges are often very regular and at the same point on the cycle. This difference between the positive and negative discharges reveals whether the protrusion is on the busbar or chamber wall. Streamer discharges will not lead to breakdown.

Leaders follow further increases of the applied voltage, and appear every few cycles as large discharges on the positive side. They propagate in steps until either they become extinguished, or reach the other electrode and cause complete breakdown. Figure 1 shows a typical display of PD from a protrusion on the busbar at this stage of the discharge process. Leaders are the precursors of breakdown, and there is always a risk of failure when they are present.

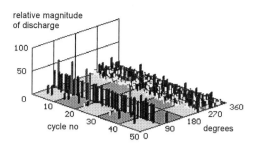

Figure 1. Busbar corona streamer and leader phase.

Free Metallic Particle

A particle lying on the chamber floor becomes charged by the electric field, and if the force on it exceeds that due to gravity it will stand up and dance along the chamber floor. This generates a discharge pulse each time contact is made with the floor, and the particle assumes a new value of charge. The pulses occur randomly over the complete power frequency cycle, but their peak amplitudes follow the phase of the voltage. A typical PD pattern for a free metallic particle is shown in Figure 2. At higher voltages the particle will start to jump towards the busbar, and after several cycles may reach it. Two factors combine to make this an especially serious condition which often leads to breakdown; (i) as the particle approaches the busbar it discharges and generates a voltage transient which increases still further the stress at its tip, and (ii) breakdown can occur by leader propagation well before any space charge has had time to develop and shield the tip.

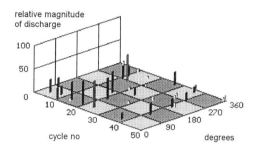

Figure 2. A typical free particle signal pattern.

Floating Electrode

This arises if the contact to, for example, a stress shield deteriorates and sparks repetitively during the voltage cycle. The sparking is energetic because the floating component usually has a high capacitance, and this degrades the contact further. Metallic particles are produced, which may lead to complete breakdown. The discharges are concentrated on the leading quadrants of the positive and negative half-cycles, and have a constant amplitude which does not vary with the applied voltage. Often the gap is asymmetrical, and sparks over at a different voltage on the positive and negative half-cycles. Then a different charge is trapped on the floating

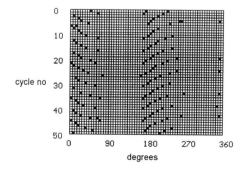

Figure 3. A typical floating electrode signal pattern.

component, and this gives rise to the characteristic wing shaped patterns shown in the plan view of Figure 3. These provide a positive confirmation that the defect is a floating component.

Defect Classification

PD monitoring systems can generate a large amount of data and computer-aided interpretation and classification of defects can be of great assistance when analysing this information. Interpretation is based on the analysis of statistical parameters extracted from the data, particularly those relating to the amplitude, repetition rate and point on wave of the detected UHF signals. Neural networks are well suited to the pattern recognition problem presented by the data in Figures 1-3. A neural network consists of layers of neurons, which are interconnected by links. Each output of the neural network represents the probability that the input data corresponds to a particular type of defect. During system training using a database of real PD patterns from known defect types, the links between neurons are automatically strengthened or weakened on the basis of the required output. The accuracy of defect classification can exceed 95% using this strategy[2].

A PD monitoring system is therefore able to identify the defect type, and this automated classification procedure is already being used by utilities in the UK. Eventually, an expert system will be included to provide the engineer with an assessment of the probability that a particular discharge will lead to breakdown. This is a more difficult task and while work continues, a display of the summary output table from the neural network can provide the engineer with more concise information on which to base a judgement.

SITE TEST PROCEDURES FOR GIS

The problem of how best to undertake the site commissioning test for new GIS has been studied in depth by CIGRE JWG 33/23.12 'Insulation co-ordination of GIS: return of experience, on site tests and diagnostic techniques'. Its main conclusion[3] is that *the recommended dielectric on site test procedure is a high voltage AC test together with a sensitive partial discharge measurement. For GIS with insulation test level chosen in IEC 694, the high AC voltage test level on site has to be linked to the LIWL rather than to the PFWL of the GIS even if this results in the AC on site test level being close to the PFWL.*

The reasoning is that the purpose of the test is to detect 'critical defects', which are defects of a size likely to affect the specified LIWL and PFWL voltages. Extensive series of tests made in a number of laboratories have shown the critical sizes to be:

Free metallic particles	2-5 mm long
Protrusion from the HV conductor	1 mm height
Particle on barrier surface	2 mm long

All these defects produce PD before breakdown occurs. An important part of the CIGRE studies has therefore been *to find the AC voltage level that will cause the critical defects to generate partial discharges at a level measurable on site.* This has led to the adoption of test voltages at which the critical defects will produce PD levels of 1-10 pC. The highest permissible PD level during the test has been set at 5 pC, or the equivalent signal if the measuring system is calibrated in other units.

The proposed test procedure is:

1. Conditioning with AC according to the manufacturer's recommendations
2. A PD measurement at $0.8 \times U_t$, with a highest permissible PD level of 5 pC or equivalent *
3. A 1 minute AC test at $U_t = 0.36 \times LIWL$ **

* Alternatively, if a PD measurement cannot be made, then a LI test at $0.8 \times LIWL$

**Alternatively, if the highest available test voltage is less than $0.36 \times LIWL$, then an OSI test at $0.65 \times LIWL$

The UHF and acoustic techniques are seen as being the most promising for use in this test procedure, and future CIGRE work will concentrate on developing means of calibrating them so it can be shown that the highest permissible level of discharge is not exceeded.

CALIBRATING UHF MEASUREMENTS

The path of the signal from the defect to the UHF coupler is shown in Figure 4, and involves the following stages:

Excitation of resonances in the chamber - the partial discharge in SF_6 is an extremely short current pulse, which radiates energy into the chamber and sets up many modes of microwave resonance. Studies have shown the amplitude of the microwave field to be dependent on the length of the PD current path, and its position and orientation in the chamber.

Propagation of the signal - the UHF field intensity is reduced by reflections from discontinuities such as barriers, bends, junctions and changes in the diameter of the vessel. It also undergoes dispersion, and to a small extent attenuation due to the skin effect.

UHF coupler response - the coupler acts as an antenna, producing a voltage at its output terminal in response to the UHF field incident upon it. The efficiency of this conversion, both in terms of sensitivity and bandwidth, needs to be known.

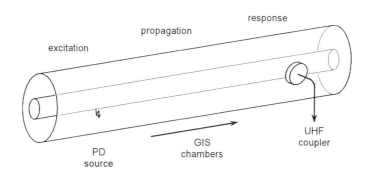

Figure 4. Stages in the transfer of energy from a defect to the UHF coupler.

All the above must be considered before an accurate calibration can be made, and this presents a challenge which is currently being studied. It is probable that some approximations will be made and a practical calibration technique established, but this remains to be seen. For the present, experiments in which SF_6 filled test cells containing defects producing known levels of PD have been placed in GIS, and the UHF coupler signals measured. In one test using a needle electrode in the cell, the onset of corona on the negative peak of voltage was measured by a conventional discharge detector as being less than 1 pC. The cell was then placed in a typical GIS section and the same pattern of discharge obtained from the UHF coupler, even though it was many metres away. Both this and other tests have demonstrated that the sensitivity required for the proposed CIGRE test procedure can be met under site conditions without difficulty.

Coupler Calibration

Even before a complete calibration is available, it is important to be able to measure the characteristics of the UHF couplers now being fitted to GIS. This enables the couplers to be designed both in terms of sensitivity and bandwidth to optimise the performance of a monitoring system fitted to a particular substation.

A recently developed coupler calibration system[4] is shown in Figure 5. A voltage step of risetime < 50 ps is applied to the input of a gigahertz transverse electromagnetic cell (GTEM), and propagates as a step electric field towards the open end of the cell. As the field passes over the coupler aperture in the top wall of the GTEM, the incident electric field is measured using a monopole probe with a known response. The voltage signal from the probe is converted to the frequency domain using a fast Fourier transform (FFT), and stored as a reference. The UHF coupler is then fitted over the aperture, and the same step field applied. Having measured the voltage output from the coupler, it is again converted by a FFT. Dividing this frequency domain signal by the reference signal previously stored allows the transfer function of the UHF coupler to be calculated. The most useful measure of the sensitivity of a UHF coupler is its voltage output per unit incident electric field; i.e., an effective height, H_e. This quantity can be interpreted as the height of an equivalent monopole probe located on the axis of the coupler, whose output voltage is simply its height multiplied by the amplitude of the incident electric field.

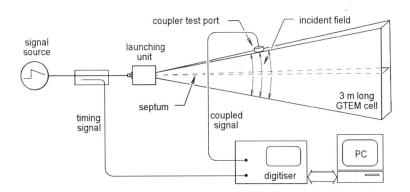

Figure 5. Block diagram of the UHF coupler calibration system.

Since the coupler test aperture is located at the mid-point of the GTEM, a reflection-free window of 10 ns is available for the measurement before the reflection from the open end of the cell returns. Consequently, most of the coupler response to the step excitation should be contained within this 10 ns period for an accurate measurement to be obtained. This requirement has been satisfied by all of the couplers that have been tested on the system.

The signal processing used to determine the frequency response of the coupler can extend the measurement bandwidth beyond the normal limits of the test equipment[5]. The basis of the procedure is to use identical test equipment and cables for both the incident field measurement and the coupler response measurement. Using this approach, a measurement system bandwidth of 2 GHz has been achieved.

An important aspect governing the performance of a coupler is its mounting configuration, because this determines how well the coupler responds to UHF fields inside the GIS chambers. Couplers are normally located in recesses, or externally on the GIS, which makes them more remote from the UHF fields. They should therefore be tested on a port that replicates their mounting configuration on the GIS.

As an example, Figure 6 shows an external coupler designed to fit a 90 mm diameter window on a 132 kV GIS. The cut-off frequency of the port is 1950 MHz, and relatively small lengths cause significant attenuation below this frequency. To investigate this, the coupler sensitivity was measured for port lengths d of 10, 24 and 61 mm. The results are given in Figure 7, which shows the attenuation of the signal increasing for higher values of d, particularly at lower frequencies.

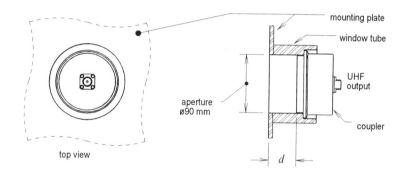

Figure 6. Mounting an external coupler for testing at different port lengths.

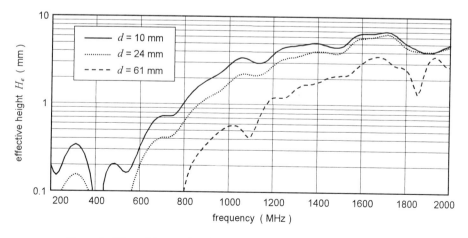

Figure 7. Effect of the port length on the frequency response of the coupler.

Specifying UHF Couplers

When preparing the specification for a new GIS with integral couplers, the coupler response should be specified to ensure that the required sensitivity of the complete monitoring system will be achieved. A specification in terms of effective height H_e over the frequency range of interest has been prepared by the National Grid Company[6], and an example is shown in Table 1. Here the minimum effective height $H_{e\,min}$ is defined as the sensitivity that must be exceeded over at least 80 % of the frequency range, and the mean effective height $\overline{H}_e$ is the sensitivity that must be exceeded by the average value of H_e over the full frequency range.

Table 1. Coupler specification for a 420 kV GIS

frequency range	minimum effective height $H_{e\,min}$	mean effective height $\overline{H}_e$
500 - 1500 MHz	2.0 mm	6.0 mm

The use of two conditions rather than simply specifying $\overline{H}_e$ ensures that the couplers are broadband. If the minimum value was not specified, the average value could be achieved by using a very sensitive, but narrow band coupler. The specification requires that the calibration is carried out with the coupler mounted on a replica of the GIS mounting arrangement.

SERVICE EXPERIENCE

On-line partial discharge monitoring (PDM) systems have been installed in a number of 400 and 500 kV GIS, and are proving a valuable means of detecting incipient failures. Most experience has been gained with the PDM at the ScottishPower 400 kV GIS at Torness, where it has been reported that early warnings of developing problems have enabled remedial action to be taken. It is said by ScottishPower that at least 3 in-service failures and possible disruptions of supply have been prevented in this way.

At present four PDM systems are operating on GIS in the UK, two in Singapore, and one each in Hong Kong and the USA. New PDM projects are under way in Korea, Malaysia and Brazil, and it expected that the number of installations will increase significantly as more utilities recognise the cost benefits they can bring.

Even prior to the publication of the CIGRE report of February 1998, several GIS in the UK and elsewhere had been commissioned with the aid of PDM systems; indeed the very first UHF measurements were made during the commissioning tests undertaken at Torness GIS in 1984. Such measurements are now made much more conveniently, because PDM software dedicated to the commissioning test enables data from up to 12 UHF couplers in the test section to be displayed simultaneously, and updated every few seconds. The data is also stored as a permanent record of the test.

Should the test voltage be provided by a variable frequency resonant test set, a low voltage signal from the test set enables the PDM to lock on to the test frequency and maintain the correct phase relationship in the data displays.

It is usual for the test voltage to be raised in two or three stages, and held for some minutes at these intermediate levels before being increased for the one minute test. During this time it is an education to see the conditioning processes in action; bursts of activity while a protrusion burns off, a patch of oxide is removed, or even a dust-like particle moved away or being destroyed. Then after a minute or so, the test section is usually completely free of discharges until the voltage is raised to the next level.

Unfortunately on occasions the defect is found to be permanent, whereupon the test may be suspended while it is removed, or the voltage raised with the possibility of causing a test flashover. Whatever the decision, the test engineer makes it knowing the type of defect present. Normally the GIS will be opened, but not before the defect is located. This is often done by a time-of-flight measurement of signals arriving at couplers on either side of the defect, but if the discharge is large enough an acoustic measurement is a simpler alternative.

At the conclusion of the test, measurements are often made at the service voltage, or just above, to ensure that the test section contains no measurable discharges and is in the best possible condition before handing it over to the utility.

REFERENCES

1. M. D. Judd and O. Farish, High bandwidth measurement of partial discharge current pulses, *IEEE Int. Symp. on Electrical Insulation (ISEI)*, Washington (1998)

2. J. S. Pearson, O. Farish, B. F. Hampton, M. D. Judd, D. Templeton, B. M. Pryor and I. M. Welch, Partial discharge diagnostics for gas insulated substations, *IEEE Trans. Dielectrics and Electrical Insulation*, Vol. 2, No. 5, pp. 893-905, (1995)

3. CIGRE Joint Working Group 33/23.12, Insulation co-ordination of GIS: Return of experience, on site tests and diagnostic techniques, *ELECTRA*, No. 176, pp. 67-97 (1998)

4. M. D. Judd, O. Farish and P. F. Coventry, UHF couplers for GIS - sensitivity and specification, *10th Int. Symp. on High Voltage Engineering (ISH)*, Montreal (1997)

5. J-Z. Bao, J. C. Lee, S-T. Lu, R. L. Seaman and Y. Akyel, Analysis of ultra-wide-band pulses in a GTEM cell, *Proc. Soc. Photo-optical Instrumentation Engineers (SPIE)*, Vol. 2557, pp. 237-248 (1995)

6. The National Grid Company plc, *Capacitive couplers for UHF partial discharge monitoring*, Technical Guidance Note TGN(T)121 Issue 1 (1997)

DISCUSSION

A. BULINSKI: (1) What is the effect of external interference on the sensitivity of the UHF technique? (2) What is the PD source location capability of this technique?

B. HAMPTON: (1) In a GIS fed by cables there is virtually no interference. However, when it is connected to overhead lines, interference from air discharges may be fed into the chambers. The frequency of this is mostly below 300 MHz, and it is rejected by fitting high-pass filters of, say, 500 MHz at the inputs of the monitoring system. Other interference might be seen from broadcast transmissions, where the antenna is close to the GIS. In exceptional cases this may need to be rejected by notch filters. (2) The most accurate means of locating a PD is by time-of-flight, using the signals from couplers on either side of the discharge. A two-channel, digital oscilloscope of 1 GHz bandwidth gives excellent results.

H. OKUBO: In Japan, Chubu Electric Power Company has developed the long distance 275 kV GIL. Before the construction, they constructed a test line to measure the signal attenuation and sensitivity characteristics of the UHF sensor. I myself have made measurements with Mitsubishi Electric Corporation. We obtained the results of very low attenuation rate, say, 3-5 dB/100 m, and even below 1 pC PD could be detected by the UHF sensor at 100 m distance away from the PD source. What do you think of this sensitivity compared to the case of GIS, requiring an UHF sensor every 20 m?

B. HAMPTON: The PD signals in a GIS are lost mostly by diversion at the junctions, and reflections at the mismatches caused by barriers, bends, changes in chamber diameter, and so on. Compared with these, the attenuation due to the skin effect is very low. In a GIL the signal does not suffer these losses to anything like the same extent, and so the couplers may be placed much further apart and still retain a good sensitivity.

DIELECTRIC PROPERTIES OF SMALL GAS GAP IN SF_6/N_2 MIXTURES

M. Ishikawa,[1] T. Kobayashi,[1]
T. Goda,[2] T. Inoue,[2] M. Hanai,[2] T. Teranishi[2]

[1]The Tokyo Electric Power company, Inc.
Tokyo, Japan
[2]Toshiba Corporation
Kawasaki, Japan

INTRODUCTION

Gas-insulated transformers (GIT) using SF_6 gas as an insulating medium offer the advantages of making substations nonflammable and reducing the space required. However, the possible effects of SF_6 gas on the environment have become a topic as "GREENHOUSE" gas[1,2]. Therefore, it is important to reduce SF_6 use in the future, and SF_6/N_2 or N_2 are considered to be suitable substitutes[3].

GIT has many small gas gaps in its dielectric structure. For example, turn-to-turn insulation of windings form wedge-shaped small gas gaps between insulated conductors. Such small gas gaps create dielectric weak points due to electric field concentration. Dielectric properties of small gas gaps are one of the important factors in the dielectric design of GIT.

In this research, we investigated the impulse dielectric properties of small gas gap in SF_6/N_2 mixtures experimentally.

EXPERIMENTAL SET-UP

The experimental setup is shown in Figure 1. The chamber has two inlets for SF_6 and N_2 which are to be filled separately. An electric fan is installed in the chamber to mix SF_6 and N_2. First, the chamber is filled with SF_6 and N_2 to a fixed pressure ratio. Then, the gas is agitated by the fan for 15 minutes. We equipped the chamber with an image intensifier (I.I.) camera to observe partial discharges occurring in a test sample.

The structure of the test sample is shown in Figure 2. A rod electrode is put on two earth electrodes with flat surfaces covered with a 1mm-thick insulator sheet. The rod electrode is covered with polyethylenterephthalate film (PET, 2mm thick) or kraft paper (KP, 1mm thick) at the straight part and has spheres at both ends. The diameter of the rod electrode is ϕ15mm. The earth electrodes are molded with epoxy resin to prevent wedge-shaped gas gaps forming between the earth electrodes and the insulator sheet.

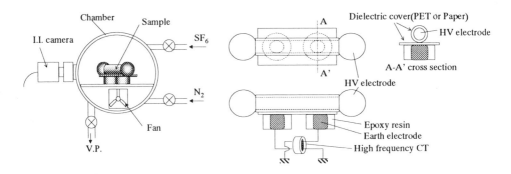

Figure 1. Experimental apparatus　　　　**Figure 2.** Test sample

A positive lightning impulse voltage (1.2/50μs) was applied to the rod electrode. The partial discharge inception voltage (PDIV) was measured with a high-frequency CT by using the current method. Two earth electrodes were used to make partial discharge measurements more sensitive by canceling the charging current.

Gas pressure was 0.1, 0.2, 0.35, 0.5MPa (absolute value). SF_6 content in N_2 was 0, 0.05, 0.15, 0.25, 0.5, 1 in case of PET covered model, 0, 0.05, 0.15, 0.5, 1 in case of KP.

RESULT AND DISCUSSION

PDIV characteristics of various gas pressure and SF_6 content measured for PET covered samples are shown in Figure 3. Each plot shows the average value of five data under the same test conditions.

A positive synergism was observed in the PDIV characteristics in SF_6/N_2 mixtures such as the uniform-field dielectric properties in the mixture[4]. Small contents of SF_6 to N_2 lead to a remarkable increase of PDIV at a higher gas pressure. And, PDIV increase has a tendency toward saturation at a SF_6 content higher than 0.5. On the other hand, At a lower pressure, the synergism was limited compared to the higher pressure. PDIV characteristics measured for the KP-covered samples are shown in Figure 4. There is a positive synergism for KP-covered samples as there is for PET covered samples.

264

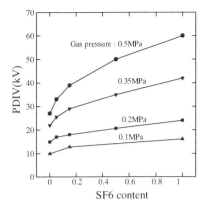 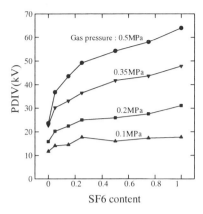

Figure 3. PDIV characteristics of PET sample   **Figure 4**. PDIV characteristics of KP sample

Light emissions of partial discharges observed with the I.I. camera at PDIV are shown in Figure 5. The partial discharge was brighter in N_2 than SF_6 at 0.1MPa. On the other hand, The region of light emissions of partial discharges become limited at 0.35, 0.5MPa. Although PDIV characteristics showed strong synergism at higher pressures, no clear difference was observed between N_2 and SF_6

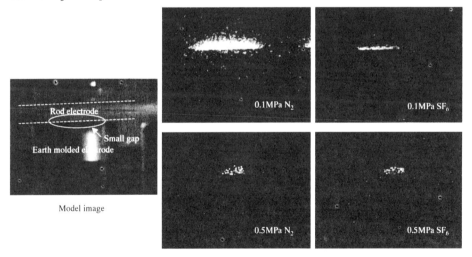

Figure 5. The image of partial discharge

Breakdown properties of KP samples are shown in Figure 6. Breakdown strength of PET samples are so high that breakdown properties of PET samples could not be measured. Breakdown voltages of KP samples are considered to be determined by the penetrating strength of kraft paper, that is solid breakdown. In spite of this a positive synergism is observed in the breakdown properties of KP samples as with PDIV characteristics. It is considered that breakdown voltages of KP samples were influenced by the properties of the dielectric gas, because KP has gas permeability.

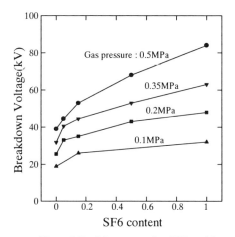

Figure 6. Breakdown properties of KP model

CONCLUSION

We investigated dielectric properties of small gas gaps formed by insulation covered electrodes in SF_6/N_2 mixtures. The following results are obtained.

(1) PDIV characteristics show a positive synergism as observed under uniform field conditions.

(2) The synergism is remarkable in the region of small SF_6 content and at higher pressures.

(3) Breakdown properties of kraft paper-covered samples also show a positive synergism.

REFERENCES

1. M.Maiss and I.Levin, Global increase of SF_6 observed in the atmosphere, *Geophysical Research Letters*, Vol.21, No.7, pp.569-572(1994)
2. L.G.Christophrou, J.K.Olthoff and R.J.Van Brunt, Sulfer Hexafluoride and Electric Power Industry, *IEEE Electrical Insulation Magazine*, Vol.13, No.5, pp.20-24(1997)
3. L.G.Christophrou and R.J. Van Brunt, SF_6/N_2 Mixtures, *IEEE Transactions on Dielectric and Electrical Insulation*, Vol.2, No.5, pp.952-1003(1995)
4. D.J.Skipper and P.I.McNeall, Impulse-Strength Measurement on Compressed-Gas Insulation for Extra-HV Power Cables, *Proc. IEE*, Vol.112, pp.103-108(1965)

DISCUSSION

Y. QIU: How small was the gas-gap length? Could you control the gas-gap length in your experiments?

T. GODA: We cannot determine the gas-gap length. Therefore, we can say that the discharge path is important and PDIV is determined by the Paschen minimum.

H. I. MARSDEN: How do you control moisture in your experiment? Can you address the role of moisture with regard to aging in a transformer?

T. GODA: In this experiment the test model is dried at 100 °C for 72 h. The humidity level is quite low. Therefore, the influence of moisture is not observed and there is no aging effect for the test model.

DISCRIMINATION OF STREAMER/LEADER TYPE PARTIAL DISCHARGE IN SF6 GAS BASED ON DISCHARGE MECHANISM

H.Okubo,[1] T.Takahashi,[1] T.Yamada[1] and M.Hikita[2]

[1] Dept. of Electrical Engineering
Nagoya University
Nagoya, 464-8603, Japan
[2] Dept. of Electrical Engineering
Kyushu Institute of Technology
Kitakyushu, 804-8550, Japan

INTRODUCTION

Since SF6 gas insulated power apparatus like gas insulated switchgears (GIS) have been widely used in high and middle voltage substations for electric power transmission and distribution, it is strongly needed to enhance the operational reliability of the apparatus.[1] For example, insulation defects like metallic particles in GIS would be a cause of partial discharge (PD) and could lead to dielectric breakdown (BD).[2] Thus, for the stable operation of GIS, it is very important to diagnose the insulation condition in service and to predict the BD using PD detection techniques.[3-5] Therefore, it is crucial to clarify the PD mechanism in SF6 and to develop the BD prediction technique in GIS. Under ac voltage application, however, space charges generated by PD show complicated behavior and have an influence to PD themselves, due to the temporal change of the instantaneous value and the polarity of applied voltage.[6] Thus, it is important to investigate the PD mechanism under ac voltage application. From these points of view, we have investigated the PD characteristics under ac voltage application in SF6 gas.

This paper describes, firstly, the temporal change of PD characteristics under long time voltage application until BD, and secondly, the discrimination technique of leader type discharge from PD using simultaneous measurement of current and light intensity waveforms of single PD. Finally, from these results and discussions, we conclude that the detection of the specific PD signal might be a promising technique for BD prediction.

LONG-TERM MEASUREMENT OF PARTIAL DISCHARGE AND THE PROCESS TO DIELECTRIC BREAKDOWN

Experimental

Figure 1 shows a model GIS and a test circuit for a long-term PD measurement. As a PD source, we used a needle electrode made of aluminum with 5mm in length and 0.25mm in diameter fixed on the high voltage conductor in the model GIS which simulated the severest condition of metallic particle in GIS. A plane electrode was placed 7.0mm below the needle tip to have a needle-plane electrode configuration. Phase-resolved PD characteristics were measured by a computer-aided PD measuring system through CR detection impedance. SF6 gas pressure in the model GIS was kept constant at 0.4MPa under room temperature condition.

Temporal Change of PD Characteristics

We applied ac 55kVrms, at which PD from needle electrode was generated constantly but BD did not occur. With keeping the applied voltage at 55kVrms, we started the measurement of PD characteristics until BD using the computer-aided phase-resolved PD measuring system for a long time.

Figure 2 shows a typical result of temporal change of PD behavior for time to BD t_B=149min. In Fig. 2, the charge magnitude of PD increases with time under the constant voltage application. Especially, as can be seen in Fig. 2 (f) at the stage just before the BD, sharp PD pulse with relatively large magnitude of charge is generated at the phase region of positive applied voltage peak. This fact would indicate that the growth of the magnitude of PD charge and the PD behavior have a close relationship to BD.

In order to clarify the reason of the pattern change of PD characteristics under constant voltage application, we observed the shape of needle tip during voltage application. Figure 3 shows the temporal change of needle tip radius for different exposing time of PD under constant ac voltage application. As seen in Fig. 3, the radius of the needle tip increases with exposing time of PD. The reason would be that the needle tip was gradually worn down by the tremendous number of PD.

On the other hand, the BD and positive PD inception voltage characteristics as a function of gas pressure for different needle tip radius are obtained as in Fig. 4. As increasing gas pressure, the BD voltage (BDV) once increases to the maximum, suddenly decreases to the positive PD inception voltage level, and then increases again. This characteristics is due to the corona stabilization effect of needle tip PD and called as "N characteristics". The pressure having the maximum BDV decreases with increasing the tip radius. The reason of this change would be that the leader discharge inception voltage decreases with the increase of the tip radius.

Consequently, it is obvious that the needle tip radius changes along with PD exposing time leading to the change of N-characteristics and the decrease of the leader transition voltage, and the inception of the leader discharge causes the final BD. From this point of view, in order to predict the BD in GIS, it is worthy to detect and discriminate leader type discharge from streamer discharge, under normal operating condition of GIS.

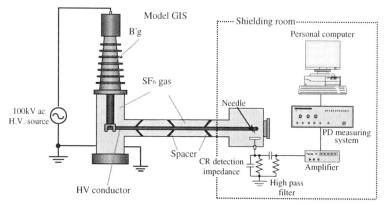

Figure 1. Experimental setup for long-term PD measurement.

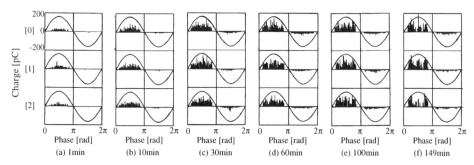

Figure 2. PD phase characteristics from PD inception to BD during long time ac voltage application.

φ=0.25mm, g=7mm, Applied voltage Va=55kVrms, Time to BD t_B=149min, Gas pressure P=0.4MPa.

(a) 10min (b) 60min

(c) 149min

Figure 3. Change of neelde tip shape for different time of ac voltage application under PD condition. (Va=55kVrms)

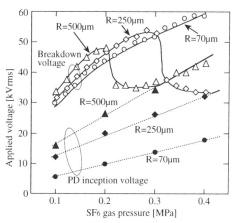

Figure 4. Breakdown voltage and positive PD inception voltage characteristics as a function of SF6 gas pressure.

DISCRIMINATION OF LEADER TYPE PARTIAL DISCHARGE

Experimental

Figure 5 shows an experimental circuit for simultaneous measurement of a single PD under ac voltage application. We used a needle electrode with tip radius r=500μm and length ℓ=20mm, made by stainless steel, fixed on the high voltage conductor of model GIS to simulate a metallic particle in GIS. Ac voltage of 60Hz was applied to the needle to generate PD in SF6 gas. A plane electrode with a diameter R=30mm was set 20mm below the needle electrode.

PD pulse current signal detected by the plane electrode was fed into a high speed digital oscilloscope (Sampling rate : 10GS/s, analog band width : 750MHz) through a matching circuit having broadband response, 1MHz to 1GHz, or more.

PD light intensity was observed using dielectric mirrors, optical lenses, two types of photo multiplier tubes (PMT-B and -R) and a digital oscilloscope (5GS/s, 750MHz) so that we could observe PD light intensity waveforms for two different spectral ranges, 420nm~540nm and 630nm~700nm, respectively.

Moreover, light images of PD around the needle tip were recorded with lenses, an image

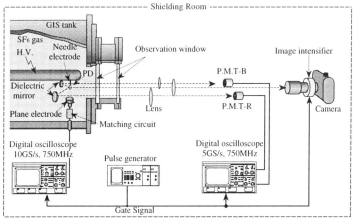

Figure 5. Simultaneous measurement of current pulse, light intensity waveform and light image of a single PD in SF6.

intensifier (I.I.) and a still camera. The I.I. has gate function with the width of 100ns at an arbitrary phase of applied ac voltage and enables to enhance the light intensity up to 5.3×10^6 times with wavelength at 180~700nm. We constructed the simultaneous measuring system to make it possible to observe the single PD current pulse and corresponding light intensity waveform and light image simultaneously by controlling the I.I. and the digital oscilloscopes with the gate signal synchronized with applied ac voltage phase.

Discrimination of Leader Discharge from Streamer Discharge by Simultaneous Measurement of PD Current Pulse, Light Intensity and Light Image

Figure 6 shows a typical current pulse and light intensity waveform and light image of a corresponding single PD pulse occurring at around 90° phase angle of applied voltage $V_a=40kV_{rms}$ at gas pressure P=0.3MPa.

It is obvious from Fig. 6 that the PD current waveform in SF_6 has very short rise time T_r : $T_r\approx0.8ns$. The shape of light intensity waveform has also very short rise time and similar to the current pulse waveform. Moreover, the PD light emission image shows that the PD extends less than 1mm length from the needle tip. From the result of the electric field calculation, it is revealed that PD shown in Fig. 6 extends shorter than the length to the point where the electric field strength is equal to the critical electric field strength of SF_6 gas; E_{cr} ($E_{cr}/P=89V/m\cdot Pa$). Thus, we can conclude that the PD shown in Fig. 6 is the streamer type discharge.[7]

Figure 7 shows the typical example of the PD current pulse waveform and corresponding light intensity waveforms of PD with double peaks. It is very interesting that PD having double peaks indicated in Fig. 7 was obtained depending on the gas pressure and applied voltage condition. This fact will be discussed in detail later. From Fig. 7, we can find the similarity of the current

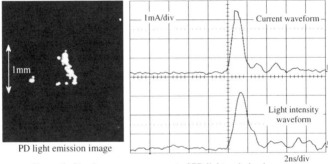

Figure 6. Simultaneous measurement of PD light emission image, PD current pulse and PD light intensity waveforms. ($V_a=40kV_{rms}$, P=0.1MPa)

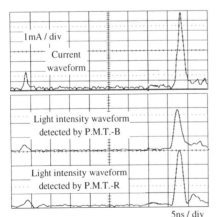

Figure 7. Light wavelength characteristics of double peaks PD current pulse in SF_6 gas. ($V_a=40kV_{rms}$, P=0.4MPa)

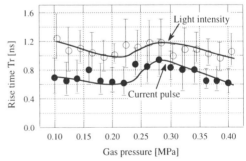

Figure 8. Rise time Tr of PD current pulse and PD light intensity waveforms for different gas pressures. ($V_a=40kV_{rms}$, 90°)

pulse and the light intensity waveforms, but easily recognize the difference of intensity between PMT-B and PMT-R. Now, we define the value of the peak intensity of the smaller pulse (first pulse) measured by PMT-R and PMT-B as R1 and B1, the larger pulse (second pulse) measured by the PMT-R and the PMT-B as R2 and B2, respectively. It is seen from Fig. 7 that the ratio (R2/B2) is larger than (R1/B1). Similar measurement were repeated 17 times and showed the value 1.4 as an average ratio (R2/B2)/(R1/B1). The ratio (R1/B1) and (R2/B2) mean the ratio of the red and blue light band component of PD light emission for the first pulse and second one, respectively. In general, it is known that the leader discharge in SF_6 gas is accompanied by longer light wavelength component compared with streamer discharge.[7-9] From this point of view and the result of the average of the ratio (R2/B2)/(R1/B1)=1.4, we could conclude that the first peak of the waveform is the streamer and the second peak is the leader discharge in Fig. 7.

Figure 8 shows the gas pressure dependence of rise time of PD current pulse and PD light intensity waveforms at V_a=40kVrms. As seen in Fig. 8, rise time of PD current pulse becomes longer at the gas pressure around 0.3MPa. This would be because leader discharge having a longer rise time of current pulse were generated at around 0.25~0.3MPa.

Pressure and Applied Voltage Dependence of Leader Discharge Transition

The measurements mentioned above were carried out for different gas pressure and applied voltage. Figure 9 shows the obtained generation ratio of the double-peak current pulse for different pressure and applied voltage. It should be noted that the measurements of the current waveforms were focused on PD occurred at 90° voltage phase angle with the detection sensitivity of 0.5mA corresponding to 0.5pC of charge. To distinguish the double peaks characteristics from burst corona, we took only the waveforms with time interval 20~100ns as the double-peak current pulse.

As can be seen in Fig. 9, at the low pressure region P≤0.2MPa, the generation probability of the current pulse waveform with double peaks is less than 0.1% irrespective of applied voltage. On the other hand, at the high pressure region P≥0.2MPa, the current pulses with double peaks occur more frequently with increasing the applied voltage approaching to the BDV at each gas pressure. Thus, it is pointed out that there is a double-peak current pulse region and the generation probability depends highly on the applied voltage and gas pressure.

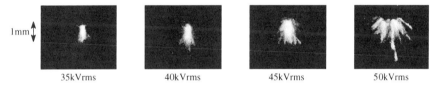

Figure 10. PD light emission image for different applied voltage. (P=0.3MPa, 80°~100°, 60 ac cycles)

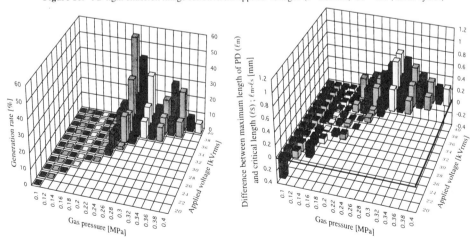

Figure 9. Generation rate of double-peak current pulse for different gas pressure and applied voltage. (90°)

Figure 11. Difference between maximum distance of PD light image ℓ_m and critical distance ℓ_s as functions of gas pressure and applied voltage.

Next, we obtained the maximum extension length of PD from the measurement of the summing up PD light emission image for 1sec, as shown in Fig. 10. We can define the PD maximum extension length measured at different gas pressure and applied voltage as ℓ_m. We also calculated the length ℓ_s which was the length from the needle tip to the point that electric field strength was equal to the critical electric field E_{cr}. From these results, we can obtain Fig. 11 which shows the difference between maximum length ℓ_m and critical length ℓ_s ; $\ell_m - \ell_s$. Note that the value $\ell_m - \ell_s \leq 0$ means that the discharge extends only within the region of the critical electric field E_{cr}, and $\ell_m - \ell_s > 0$ means that the discharge extends beyond the region of the electric field E_{cr}.

As shown in Fig. 11, it can be said that PD extends the length less than the critical length in the gas pressure region P<0.18MPa, on the other hand, longer than the critical length in the gas pressure region P≥0.18MPa. Furthermore, the extension length of PD increases with the increase of the applied voltage for each gas pressure. Generally, the leader discharge extends longer distance to the region of the electric field less than E_{cr} and the streamer discharge within the region of the electric field more than E_{cr}.[7] From these discussions and the evidence of the measurement, the leader discharge may occur at the gas pressure region P≥0.18MPa in Fig. 11.

The comparison between the Fig. 9 and Fig. 11 gives us the similarity of the pressure-voltage region of double-peak current pulse generation and the leader discharge transition. That is to say, the double-peak current waveform could be the evidence of the occurrence of the leader discharge. Generally, the leader discharge occurs after the streamer discharge generation sequentially.[7] Therefore, the first small peak of the current waveform shown in Fig. 7 can be considered as the streamer discharge and the second large peak as the leader discharge, respectively. The time interval between the first peak and the second peak means the leader transition. These evidence give us a deep consideration that the detection of double-peak current pulse shows the leader type discharge generation and could lead to BD prediction in GIS.

CONCLUSIONS

In this paper, we described the measurement results of PD characteristics and transition process from PD to BD under operating condition of GIS.

Firstly, we investigated the long-term temporal change of PD characteristics. As a result, we could find that the PD characteristics temporally changed and finally PD led to breakdown. From this result, we could pointed out that the needle tip might be worn by PD, that could cause the change of breakdown characteristics due to the increase of the possibility of the leader discharge generation. Thus, we concluded that the breakdown followed by the long-term PD occurrence was due to the temporal increase of leader generation rate and it suggested the importance of the measurement of PD, especially the leader discharge generations.

Moreover, we introduced the simultaneous measurement of PD with electrical and optical method to distinguish the leader type discharge from a lot of PD and we could find the similarity of two parameters; the double-peak current pulse generation rate and the leader discharge transition. By applying this finding fact, we pointed out that the detection of the double-peak current pulse of PD might be a promising technique to predict BD in GIS.

REFERENCES

1. S.Yanabu, Y.Murayama and S.Matsumoto, SF_6 Insulation and its Application to HV Equipment, *IEEE Trans. on Elect. Insul.*, Vol.26, No.3, pp.358-366 (1991).
2. R.Baumgärtner, *et al.*, Partial Discharge - Part IX: PD in Gas-Insulated Substation - Fundamental Considerations, *IEEE Elect. Insul. Magazine*, Vol. 7, No. 6, pp. 5-13 (1991).
3. T.Utsumi, *et al.*, Preventive Maintenance System with a Different Gas Injecting Facility for GIS, *IEEE Trans. on Power Delivery*, Vol. 8, No. 3, pp. 1107-1108 (1993).
4. H.Muto, *et al.*, Frequency Spectrum Due to Standing Waves Excited by Partial Discharges in a GIS, *Proc. of the 10th Int. Symp. on High Voltage Eng.*, Vol. 4, pp. 179-182 (1997).
5. M.D.Judd, O.Farish and J.S.Pearson, UHF Couplers for Gas-insulated Substations: a Calibration Technique, *IEE Proc. of Sci. and Meas. Tech.*, Vol. 144, No. 3, pp. 117-122 (1997).
6. R.J.Van Brunt, *et al.*, Mechanisms for Inception of DC and 60-Hz AC Corona in SF_6, *IEEE Trans. on Elect. Insul.*, Vol. 17, No. 2, pp. 106-120 (1982).
7. I.Gallimberti and N.Wiegart, Streamer and Leader Formation in SF_6 and SF_6 Mixtures under Positive Impulse Conditions: II. Streamer to Leader Transition, *J. Phys. D: Appl. Phys.*, Vol.19, pp.2363-2379 (1986).
8. W.Pfeiffer, *et al.*, Analysis of Prebreakdown Phenomena in SF_6 in Non-uniform Gaps for Very Fast Transient Voltage Stress Using Electro-optical Diagnostics, *11th Int. Conf. on Gas Dis. and Their Appl.*, pp.II-92-95 (1995).
9. V.Zengin, *et al.*, Analysis of SF_6 Discharge by Optical Spectroscopy, *Gaseous Dielectrics VI*, pp.595-599 (1991).

DISCUSSION

R. ARORA: (1) For leader PD in free gas the gap distance of 20 mm is perhaps too short. In my opinion, it may give rise only to stable streamers and unstable leaders just before the breakdown. (2) Some solid dielectric support to the H. V. electrode must have been provided in the experimental setup. Was the possibility of surface discharge (tracking) checked? Surface discharge or tracking is normally leader discharge.

T. TAKAHASHI / H. OKUBO: (1) Not only the streamer PD but also the leader PD could be generated in the electrode configuration with highly non-uniform fields as we used. The generation and transition of the leader PD is reported by Gallimberti et al. (Ref. 7). We confirmed the "leader" PD generation with optical and electrical measurements shown in this paper. (2) We did not use solid dielectrics around the PD source. Therefore, the surface discharge is out of consideration in this paper.

A. GOLDMAN: Have you an idea of the temperature reached by the gas inside the discharge paths under your different working conditions? Effectively, the leader formation implies a temperature high enough to eliminate (or, at least, to strongly reduce) electron attachment, that will result in an increased electrical conductivity of the gap, as has been shown in the paper presented at this symposium by Champain et al.

T. TAKAHASHI / H. OKUBO: As you pointed out, the temperature inside the leader channel is high enough to increase the electrical conductivity. This could result in the difference of the ratio of the red component to the blue component of light emission by the leader PD and by the streamer PD shown in our paper. This difference could be due to the difference of the degree of ionization. The SF_6 molecules in the leader channel are sufficiently ionized into S^+ and F^+ to emit red component of light emission larger than in the streamer channel.

THE INFLUENCE OF ACCELERATED PARTIAL DISCHARGES TESTS ON THE EFFECTS ON EPOXY COMPOSITE SURFACE IN SF6

Helena Słowikowska, Tadeusz Łaś, Jerzy Słowikowski

High Voltage Department
Electrotechnical Institute
Warsaw, Poland

INTRODUCTION

As it is well known, the surface strength of insulators plays an essential role in the reliability of insulating systems with SF6. Partial discharges (PD) decides to a great degree on initiation of slowly developing defects which impair the electric surface strength.
Checking the possibilities of selecting parameters of accelerated ageing in surfaces of epoxy composites in gaseous SF_6 medium at ac voltages was the goal of the measurements mentioned above.

The tests were conducted at a pressure of about 1 bar in order to compare the results with those of tests made in air atmosphere.

The waveshapes of mean PD currents **I**, and absolute values of the max. charge **q**, which occurred in positive or negative half-cycles of the sinusoidal voltage were recorded and cumulated charge **Q** during measurements.:

$$Q = \int_o^{tn} I \, dt \qquad [1]$$

where: t_n - exposure time
I - acc. to IEC 270, equal to
$$I = 1/T \,[\, |q_1| + |q_2| + \ldots\ldots |q_n|\,] \qquad [2]$$
T - period of the ac waveform).

Scanning profilometry (measurements of the changes in the dielectric`s surface), surface resistance measurements and gas chromatography analysis (concentration of SF_6 by-products) were used to quantitatively evaluate the effect of PD action. Qualitative evaluation was made by means of optical and scanning microscopy monitoring (between others changes in the shape of the point electrode).

EXPERIMENTAL SETUP

The partial discharges were generated in a sealed vessel, made of stainless steel, with a volume of 300 ml. The upper (HV) point electrodes were used, made of stainless steel with curvature of 53 µm mounted on a structure which enable precise regulation of the distance d between point electrode and tested plate of dielectric. The distance from 0.5 to 5.0 mm was kept during the tests. The specimens, to which the presented here results refer, were epoxy resin composite planar sheets of thickness 2 mm, based on besphenol A and epichlorhydrite with alumina filler [1]. The experiments were performed with SF6 of 99.9% purity introduced at a pressure of 140 kPa in chamber which was evacuated before. The amount of residual air in the vessel in initial state was ≤ 0.2.%. This level of air linked to a practical level in equipments [2]. Before initiating the tests, the samples were dried at the 90° C during 24 hours.

The tests were carried out under sinusoidal voltage. During the exposure, PD activity was measured by means of a conventional PD detector (Tettex 9126) and stored in PC memory via RS232 interface. The following PD parameters were determined : average current I, maximum value of PD pulses charge q (±) in plus and minus half-cycles.

METHODS USED IN RESEARCH AFTER TESTS

- Complex measurements of roughness parameters: scanning profilometer with 32-bits computer IBM PC and Turbo Pascal 6.0 software was used. The profilograms, maps and 3-dimentional upturned views were made. On this base surface measurements such as : decrement volume (v), (volume of material removed), the area of decrement (a) and maximum depth of decrements (d max.) were calculated by computer analysis [3,4].
- Surface resistance R_s - was measured using the Keithley 617 electrometer with DC 100 volts. Parallel 5 mm long brass electrodes with a distance of 0.5 mm between them were moved on the epoxy samples from outside of the area to the central point of exposure.
- Gas chromatograph Carlo Erba , model Vega 6000 housing column packed with Poropack Q was used. Samples of gas, volume 1 ml were taken from the cell, during the tests and directly after the tests.
- Microscope observations were made with optical and electronic scanning microscope, using 150x and 300x magnification.

CHARACTERIZATION OF EXPOSURE AND ITS EFFECTS

Parameters of exposures

Tests were made at three distances between point electrode and sample surface (0.5; 2.0 and 5.0 mm) at three voltage values (5 ;7 and 10 kV) and two frequences (50 and 1000 Hz). Exposure times at 50 Hz were within 48h and 120h, in one case also 428h while at 1000 Hz within 2.4h and 6h.

An initial working hypothesis was assumed that similar exposure hazards should occur at 1000 Hz and 50 Hz if the ratio of exposure time were 1:20 respectively.

Values of the maximum (apparent) charge (positive and negative) q (±) fluctuated within different limits depending on test conditions where the lowest value (sample No.2, Table.1 - 0.5mm, 5 kV, 50 Hz) was 3 pC while the highest one (sample No.3, Table.1 - 0.5mm, 7 kV, 50Hz) was 1500 pC. In one case (sample No 0.5mm, 7kV, 50Hz) the characteristic was unstable; break-through streamer 10000 pC appeared periodically.

Starting from the sample No.3 the following values were recorded: mean current I, max. charge q(±) and total charge Q during the whole duration of the test. The value of Q, according to the erlier work was assumed to be the measure of exposure at given test conditions [5]. Depending on test conditions and exposure time the value of Q was within 3.8 mC and 340 mC. Exemplary PD characteristics as a function of test duration are shown in Fig.1.

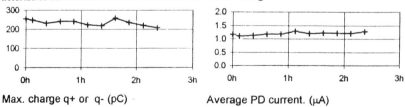

Max. charge q+ or q- (pC)   Average PD current. (µA)

Figure 1. Exemplary PD characteristic

Table 1. presents exposure conditions and a description of the appearance on the surface of the samples exposed.

Table 1. Characterization of exposure and its effects

(Notation: d-distance between point electrode and sample surface, U- voltage,
f- frequency of voltage, t- exposure time, q(±) - max. apparent charge
(positive and negative), Q- total apparent charge)

d mm	U kV	f Hz	No	PD characterization	Changes in sample surface
0.5	5	50	1	t = 114 h q (±) = 10 ÷ 100 pC,	Droplets at exposure center, single bubbles (blisters) around the center. Growth of roughness of the upper layer in both areas revealing filler grains (Fig.4 a,b,c). Drop in R_S at exposure center down to 10^{10} ohm (Fig.4).
	7	50	2	t = 120 and 480 h Unstable characteristic; q(±) reaches 10 000 pC intermittently. In the remaining time intervals q (±) = 3 ÷ 35 pC.	Together with a growth in roughness at exposure center and its surrounding - intense deep-seated defects (Fig). Drop in R_S down to 10^6 ohm (Fig 4).
2.0	10	1000	5	t = 2.4 h, q(±) = 800 ÷ 1500 pC; $Q_{2,4}$ = 3.8 mC.	Droplets and carbonized particles present - more and more frequent at exposure center.
		50	4	t = 48h, q(±) = 50 ÷ 400 pC; Q = 80 mC,	
				t = 120h, q(±) = 100 ÷ 1000 pC; Q = 340 mC,	A spot of liquid with carbonized intrusions at exposure center (Fig.3a). numerous droplets around the center. Increased surface roughness below droplets (Fig.3b. surface after washing)
5.0	10	1000	5	t = 2.4 h, q(±) = 500 ÷ 900 pC; Q = 5.2 mC,	Single droplets and carbonized points at the center. Drop in R_S down to 10^{10} ohm.
		50	6	t = 48 h, q(±) = 210 ÷ 260 pC; Q = 33 mC,	Exposure effect analogous as that in sample No 4 after 120h but less intensive.
		1000	7	t = 2.4 h, q(±) = 210 ÷ 260 pC; Q = 10.5 mC,	Exposure effect analogous as that in sample No.5, after 2.4 h but less intensive.

TEST RESULTS

Surface disturbances in the form of roughness (Fig.2 a and b) occured in all samples where no break-through streamer discharges were found.

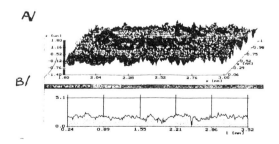

Figure 2. Surface disturbances after stable PD (A -upturned view, B- profilogram)

That disturbance was of fuzzy nature without occurrence of a larger material defect at exposure center. Presence of fluid droplets and carbonized particles was established in each of the cases mentioned above. With increased exposure, the presence of fluid droplets formation was observed (Fig.3a,b).

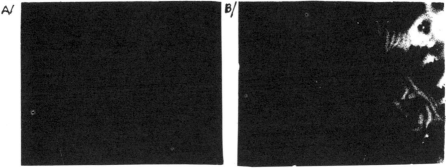

Figure 3. Surface changes after PD action : A) fluid drops and carbonized particles,
B) material defects under the drops (visible after washing)

More intensive effect, with the presence of bubbles (blisters),in the case of epoxy composite with SiO_2 filler, was observed by Chu[6].

Electrode distance, value of voltage and, first of all, intensity of exposure, whose measure in the cases concerned was Q, decided on the amount of effects. Quantitative evaluation of the effect of voltage and electrode distance on exposure results proved to be difficult to assess, because of limited possibilities of quantitative determination of PD effects. More precise data gave the changes of R_s; it was found, between others, that the distance **d,** increased from 0.5 mm to 5 mm (at similar values of Q), caused increased area with lower value of R_s (comp. Fig.4, curve 1 and 2) which can be ascribed to the enlarged area of reactions in gas, which - in the second case -are more intensive.

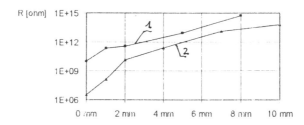

Figure 4. Surface resistance of R_S as a function of radial distance from the center.
(conditions: 1- 0.5mm,5 kV,50 Hz,120h; 2- 5.0mm, 7 kV,50 Hz, 120h).

An interesting is the fact of accumulation of carbonized particles at exposure center first of all. The presence of those particles was the consequence of erosion of the point electrode (comp. Fig.5 a and b)

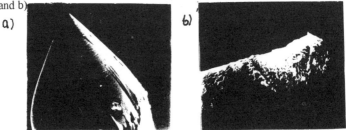

Figure 5. The erosion of the point electrode after PD action

Most probably it decides on the occurrence of a minimum in R_s at exposure center. Increase the content of CO_2 in gas (fig.6) should also be ascribed to electrode erosion, according to observations which were made in this respect by Siddigangappa a. Van Brunt [6].

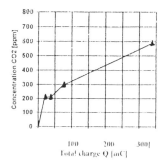

Figure 6. Increase the content of CO_2 as a function of Q.

Relation between concentration of durable decomposition products of SF_6 ($SOF_2 + SO_2F_2$) and the low values of total charge Q, was established (Fig.7). The form of the relation is near linear.

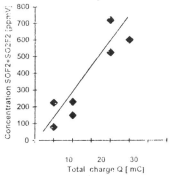

Fig. 7. Sum of concentration SOF_2 and SO_2F_2 as a function of Q

The hipothesis concerning PD exposure accelerated by 20 times as a result of voltage frequency increased by 20 times (from 50 Hz to 1000 Hz) (level of tests No. 4 and 5 as well as 6 and 7, Table 1), was not confirmed. The ratio of Q after exposure times selected in relation of 1:20 was not 1:1 but 15.4; 16.2 at distance d=2 mm and 3.2 at a distance d= 5.0 mm.

In the case of occurrence of break-through streamer ($q(\pm) \approx 10000$ pC - see test no.2, Table.1) a significant defect of material volume was established at exposure center, similarly as was observed by authors in the case of PD in air at the same distance of electrodes and ac frequency 50 Hz (d=0.5 mm/f=50 Hz) but at a lower voltage (4 kV) [7,4] In this case, the intensive development of SF_6 by-products was also observed (Table.2).

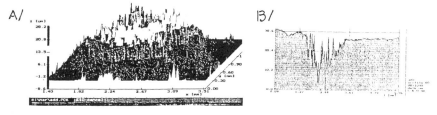

Figure 8. Upturned view (A) and profilogram (B) of the surface after unstable PD (see poz.2 in Table 1)

Table 2. The influence of the presence of streamers on the surface
effects and concentration of SF_6 by-products
(The mean condition of the test: d = 0.5 mm, f = 50 Hz, t = 120 h.)

Sample	Profilometry parameters			SF_6 by-products content [vpm]			
	a [mm^2]	V [$mm^3 \times 10^{-5}$]	d_{max} [μm]	CF_4	SOF_2	SO_2F_2	CO_2
1- stable PD charact. (see No.1- Table.1)	0.07	16.7	1.9	<50	199	<100	< 50
2 - unstable PD charact.*/ (see No.2 - Table 1)	2.3	1232	31.6	440	18800	38800	2720

*/ breakdown strimers present occasionally

CONCLUSIONS

With occurrence of partial discharges of max. apparent charge not exceeding 1500 pC, the PD effect on dielectric surface is reduced to chemical action of SF_6 decomposition products. Apart of disturbance to the surface of epoxy composite which manifests itself by increase in roughness, this action causes decrease in surface resistance R_S. A significant role in this decrease is played by condensation of liquid by-products of SF_6, inclusive deposition of conducting particles created by erosion of the point electrode. Deposition of the latter occurs first of all at exposure center. Both effects are observed in the presence of air in gas below 0.5 %$_{vol.}$

In the electrode system applied, the expected intensification of PD exposure with respect to the voltage of 50 Hz did not occur, , at a voltage of 1000 Hz. This was caused by considerable decrease in pulse numbers occuring during voltage cycle.

Occurrence of break-through streamers with charges ≥ 10 000 pC caused surface disturbance depending on in-depth-volume-defects of epoxy composite at exposure center. This phenomenon is accompanied by a significantly more intensive evolution of gaseous SF6 decomposition products than in the case discussed earlier.

ACKNOWLEDGMENTS

Investigations were conducted in the framework of cooperation between the Electrotechnical Institute in Warsaw (Poland) and the National Institute of Standards and Technology (USA), financed by the Maria Skłodowska-Curie Fund II.

REFERENCES

1. R.Rygał.,H.Latour-Słowikowska, J.Słowikowski, D.Paziewska.,*The Influence of the filler characteristics on the surface resistivity of epoxy composition applied in SF6 insulation systems*, Conf. Record of the 1992 IEEE Intern. symp. on Electr. Insul; F-16, pp.279-282 (1992).
2. G.Mauthe and aths.*SF_6 recycling guide; re-use of SF6 gas in electrical power equipment and final disposal*, TF 23.10.01 Electra,No. 173, August,1997.
3. H.Słowikowska,,T.Łaś, J.Słowikowski,J.R.Van Brunt,*Modification of cast epoxy resin surfaces exposure partial discharge*,VII Intern. Symp."Gaseous Dielectrics", USA, 1994.
4. H.Słowikowska,J,Słowikowski,T.Klimczak,J,Dukaczewski,A,Winiarska,*Contribiution to the Qualitative and Quantitative Estimation of the PD Effect on the Surfac3e Layer of Epoxy resins and Composites*, VII Intern. Conf. Dielectr. Mat. Meas. a. Appl., UK, pp.23-26(1996).
5. R.J.Van Brunt, P.von Glahn, T.Łaś ,*Nonstationary behavior of partial discharge during discharge indused ageing of dielectrics*, IEE Proc-Sci. Meas.Techn.,Vol.142,No.1,pp 37-45.
6. M.C. Siddagangappa and R.J. Van Brunt, *Decomposition products from corona in SF_6/N_2 and SF_6/O_2 Mixtures*, Proc. 8th Int. Conf. on Gas Discharges and their Applications, Leeds University Press, pp. 247-250, 1985.

INFLUENCE OF VOID GEOMETRY AND BULK DIELECTRIC POLARIZATION UPON PD TRANSIENTS

I. W. McAllister and G. C. Crichton

Department of Electric Power Engineering
Technical University of Denmark
DK-2800 Lyngby, Denmark

INTRODUCTION

Through a field-theoretical approach, a physically valid theory of partial discharge transients has been developed. The theory is based upon the concept of the charge induced upon the detecting electrode by the partial discharge. Mathematically, this induced charge is composed of a component associated with the actual space charge in the void, and one related to changes in the bulk dielectric polarization brought about by the space charge.

The charge induced by a partial discharge on the detecting electrode can be evaluated using either the λ function, or the ϕ function.[1,2] These functions take account of the dielectric polarization implicitly (λ) or explicitly (ϕ). Hence by using these two functions it becomes possible to identify the influence of the dielectric polarization upon the induced charge. The effect is associated with the change in polarization $\delta\vec{P}$ due to an incremental change in the electric field strength $\delta\vec{E}$. The source of the latter is the partial discharge space charge.

In the present paper, the influence of the void geometry upon $\delta\vec{P}$ is examined for a heterogeneous bulk dielectric system. It is shown that the component of the induced charge due to $\delta\vec{P}$ may increase or decrease depending upon the ratio of the dielectric permittivities and within which dielectric the void is located. The magnitude of this effect is also strongly dependent upon the prolateness/oblateness of the void geometry. This behaviour is reflected directly in the Poissonian induced charge, and consequently in the recorded partial discharge transient.[3]

POISSONIAN INDUCED CHARGE

The induced charge can be described in terms of a Poissonian and a Laplacian component.[3] The Poissonian induced charge is that component of the induced charge which is rigidly linked to the space charge source, and which together with this source gives rise to the Basic Poisson Field.[2] Mathematically, the final value of the Poissonian induced charge q, due to a partial discharge, can be resolved into two components:

$$q = q_\mu + q_P \tag{1}$$

The induced charge component q_μ is directly associated with the space charge in the void; q_p represents the induced charge related to the change in dielectric polarization ($\delta\vec{P}$) brought about by the presence of this space charge.[2] With reference to induced charge, the effect of the void wall charges can be considered equivalent to that of an electric dipole of moment $\vec{\mu}$ located within the void.[1] The Poissonian induced charge arising from a dipole is given by

$$q = -\vec{\mu}\cdot\vec{\nabla}\lambda \qquad (2)$$

where λ represents the proportionality factor between the charge in the void and the induced charge on the detecting electrode. The λ function is a solution of the general Laplace equation[2]

$$\vec{\nabla}\cdot(\varepsilon\vec{\nabla}\lambda) = 0 \qquad (3)$$

in which ε denotes permittivity. The boundary conditions are $\lambda = 1$ at the detecting electrode, and $\lambda = 0$ at the surfaces of all other electrodes. In addition, the following condition must be fulfilled at all dielectric interfaces, such as the walls of voids; viz.

$$\varepsilon_+(\partial\lambda/\partial n)_+ = \varepsilon_-(\partial\lambda/\partial n)_- \qquad (4)$$

where λ is differentiated in the direction normal to the interface, and the signs + and − refer to each side of the interface, respectively. Any method of solving Laplace's equation can be used to determine λ.

If, however, the dimensions of the void are such that $\vec{\nabla}\lambda$ may be assumed constant within the void, then we can introduce another function, λ_0. This function, which is derived for the same boundary conditions but with the void absent, is designated the unperturbed λ function. As both these functions are solutions of Laplace's equation, then by mathematical analogy with electrostatic fields, the relation between the λ and the λ_0 function is given by

$$\vec{\nabla}\lambda = h\vec{\nabla}\lambda_0 \qquad (5)$$

For the voids under consideration, the parameter h is a scalar which depends on the void geometry and the relative permittivity of the bulk medium. Following the introduction of λ_0, the Poissonian induced charge on the detecting electrode may be expressed as

$$q = -h\vec{\mu}\cdot\vec{\nabla}\lambda_0 \qquad (6)$$

The component of the Poissonian induced charge related to the void space charge *alone* may be obtained from

$$q_\mu = -\vec{\mu}\cdot\vec{\nabla}\phi \qquad (7)$$

where ϕ, another proportionality factor, is a solution of the reduced Laplace equation[2]

$$\nabla^2\phi = 0 \qquad (8)$$

The boundary conditions are $\phi = 1$ at the detecting electrode, and $\phi = 0$ at the surface of all other electrodes. Hence from (1), (6) and (7), the polarization component q_p of the Poissonian induced charge may be expressed as

$$q_p = -\vec{\mu}\cdot(h\vec{\nabla}\lambda_0 - \vec{\nabla}\phi) \qquad (9)$$

HETEROGENEOUS DIELECTRIC SYSTEM

The influence of $\delta\vec{P}$ upon the induced charge will be examined for a void located in a planar heterogeneous dielectric geometry. With this simple system, it is possible to illustrate the basic consequences associated with $\delta\vec{P}$.

We consider a planar electrode geometry with a two layer dielectric. If in rectangular coordinates, the electrodes are represented by $z = 0$ and $z = d$, then the dielectric interface is taken as $z = s$, with $s < d$. The permittivity of the upper dielectric is ε_2 for which $s \leq z \leq d$, while while that of the lower is ε_1, for which $0 \leq z \leq s$.

If the lower electrode is used as the detecting electrode, then the boundary conditions for the λ_0 function are $\lambda_0 = 1$ for $z = 0$ and $\lambda_0 = 0$ for $z = d$. Hence the λ_0 functions of the two media are given by

$$\lambda_{01} = \frac{\varepsilon_1(d - s) + \varepsilon_2(s - z)}{\varepsilon_1(d - s) + \varepsilon_2 s} \tag{10}$$

for $0 \leq z \leq s$, and

$$\lambda_{02} = \frac{\varepsilon_1(d - z)}{\varepsilon_1(d - s) + \varepsilon_2 s} \tag{11}$$

for $s \leq z \leq d$, where the λ_0 subscripts, 1 & 2, refer to the lower and upper regions, respectively.

On differentiating with respect to z, we obtain the relevant expressions for the associated λ_0 gradients:

$$\vec{\nabla}\lambda_{01} = \frac{-\varepsilon_2 \vec{e}}{\varepsilon_1(d - s) + \varepsilon_2 s} \tag{12}$$

and

$$\vec{\nabla}\lambda_{02} = \frac{-\varepsilon_1 \vec{e}}{\varepsilon_1(d - s) + \varepsilon_2 s} \tag{13}$$

where $\vec{e}$ is a unit vector in the positive z direction.

For a homogeneous system, $\lambda_0 = \phi$. Thus for a planar system we have

$$\vec{\nabla}\lambda_0 = \vec{\nabla}\phi = -\frac{\vec{e}}{d} \tag{14}$$

Both (12) and (13) reduce to this expression for $\varepsilon_1 = \varepsilon_2$.

INDUCED CHARGE COMPONENT DUE TO $\delta\vec{P}$

To undertake a comparative assessment of the influence of the void geometry upon PD transients we will consider the dipole $\vec{\mu}$, associated with the charge accumulated at the void wall, to be a constant in this study. Furthermore, it will be assumed that the void is located more than 10 times its greatest linear dimension from the dielectric interface, such that the $\vec{\nabla}\lambda$ distribution within the void remains uniform; i.e. the existence of this interface does not perturb $\vec{\nabla}\lambda$ in the void. This assumption implies that the concept of h is valid and that (5) may be employed.

Variation of q_{Pn}

With respect to the component of the induced charge related to $\delta\vec{P}$, we have upon combining (6) and (9)

$$\frac{q_{Pn}}{q_n} = \frac{-\vec{\mu}\cdot(h_n\vec{\nabla}\lambda_{0n} - \vec{\nabla}\phi)}{-\vec{\mu}\cdot h_n\vec{\nabla}\lambda_{0n}} \qquad (15)$$

where q_n is the Poissonian induced charge of the heterogeneous system, with $n = 1,2$ depending in which dielectric medium the void is located.

On account of the planar geometry, the dipole moment is directed either away from or towards the coordinate origin. However, without loss of generality, we select

$$\vec{\mu} = \mu\vec{e} \qquad (16)$$

On performing the vector operations, (15) simplifies to give

$$\frac{q_{Pn}}{q_n} = 1 - \frac{d\phi/dz}{h_n d\lambda_{0n}/dz} \qquad (17)$$

From (17) it is clear that the polarity of q_{Pn}/q_n is dependent upon whether

$$\frac{d\phi/dz}{h_n d\lambda_{0n}/dz} \gtrless 1 \qquad (18)$$

Hence using (12), (13) and (14) we obtain for a void in medium 1

$$\frac{q_{P1}}{q_1} = 1 - \frac{\varepsilon_1(d-s) + \varepsilon_2 s}{\varepsilon_2 h_1 d} \qquad (19)$$

and for a void in medium 2

$$\frac{q_{P2}}{q_2} = 1 - \frac{\varepsilon_1(d-s) + \varepsilon_2 s}{\varepsilon_1 h_2 d} \qquad (20)$$

As both (19) and (20) contain h, this implies that q_{Pn} is dependent upon the void geometry.

Influence of Void Geometry

The influence of the void geometry upon q_{Pn} will be studied with reference to spheroidal voids, for which $\vec{\nabla}\lambda$ is parallel to the axis of rotation. For such voids, h is given by[4]

$$h = \frac{K\varepsilon_r}{1 + (K-1)\varepsilon_r} \qquad (21)$$

where ε_r is the relative permittivity of the relevant bulk dielectric and K is a dimensionless parameter

For the present situation, we have for an oblate spheroidal void[4]

$$K = \frac{u^3}{(u^2 + 1)(u - \arctan u)} \qquad (22)$$

The variable u is given by

$$u = \sqrt{(b/a)^2 - 1} \qquad (23)$$

where a and b are the semi-axes of the oblate spheroid such that $b/a > 1$, i.e. a is the axis of rotation.

For a prolate spheroidal void ($b/a < 1$), the associated expression for K is[4]

$$K = \frac{2v^3}{(1 - v^2)\left[\ln\left(\frac{1+v}{1-v}\right) - 2v\right]} \qquad (24)$$

where

$$v = \sqrt{1 - (b/a)^2} \qquad (25)$$

and a is again the axis of rotation. Following the evaluation of the K values, the corresponding h values can be derived and thereafter inserted into (19) and (20).

The variation of q_{P1}/q_1 with a/b is shown in Figure 1 for selected values of $\varepsilon_2/\varepsilon_1$. If we take the homogeneous ($\varepsilon_2 = \varepsilon_1$) case as the reference condition, then from Figure 1 it is clear that, when $\varepsilon_2/\varepsilon_1 > 1$, q_{P1} is increased. The prolate void exhibits the greater *relative* effect. For $\varepsilon_2/\varepsilon_1 < 1$, there is a reduction in q_{P1}. This can be so marked that the *polarity* of q_{P1} is reversed. Here, the prolate void displays the greater reductions, both in *relative* and *absolute* terms.

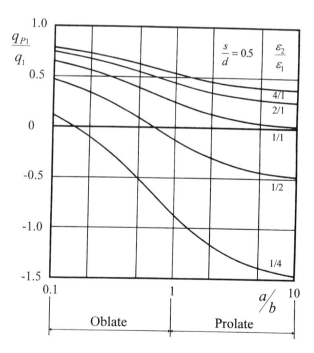

Figure 1. Variation of the polarization component of the Poissonian induced charge for spheroidal voids in a heterogeneous dielectric system.

With reference to Figure 1, if we interchanged the ε subscripts 2 and 1 and considered the void to be in medium 2, then this Figure would display the variations of q_{p2}/q_2 as a function of a/b.

Finally, it is interersting to note that a $(q_{p1}/q_1 = 0)$-condition exists for a range of values of the permittivity ratio; see Figure 1. The actual values depend on the functional form of the *homogeneous* $(\varepsilon_2 = \varepsilon_1)$ curve, and consequently are subject to void shape. For oblate voids, the *changes* in permittivity ratio required to bring about this zero condition are relatively large. In contrast, for prolate voids small changes in the ratio readily give rise to a zero-polarization component. From (19), it can be deduced that for prolate voids the ratios lie in the range $0.6 < \varepsilon_2/\varepsilon_1 < 1$.

CONCLUSIONS

Changes in the polarization $(\delta\vec{P})$ of the bulk dielectric arising from a partial discharge can significantly affect the magnitude of the polarization component of the Poissonian induced charge. For a two dielectric system, this influence is dependent not only upon the ratio of the dielectric permittivities and within which medium the void is located, but also on the void geometry. This latter parameter also affects the values of the permittivity ratio leading to the $(q_{Pn}/q_n = 0)$-condition.

REFERENCES

1. A. Pedersen, G. C. Crichton, and I. W. McAllister, The theory and measurement of partial discharge transients, *IEEE Trans. Elect. Insul.* 26:487 (1991).
2. A. Pedersen, G. C. Crichton, and I. W. McAllister, The functional relation between partial discharges and induced charge, *IEEE Trans. Diel. & Elect. Insul.* 2:535 (1995).
3. A. Pedersen, G. C. Crichton, and I. W. McAllister, Partial discharge detection: Theoretical and practical aspects, *IEE Proc. - Science, Measurement and Technology*, 142:29 (1995).
4. G. C. Crichton, P. W. Karlsson, and A. Pedersen, Partial discharges in ellipsoidal and spheroidal voids, *IEEE Trans. Elect. Insul.* 24:335 (1989).

PARTIAL DISCHARGE INCEPTION AND BREAKDOWN CHARACTERISTICS IN GAS MIXTURES WITH SF6

H. Okubo,[1] T. Yamada,[1] T. Takahashi,[1] and T. Toda[2]

[1]Department of Electrical Engineering
Nagoya University, Nagoya, 464-8603, Japan
[2]Electric Power Research & Development Center
Chubu Electric Power Co., Inc., Nagoya, 459-8522, Japan

INTRODUCTION

SF6 is the most commonly used as an insulating gas in electric power apparatus such as gas insulated switchgears (GIS), due to its superior insulation strength.[1,2] However, it is known that breakdown strength of SF6 under nonuniform electric field like metallic particle condition is extremely susceptible leading to be lower breakdown voltage. Furthermore, from the view point of environmental protection, as SF6 has strong greenhouse effect, the use of SF6 will be strongly controlled. Thus, it is needed to develop the alternative dielectric gas or gas mixtures having better insulating characteristics and no greenhouse effect.

For these reasons, insulation characteristics of various kind of gas or gas mixture have been extensively studied.[1-3] A lot of studies on gas mixtures have been done for breakdown characteristics under lightning impulse voltage application on (quasi-)uniform field,[4,5] while very few research on breakdown (BD) characteristics based on partial discharge (PD) behavior under nonuniform field in gas mixtures have been reported.

From these backgrounds, we carried out the experiment on PD and BD characteristics in gas mixtures by applying ac HV. We employed N_2 as environmentally acceptable gas, in addition to CO_2 and CF_4, greenhouse effect gases, because of the comparison with SF_6/N_2. This paper describes partial discharge inception voltage (PDIV) and breakdown voltage (BDV) characteristics of gas mixtures which could be alternative to SF6. As a result, in low concentration rate of SF6, PDIV in gas mixtures showed small dependence on mixture rate of SF6, while BDV showed the significant influence of SF6 mixture rate. This means that PDIV in gas mixtures mainly depends on the gas having lower insulation strength, but BDV depends strongly on the PD behavior, that is, the corona stabilization effect of SF6. Moreover, we also showed that BDV in gas mixtures with low concentration of SF6 is not dependent on the kind of gas mixed with SF6. Then, we indicated that mixture of only 1% SF6 drastically enhance the BDV of each pure gas. These results show that SF_6/N_2 gas mixtures will be one of the expected insulation gas for SF6 substitute for HV power apparatus.

EXPERIMENTAL

Figure 1 shows a 66kV model GIS and the test circuit for measuring PD. A needle electrode with tip radius 500μm and length 20mm was fixed on the high voltage conductor

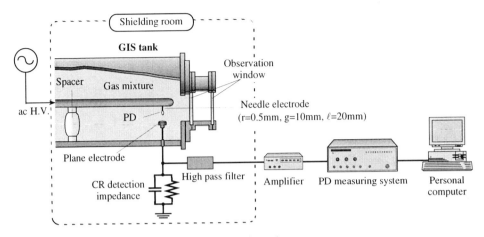

Figure 1. Experimental setup.

to simulate a metallic particle condition in GIS. The gap length of needle-plane electrodes was 10mm. PD was generated at the tip of the needle by applying ac 60Hz voltage to the conductor in GIS. GIS filled in different gas mixtures. The PD current pulses detected at the plane electrode were fed into PD measuring system through CR detection impedance, high pass filter and amplifier. The system allows to storage continuously the signal of PD charge and the PD generation phase of applied ac voltage.[6,7]

All experiments were carried out under room temperature. Mixture ratio of gas mixture was determined by partial pressure of each gas.

PARTIAL DISCHARGE INCEPTION AND BREAKDOWN CHARACTERISTICS

Effect of Gas Pressure

Figures 2 and 3 show positive and negative PDIV and BDV as a function of gas pressure in SF_6/N_2. PDIV in SF_6/N_2 has linear property for gas pressure, increasing as SF_6 mixture rate increases. On the other hand, BDV in SF_6/N_2 shows nonlinear property for gas pressure, that is, peaking characteristics at a certain gas pressure. This phenomenon is caused by electron attachment of electronegative gas, known as "N-characteristics".[8] BDV in pure N_2, however, shows linear property because of its non-electronegativity.[2] Mixing 1% SF_6 into pure N_2, the BDV shows the peaking characteristics at around 0.28MPa as shown in Fig. 3. Variation of mixture rate of SF_6 from 1% to 100% tends to shift the gas pressure toward low pressure at which the peaking characteristics appears.

Figures 4 and 5 show positive and negative PDIV and BDV as a function of gas pressure in SF_6/CO_2. It is obviously similar to SF_6/N_2 that PDIV in SF_6/CO_2 have linear property for gas pressure, while BDV in SF_6/CO_2 and pure CO_2 have nonlinear property. The nonlinear property of pure CO_2 would result from dissociative attachment process of $(CO_2 + e^- \rightarrow CO + O^-)$.[9,10]

Moreover, Figures 6 and 7 show positive PDIV and BDV as a function of gas pressure in SF_6/CF_4. The PDIV and BDV in SF_6/CF_4 have similar dependence on gas pressure as SF_6/N_2 and SF_6/CO_2. BDV characteristics in pure CF_4 show smaller peaking characteristics than that in SF_6, because the electron attachment coefficient of CF_4 is much smaller than that of SF_6.[10] The nonlinear property of SF_6/CF_4 containing a small amount of SF_6 does not appear remarkably unlike SF_6/N_2 or SF_6/CO_2 because a small amount of SF_6 have little influence of electronegative CF_4.

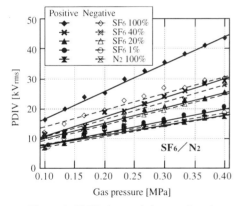

Figure 2. PDIV characteristics as a function of gas pressure in SF₆/N₂ gas mixtures with percentages of SF₆ in the mixtures.

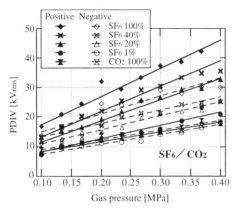

Figure 4. PDIV characteristics as a function of gas pressure in SF₆/CO₂ gas mixtures with percentages of SF₆ in the mixtures.

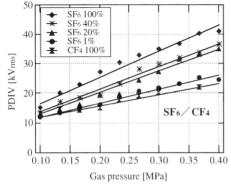

Figure 6. Positive PDIV characteristics as a function of gas pressure in SF₆/CF₄ gas mixtures with percentages of SF₆ in the mixtures.

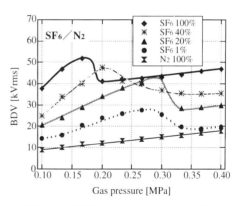

Figure 3. BDV characteristics as a function of gas pressure in SF₆/N₂ gas mixtures with percentages of SF₆ in the mixtures.

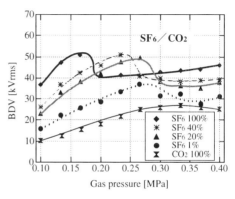

Figure 5. BDV characteristics as a function of gas pressure in SF₆/CO₂ gas mixtures with percentages of SF₆ in the mixtures.

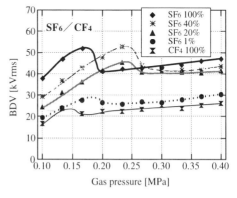

Figure 7. BDV characteristics as a function of gas pressure in SF₆/CF₄ gas mixtures with percentages of SF₆ in the mixtures.

Effect of Mixture Rate of SF6

Figures 8 and 9 show positive and negative PDIV and BDV as a function of mixture rate in SF_6/N_2. PDIV characteristics as a parameter of gas pressure have relative similar characteristics. Each PDIV characteristics shows a small increase in about 0~20% mixture of SF_6 and then saturation in about 20~90% mixture of SF_6, indicating almost the same PDIV between positive and negative cycle. In about 90~100% mixture of SF_6, different characteristics between the positive and negative appear, in other words, a positive PDIV increases remarkably, while a negative PDIV remains almost constant. On the other hand, it is clear from Fig. 9 that BDV characteristics are notably different in gas pressure. These different characteristics result from the nonlinear property as shown in Fig. 3. In addition, it should be noted that every BDV characteristics for different gas pressure show drastic increase with containing a small amount of SF_6.

Figures 10 and 11 show positive and negative PDIV and BDV as a function of mixture rate in SF_6/CO_2, respectively. Apparently, the PDIV and BDV characteristics in SF_6/CO_2 are similar to those of SF_6/N_2. The positive PDIV in SF_6/CO_2 containing about 90~100% mixture of SF_6 have no drastic increase, different from the case of SF_6/N_2 as shown in Fig. 8. From this result, we consider that electron attachment of SF_6 and dissociative attachment of CO_2 interact one another in SF_6/CO_2, compared with electron attachment of SF_6 alone in SF_6/N_2.

As a result, we made clear that BDV in gas mixtures increase more drastically than PDIV by mixing a small amount of SF_6. This result is interpreted that by mixture with SF_6 the changes of PD characteristics affect corona stabilization effect remarkably.

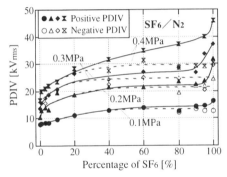

Figure 8. PDIV characteristics as a function of mixture rate of SF_6 in SF_6/N_2 gas mixture at different gas pressure.

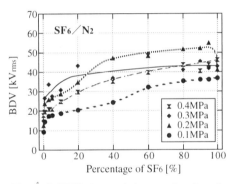

Figure 9. BDV characteristics as a function of mixture rate of SF_6 in SF_6/N_2 gas mixture at different gas pressure.

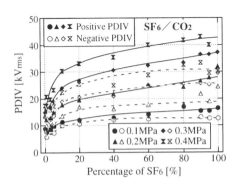

Figure 10. PDIV characteristics as a function of mixture rate of SF_6 in SF_6/CO_2 gas mixture at different gas pressure.

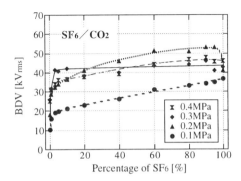

Figure 11. BDV characteristics as a function of mixture rate of SF_6 in SF_6/CO_2 gas mixture at different gas pressure.

ϕ-q-n CHARACTERISTICS OF PARTIAL DISCHARGE

Partial Discharge in Pure Gases

Figures 12(a)~(c) show ϕ-q-n characteristics of PD occurring in pure SF_6, N_2 and CO_2 (applied voltage: just above positive PDIV of each pure gas, gas pressure: 0.1MPa). The ϕ-q-n characteristics show the summing data of phase, charge and number of PD occurring for 300 cycles of ac applied voltage. The ϕ-q-n characteristics of SF_6 in positive cycle show that PD with large charge at around PD inception phase are followed by only PD with relatively small charge. This is considered to result from temporary decline of corona stabilization effect, namely, decrease of space charge around the needle tip in turnover from negative cycle to positive cycle due to ac voltage application. After occurrence of PD with large charge, only PD with relatively a small charge occur due to the recovery of corona stabilization effect. On the other hand, PD with relatively small charge occur in negative cycle. This difference between positive and negative PD corresponds to different formation mechanism of discharge; streamer PD in positive, Trichel PD in negative.[9]

In positive cycle of the ϕ-q-n characteristics shown in Fig. 12(b), PD in pure N_2 occur only at 60°, PD inception phase. The charge of PD in N_2 is over 10 times larger than that in pure SF_6. The PD with large charge are caused by easy generation of electron avalanche in nonattaching gas.[2]

The ϕ-q-n characteristics shown in Fig. 12(c) show that PD in pure CO_2 is relatively similar to that in SF_6 in positive cycle, while a large number of PD with much more charge generate in negative cycle. The generation number of negative PD is about 22 times larger than that of

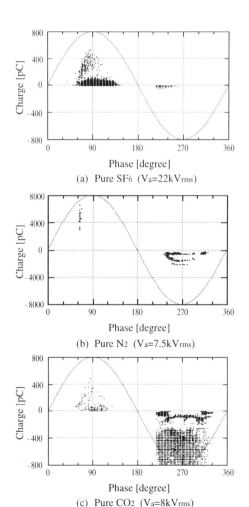

Figure 12. ϕ-q-n characteristics in different gases at 0.1MPa.

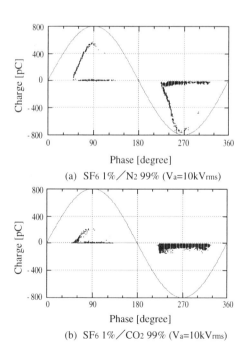

Figure 13. ϕ-q-n characteristics in the mixture of 1% SF_6 gas／different pure gases at 0.1MPa.

positive PD and the whole charge of negative PD reaches 200 times larger than that of positive PD. Such a large negative PD result from the easy extension of PD because the dissociative attachment of CO_2 works only just at the near area of the needle tip by collision with electrons and photons with high energy.[9,10]

Partial Discharge in Gas Mixtures

Figures 13(a) and (b) show φ-q-n characteristics of PD occurring in SF_6/N_2 and SF_6/CO_2 containing 1% SF_6. PD charge in $SF_6(1\%)/N_2(99\%)$ in Fig. 13(a) drastically reduce, compared with the charge in pure N_2 shown in Fig. 12(b). The φ-q-n characteristics in $SF_6(1\%)/N_2(99\%)$ is relatively similar to that of pure SF_6 shown in Fig. 12(a). In the same way, the φ-q-n characteristics of $SF_6(1\%)/CO_2(99\%)$ is relatively similar to that of pure SF_6, in other words, the large charge of negative PD in pure CO_2 reduced drastically by mixing 1% SF_6 into pure CO_2.

Consequently, φ-q-n characteristics in 1% SF_6 gas mixture changed drastically close to the characteristics in pure SF_6. The significant change of the PD characteristics would have strong influence on breakdown strength.

CONCLUSIONS

We examined the PD and breakdown characteristics in SF_6/N_2, SF_6/CO_2, SF_6/CF_4 mixtures. In this experiment, it was clarified that by mixing a small amount of SF_6, the PDIV has small dependence on mixture rate, but the PD behavior change drastically close to that in pure SF_6, and thus the BDV increases dramatically. This dependence on mixture rate of SF_6 is caused by PD generation mechanism, in other words, PD in the gas mixture of nonattaching or weakly attaching gas with SF_6 acts with mutual interaction of corona stabilization effect.

From the above consideration, the dielectric properties of gas mixtures consisting of strongly attaching gas and nonattaching gas like SF_6/N_2 are almost similar to that of gas mixture consisting of strongly attaching gas and weakly attaching gas. Therefore, the gas mixture with very commonplace N_2 with lower liquified temperature, high chemical stability and no greenhouse effect is promising gas dielectrics applicable in future power apparatus.

REFERENCES

1. S. Okabe, et al., Recent Development in Diagnostic Techniques for Substation Equipment, *CIGRE, Session Paper* 15/21/33-08, (1996).
2. L. G. Christophorou, et al., SF_6/N_2 Mixtures, *IEEE Trans. on Elect. Insul.*, Vol. 2, No. 5, pp. 952-991, (1995).
3. K. Mardikyan, et al., AC Breakdown Strength of N_2, SF_6 and a Mixture of N_2+SF_6 Containing a Small Amount of SF_6, *1996 IEEE Int. Symp. on Elect. Insul.*, June 16-19 Vol. 2, pp. 763-765, (1996).
4. H. L. Marsden, et al., High Voltage Performance of a Gas Insulated Cable with N_2 and N_2/SF_6 Mixtures, *10th Int. Symp. on High Voltage Eng.*, Vol. 2, pp 9-12, (1997).
5. T. B. Diarra, et al., N_2-SF_6 Mixtures For High Voltage Gas Insulated Lines, *10th Int. Symp. on High Voltage Eng.*, Vol. 2, pp 105-108, (1997).
6. M Hikita, et al., Partial Discharge Measurements in SF_6 and Air using Phase-resolved Pulse-height Analysis, *IEEE. Trans. on Elect. Insul.*, Vol. 1, No. 2, pp. 276-283, (1994).
7. T. Kato, et al., Partial Discharge Behavior and Breakdown Prediction in SF_6 Gas, *Proc. of 4th Int. Conf. on Prop. and Appl. of Dielect. Materials*, (1994).
8. J. K. Nelson, Modification of Corona Stabilization in Electronegative Gases by Low Ionization Potential Additives, *Gaseous Dielectrics IV*, pp. 175-183, (1984).
9. M. S. Bhalla, et al., Measurement of Ionization and Attachment Coefficients in Carbon Dioxide in Uniform Fields, *Proc. Phys. Soc.*, Vol. 76, pp. 369-377, (1960).
10. L. G. Christophorou, et al., Basic Physics of Gaseous Discharges, *IEEE Trans. on Elect. Insul.*, Vol. 25, No. 1, pp. 55-74, (1990).
11. J. Liu, et al., Simulation of Corona Discharge-Negative Corona in SF_6, *IEEE, Trans. on Dielect. and Elect. Insul.*, Vol. 1, No. 3, pp. 520-529, (1994).

DEPENDENCE OF PARTIAL DISCHARGE AND BREAKDOWN CHARACTERISTICS ON APPLIED POWER FREQUENCY IN SF_6 GAS

T.Takahashi,[1] T.Yamada,[1] M.Hikita[2] and H.Okubo[1]

[1] Dept. of Electrical Engineering
Nagoya University
Nagoya, 464-8603, Japan
[2] Dept. of Electrical Engineering
Kyushu Institute of Technology
Kitakyushu, 804-8550, Japan

INTRODUCTION

In recent years, SF_6 gas is extensively utilized as an insulation gas for high voltage power apparatus, such as gas insulated switchgears (GIS), transformers, etc.[1] One of the most important techniques to keep high reliability of such apparatus is an insulation diagnostic technique. In GIS, insulation defects like metallic particles would cause the partial discharge (PD) and may lead to breakdown (BD).[2] Therefore, it is very important for the stable operation of GIS to detect and measure PD based on its discharge mechanism in SF_6 gas. However, under ac voltage application the temporal change of the instantaneous voltage values and the polarity makes the space charge behavior generated by PD so complicated that there is much difficulty to clarify the PD mechanism. Until now, a few study has tried to investigate the PD mechanism with consideration of space charge behavior under ac voltage application.[3]

We have been investigating the PD mechanism under ac voltage application.[4,5] In this paper, we describe the measurement results of PD characteristics as a function of the applied power frequency with high speed optical and electrical measurement technique. Based on the results, we discuss the PD mechanism with the consideration of the space charge behavior.

EXPERIMENTAL

Figure 1 shows a model GIS and a testing circuit with a variable power frequency high voltage source (f=60~600Hz). As a PD source, a needle electrode was set on the high voltage conductor of the model GIS to simulate the metallic particle condition in GIS. The needle electrode was made of stainless steel with 20mm in length and 500μm in tip radius. A plane electrode was placed 10mm below the needle electrode to consist the needle-plane electrodes. PD current pulse signal flowed into the plane electrode was measured by a high speed digital oscilloscope (analog band width : 1GHz, sampling rate : 5GS/s) through a matching circuit having wide-band frequency response 1MHz to 1GHz.[4] Moreover, a trigger signal synchronized with the applied ac voltage phase was inputted to the oscilloscope so that we could observe a single PD current pulse occurring at a designated phase angle of applied ac voltage. In addition, we observed the corresponding light images of PD with optical method.[5] The light emission image of PD was enlarged by lenses set inside and outside GIS, intensified by an image intensifier (I.I.) and recorded by a still camera. Here, a gate signal synchronized with the applied ac voltage was led into the I.I. so that we could

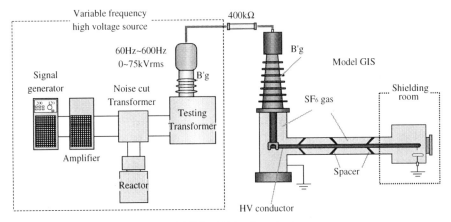

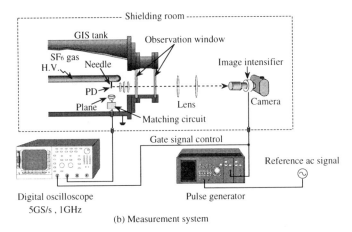

Figure 1. Variable frequency high voltage source and experimental setup for measuring PD characteristics.

also observe the light image of PD occurring at a designated phase angle region of ac applied voltage.

The GIS was filled with SF6 gas and all experiments described in this paper were carried out in room temperature.

POWER FREQUENCY DEPENDENCE OF PARTIAL DISCHARGE INCEPTION AND BREAKDOWN

Figure 2 shows PD inception and breakdown voltage characteristics as a function of gas pressure at different applied power frequency (f=60Hz and 400Hz). It can be seen in Fig. 2 that at f=60Hz, breakdown voltage (BDV) once increases to pressure P_m at which BDV is maximum, suddenly decreases to pressure P_c at which BDV becomes equal to positive PD inception voltage (PDIV), and increases again with increase of gas pressure. This nonlinear property also appears in f=400Hz. Comparison of BDV characteristics at f=60Hz with at 400Hz reveals that at the pressure region $P \leq P_m$, BDV shows nearly equal values for these two frequencies, while the difference of BDV for f=60Hz and f=400Hz emerges at $P \geq P_m$.

Generally speaking, breakdown is caused by the extension of the streamer discharge at $P \leq P_m$, whereas at $P \geq P_m$, breakdown resulted from the transition of streamer to leader discharge and extension of leader discharge.[6] From these points of view, at $P \leq P_m$, breakdown does not occur until the electric field distribution becomes a certain condition. Thus, at $P \leq P_m$, breakdown characteristics

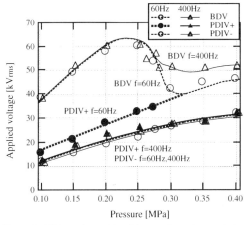
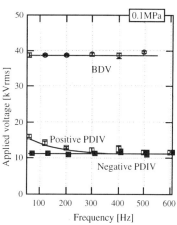

Figure 2. PD inception and breakdown voltage characteristics as a function of gas pressure at different power frequency.

Figure 3. Power frequency dependence of positive and negative PD inception voltage and breakdown voltage. (P=0.1MPa)

is highly dependent on the instantaneous value of applied voltage, and this means that BDV is irrespective of the power frequency. On the contrary, the extension of leader discharge is strongly dependent on the space charge distribution, that is, a large amount of the space charges around the needle tip tend to suppress the extension of leader discharge. Thus, at P≥Pm, increase of the power frequency leads to decrease the travelling distance of space charges in half cycle of applied ac voltage and would result in the BDV enhancement.

As seen in Fig. 2, negative PDIV increases with gas pressure and is not dependent on the power frequency. This is because negative PD would start with the initial electron generated by field emission,[3] and then the negative PD is dependent not on the power frequency but on the instantaneous value of applied voltage. Positive PDIV shows the increase characteristics with gas pressure similarly for each power frequency, however, the values at f=400Hz are smaller than at f=60Hz. It is generally said that the initial electrons for positive PD is generated by the field-induced collisional detachment from negative ions.[3] The negative ions generated by negative PD in negative ac half cycle remain more at f=400Hz than at f=60Hz. This is because the travelling distance of the ions during half cycle are reduced with increase of power frequency. From these points of view, it can be said that the power frequency dependence of positive PDIV is caused by the change of negative space charge behavior due to the change of power frequency.

Next, we show the power frequency dependence of positive and negative PDIV and BDV characteristics in gas pressure P=0.1MPa as Fig. 3. As shown in Fig. 3, PDIV and BDV characteristics mentioned before can be seen more clearly, that is to say, BDV and negative PDIV is not dependent on the power frequency but positive PDIV is reduced with increase of power frequency and equals to negative PDIV at the frequency region f=400Hz or higher.

POWER FREQUENCY DEPENDENCE OF PARTIAL DISCHARGE CURRENT PULSE WAVEFORM

Figures 4 (a)~(d) show typical PD current pulse waveforms observed at around 40° and 210°, for f=60Hz and f=400Hz, respectively, all at P=0.1MPa V_a=30kVrms. Here, 40° and 210° mean the voltage phase angle vicinities of positive and negative PD inception at both frequency, respectively. As seen in Figs. 4 (a) and (c), the positive PD current pulse waveforms in different power frequency have different shapes, while the negative PD current pulse waveforms in different power frequency shown in Figs. 4 (b) and (d) have similar shapes, namely, the similar rise time and amplitude.

Similar measurements were carried out 30 times for every 10° phase width of applied ac voltage. Figures 5 (a) and (b) show thus obtained phase dependence of PD current pulse height at V_a=30kVrms for f=60Hz and f=400Hz, respectively. As seen in Figs. 5 (a) and (b), PD in negative half cycle have similar phase characteristics in both frequency. This is due to the PD mechanism, based on the Trichel pulse having regular occurrence and equal charges during the half cycle.[7]

On the other hand, positive PD characteristics at 30°~40° for f=60Hz is not similar to that for f=400Hz. This difference of PD characteristics on applied power frequency could be due to the difference of the space charge behavior. In low frequency as f=60Hz, space charges generated by

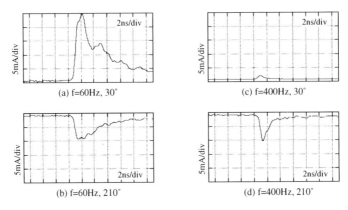

Figure 4. Typical PD current pulse waveforms. (P=0.1MPa, Va=30kVrms)

negative PD in negative half cycle no longer remain nearby the needle tip due to the drift of ions along the electric line of force. On the contrary, in high frequency as f=400Hz, space charges generated by negative PD in negative half cycle could remain nearby the needle tip at positive PD inception phase. Therefore, the relaxation of the electric field nearby the needle tip by the positive ion generated by negative PD at previous negative half cycle could have an effect at positive PD inception phase. This corona stabilization effect will suppress the extension of PD at positive PD inception phase for high frequency.

Here, it should be noted that at low applied voltage as positive PDIV, negative ions generated by the negative PD influence to the generation of positive PD, while at high applied voltage as V_a=30kVrms, positive ions generated by the negative PD tend to remain nearby the needle tip and suppress the extension of positive PD.

Figures 6 (a) and (b) show the current increase rate di/dt at the wave front of PD current pulse as a function of phase angle under the same condition as those in Fig. 5. As can be seen in Fig. 6, di/dt around positive PD inception phase at low frequency as f=60Hz is larger than those of the other positive phase regions. At high frequency as f=400Hz, di/dt at around positive inception phase is much smaller than those at low frequency as f=60Hz. This result can be due to the effect of corona stabilization at around the positive PD inception phase. In negative half cycles, di/dt shows a little bit smaller value in negative cycle than in positive half cycle, irrespective to applied power frequency. This negative PD characteristics is due to the PD mechanism based on Trichel pulses.[7]

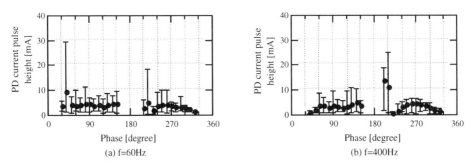

Figure 5. Phase dependence of PD current pulse height at different applied power frequency. (P=0.1MPa, Va=30kVrms)

POWER FREQUENCY DEPENDENCE OF LIGHT EMISSION CHARACTERISTICS OF PARTIAL DISCHARGE

Figures 7 (a)~(f) show the light emission of PD occurring at f=60Hz and 400Hz, at 30° (positive PD inception phase), 90° and 150° (positive PD extinction phase), respectively. Note that the applied voltage was V_a=30kVrms and the light images shown in Fig. 7 were taken by superimposed light emission of PD generated in 60 cycles of ac applied voltage and within 10° width of each

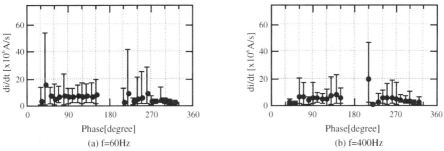

Figure 6. Phase dependence of current increase rate di/dt at different applied power frequency. (P=0.1MPa, Va=30kVrms)

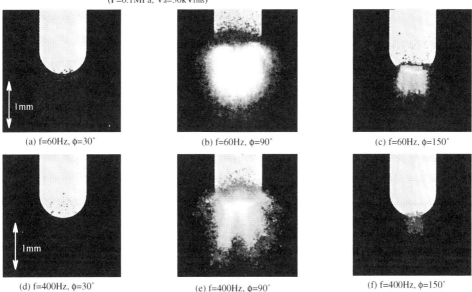

Figure 7. Light emission images of PD at different applied power frequency and phase. (P=0.1MPa, Va=30kVrms, gate width: 10°, 60 ac cycles)

observation phase. As seen in Figs. 7 (b), (e) and (c), (f), PD occurring at around 90° and 150° show brush-like luminous images for both f=60Hz and 400Hz. From this light emission images, PD at those phase region could be streamer discharge. As seen in Fig. 7 (a), PD occurring at around PD inception phase (30°) for f=60Hz could also be streamer discharge due to the brush-like luminous image, however, PD occurring for f=400Hz at around positive PD inception phase (30°) shown in Fig. 7 (d) does not show brush-like but dot-like luminous image. As was seen in Fig. 5 (b) for f=400Hz at around 30° of applied voltage phase, we could not observe pulse-like current waveform. From this result, PD occurring for f=400Hz at around 30° could be positive glow discharge, which has luminous image with very short length and non-pulse-like current waveform. This difference of PD extension characteristics on applied power frequency could be due to the difference of the space charge behavior; corona stabilization is not effective at low frequency but effective at high frequency even at the positive PD inception phase.

POWER FREQUENCY DEPENDENCE OF PARTIAL DISCHARGE INCEPTION PHASE

Figures 8 (a) and (b) show the applied power frequency dependence of PD inception phase

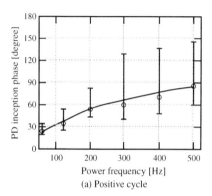
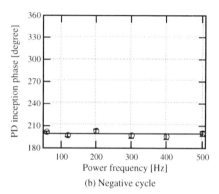

(a) Positive cycle (b) Negative cycle

Figure 8. Power frequency dependence of PD inception phase. (P=0.1MPa, Va=30kVrms)

with 30kVrms of applied voltage at positive cycle and negative cycle, respectively. As seen in Fig. 8 (b), negative PD inception phase is not dependent on applied power frequency. This means that negative PD inception is dependent only on the instantaneous value of applied voltage, and on the negative PD mechanism, that is, initial electron for negative PD is generated by the field emission from the needle tip. As seen in Fig. 8 (a), however, positive PD inception phase tends to be shifted with increase of the applied power frequency. This is due to the increase of remained space charges around needle tip with the applied power frequency; increasing the amount of space charges more relaxes the electric field strength around the needle tip at positive PD inception phase so that the inception and extension of PD could be more influenced.

CONCLUSIONS

In this paper, we investigated the effect of space charges on PD behavior and BD characteristics with changing applied power frequency from 60 to 600Hz. As a result, we could find the power frequency dependence of the breakdown voltage at the region of the leader dominant breakdown, positive PD inception voltage and positive PD inception phase. These power frequency dependence could be due to the positive PD mechanism; positive PD starts with the initial electron generated by the collisional detachment of negative ions around the needle tip, and the fact that the drift distance of the ions in half cycle of applied ac voltage reduces with the increase of the power frequency. Therefore, it could be said that the space charge behavior greatly affects the positive PD characteristics.

From these results, we pointed out that the space charge behavior should be considered for the discussion of PD and breakdown diagnostic technique for HV power apparatus.

REFERENCES

1. T.Utsumi, *et al.*, Preventive Maintenance System with a Different Gas Injecting Facility for GIS, *IEEE Trans. on Power Delivery*, Vol. 8, No.3, pp.1107-1113 (1993).
2. R.Baumgärtner, *et al.*, Partial Discharge - Part IX: PD in Gas-Insulated Substation - Fundamental Considerations, *IEEE Elect. Insul. Magazine*, Vol. 7, No. 6, pp. 5-13 (1991).
3. R.J.Van Brunt, *et al.*, Mechanisms for Inception of DC and 60-Hz AC Corona in SF_6, *IEEE Trans. on Elect. Insul.*, Vol. 17, No. 2, pp. 106-120 (1982).
4. M.Hikita, *et al.*, Phase Dependence of Partial Discharge Current Pulse Waveform and Its Frequency Characteristics in SF_6 Gas, *IEEE Int. Symp. on Elect. Insul.*, pp. 103-106 (1996).
5. H.Okubo, *et al.*, Investigation of Partial Discharge Mechanism in SF_6 Gas by Simultaneous Measurement of Current Waveform and Light Emission, *Proc. of the 10th Int. Symp. on High Voltage Eng.*, Vol. 2, pp. 177-180 (1997).
6. F.Pinnekamp and L.Niemeyer, Qualitative Model of Breakdown in SF_6 in Inhomogeneous Gaps, *J. Phys. D: Appl. Phys.*, Vol. 16, pp. 1293-1312, (1983).
7. G.W.Trichel, The Mechanism of the Negative Point to Plane Corona Near Onset, *Phys. Rev.*, Vol. 54, pp.1078-1084 (1938).

PARTIAL DISCHARGE CHARACTERISTICS OF SMALL GAS GAPS IN SF_6 GAS

M. Ishikawa,[1] T. Kobayashi,[1]
T. Goda,[2] T. Inoue,[2] M. Hanai,[2] T. Teranishi[2]

[1]The Tokyo Electric Power company, Inc.
Tokyo, Japan
[2]Toshiba Corporation
Kawasaki, Japan

INTRODUCTION

Gas-insulated transformers (GIT) using SF_6 gas as an insulating medium offer the advantages of making substations nonflammable and reducing the space required.

GIT has many small gas gaps in its dielectric structure. For example, turn-to-turn insulation of windings form wedge-shaped small gas gaps between insulated conductors. Such small gas gaps create dielectric weak points due to electric field concentration. Dielectric properties of small gas gaps are one of the important factors in the dielectric design of GIT. However, it is difficult to apply the theory of gas breakdowns in a uniform electric field directly to such small gas gaps with covered electrodes[1,2].

This is a report on an investigation of the effect of insulators and their surfaces on the partial discharge characteristics of small gas gaps.

EXPERIMENTAL SET UP

The structure of the test sample is shown in Figure 1. A rod electrode is put on two earth electrodes with flat surfaces covered with a 1mm-thick insulator sheet. The rod electrode is covered with insulator at the straight part and has spheres at both ends. The diameter of the rod electrode is ϕ

15mm. The earth electrodes are molded with epoxy resin to prevent wedge-shaped gas gaps forming between the earth electrodes and the insulator sheet.

A positive lightning impulse voltage (1.2/50μs) was applied to the rod electrode. The partial discharge inception voltage (PDIV) was measured with a high-frequency CT by using the current method. Two earth electrodes were used to make partial discharge measurements more sensitive by canceling the charging current. The absolute pressure of SF_6 gas in the test was 0.225 or 0.5MPa.

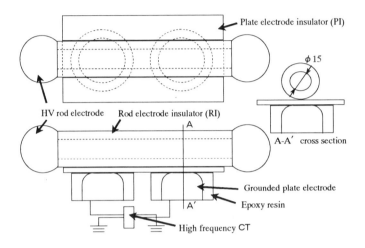

Figure 1. Test sample

Figure 2 shows the five kinds of insulator structures of test samples. Each consisted of a plate electrode insulator (PI) and a rod electrode insulator (RI). In structure 1, PI was fabricated from a press-board, which was 1mm thick, and RI of a roll of kraft paper (KP), which had a wall thickness of 1-6mm. In structure 2, PI was a laminate, which was 1mm thick, of polyethyleneterephthalate (PET) sheets and RI a roll of PET sheets, which had a wall thickness of 1-6mm. PI and RI in structure 3 were made by covering PI and RI in structure 1 with PET film. Structure 4 was nearly the same as structure 1, with only PI covered with PET film. Structure 5 was also nearly the same as structure 1, with only RI covered with PET film. The PET film was 0.025mm thick.

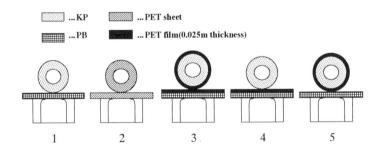

Figure 2. Insulation structures

TEST RESULTS

Figure 3 shows the PDIV characteristics of structure 1 through 3. Each plot shows the average value of five samples in each case. The applied voltage was a positive lightning impulse.

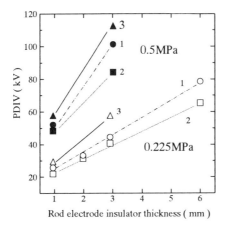

Figure 3. PDIV characteristics

With the sample (structure 1) made of KP and PB insulators, PDIV values were higher than with the sample (structure 2) made of PET insulation. Covering the KP and PB insulators with PET film (structure 3) tended to increase partial discharge inception voltages. These results shows that PDIV characteristics of small gas gaps with covered electrodes depend on the materials of insulators and their surface conditions.

Figure 4 shows the PDIV characteristics obtained for the samples in structure 1 through 5 under impulse voltages, positive and negative (structure 2 with positive only). The wall thickness of the rod electrode insulators was 1mm. The test gas pressure was 0.225 MPa.

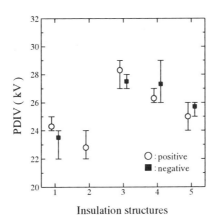

Figure 4. Influence of insulation surface materials

303

The PDIV values were highest with structure 3, both PB and KP insulators covered with PET film, and second highest with structure 4, with only the PB insulator covered with PET film. The PDIV values of structure 5, only KP insulator covered with PET film, were much the same as those of structure 3. With any structure, the polarity of impulse voltage had no effect. This is probably because the wedge-shaped small gas gap formed between two electrodes is nearly symmetrical.

DISCUSSION

Figure 5 shows the partial discharge inception electric field (PDIE) of the five samples based on the test results in Figure 4. The PDIE values shown here are those obtained on RI surfaces 10 degrees off the perpendicular on contact between RI and PI. The electric field distribution in the wedge-shaped gas gap come into contact with the PDIE curve of SF_6 gas determined by the Paschen's curve nearly at this point, from where partial discharges are supposed to occur[34].

The dielectric constants of the insulators used for calculation were 3.2 (PET), 3.4 (PB) and 1.85 (KP).

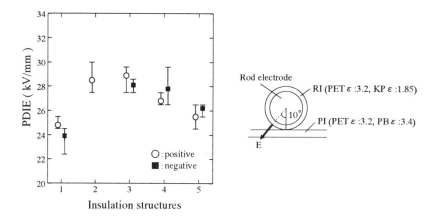

Figure5. Field evaluation of structures

Partial discharge inception field values differed little between structure 2, all PET, and structure 3, KP and PB insulators covered with PET film. With structure 2 and 3, the RI surface field values at the points mentioned above were almost 1.4 times the theoretical gas breakdown field values. It was reported that PDIE in the small gas gaps with covered electrodes become higher than the theoretical breakdown field of SF_6 gas because of the lack of initial electrons[2]. The results obtained in this experiment were coincident with the former results[2]. The PDIV values were lower with sample (structure 4), only the PB covered with PET film, further with sample (structure 5), only the KP covered with PET film, and lowest with sample (structure 1), without a PET film covering.

The electric field strength in the small gas gaps is determined by the dielectric constant of insulation. So, PDIV values of samples with KP and PB insulators covered with a PET film (structure 3) became higher than those of the samples with PET (structure 2) because the dielectric constant of KP (ε =1.85) is lower than that of PET (ε =3.2). On the other hand, with PB and KP insulators without PET film coverings, PDIE values were lower than with PET insulators. This can possibly be ascribed to the roughness of PB and KP surfaces in contrast to the smooth PET surfaces. Covering PB with PET film was more effective to increase PDIV than covering KP with PET film. This is probably because PB has much more rough surface than KP.

CONCLUTION

Using several insulating materials, an investigation was carried out of small gas gaps formed between covered electrodes, discovering the following:

(1) The PDIV characteristics of small gas gaps between insulators are affected by the materials of insulators and their surface conditions.

(2) The PDIE values of small gas gaps between insulators covered with PET film are almost 1.4 times the theoretical gas breakdown field values.

(3) The PDIE values of small gas gaps between PB or KP insulators are lower because of surface roughness. This is more remarkable with PB surface.

REFERENCES

1. A.Inui, T.Teranishi, H.Okubo and S.Yanabu, Partial Discharge Characteristics of Wedge Gap in SF_6 Gas-insulated Transformers, *7th Int. Symp. on High Voltage Engineering*, 31.10 (1991).
2. T.Ishii, M.Hananmura, M.Hanai, K.Toda, T.Teranishi, H.Murase and S.Yanabu, Insulating Characteristics of Small Gas Gap under Lightning Impulse Voltage, *6th Int. Symp. on Gaseous Dielectrics*, 49 (1990).
3. A.Inui, T.Inoue, T.Teranishi, H.Murase, I.Ohshima and K.Toda, *Dielectric Characteristics of Static Shield for Coil-end of Gas-insulated Transformer*, IEEE Transaction On Electrical Insulation, Vol.27, No.3, 572 (1992).
4. A.Inui, S.Yamada, T.Teranishi, H.Murase, T.Teranishi and I.Ohshima, Impulse Insulation Characteristics of Composite Insulation System Having a Wedge Gap in SF6 Gas, *T. IEE Japan*, Vol.112-A, No.2, 106 (1992), Japanese.

CHARACTERISTICS OF PARTIAL DISCHARGES ON A DIELECTRIC SURFACE IN SF_6-N_2 MIXTURES

Xiaolian Han, Yicheng Wang, Loucas G. Christophorou, Richard J. Van Brunt

National Institute of Standards and Technology
Electricity Division, Gaithersburg, MD 20899

INTRODUCTION

An important tool for improving the reliability of HV-insulation systems relies on partial discharge (PD) measurements. The assessment of the insulation failure of HV equipment using PD measurements requires an interpretation of the PD measurements themselves. The statistical characterization of pulsating PD signals has been shown[1-3] to play an important role in understanding PD phenomena. The development of diagnostic systems which utilize statistical data on partial discharges for pattern recognition is important for the identification of types of defects in electrical insulation.

This paper presents the results of a recent investigation of pulsating PD phenomena occurring in point-dielectric gaps under an alternating voltage in SF_6-N_2 mixtures. The dielectric is a cast epoxy resin with Al_2O_3 filler. A computer-based PD recording system has been utilized which captures the amplitudes and time of occurrences of "all" pulsating PD events above a certain amplitude within an arbitrary time period. The data have been collected over sufficiently long time periods to make stochastic analysis possible. From the analysis of the stochastic behavior, the differences of the time-resolved histograms of PD in SF_6-N_2 mixtures and air have been obtained. Due to the strong electronegative character of SF_6 and its tendency to decompose during a discharge, the PD patterns in SF_6-N_2 mixtures are notably different from those in air. The measured PD pulse rate and average current show that partial discharges in SF_6-N_2 mixtures in the presence of an epoxy surface are non-stationary, that is they vary with time. Measurements made over a twenty-four-hour period show that there are different stages of the partial discharge behavior. These results are presented and briefly discussed.

EXPERIMENTAL CONDITIONS

The schematic diagram of the experiment is shown in Fig. 1. A zero-crossing detector is utilized to obtain the time of occurrences of the partial discharges, and digitized data are collected by the computer. The PD was generated between the point and the dielectric epoxy by applying a continuous sinusoidal alternating voltage at 60 Hz to a point stainless steel electrode. The tip radius is about 50 μm and the gap distance is 0.5 mm ±0.01 mm. The insulating gas mixture consists of 10%SF$_6$+90%N$_2$ at a pressure of 100 kPa.

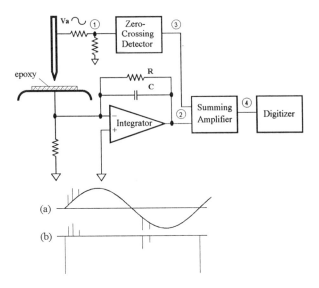

Fig. 1. Experimental arrangement for the study of PD characteristics and the measurement of the time-resolved pulse amplitudes.
(a). A typical waveform superimposed by the voltages from test point ① and ②.
(b). A waveform seen at test point ④ which is the sum of the voltages from test point ② and ③.

EXPERIMENTAL RESULTS AND DISCUSSION

Non-stationary Behavior

Figure 2(a) shows the measured 24-hour PD average current at a voltage of 2.30 kV, and Fig. 2(b) shows the measured relative PD pulse rate over the same time period. Over this time period, both the PD average current and the pulse rate show distinct patterns of behavior. According to the International Electrotechnical Commission (IEC) Standard 270,[4] the average current I is given by

$$I = \frac{1}{T_{ref}}(|q_1|+|q_2|+...+|q_i|)$$

which is the sum of the absolute values of the individual PD amplitudes q_i during a certain time interval T_{ref}, divided by this time interval. In this paper the time interval used for the average current is one second. Figure 3 shows individual PD (positive and negative) amplitude distributions. All data were recorded after an initial 10-hour conditioning period during which no measurements were made because the inception voltage varied randomly by several thousand volts. The data shown in Figs. 2 and 3 were taken at a rate of one point per second for the first 2 minutes of each consecutive 10-minute period. Pulses with amplitudes lower than 3.0 pC (positive) and 3.6 pC (negative) were not recorded.

The PD (positive and negative) pulse amplitude distributions in Fig. 3 at stages 1, 3, and 5 are quite different from those at stages 2 and 4. The former is characterized by a widely scattered pulse-amplitude distribution while the latter is characterized by a narrow pulse amplitude distribution. It is clear that the PD patterns of stages 2 and 4 are different

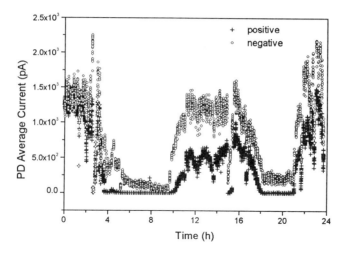

Fig. 2(a). PD average current measured for 24 hours with a gap distance of 0.5 mm ±0.01 mm, a frequency of 60 Hz, and a voltage of 2.30 kV.

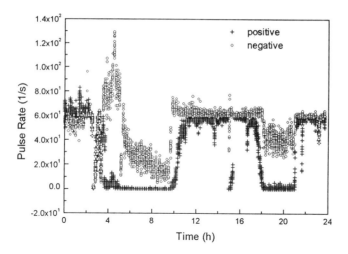

Fig. 2(b). PD pulse rate measured for 24 hours with a gap distance of 0.5 mm ±0.01 mm, a frequency of 60 Hz, and a voltage of 2.30 kV.

from those of stages 1, 3 and 5. According to the charge distributions in Fig. 3, the positive partial discharges cease to exist after about 4 hours and the negative partial discharges shift to lower values. A similar cessation of positive discharge has been observed in air exposed to PD for 2.9 hours by Van Brunt et al. by applying an alternating voltage to a point electrode touching a dielectric surface.[3] For the data in Fig. 3, the partial discharges are observed to appear again after about 10 hours (stage 3). The pattern in stage 3 is similar to stage 1, but after about 18 hours the pattern shifts to that of stage 2. The data in Figs. 2 and 3 show that the partial discharges on the epoxy surface exhibit correspondingly non-stationary behavior manifested by different stages which reoccur upon the continuous recording of the PD pulses.

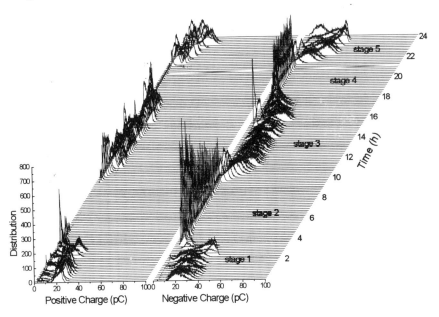

Fig. 3. Progression of individual PD (positive and negative) amplitude distributions measured for 24 hours with a gap distance of 0.5 mm ±0.01 mm, a frequency of 60 Hz, and a voltage of 2.30 kV.

Surface Condition

In Fig. 4(a) is shown the concave profile of the epoxy surface after exposure to partial discharges for 36 hours in a 10%SF$_6$+90%N$_2$ mixture. Interestingly, within the concave are seen protrusions (shown as a and b in the figure) of about 1 μm to 3 μm high and about 10 μm wide. In Fig. 4(b) is shown the condition of the epoxy surface, amplified by a factor of 1000. Evidently, there is a clustering of crystalline-like structure on the epoxy surface.

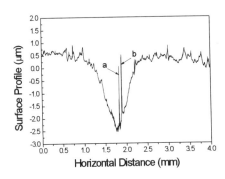

Fig. 4(a). Profile of the epoxy surface after exposure to partial discharges for 36 hours in a mixture of 10%SF$_6$+90%N$_2$. The protrusions, indicated in the figure by *a* and *b*, are about 1 μm to 3 μm high and about 10 μm wide.

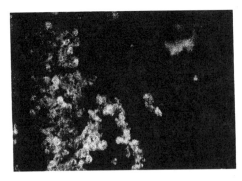

Fig. 4(b). Condition of the epoxy surface, corresponding to a small area in the profile of Fig. 4(a), magnified 1000 times, after exposure to partial discharges for 36 hours in 10%SF$_6$+90%N$_2$. The white color shows the crystalline-like structure formed.

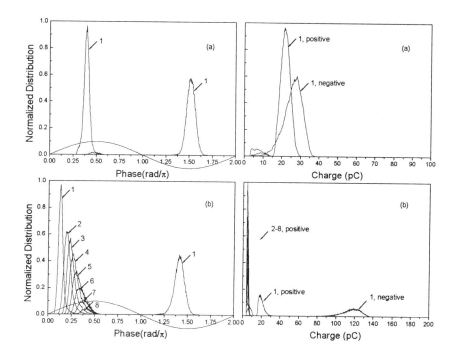
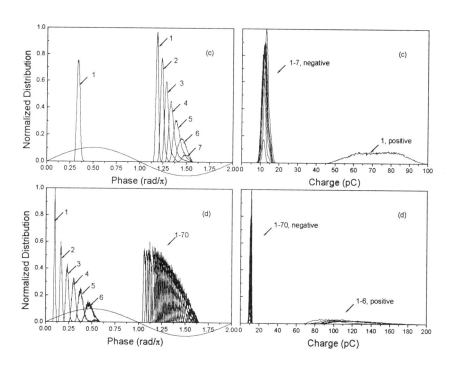

Fig. 5. Individual PD phase and amplitude distributions (gap distance = 0.5 mm ±0.01 mm, frequency 60 Hz). (a) 10%SF_6+90%N_2 mixture, voltage = 2.28 kV; (b) 10%SF_6+90%N_2 mixture, voltage = 3.18 kV; (c) air, voltage = 1.79 kV; (d) air, voltage = 3.00 kV. Numbers (*1, 2, 3, ...*) indicate pulse sequence within the positive half cycle or negative half cycle of the sinusoidal alternating voltage.

Voltage Dependence of Statistical Distribution

In Figs. 5(a) and 5(b) are shown representative PD phase and charge distributions for the 10%SF_6+90%N_2 mixture at two values of the applied voltage. Similar data are shown in Figs. 5(c) and 5(d) for air. Although for the long periods the partial discharges exhibit non-stationary behavior, for relatively short time intervals, stable signals can be obtained which allow the study of the voltage dependence of the statistical distributions. The data presented here were taken after one-hour conditioning exposure to partial discharges. For the 10%SF_6+90%N_2 mixture, the number of pulses increases in the positive cycle and stays the same in the negative cycle when the voltage is increased. While the same behavior in the positive cycle is exhibited by the PD in air,[5] the number of pulses in the negative cycle increases for air as well. For both gaseous dielectrics, the larger the difference is in the number of pulses between the positive and negative cycles, the larger is the observed difference in the amplitudes of the individual pulses between positive and negative cycles. These types of behavior and the differences between the SF_6-N_2 mixture and air are likely to be due to the different electron attachment and detachment processes in these two gases. In this regard, it is interesting to note that when the PD reappears in the SF_6-N_2 mixtures after a few cycles with no pulses, a PD always first occurs in the negative cycle followed by a PD in the positive cycle. This suggests that the second discharge is triggered by electrons detached from negative ions formed during the preceding pulse.

CONCLUSIONS

Changes of the dielectric surface induced by partial discharges modify the statistical behavior of the partial discharges themselves. Positive partial discharges reappear even after they have been extinguished for hours. Consideration of the time development of partial discharges is thus necessary for defining the PD patterns for PD pattern recognition and detection.

ACKNOWLEDGEMENTS

The authors are grateful for the epoxy surface condition measurement performed by J. Song and his co-workers of the Precision Engineering Division of NIST.

REFERENCES

1. R. J. Van Brunt, Stochastic properties of partial-discharge phenomena, *IEEE Transactions on Electrical Insulation*. 26:902(1991).
2. R. J. Van Brunt, P. von Glahn, and T. Las, Anomalous stochastic behavior of partial discharge on aluminum oxide surface, *Journals of Applied Physics*. 81:840(1997).
3. R. J. Van Brunt, P. von Glahn, and T. Las, Non-stationary behavior of partial discharge during discharge induced ageing of dielectrics, *IEE Proceedings of Science Measurement Technology*. 142:37(1995).
4. *Partial Discharge Measurement,* International Electrotechnical Commission, IEC Standard, Publication 270, 47(1981).
5. Y. Wang, X. Han, and R. J. Van Brunt, Digital recording and analysis of partial discharges in point-dielectric gaps, in: *1998 IEEE International Symposium on Electrical Insulation,* Washington, D.C., USA (1998).

ANALYSIS OF THE DEGRADATION OF POLYETHYLENE IN AIR USING ELECTRICAL AND PHYSICAL DATA

John Horwath and Daniel L. Schweickart

Air Force Research Laboratory
Propulsion Directorate
Wright-Patterson Air Force Base, OH 45433

Yicheng Wang

Electricity Division, National Institute of Standards and Technology
Gaithersburg, MD 20899

ABSTRACT

Discharge currents and degradation of polyethylene from high dc electric field stress in air were investigated. A positive point-to-plane configuration was used with thin polyethylene samples on the cathode surface. Voltage, spacing, and relative humidity were independent variables. The surface chemistry and resistivity of the samples evidence changes as degradation occurred. This changing boundary condition affected the gaseous discharge. Both electrical and physical measurement techniques have been utilized to characterize the degradation. Electrical measurements include surface resistivity measurements, pulse train recording of partial discharges, and continuous current measurements. Physical measurements include X-ray photoelectron spectra (XPS), atomic force microscopy (AFM) roughness measurements, and Fourier transform infrared (FTIR) spectroscopy spectra.

INTRODUCTION

High voltage engineering has been dependent on material performance since the beginning of commercial electrical power. Higher electrical voltages, cost, and greater service life have been driving dielectric development since the late 1800's. To assist in the development of dielectrics, various tests have been devised to understand basic phenomena or to act as quality control checks. The increasing sophistication of the tests has largely followed the advancement of electronics over the last sixty years. With the availability of increasing computing power and of faster analog-to-digital convertors to record incipient failure transients, new possibilities in dielectric testing are possible. In particular, more sophisticated pulse train recorders suitable for dielectric research can be designed and built. While such pulse train recorders are suitable for charge transfer measurements in gaseous, liquid, or solid dielectrics, this effort focuses on polyethylene in air environments. Electrical discharges that occur in a hybrid solid and gaseous insulation systems are complex phenomena. Explanation of an insulation degradation mechanism, given an experimental result, can be problematic. The use of a pulse train recorder and subsequent stochastic analysis of data may further the understanding of these basic mechanisms. Such knowledge may improve the testing and design of high voltage polymer systems.

MEASUREMENT TECHNIQUE

Measurements were taken in a vented cylindrical chamber of ≈5 liter capacity. Air was pumped through the chamber at the rate of 5 liters per minute. Pressure was held at one atmosphere for all tests. Ambient room temperature averaged 23.8 °C.

The electrical measurements involved both dc and pulse measurements. A simplified block diagram of the measurement system is shown in Figure 1. A point-to-plane gap was used. The anode was machined from a 6 mm diameter stainless steel rod. The anode had a 64 degree cone shaped point with a tip radius of ≈60 μm. The cathode was a polished disk plane of 5 cm diameter. The plane of nickel alloy (Kovar) was covered by 63 mm diameter polymer disks.

Long term exposure to high voltage will permanently degrade the surface of the polymer and alter the electrical measurements. Unless otherwise noted, measurements were taken over ten minute periods for each setting of spacing, voltage, and relative humidity. To minimize such aging effects for the ten minute tests, new disks were used before one hour was exceeded on a disk. Tests were randomized within the hour exposure. The polymers used for these tests were low density polyethylene (LDPE) and high density polyethylene (HDPE). Disks were cut from sheet stock. The LDPE sheet was 0.8 mm in thickness and the HDPE sheet was 0.83 mm in thickness.

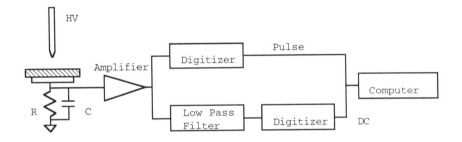

Figure 1. Schematic diagram of the PD detection system.

The dc current is measured via the 430 kΩ shunt resistor. The voltage at the resistor is low-pass filtered with a cutoff frequency of 30 Hz. The filtered voltage signal is sampled every ten seconds during a ten minute test run.

The pulse current, (I_p), is given by

$$I_p = \frac{\sum_{n=1}^{k} n_i Q_i}{t} \quad (1)$$

where i is the channel, n_i is the number of counts per channel, Q_i is the charge per channel, k is the total number of channels, and t is the measurement time. Both time and amplitude were recorded for each pulse. A low level cutoff of twenty-five millivolts was used. A maximum analog voltage of five volts could be digitized. With a capacitance of 0.1 nF, the measurable charge amplitude of a pulse ranges from 2.5 pC to 500 pC. The digital partial discharge (PD) detection system has been described in detail by Wang[1]. Pulses are detected by the RC (C = 0.1 nF and R = 430 kΩ) circuit in front of a high impedance unity-gain amplifier. The digitizing system uses a multifunction data acquisition board to acquire the pulse current waveform continuously at a sample rate of 500 k-samples/s and a resolution of 12 bits. The board cyclically fills a circular buffer that holds 1 million data points. The buffer is divided into two equal halves so that data in one half of the buffer can be transferred, analyzed, and saved while the other half is being filled. The system consecutively acquires waveforms of one second duration.

ELECTRICAL MEASUREMENTS

Voltage dependency

The relationship between voltage and dc current for typical samples is shown in Figure 2. The gap spacing was one centimeter and relative humidity was 50 percent. The dc current, (I_{dc}), generally increases with the voltage. An exponential current growth trend is shown, as might be expected in a gap with no dielectric layer.

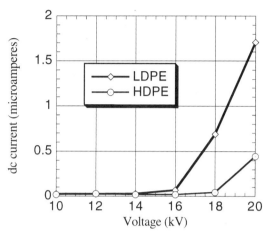

Figure 2. dc current versus voltage.

Pulse current versus voltage is shown in Figure 3 for a gap spacing of one centimeter and 50 percent relative humidity. In contrast to the dc measurements, the pulse current for both polyethylenes evidences a very nonlinear voltage dependency. As voltage is increased, the pulse current reaches a maximum then decreases in magnitude. Point-plane gaps without insulators also show a similar non-monotonic pulse current behavior with respect to increasing voltage[2]. For both LDPE and HDPE, the dc current is at least one order of magnitude larger than the pulse current for the range of voltages shown. The pulse current can be attributed to the electron transport while the dc current results from the transport of both electrons and ions across the gap.

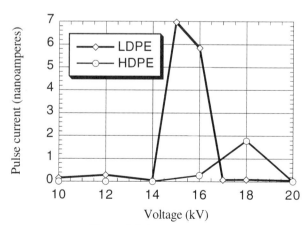

Figure 3. Pulse current versus voltage.

Spacing considerations

Spacing effects for dc currents are shown in Figure 4. The voltage was 20 kV and relative humidity was 50 percent. Both the low density and high density polyethylene dc currents show that the maximum current does not occur at the closest spacing. As the ratio of gap spacing to cathode diameter increases, electric field geometry begins to affect the dc current magnitude. Charge carriers (free electrons and ions), which make up the dc drift current, follow field lines that are not intercepted by the dielectric layer. These field lines increase in number as the gap widens. However, with larger spacings the electric field strength is reduced resulting in decreasing current after the peak.

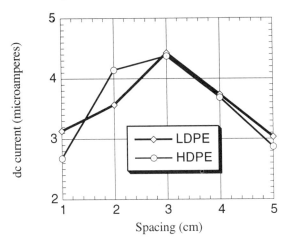

Figure 4. dc current versus spacing.

The pulse current is not strongly correlated with spacing. The pulse current is strongly correlated with type of polyethylene. LDPE shows a significantly higher pulse current than HDPE.

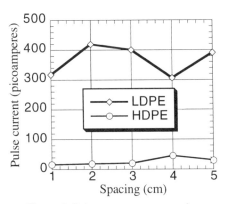

Figure 5. Pulse current versus spacing.

Humidity effects

The spacing for the humidity tests was 1 cm and the voltage was 20 kV. Humidity effects are complex and a complete description is difficult. The dc current versus relative humidity is shown in Figure 6. Humidity can cause a clustering of water molecules around ions[3]. The clustering can cause a decrease in ion mobility and subsequent current[4]. Pulse current versus humidity is shown in Figure 7. The pulse current is strongly affected by humidity. At the lower relative humidities of 25 and 28 percent only three or four pulses were recorded per ten minute test. At the higher relative humidities of 78 and 79 percent at least 123,000 pulses were recorded per ten minute test.

Surface resistivity measurements

Surface resistivity measurements were taken with a high resistance meter. Parallel plate probes 1 mm apart with a 5 mm length were used. Surface resistivity measurements on polymers can be highly susceptible to ambient humidity[5]. Measurements were taken in a shielded enclosure at two percent relative humidity. LDPE showed an initial surface resistivity of $6.3 \times 10^{16} \Omega$ and HDPE showed an initial surface resistivity of $8.3 \times 10^{16} \Omega$. Surface resistivity can influence the type of gaseous discharge; typically as surface resistivity of a polymer decreases, the discharge may have a decreased pulse component[6]. However, the small difference in surface resistivity between the two polyethylenes does not agree with this trend when viewing Figure 5.

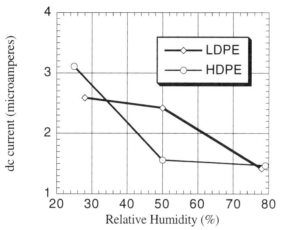

Figure 6. dc current versus relative humidity.

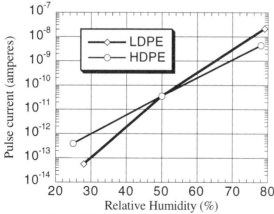

Figure 7. Pulse current versus relative humidity.

CHEMICAL MEASUREMENTS

The XPS results show an increase in surface oxidation. HDPE unexposed to high voltage had a surface oxygen to carbon ratio of 0.026. HDPE exposed to 20 kV for seven hours had an average surface oxygen to carbon ratio of 0.30.

FTIR results show an increase in two carbon and oxygen functional groups at the surface of HDPE after seven hours at 20 kV with a 5 mm spacing. An absorbance was observed at 1718 cm^{-1}. Absorbances in this region are typical for carbonyl groups (C=O). Another absorbance was observed in the 1270 cm^{-1} region. This absorbance is consistent with C-O groups.

PHYSICAL MEASUREMENTS

AFM results show an increase in surface roughness of HDPE upon exposure to high voltage. The standard deviation of surface height in a 400 μm^2 area unexposed to high voltage is 77.5 nm. The standard deviation of surface height in the same size area after exposure to 20 kV for seven hours is 113 nm.

DISCUSSION AND CONCLUSIONS

Understanding the basic phenomena of gaseous discharge and polymer degradation is essential for development of valid industrial dielectric testing and high voltage design of polymer insulation systems. This investigation shows that several variables affect the gaseous discharge with resulting physical and chemical changes. Physical and chemical measurements show erosion and oxidation. The discrepancy between pulse current activity and surface resistivity may be due to surface morphology. Gaseous discharge measurements show voltage and relative humidity dependencies. For these experiments, the decrease of dc current with increasing humidity seems complementary to the increase of pulse current with increasing humidity. The magnitude of the pulse current is also strongly dependent on voltage, and in general much smaller than the dc current. The latter observation may have significant implications in improving methods for quantifying partial discharge. Currently only the apparent charge associated with the pulse current is used in quantifying partial discharge[7].

Continuing efforts will provide additional data for verification of these trends, by testing a greater number of samples in various environments. Future research will include surface profile measurements to calculate volume erosion and longer term testing over several weeks. A model for polymer degradation processes using data from both electrical and physical measurements will be developed.

REFERENCES

1. Y. Wang, "New Method for measuring statistical distributions of partial discharge pulses," *J. Res. Natl. Inst. Stand. Technol.* 102, 569 (1997).
2. G. Buchet, M. Goldman, and A. Goldman, "Décharges dans les Gaz: Sur la Nature Courant Permanent dans les Décharges Couronne Pointe-plan en Tension Continue," *C. R. Acad. Sci. Paris*, vol. 263B, pp. 356-359, (1966).
3. R. Messaoudi, A. Younsi, F. Massines, B. Despax, and C. Mayoux, "Influence of Humidity on Current Waveform and Light Emission of a Low-frequency Discharge Controlled by a Dielectric Barrier," *IEEE Transactions on Dielectrics and Electrical Insulation*, vol. 3, no. 4, pp. 537-543, Aug. (1996).
4. I. A. Metwally, "Factors Affecting Corona on Twin-Point Gaps under dc and ac HV," *IEEE Transactions on Dielectrics and Electrical Insulation*, vol. 3, no. 4, pp. 544-553, Aug. (1996).
5. D. K. Das-Gupta, "Electrical properties of surfaces of polymeric insulators," *IEEE Transactions on Dielectrics and Electrical Insulation*, vol. 27, no. 5, pp. 909-923, Aug. (1992).
6. C. Hudon, R. Barnikas, and M. R. Wertheimer, "Spark-to-glow discharge transition due to increased surface conductivity on epoxy resin specimens, " *IEEE Transactions on Dielectrics and Electrical Insulation*, vol. 28, no. 1, pp. 1-8, Feb. (1993).
7. "Partial Discharge Measurements," International Electrotechnical Commission, IEC Standard, Publ. 270, 1981.

DIGITAL ANALYSIS OF PD SOURCES IN GAS INSULATED SWITCHGEAR SUBSTATION

A. H. Mufti, W. M. Al-Baiz, A. O. Arafa

High Voltage Laboratory
Electrical & Computer Engineering Dept.
Faculty of Engineering
King Abdulaziz University
P.O. Box 9027, Jeddah 21413
Saudi Arabia

INTRODUCTION

It has been revealed that imperfection such as microscopic protrusions on the interior of the Gas Insulated Switchgear systems (GIS) metal work, foreign particles, components that float in potential and deterioration of solid insulation reduces the insulation level of switchgear[1].

During normal operation, free conducting particles tend to bounce along the bottom of the enclosure. The individual bounce amplitude depends on the particle size, shape, the potential applied to the system and the random phase at which the particle lands[2,3]. The microscopic surface structure of the conductor causes local enhancement of electric stress, which in turn leads to a premature breakdown of the gas medium[1]. Components that float in potential can also be considered as discharge source in GIS system. The most common type of floating components is either spacer insert or corona shields[4]. As a result of the above discharge sources, a method of inspecting the inside condition of GIS before insulation failures can take place as an objective which, if realized, can lead to improving its performance and reliability of equipment.

The aim of this paper is to analyse the discharge sources in gas switchgear system. The measurements are recorded in terms of discharge magnitude and pulse count rate number as a function of phase angle. Various three dimensional relationship pattern of partial discharge sources are obtained under the effect of gas pressure, electrode gap and source location. The particle type and mode under different applied voltage are discussed.

APPARATUS AND EXPERIMENTAL TECHNIQUE

The measurements are carried out using a transparent, Plexiglas cylindrical pressure vessel of 130 mm internal diameter and 520 mm height. The needle-to-plane configuration is used to present the discharges from high voltage side and earthed side with a gap of

10 mm and 20 mm. The concave electrode with an interelectrode gap separation of 26.6 mm is used to present the various discharge mode of wire and sphere particles of different size and type. The SF_6 gas used is of commercial purity i.e. 99.8% pure. The pressure of 0.34 MPa and 0.5 MPa are used in the experiment.

The high voltage source is a 60 Hz, 100 kV test transformer. A current limiting water resistor is connected in series with the high voltage supply. The partial discharge digital analyzer system is used with coupling quadripole and coupling capacitor. The Partial discharge calibrator which is used corresponds to major international standards. The measurements are recorded in terms of charge and number of pulses for different discharge types. The a.c. voltage has been increased slowly to observe the appearance of electrical discharge pulse on (CRO) up to active point of inception voltage slightly higher than the inception voltages to determine relative magnitude of partial discharge and print out. Without the test geometry in the circuit, the system had a partial discharge level of < 5 pC for voltages of 100 kV.

RESULTS AND DISCUSSION

The experimental result presents various discharge data by computer-aided digital processing. The discharge magnitude and the discharge position are related to the power frequency cycle. For this purpose the power frequency cycle is divided in several windows $\varnothing$. Also the momentary value in kV of the test voltage at which the discharge has occurred is registered. The discharge distribution $Hn(\varnothing,q)$ present the relationship between the discharge count rate n and the discharge magnitude q as a function of the phase angle $\varnothing$ of the voltage cycle. This three dimensional relationship is used to evaluate the partial discharge and may yield a better discrimination performance.

The needle-to-plane geometry offers a convenient method of studying non-uniform field discharges. Figure 1 shows the discharge of a needle fixed on the high voltage side with a gap of 20 mm and a pressure of 0.34 MPa. The maximum discharge magnitude Hqmax($\varnothing$) at 20 kV records a value of 11 pC around the positive half-cycle peak and 8.8 pC around the negative half-cycle peak. The mean pulse height distribution Hqn ($\varnothing$) records a value of 7.7 pC which represents the average value of the discharges. As the pressure is increased to 0.5 MPa, the rate of the electron production decreases. Both the ion diffusion and the photoionization mean free path are reduced whereas the voltage is increased almost linearly with the pressure[5,6]. The discharge is observed at 24 kV and Hqmax($\varnothing$) records a value of 10 pC around the positive peak and 6.9 pC around the negative peak as shown in figure 2. The mean pulse height distribution Hqn($\varnothing$) records a value of 7.1 pC. The discharge count rate distribution Hn ($\varnothing$, q) show the effect of the pressure on the discharge distribution pattern.

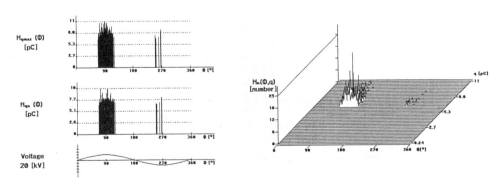

Figure 1. The discharge source is a needle fixed on the high voltage side with a gap of 20 mm and a pressure of 0.34 MPa.

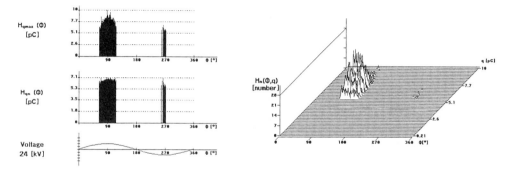

Figure 2. The discharge source is a needle fixed on the high voltage side with a gap of 20 mm and a pressure of 0.5 MPa.

The gap distance is been reduced to 10 mm and the vessel is pressurised to 0.34 MPa. Hqmax(∅) reaches a value of 13.5 pC around the positive half cycle peak and 14 pC around the negative half cycle peak at a threshold voltage of 14 kV as shown in figure 3. The decrease of the gap distance causes an increase in the electric field at the electrode point[5]. The Hqn(∅) records a value of 10.5 pC. The Hn(∅,q) shows the discharge distribution of the reduced gap. This gives a clear detection performance which may help to identify different discharge pattern.

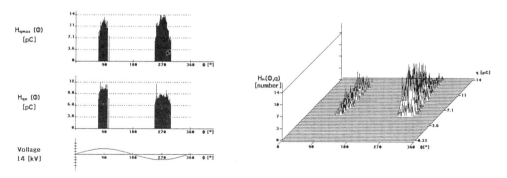

Figure 3. The discharge source is a needle fixed on the high voltage side with a gap of 10 mm and a pressure of 0.34 MPa.

The partial discharge location in GIS systems has a great concern from a practical point of view. In this study, the needle is connected to the earth side with a gap of 10 mm and a pressure of 0.34 MPa. The onset voltage is observed at a threshold value of 18 kV. The Hqmax(∅) records a value of 16 pC around the positive half cycle peak and 6.8 pC around the negative half cycle peak as shown in figure 4. The Hqn(∅) records a value of 9 pC. It can be clearly seen from figure 3 and figure 4 that the discharge patterns have different characteristics. This yields to the recognition and classification of partial discharge locations.

Wire particles produce high field intensification and the particles exist in an infinite variety of size and shapes so that any study of particles with the initiation of discharges in gases comes by nature[2,3]. The particle length and diameter are important factors in lowering the dielectric withstanding capabilities[1]. Figure 5 shows different patterns of partial discharge of a wire particle of 10 mm length and 1 mm diameter with a pressure of 0.5 MPa. For the particle lies on the grounded electrode, the maximum discharge magnitude Hqmax(∅) records a value of 1.9 nC around the positive half cycle peak and 1.5 nC around the negative half cycle peak at 38 kV. The Hqn(∅) records a value of 1.5 nC. As

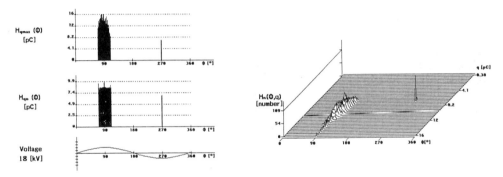

Figure 4. The discharge source is a needle fixed on the earth side with a gap of 10 mm and a pressure of 0.34 MPa.

the voltage is increased, the electrostatic force acting on the particle exceeds the gravitational force[7]. The particle moves on the grounded electrode gap. The Hqmax(∅) records a value of 2.2 nC on the positive half cycle and 1.6 nC on the negative half cycle. The Hqn(∅) records a value of 1.9 nC. The particle starts to bounce as the voltage reaches 50 kV and the Hqmax(∅) records a value of 1.1 μC on the positive half cycle. The above discharge pattern shows a clear discrimination performance of a particle under three modes of activities.

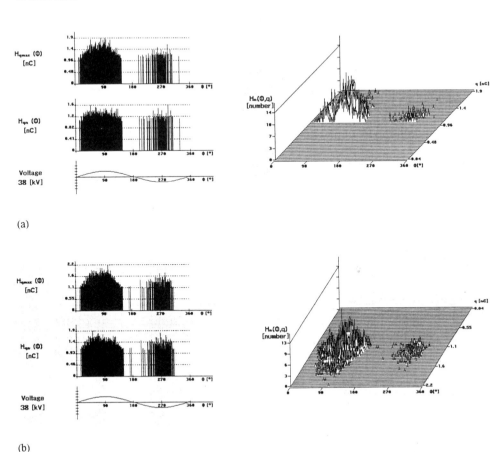

(a)

(b)

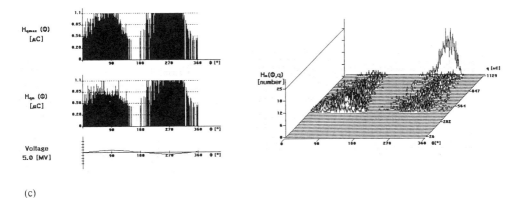

(c)

Figure 5. The discharge source is a metallic wire particle of 1 mm diameter and 10 mm length with a pressure of 0.5 MPa. (a) The particle lies on the grounded electrode, (b) the particle moves on the grounded electrode, (c) the particle bounces in the inner electrode gap.

Free spherical particles have a low field intensification. Their breakdown characteristics are not similar to those of the wire particles[8]. In this experiment a spherical particles of 3.18 mm diameter is used with an electrode distance of 20 mm with a pressure of 0.34 MPa. As the voltage increases to 45 kV, the electrostatic force acting on the charged sphere exceeds the gravitational force and the sphere is lift off the electrode surface. The maximum pulse height distribution records a value of 216 pC on both positive and negative half cycles as shown in figure 6.

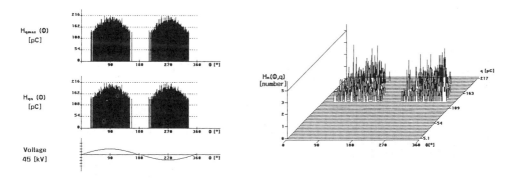

Figure 6. The discharge source is a metallic sphere particle of 3.18 mm diameter with a pressure of 0.34 MPa. The particle is lifted off the grounded electrode.

The use of partial discharge digital recording system has become very popular in recent years[9,10]. The evaluation of the measuring information permits a better discrimination performance of PD. The discharge distribution, the discharge count rate and the discharge magnitude show the effect of pressure, gap and locations. As the pressure is increased, the discharge count rate increases. When the gap length is reduced, the discharge count rate number increases particularly on the negative half cycle under the effect of the electric field increase at an electrode point. The effect of the discharge source location on the pattern is clearly seen in terms of discharge count rate number. However, the maximum discharge magnitude and the mean pulse height distribution of the needle on the earthed side records higher values than those of the needle on the high voltage side.

323

In case of metallic particle, the particle size and mode are the main factors that influence the discharge count rate and the discharge magnitude on both positive and negative half cycles. This can be observed when the particles mode is changed from laying on the lower electrode to a moving mode. However, in the bouncing mode, the partial discharge pulse pattern covers the whole cycle with a large increase in pulses during a negative half cycle.

The partial discharge pattern of the spherical particle is also affected by the particle mode under the applied voltage stress. The discharge pulse pattern of a lifted off mode covers the whole positive and negative half cycles and appears identical. More detection data can be obtained for more systematic analysis of the discharge sources.

CONCLUSIONS

The conventional and digital analysing detection techniques may help in recognizing different partial discharge sources in the GIS systems. The various partial discharge pattern activities can be obtained from fixed and conducting particles at different mode and location. The PD magnitude and the pulses count rate around the positive and negative half cycles can give appropriate method of classification. The use of a partial discharge digitizer detection system offers the opportunity to store the various pattern of discharge pulses.

REFERENCES

1. H.Ryan, D.Lightle, D.Milne, Factors influencing Dielectric Performance of SF_6 insulated GIS, IEEE Trans. Power Apparatus and Systems, Vol. PAS-104, No. 6, pp 1527-1535, (1985).
2. C.Cooke, R.Wooton, A.Cookson, Influence of Particles on as and dc Electrical Performance of Gas Insulated Systems at Extra-High-Voltage, IEEE Trans. Power Apparatus and Systems, Vol. PAS-96, No.3, pp 768-777, (1977).
3. T.Hattori, M.Honda, H. Aoyogi, N. Kobayashi, K. Tevasaka, A Study on Effects of Conducting Particles in SF_6 Gas and Test Methods for GIS, IEEE Trans. Power Delivery, Vol. 3, No. 1, pp. 197-204, (1988).
4. R.Schurer, K.Feser, The Effect of Conducting Particles Adhering to Spacers in Gas Insulated Switchgear, Intern. Symp. on High Voltage, Vol. 2, pp. 165-168, (1997).
5. H.Anis, K.Srivastava, Pre-Breakdown Discharges in Highly None Uniform fields in Relation to Gas Insulated Systems, IEEE Trans. Electr. Insulation, Vol. EI-17, No. 2, (1982).
6. O.Ibrahim, O. Farish, Positive-Point Corona Phenomena in Sulfur Hexafluoride, IEE Colloquim on Corona Discharge, Digest No. 1979/73, (1979).
7. H.Parekh, K.Srivastava, Breakdoen Voltage in Particle Contaminated Compressed Gases, IEEE Trans Electr. Insul., Vol. EI-14, No. 2, pp 101-106, (1979).
8. A.Cookson, O.Farish, G.Sommerman, Effect of Conducting Particles on ac Corona and Breakdown in Compressed SF_6, IEEE Summer Power Meeting, No. 71, TP-508-PWR, pp 1329-1338, (1971).
9. E.Gulski, Computer-Aided Recognition of Partial Discharges Using Statistical Tools, Delft University Press, (1991).
10. E.Gulski, Application of Modern PD Detection Techniques to Fault Recognition in the Insulation of High Voltage Equipment, Ninth Inter. Symp. on High Voltage, 5642-1, (1995).

EXPERIMENTAL ESTIMATION OF SCHWAIGER FACTOR LIMIT (η_{LIM}) IN ATMOSPHERIC AIR

Ravindra Arora, Sunil Prem

Department of Electrical Engineering
Indian Institute of Technology
KANPUR - 208016 (India)

INTRODUCTION

Behaviour of dielectrics strongly depend upon the type of field these are subjected to. No stable Partial Discharge (PD) occur under uniform and weakly non-uniform field conditions. The transition from weakly to extremely non-uniform field in a gaseous dielectric accompanies with Partial Discharges (PD) or corona inception.

The field configurations mentioned above can be analytically described on the basis of the factor 'η', introduced by Schwaiger in 1922. The Schwaiger factor, also known as the "degree of uniformity of a field" is defined as the ratio of mean to maximum field intensities in a given configuration.

$$\text{Schwaiger Factor, '}\eta\text{'} = E_{mean}/E_{max}$$

The value of η lies between 0 and 1 ($0 < \eta \leq 1$). For an uniform field, η is equal to 1. Lower values of η within a limit represent weakly non-uniform fields. The lower limiting value of η, known as "η_{lim}", represents the transition from weakly to extremely non-uniform field.

The value of η_{lim} depends upon the behaviour of the dielectric and its physical conditions such as pressure, temperature and purity. It may also depend upon the type of voltage applied and the size of the electrodes and their conditions it is measured with.

In this work the value of Schwaiger Factor Limit (η_{lim}) for atmospheric air was determined experimentally in the laboratory with ac power frequency voltage. The measurements were carried out with the help of a PD detector and a high sampling rate digital oscilloscope. Sets of spherical electrodes of different size were used to produce variation in field configuration with gap distance which gave different values of η. The maximum value of η for which stable PD inception could be measured, was carefully determined. Due consideration was made to the effect of field distortion due to connecting shanks and grounding of the lower electrode for the estimation of Schwaiger factor.

Gaseous Dielectrics VIII, Edited by Christophorou
and Olthoff, Plenum Press, New York, 1998

CLASSIFICATION OF ELECTRIC FIELDS

The electric field between two electrodes is generally classified into two categories, that is Uniform and Non-uniform fields. A field is uniform only when the Schwaiger factor is equal to one. Any value of η lower than one makes the field to be non-uniform. On the basis of the occurrence of the PD phenomenon, a novel concept of 'Weakly Non-uniform Field' was introduced. A non-uniform field is categorised as weakly non-uniform if no stable PD occur before the breakdown. Hence under the given field conditions, if the PD inception voltage U_i is equal to the breakdown voltage U_b, the field is said to be weakly non-uniform. On increasing the non-uniformity in the field a stage comes when stable PD occur before the breakdown. Under such conditions the PD inception voltage is lower than the breakdown voltage. This field is classified as the 'Extremely Non-uniform'. The more is the field non-uniform, the difference between U_i and U_b increases.

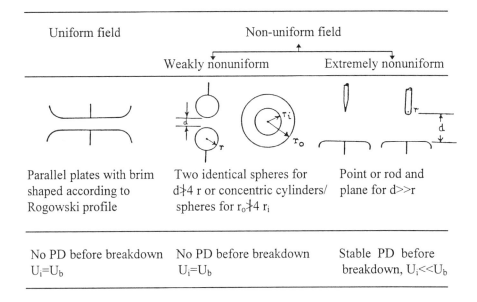

Fig.1 : Classification and typical configurations of electric fields

The transition from weakly to extremely non-uniform fields can be analytically related with the Schwaiger factor. The value of η_{lim} represents this transition under the given conditions of the dielectric. Like in uniform field, under weakly non-uniform field configurations too no PD occur. The knowledge of exact value of η_{lim} under given conditions is useful for the design of equipment.

EXPERIMENTAL SETUP

The high voltage supply was obtained from a 100 kV, 50 kVA, ac power frequency, PD free test transformer. The high voltage dome of this transformer was connected to the other dome on the top of a 1.1 nF, 100 kV PD free coupling capacitor with the help of a 7.5 cm diameter aluminium pipe. The test object was connected to this dome with the help of a 2.5 cm diameter flexible steel conduit. A PD detector, having sensitivity of 1.0 pC, was connected to the low voltage arm of the coupling capacitor. The entire set up was ensured to be PD free up to 70 kV (rms) with this detector.

Electrode Configuration for Investigation

As shown in Fig.1, a sphere-sphere electrode configuration is a classical example of weakly non-uniform field for a limited gap distance. The Schwaiger factor for these electrodes is possible to be analytically estimated for different gap distance.

For the construction of the spheres, standards laid down by the Bureau of Indian Standards[1] and the British Standards[2] were followed. Five pairs of twin stainless steel spheres of diameters 9.92, 15.0, 20.0, 25.4 and 31.7 mm were prepared for investigations. Arrangement was made to mount them on an ebonite stand with the help of circular cross section shanks and their cylindrical extensions.

CALCULATION OF SCHWAIGER FACTOR

For the estimation of maximum field intensity E_{max} between two spherical electrodes a number of analytical solutions are proposed in the literature. The method suggested by Carter and Lob[3] was adopted. Two conditions of the gap, the symmetrical in which the spheres are assumed to carry equal and opposite charges and the unsymmetrical in which one sphere is grounded, are considered separately.

Symmetrical Gap

The calculation of electric field in a gap between two identical spheres by means of dipolar coordinates (u, v, w) is obtained by the rotation of two orthogonal systems of coaxial circles about the line of symmetry passing through real limiting points of one system. The solution of Laplace's equation in the form of Legendre's equation gives rise to an expression for the maximum value of potential gradient in the dielectric at the surface of the spheres to be ;

$$E_{max} = \frac{U}{a} \cosh^3 u_1/2 \sum_{m=0}^{\infty} \frac{\sinh(m+1/2)u_1}{\cosh^2(m+1/2)u_1}$$

and the expression for mean field intensity is ;

$$E_{mean} = \frac{U}{2a} \coth u_1/2$$

where U is the magnitude of the applied voltage and 'a' and 'u_1' are related to the geometry of the gap as shown in the Fig.2.

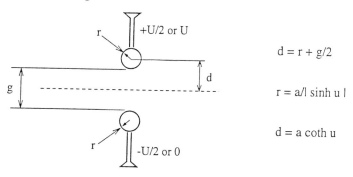

$d = r + g/2$

$r = a/|\sinh u|$

$d = a \coth u$

Fig.2 : Symmetrical Sphere Gap Geometry

The surfaces $u = u_1$, $u = -u_1$ (where $u_1 > 0$) are the identical non-intersecting spheres, the electrodes. The potential over each of these spheres is constant and it is considered to be $+U/2$ on $u = u_1$ and $-U/2$ on $u = -u_1$. The origin and the value of 'a' are chosen such that any two given non-intersecting spheres are included in the system.

Knowing the mean and the maximum field intensities, the relation for Schwaiger factor works out to be ;

$$\eta = \frac{E_{mean}}{E_{max}} = 1 \bigg/ \left[2\sinh u_1/2 \cosh^2 u_1/2 \sum_{m=0}^{\infty} \frac{\sinh(m+1/2)u_1}{\cosh^2(m+1/2)u_1} \right]$$

Unsymmetrical Gap

This is the condition when one of the sphere (lower) is grounded. Potential on one sphere is U and on the other it is equal to zero. The expression for maximum potential gradient in this case works out to be ;

$$E_{max} = \frac{2U}{a} \cosh^3 u_1/2 \sum_{m=0}^{\infty} \frac{\sinh(2m+1/2)u_1}{\cosh^2(2m+1/2)u_1}$$

Thus the expression for Schwaiger factor taking in to account the grounding of the lower sphere is ;

$$\eta = \frac{E_{mean}}{E_{max}} = 1 \bigg/ \left[4\sinh u_1/2 \cosh^2 u_1/2 \sum_{m=0}^{\infty} \frac{\sinh(2m+1/2)u_1}{\cosh^2(2m+1/2)u_1} \right]$$

The values of η calculated by the above expression for 9.92 mm diameter spheres at different gap distance settings of the measurement are given in Table 1. These represent the values for the unsymmetrical, isolated configuration of the spheres when one sphere is grounded.

Effect of Connecting Shanks

The spheres were connected with the help of metallic shanks for mounting them on the stand. As concluded by Binns[4] and Kuffel et al.[5], the net-effect of shanks is to make the field more uniform. Hence, to determine the exact-value of η, the effect of shanks on the field must be taken in to account.

Pillai and Hackam[6] have compared the values of η for sphere-sphere gaps under the two conditions, that is, for isolated spheres and for spheres having cylindrical shanks. According to their estimation, if 'r' is the radius of the spheres and 'd' the distance between them, then for d/r=1, the value of η for spheres with shanks is about 2% higher than for spheres without shanks. Similarly, for d/r=10, η is increased by about 9%. Considering these investigations, a linearly varying correction factor has been applied to the values depending upon the d/r ratio of the electrode configuration. For the 9.92 mm spheres, the corrected values are given in Table 1.

MEASUREMENTS AND RESULTS

For all the five pairs of twin spheres breakdown voltages for increasing gap distance settings were measured beginning with a gap of a few mm under normal room temperature and atmosphereic pressure conditions. The PD detector was kept on concurrently to detect any stable PD before the breakdown. A sample of such measured results are given in Table 1. The gap distance, at which stable PD could be first detected, was extra carefully measured. The breakdown and PD inception voltages are plotted for increasing gap distances in Fig. 3 & 4 for two sets of electrodes out of five investigated.

Table 1 Values of U_i, U_b, and η for different gap settings for 9.92 sphere gap

Ser No	Gap d in mm	U_i kV(peak)	U_b kV(peak)	Schwaiger Factor (η)		
				Calculated Value	Percentage Correction	Corrected Value
1.	5.0	No PD	13.7	0.656666	2.15	0.67143
2.	10.0	No PD	25.5	0.425045	3.0	0.437846
3.	13.0	No PD	28.1	0.344838	3.45	0.356735
4.	14.0	No PD	29.1	0.323992	3.6	0.335704
5.	15.0	No PD	29.7	0.305377	3.75	0.316829
6.	20.0	No PD	32.8	0.236302	4.5	0.246992
7.	22.0	33.7	34.1	0.216443	4.8	0.226890
8.	25.0	33.9	35.1	0.192073	5.25	0.202215
9.	30.0	36.2	37.3	0.161537	6.00	0.171288
10.	35.0	36.5	39.0	0.139261	6.75	0.148720
11.	40.0	37.9	43.0	0.122325	7.50	0.131559
12.	45.0	38.5	45.0	0.109028	8.25	0.118082
13.	50.0	38.7	48.5	0.093193	9.00	0.107227
14.	60.0	38.9	53.2	0.082148	9.75	0.090833
15.	70.0	38.9	60.5	0.070526	10.5	0.079049

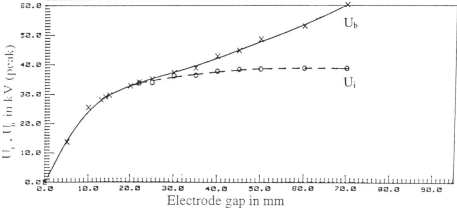

Fig. 3: PD inception U_i and Breakdown voltage U_b characteristics for 9.92 mm dia spheres

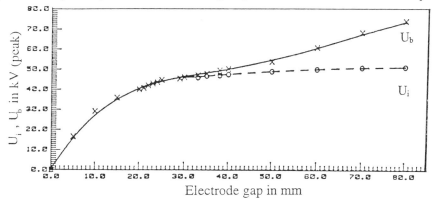

Fig. 4: PD inception U_i and Breakdown voltage U_b characteristics for 15.00 mm dia spheres

Fig.3&4 show typical characteristics of the transition from weakly to extremely nonuniform field. No PD could be measured in weakly nonuniform fields. In the extremely nonuniform field conditions, stable PD were measured. The difference between U_i and U_b increases as the nonuniformity in the field is increased and PD activity takes place in a wider range of applied voltage.

The calculated and corrected values of Schwaiger factor at the transition (η_{lim}) for the five pairs of twin spherical electrodes are given in Table 2.

Table 2 : Values of Schwaiger factor limit determined for different size of electrodes

Sr. No.	Sphere Diameter D (mm)	Schwaiger Factor Limit (η_{lim})
1.	9.92	0.246992
2.	15.0	0.248775
3.	20.0	0.240765
4.	25.4	0.239573
5.	31.7	0.243321

As seen from the Table 2, the values of Schwaiger factor limit in atmospheric air measured with different size of the spherical electrodes, do not have any appreciable difference. These lie within a narrow range between 0.24 and 0.25. Hence it can be concluded that there is no affect of size of the electrodes investigated on the value of η_{lim}. A rounded value of η_{lim}, convenient to remember, of 0.25 is proposed for Schwaiger factor limit for atmospheric air for ac power frequency voltage.

As seen in Figs. 3&4 and Table 1, the value of η_{lim} for all the electrodes was measured at a gap distance just above twice the diameter of the twin spheres. Hence, it can be safely concluded that in atmospheric air the field in the gap between two identical spheres, one of them grounded, can be considered to be weakly nonuniform up to a gap distance twice the diameter of the spheres.

REFERENCES

1. IS : *Method for Voltage Measurement by Sphere Gaps* (One Sphere Earthed), ISI (3rd Print) July 1982.
2. BS358 : British Standard.
3. Carter, G.W. and Loh, S.C, *The Calculation of the Electric field in a Sphere Gap by Means of dipolar coordinates*, Proc. IEE, Vol 106 C, (1959) Page 108-111.
4. Binns, D.F., *Calculation of the Field Factor for a Vertical sphere Gap Taking Account of The Surrounding Earthed Surfaces*, IEE Vol 112, (1975) page (1575-82).
5. Kuffel E and Husbands A.S., : *Influence of nearby Earthed objects and of the polarity of the voltage on the Direct Voltage Breakdown of Horizontal Sphere Gaps*, Proc. IEE, Vol 108 A, (1961), Page 302.
6. Pillai A.S., Hackam R., *Electric Field and Potential distributions for unequal spheres using Symmetric and Asymmetric Applied Voltage*, IEEE Trans. EI, Vol. EI-18, No. 5, (Oct. 1983).

SECTION 5: HIGH PRESSURE GAS DIELECTRICS / SF$_6$-N$_2$ MIXTURES

SF_6/N_2 Mixtures for HV Equipment and Practical Problems

Tatsuo Kawamura[1], Satoshi Matsumoto[2], Masahiro Hanai[2] and Yasufumi Murayama[2]

[1] Shibaura Institute of Technology (Tokyo, Japan)
[2] Toshiba Corporation (Kawasaki, Japan)

Introduction

Since SF_6 gas has a high dielectric performance and interrupting capability, it is used widely in the Electric Power Industry. Recently, it is recognized that SF_6 gas is a greenhouse gas and therefore, SF_6/N_2 gas mixtures are proposed as a replacement of pure SF_6.

In this paper, firstly the general scope of SF_6 and the state-of-the-art of GIS are reviewed.

Secondly, the dielectric strengths of SF_6 and N_2, physical properties and heat transfer characteristics are reviewed from a practical point of view.

Thirdly, the difficulties in practical use, related to recycling technique of used SF_6/N_2 gas mixtures, will be discussed.

It is concluded that pure SF_6 gas can be continuously used in the Electric Power Industry by establishing an adequate recycling system, reducing emission levels appropriately.

Keywords

gaseous dielectrics, gas mixtures, gas recycling, SF_6, SF_6/N_2 gas mixtures, gas reclaiming equipment

State-of-the-art of GIS and General Scope of SF_6 Gas

In Japan, size reduction of substations is quite important because of narrow space and high land cost in urban areas. On the other hand, aspect around the substation is requested to be in harmony in the suburbs. By applying GIS, these problems are solved, and noise from the outdoor equipment is minimized.

SF_6 gas has a high dielectric performance and high current interrupting capability, therefore, it is widely used in substation equipment such as GIS and Gas Insulated Transformers.

Recently, SF_6 gas has been shown to be a greenhouse gas[1]. COP3(United Nation Framework Convention on Climate Change, The 3rd Session, Conference of the Parties held in Kyoto, Japan) has decided to reduce the emission quantity of greenhouse gases in 2010 less than in 1990. Under the protocol, SF_6 gas is intended for reduction, therefore corrective action is needed against usage and exhaust of SF_6 gas worldwide.

This paper reports about three possibilities to reduce usage and exhaust of SF_6 gas;
(1) Replacement by environmentally friendly gases
(2) Reduction of leakage, optimization of recycling techniques
(3) Reduction of usage

Potential gas or gas mixtures to substitute SF_6

Requirements of the medium are summarized as follows;
a) Performance : dielectric strength, current interrupting capability, heat transfer characteristics
b) Treatment : chemical stability and inertness with respect to conducting and insulating materials, low liquefaction temperature
c) Safety : non-toxicity, non-flammable gas
d) Profitability: valid cost, stable supply
e) Environmental harmony: non-destructive for ozone layer (Ozone Depleting Potentials equals zero), less greenhouse effect (Global Warming Potentials nearly equals zero or low.)

Relevant physical properties and heat transfer characteristics of alternative gases are shown in Table 1. Toxicity, ODP and GWP are shown in Table 2.

If the environmental problem is given high priority, the SF_6 substitute will restrict to a gas of simple composition, such as nitrogen, air and CO_2. However, these gases have a dielectric strength lower than SF_6. Moreover, gas with a dielectric strength higher than SF_6 and boiling point lower than SF_6 has not been found until now.

High pressure nitrogen has a possibility as a SF_6 substitute. Fig.1 shows the dielectric strength of SF_6 and N_2 for a uniform electric field[2]. The figure shows that SF_6 possesses a few times higher dielectric strength. This means that a higher operational pressure must be used resulting in a more severe mechanical stress of the tank. This will result in higher manufacturing costs, lower safety standard and additional difficulties to handle the higher pressure.

Therefore, pure high pressure nitrogen has little possibility as a SF_6 substitute for GIS.

Table 1. Physical Properties of Gases

compound	Molecular Weight	Density kg·m-3	Boiling Point °C	Specific Heat (at 298K, 1atm) cal·g-1·K-1	Thermal Conductivity W·m-1·K-1	Viscosity (at 290K) μPa·s	Dielectric Strength % of SF6	Critical Density kg·m-3	Critical Temperature °C	Critical Pressure MPa
SF6	146.06	6.517	-64	0.157	1.55E-02	15.13*	100	725	45.64	3.767
He	4	0.1785	-269	1.242	1.50E-01	19.42	2~6	69.3	-267.96	0.229
Ne	20.2	0.9	-246	0.246	4.93E-02	30.99	1~2	484	-228.75	2.624
Ar	39.9	1.784	-186	0.125	1.77E-02	22.09	4~10	531	-122.35	4.862
H2	2.02	0.0899	-253	3.330	1.75E-01*	8.73	20	31	-239.95	1.297
N2	28.01	1.25	-196	0.248	2.38E-02*	17.40	34~43	311	-146.95	3.394
Air	29	1.293	-190	0.241	2.62E-02	18.20	37~0.40	350	-140.65	3.768
N2O	44.01	1.964	-89	0.208	1.74E-02	14.49	50	450	36.45	7.263
CO2	44.01	1.977	-79	0.202	1.42E-02*	14.53	32~37	460	31.05	7.395
SO2	64.06	2.926	-10	0.151	9.60E-03	12.60	52~100	520	157.25	7.871
CF4(FC-14)	88.01	3.927	-128	0.169	1.85E-02		39~62	626	-45.5	3.746
C2F6(FC-116)	138.02	6.158	-78	0.182*	1.70E-02	17.2**	67~90	612	24.3	2.981
C3F8(FC-218)	138.02	6.158	-38				40~112		72	2.680
C3F6	150	6.692	-29				94~106			
2-C4F6	162.04	7.229	-25				210~230			
c-C4F8(FC-C318)	200.04	8.925	-8	0.19**	1.23E-02	11.7**	111~180	620	115.3	2.785
CCl3F(CFC-11)	137.4	6.130	24	0.135***	7.90E-03	10.70	120~160	554	198	4.403
CCl2F2(CFC-12)	120.9	5.083	-30	0.145	9.70E-03	12.30	92~116	558	112	4.119
CClF3(CFC-13)	104.5	4.662	-81	0.158	1.22E-02	14.50	47~58	581	28.8	3.874
CCl4	154	6.871	78	0.129	6.73E-03	10.30	133~232	558	283.25	4.559
CHClF2(HCFC-21)	102.9	4.591	9	0.140	8.60E-03	11.4**	52~78	522	178.5	5.168
CHClF2(HCFC-22)	86.5	3.859	-41	0.157	1.05E-02	12.70	37~56	525	96	4.982
CCl2FCClF2(CFC-113)	187	8.343	48	0.161*****	7.56E-03	10***	198~250	576	214.1	3.413
CClF2CClF2(CFC-114)	170.9	7.625	4	0.170	1.04E-02	11.20	100~145	582	145.7	3.756
C2ClF5(CFC-115)	154.5	6.893	-39	0.164	1.26E-02	12.50	92~116	596	80	3.119
CF3CN	95	4.238	-62				134~140			
C2F5CN	145	6.469	-32				180~185			
C3F7CN	232.5	10.373	-2				220~233			
CF3NO2	115	5.131	-31				134			
CF3SF5	196	8.745	-20				129~155			
CF3NSF2	153.07	6.829	-6				205			
SOF2	86.1	3.841	-44				100~142			
CF3OCF3	154	6.871	-59				73~84			
CF3SCF3	170	7.585	-22				150			
NF3	71	2.980	-129	0.232				552	-39.26	4.531
SeF6	192.26	8.578	-64				110			
Notes				*: at 25°C, 0atm **: at 30°C, 0atm ***: at 30°C ****: at 60°C, 1atm *****: at 0°C, 1atm	*: at 0°C	*:14.2+5.49E-2T (T:°C) **: at 25°C, 1atm ***: at 25°C 0.1atm				

Table 2. Other Properties of Gases (Cont.)

Compound	Toxicity	ODP	GWP	Ban on Production
SF_6	Non-toxic		24900	
He	Non-toxic			
Ne	Non-toxic			
Ar	Non-toxic			
H_2	Non-toxic			
N_2	Non-toxic			
Air	Non-toxic	0		
N_2O	Non-toxic LCLo(mouse)=1500ppm		320	
CO_2	LCLo(rat)=657190ppm/15min		1	
SO_2				
CF_4	Low toxicity LCLo(rat)=895000ppm/15M	0	6300	
C_2F_6	Non-toxic		12500	
C_3F_8	Non-toxic LC50(rat)=6100ppm/4H			
C_3F_6	Toxic			
$2-C_4F_6$	High toxicity			
$C-C_4F_8$	Non-toxic		9100	
CCl_3F		1	4000	A.D.1996
CCl_2F_2	TWA=1000ppm	0.93	8500	A.D.1996
$CClF_3$	Low toxicity		11700	A.D.1996
CCl_4	Toxic TWA=10ppm	1.11	1400	A.D.1996
$CHCl_2F$	TWA=1000ppm			A.D.2020
$CHClF_2$	TWA=1000ppm	0.12	1700	A.D.2020
CCl_2FCClF_2	TWA=1000ppm	0.84	5000	A.D.1996
$CClF_2CClF_2$	TWA=1000ppm	0.71	9200	A.D.1996
C_2ClF_5	Relatively non-toxic	0.39	9300	A.D.1996
CF_3CN	High toxicity			
C_2F_5CN	High toxicity			
C_3F_7CN	Toxic			
CF_3NO_2	Toxic TWA~100ppm			
CF_3SF_5				
CF_3NSF_2				
SOF_2	Highly irritating LC50(rat)=1920mg/kg			
CF_3OCF_3	LD50 of $C_2F_6OC_2F_6$=1260ppm/kg			
CF_3SCF_3				
NF_3	Toxic TWA=10ppm		9720	
SeF_6	Toxic TWA=400ug/m3 LCLo=10ppm			

ODP:Ozone Depleting Potentials   GWP:Global Warming Potentials

SF_6/N_2 gas mixture is the most promising gas mixture as reported in the NIST technical report[3]. Dielectric Performance of SF_6/N_2 Gas Mixtures is higher than that of pure nitrogen[4]. If we use a 50% SF_6/50%N_2 Gas Mixtures, for example, the total amount of SF_6 gas is half at the same pressure of a conventional GIS. The dielectric strength will however be lower than that of pure SF_6. The size of the tank must be two-three times larger. Total amount of SF_6 gas is nearly the same as the conventional pure SF_6 type. In addition, recycling of the mixture gas is difficult, since separation and recovery for mixture gases is more difficult.

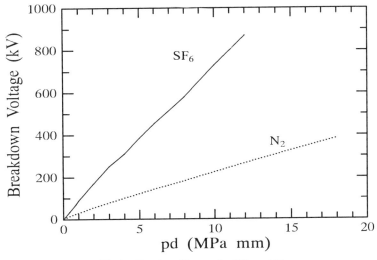

Fig.1 Paschen Curves for SF_6 and N_2

The other problem is current interruption performance. Fig.2 shows the test result of small capacitive current interruption. Arc-extinguishing performance of nitrogen for puffer-type gas circuit breaker is about half of SF_6 because of a low pressure buildup, a low peak value of puffer pressure, low cooling power and little electronegativity of the gas. Therefore, dielectric recovery characteristics is low compared to that of SF_6.

Small capacitive current interrupting performance of nitrogen is increased by mixing with SF_6, for example, by using $15\%SF_6/85\%N_2$ mixtures, as shown in Fig.2. Its performance is much better than that of pure nitrogen, but lower than that of pure SF_6.

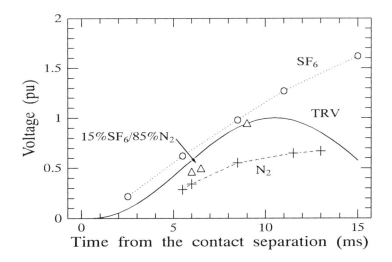

Fig.2 Small Current Interruption Performance

Other promising gases or mixtures are
- (a) SF_6/CO_2, SF_6/Air
- (b) Ternary Gas Dielectrics ($SF_6/N_2/CO_2$ etc.)
- (c) Vapor-Mist Dielectrics

These compositions have comparable problems from the practical point of view.

How to reduce the leakage of SF_6 gas from the equipment

We have two methods to reduce the leakage of SF_6 gas.

One is development of adequate recycling techniques. The other is development of new equipment using a smaller amount of SF_6 gas.

Recycling

(1) Recycling Technique

The usage of SF_6 gas is classified by the following two types.
- (a) possible or easy to recycle
- (b) impossible or difficult to recycle

Gas insulated equipment using pure SF_6 is classified into type (a). However, gas insulated equipment using gas mixtures is classified into type (b) because of the separation difficulty. Mixture gas can be recycled with or without liquefaction, but such process is complicated from the practical point of view, and needs higher costs. Therefore it is questionable how much percentage mixture gas will be recycled.

In conclusion, usage of pure SF_6 gas is much better than that of mixtures from the practical point of view, and it is possible and technically easy to recycle pure SF_6.

(2) SF_6 gas Recycling Procedure

Fig.3 shows the diagram of standard assembling and test procedure related to gas filling and exhaust gas for gas insulated equipment.

At present, the recycling system has already been used during each procedure which includes equipment testing, checking and extension. In case of failure, SF_6 gas is blown off from the equipment under the atmospheric pressure. This is caused by the condition of the recycling system. Consequently, it is necessary to establish a procedure to minimize the emission from the equipment by improvement of recycling system.

A feasibility study has already been initiated [5]. The Following steps are to be undertaken;
1) Development of SF_6 reclaimer which can evacuate SF_6 gas in the equipment at the pressure of vacuum.
2) Development of elimination techniques for impurity (Air, decomposition, moisture etc.)
3) Development of purity checking technique

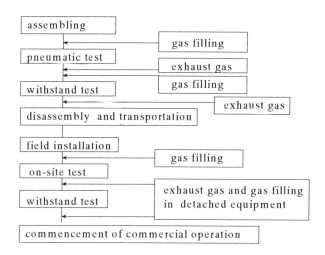

Fig.3 diagram of standard assembling and test procedure

(3)On-Site SF_6 gas Recycling System

Fig.4 shows the on-site SF_6 gas recycling system developed by Toshiba in Japan. Flowchart of gas treatment equipment is shown in figure 5. It is easy to handle on-site. The features of the system are as follows;

a) Lubrication oil free

Vacuum pump and compressor do not use lubrication oil. Therefore, there is little oil contamination in the recycled SF_6 gas.

b)High recovery percentage

By using a vacuum pump, SF_6 gas in the equipment evacuate to a pressure of 1kPa.

c)Automatic operative method

Recovery of SF_6 gas is operated automatically from high pressures to vacuum.

It is timesaving.

d)High reliability

Introduction of compressor technology based on experience of air-conditioning equipment.

e)Safety

Design concept is based on high pressure gas regulation.

Development of new equipment using a smaller amount of SF_6 gas

Electric power equipment using SF_6 gas in Japan has begun around 1965. Since then, gas insulated equipment using SF_6 gas has made remarkable progress in insulating and current interrupting technology.

Fig.4　The on-site SF$_6$ gas recycling system

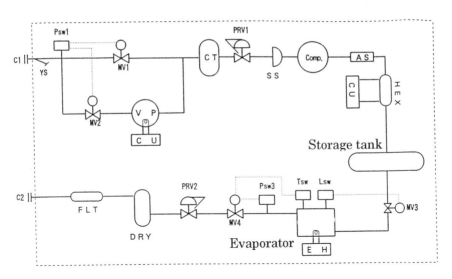

AS : After Strainer　　C : Horse Connection　Comp : Compressor
CT : Cushion Tank　　CU : Cooling Unit　　　DRY : Dryer
EH : Heater　　　　　 FLT : Filter　　　　　　HEX : Heat Exchanger
Lsw : Level Switch　　MV : Motor Valve　　　PRV : Pressure Regulating Valve
Psw : Pressure Switch　SS : Suction Strainer　Tsw : Thermometal Switch
VP : Vacuum Pump　　YS : Strainer

Fig.5　Flowchart of on-site SF$_6$ gas recycling system

Fig.6 shows the comparison of tank size for 550kV GCB's developed by Toshiba. Total gas volume in the GCB per unit is reduced to about 30% in a decade. Development of high performance equipment is quite effective to reduce the usage of SF_6 gas.

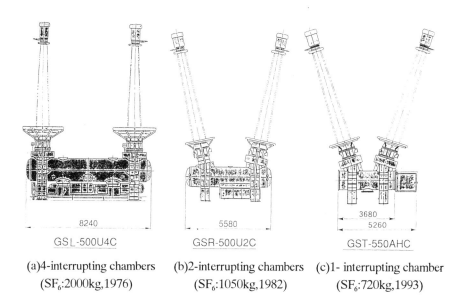

(a) 4-interrupting chambers (SF_6:2000kg,1976) (b) 2-interrupting chambers (SF_6:1050kg,1982) (c) 1-interrupting chamber (SF_6:720kg,1993)

Fig.6 The comparison of tank size and gas volume for 550kV GCB's

Conclusions

The SF_6 gas is necessary for the Electric Power Industry, therefore, it is needed to establish a SF_6 gas recycling system worldwide.

Manufacturers are trying to prevent leakage from the equipment during the manufacturing process and practical use including equipment testing and checking.

By establishing an efficient gas recycling system, recovery percentage will be 90% in 2000, 97% in 2005 during the each procedure (R&D, routine test, installation, practical use and scrapping of waste) by Japanese agreement.

References

[1] ELECTRA No.164, SF_6 and the Global Atmosphere, February 1996
[2] ELECTRA No.32, Paschen Curves for Nitrogen, Air and Sulfur Hexafluoride, January 1974
[3] NIST Technical Note 1425, Gases for Electrical Insulation and Arc Interruption: Possible Present and Future Alternatives to Pure SF_6, L.G.Christophorou, J.K.Olthoff, and D.S.Green, November 1997

[4] L.G.Christophorou et al,IEEE Trans. EI, Vol.13,No.5, 20-24 1997-9/10
[5] ELECTRA No.173, SF_6 Recycling Guide, Re-use of SF_6 Gas in Electrical Power Equipment and Final Disposal, August 1997

DISCUSSION

L. G. CHRISTOPHOROU: You have described a nice system for recycling pure SF_6. Have you tried to recycle SF_6/N_2 mixtures?

S. MATSUMOTO: No, we have not, because of technological difficulty.

H. KNOBLOCH: The small current interruption performance of SF_6/N_2 mixtures is shown in the paper. What is your experience with these mixtures concerning the arc-extinguishing capability at higher currents (1 kA to 50 kA range) at Short-Line-Fault tests and/or Terminal-Fault tests?

S. MATSUMOTO: As to the high-current interruption, we carried out the test as a trial. However, we have not enough data to discuss higher currents.

E. DOLIN: (1) In light of the goal to achieve 90% SF_6 recycling rate by the year 2000 and 97% by the year 2005, what are the assumptions about what happens to the remaining 10% and 3% of the gas? Where does it go? (2) Is the improvement in recycling rate due to advances in recycling technology or due to more widespread recycling?

S. MATSUMOTO: (1) The recycling system is not enough for preventing the release because there are some technological improvements that could be made. But the new recycling system which is described in this paper is quite effective to reduce SF_6 emission from the equipment. (2) Yes, it is. The emission rate will be decreased by establishing a recycling system in the near future.

A. H. MUFTI: The current interruption performance figure shows one example of SF_6/N_2 mixture. What are the characteristics of different percentage SF_6/N_2 mixtures?

S. MATSUMOTO: We carried out the current interruption test as a trial. We do not as yet have enough data for different percentage of SF_6/N_2 mixtures.

L. G. CHRISTOPHOROU: How does the arc interruption capability of SF_6/N_2 mixtures vary with the total pressure of the mixture?

S. MATSUMOTO: We cannot answer the question because we do not have enough data. However, the current interruption capability is supposed to be increased by increasing the total pressure of the mixture.

A. BULINSKI: From your talk I get the impression that Toshiba's position is to use SF_6 but to try to limit its leakage and increase recycling. Do you think you will be using gas mixtures or entirely different gases from SF_6 for power equipment in the future?

S. MATSUMOTO: We are now carrying out the basic research related to gas mixtures. We have not yet decided about the use of gas mixtures. However, we think that usage of pure SF_6 and establishment of pure SF_6 gas recycling system might be the best way to reduce SF_6 emissions. If a gas with a dielectric strength higher than SF_6 and boiling point lower than SF_6 is found, we will be using the gas as a SF_6 substitute.

Low SF$_6$ concentration SF$_6$ / N$_2$ mixtures for GIL

Xavier Waymel

Electricité de France
R&D Division (DER)
Les Renardières,
F-77250 Moret-Sur-Loing, France.

ABSTRACT

Environment concerns have led Electricité de France to consider the use of low-SF$_6$ content SF$_6$/N$_2$ mixtures for the development of 400 kV Gas Insulated Line technology dedicated to the underground parts of OHL links.

Intensive dielectric investigations were carried out at Les Renardières laboratories and at the Université de Pau, in order to increase literature data. The intrinsic performance of the gas under ideal conditions was evaluated first in large pure coaxial geometries. Small point-plane and rod-plane gaps were then experimented to complement the feasibility study.

With the use of pure Nitrogen, the ratings of the system could only be achieved by increasing both the diameter of the GIL structure and the gas pressure, entailing unacceptable costs.

In contrast, SF$_6$/N$_2$ mixtures containing a few percent of SF$_6$ offer acceptable compromises between dielectric performances, compacity, and costs of industrial applications : suitable choices of composition and pressure provide an alternative to pure SF$_6$ at 0.4 MPa.

INTRODUCTION

The use of SF$_6$ in high voltage industry is debated due to its involvement in global warming. Although its global contribution to the greenhouse effect has been estimated as low, it has to be widely accepted. Besides, SF$_6$ substitutes have been largely studied but none of them show better properties for both insulation and arc-breaking as well as good environmental compatibility. Every alternative to pure SF$_6$ sounds likes compromise.

In 1994, while initiating a feasibility study of Gas Insulated Line (GIL) technology for the 400 kV network, for bulk power burried links, EDF and GIL manufacturers were concerned with the environmental aspects of this alternative solution to overhead lines. Amongst technical studies, dielectric design drew attention to the potential use of pure nitrogen or SF_6/N_2 mixtures, with the lowest SF_6 content.

Literature and first investigations showed that the use of pure Nitrogen was harldy compatible with the industrial ratings of 400 kV busbars, due to the poor dielectric performances and stability, thus leading to unacceptable pipe diameters and costs. Then the interest turned to SF_6/N_2 mixtures, attempting to reduce as much as possible the amount of SF_6

This paper presents the studies of low SF_6 content SF_6/N_2 gas mixtures dedicated to the insulation of GIL. It is firstly exposed the reflections on the use of pure nitrogen for insulation. The literature review made with the University of Tennesse, Knoxville, TE, established a state-of-the-art of the gaseous insulation with nitrogen and nitrogen based mixtures. The studies conducted at EDF on pure nitrogen in large coaxial gaps are reported.

Secondly are exposed the investigations on SF_6/N_2 gas mixture insulation carried out in coaxial gap at Les Renardières labs, France, showing the influences of gas pressure and composition and wavefront time in homogeneous fields.

Thirdly, small gaps experimentations done at the Université de Pau, France, are reported. Non-homogeneous fields were investigated with an interest in the discharge development phenomenon and the conditionning process. The influences of the wavefront time, of the gas pressure and of the SF_6 content were analysed. The effect of the field homogeneity on the breakdown voltage was investigated as well.

REFLECTIONS ON PURE NITROGEN INSULATION

The gaseous dielectrics literature dealing with nitrogen based insulation is abundant on many topics. Many substitutes to pure SF_6 were studied, and experimental data were reviewed carefully, since some results appears contradictory, and it is sometimes difficult to identify all the relevant parameters of experimental configurations. Otherwise, very few experimentations were carried out with gas mixtures in large gaps. Some gases or gas mixtures offer even better performances than pure SF_6, but have undesirable chemical toxicity, or decomposition properties. Therefore, nitrogen and its mixtures with SF_6 were selected.

Literature on pure nitrogen in uniform field gives a ratio between pure SF_6 critical field and pure nitrogen critical field at the same pressure of ~2.5 at 0.5 MPa and ~3.3 at 1.5 MPa [1]. The breakdown field of nitrogen increases quite linearly up to a few bars, but efficiency stabilises progressively from 0.6 MPa on and the critical field saturates beyond 1.0 MPa [1,2].

In homogeneous field, impulse tests[1] show that the critical field of nitrogen is much higher in small structures as the pressure is increased beyond 0.8 MPa (figure 1), compared with the performances measured in industrially sized coaxial structures for impulses and AC or DC (figure 2). Therefore the similarity law for cylinders has to be considered when comparing results obtained in coaxial gaps of different sizes [5,7].

Investigations in a large coaxial gap (inner/outer electrode diameter = d/D = 185/400 mm) were conducted at EDF laboratories at Les Renardières, to increase the literature data. They show that the critical field in Lightning Impulse (LI) separates from the linear Paschen curve from 0.4 MPa, especially in positive polarity, and turns to saturation as the pressure is further increased. Numerous authors [5] explain this by the electrode surface roughness. EDF experimentations show that this phenomenon is not so dramatic for negative lightning impulse, and a crossing of the curves is observed at 0.5 MPa (figure 4). AC breakdown field follows the same shape as the most critical (positive) polarity.

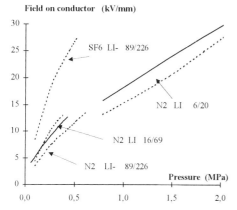
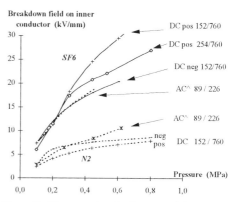

Figure 1 : Impulse breakdown field on inner conductor in coaxial geometry for SF_6 and N_2, inner/outer diameter = d/D = 89/226 mm from Cookson and al.[3], d/D = 6/20 mm from Skipper and al.[1] and d/D = 16/69 mm from Medeiros and al.[12]. Solid lines are for positive LI, dashed lines are for negative LI.

Figure 2 : Literature data for large coaxial gaps and pure SF_6 or pure nitrogen. d/D = 152/760 and d/D = 254/760 are from Cooke and al.[4], d/D = 89/226 are from Cookson and al.[3].

The ratio of AC breakdown voltage (crest) over LI breakdown voltage (positive polarity) is particularly high for pure nitrogen (about 80% at 0.6 MPa and 78% at 0.8 MPa), maybe due to the low performances in positive polarity LI.

During the experimentations, the gas conditioning effect and the scatter of the conditionned gas were unexpectedly high for positive polarity, leading to lower withstand voltages.

Moreover, some breakdown occured out of the investigation gap on a spacer shielding where the field was lower than in the investigation gap, and were attributed to roughness of the shielding : this parasitic discharges vanished when the shielding was polished. These unexplained observations seem contradictory with the other literature data[2] that give a sensitivity to electrode surface roughness lower for pure nitrogen than for pure SF_6, according to numerous experimental data and calculations with an hemispherical protrusion model by Pedersen[2].

The similarity principle was used to estimate the dimensions of a pure nitrogen insulated GIL with the help of the coaxial basic investigations, assuming insulators and sliding contacts would not produce great field enhancements. It was concluded that only the lowest lightning impulse insulation levels could be contemplated (1175 kV for 400 kV grid), with pressures above 1.0 MPa and tube diameters of at least 1.0m. These conclusions were not compatible with the economical requirements of the project.

SF_6/N_2 MIXTURES : GAS COMPOSITION AND PRESSURE CONSIDERATIONS

The attention then focussed on nitrogen based gaseous insulants, and particularly SF_6/N_2 mixtures, with low SF_6 percentage. The association of SF_6 with nitrogen takes advantage of a synergetic effect, particularly for a SF_6 content below 20 percents[2,8].

The literature for large coaxial gaps on this topic is restricted to a few papers, amongst these the one from Cookson and al.[3] (figure 3), but very little data are available between 0% (pure N_2) and 20% SF_6 content SF_6/N_2 mixtures.

The streamer criterion based on the swarm parameters measured by Aschwanden was adapted to give a simple model of the gas under ideal conditions in the similarity plots[7] (E.r as a function of P.r, with P = pressure in the vessel, E = breakdown or withstand field on the conductor, r = radius of the conductor). The area effect modelisation described by Pace and al.[7] participated in fitting the EDF experimental curves for large coaxial gaps.

The experimentations were carried out in the same geometry (d/D = 185/400mm). They showed the same saturation as that observed for pure nitrogen. The positive polarity performances in LI flattenned as the pressure was increased, whereas the negative polarity curve met saturation for higher pressures (figure 4). Consequently, curves of the two polarities cross and the positive polarity becomes critical above 0.3 to 0.5 MPa.

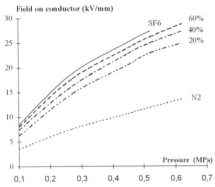
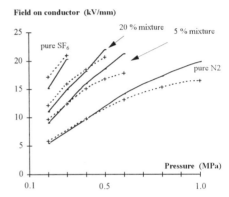

Figure 3 : Impulse breakdown field for SF_6/N_2 mixtures in negative LI for several SF_6/N_2 mixtures from Cookson and al.[3]. Solid line is for SF_6, dashed lines are for SF_6/N_2 mixture, referenced with SF_6 percentage.

Figure 4 : EDF U_{50} lightning impulse breakdown investigations in a pure coaxial gap inner/outer diameter = d/D = 185/400 mm, for pure N_2, mixtures and SF_6. Solid lines are for negative polarity, dashed lines for positive.

The impact of the gas composition experimented at EDF added information to the data from Cookson and al.[3]. The breakdown voltage substantially increases between 0% (pure nitrogen) and 5% of SF_6 concentration in N2, but the linearity of the curve in this area has not been further investigated. This increase of efficiency softens from 5% to 10%, and then stabilise from 10% to 20%. From 20% to 100% SF_6 content, EDF results are consistent with other experimentations [2,3,5] and confirm that the breakdown voltage increases almost linearly between 20% and 100% of SF_6 content.

Wavefront time influence

Experimentations in SI and AC were carried out and the same crossings were observed in SI as in LI. The SI/LI "natural" ratios are lower for mixtures than for pure nitrogen : more than 80% for N_2 at 0.6 MPa and 75% for a 10% SF_6/N_2 mixture at the same pressure. AC test results give the same conclusions for AC^/LI ratios, with about 78% for N_2 and about 71 % for 10% SF_6/N_2 mixture at 0.6 MPa. These figures may be compared with the one for pure SF_6 at 0.5 MPa, given by Diessner and al.[10] : SI/LI = 72% and AC^/LI = 62%.

This is of major impact on the voltages to be used for AC on-site dielectric tests[11] : the natural ratios of SF_6 under the usual GIS pressures are in agreement with the specified insulation levels : 630.$\sqrt{2}$ / 1425 = 62%. With SF_6/N_2 mixtures, the LI insulation level becomes predominant for dielectric dimensionning and the specified LI insulation level leads to an equivalent AC^ voltage (with natural ratios) significantly above the specified AC^ voltage. Therefore, appropriate on-site tests should be carried out at voltages consistent with the measured equivalences, and thus higher than the usual 80% of the specified levels in AC.

SMALL GAPS INVESTIGATIONS

Point-plane and rod-plane investigations were carried out to analyse the physical phenomenons associated with discharge. The conditionning of the gas and its impact on the mechanism leading to breakdown were analysed with image convertors [9].

The wavefront time influence was analysed in pure N_2 and a 10% mixture, in positive polarity and a point-plane geometry. Looking at the volt-time curve, it is observed that the breakdown voltage step between the flatten area for short fronts and the one for long fronts appears more smoothly for nitrogen than for SF_6, explaining the impact on natural ratios. The 10% mixtures is a middle case of pure nitrogen and pure SF_6 [9].

Pressure and gas composition cross influences on breakdown voltage exhibited unexpected irregularities [9]. Under LI of positive polarity, breakdown voltage increased with pressure up to 1.6 MPa and SF_6 percentage up to 15%, whereas in negative polarity, performance saturated from 0.4 MPa at the same voltage level for mixtures, regardless of SF_6 percentage. These irregularities were not observed for pure nitrogen.

The influence of the homogeneity of the gap was analysed under LI as well. It is noted that the gap length have direct effect on the breakdown voltage in negative polarity, and nearly no impact under positive polarity. On the opposite, a tip radius increase entails an increase of the breakdown voltage under LI of positive polarity, but has a limited effect under LI of negative polarity.

DESIGN AND VALIDATION OF INDUSTRIAL BUSBARS

Similar evaluation of the performances of SF_6/N_2 mixtures were conducted by GIL manufacturers in large and small gaps. All these concluded that a pure nitrogen insulated GIL was not compatible with the specifications, unless overpassing costs and dimensions. On the contrary, SF_6/N_2 mixtures lead to cost effective compromises between dielectric performances, environment respect, thermal dimensionning for buried technology.

With a 1425 kV LI rating and tube diameters limited to 700 mm, SF_6 percentages of 5% to 15% lead to consider gas pressures lower than 0.8~0.9 MPa, that is to say short into the area where performances stabilise with pressure and SF_6 percentage.

These carefull compromises between gas composition and gas pressure ensure both security margins and environment respect. The post insulators and the spacers bring additionnal non-homogeneity and field enhancement, but their contribution is usually limited to a-few-percent increase of the electrical field.

Industrial prototypes from three manufacturers (ABB, GEC-ALSTHOM and SIEMENS) associated with EDF to the project were tested to check they fullfill the specifications and estimate their security margins.

CONCLUSIONS

No substitute to pure SF_6 combines its performances in gaseous insulation and arc breaking with a better neutrality. Admittedly, the use of SF_6/N_2 mixtures in alternative to pure SF_6 in the high voltage insulation of GIL does not completely solve the environmental concerns about greenhouse effect. Pure nitrogen may be considered as the very neutral gaseous media but it proved to have poor withstand performances and it does not provide cost-effective industrial designs. Among alternative gases, SF_6/N_2 mixtures with low SF_6 percentage exhibit attractive performances, and reduce the potential environmental impact.

For Gas Insulated Line insulation, mixtures of 5 % to 15 % SF_6 in nitrogen under pressures of 0.5 to 0.9 MPa realise a satisfactory compromise between the outer tube diameter and the costs and the thermal design. The performances are then very near to those of the SF_6 under about 0.4 MPa.

One must not forget that using gas mixtures instead of pure SF_6 may complicate gas handling procedures. The remaining issue concerns the destruction and/or the separation of the components at the end of life. Cryogenic separation of the mixture currently appears to be both poorly efficient for low SF_6 percentages from 5 to 10 percent, and expensive. Alternatives techniques are contemplated by EDF in the further developments of the project.

REFERENCES

1. D.J.Skipper, and P.I. McNeal, Impulse strength measurements on compressed-gas insulation for Extra-High-Voltages power cables, *Proc. IEE 112*, 103-108.
2. L.G. Christophorou and R.J. Van Brunt, SF_6/N_2 Mixtures : Basic and HV insulation properties, *IEEE Trans on Diel. And Elect. Ins.*, vol. 2 No 5, (Oct 1995).
3. A.H. Cookson and B.O.Pedersen, Analysis of the high voltage breakdown results for mixtures of SF_6 with CO2, N_2 and air, *Third Int. Symp. on High Voltage Engineering,* 31.10 (Milan, 1979).
4. C.M. Cooke and R. Velasquez, The insulation of Ultra-High-Voltages in coaxial systems using compressed SF_6 gas, *IEEE Trans. Power Appar. Sys*, PAS-96, 1491-1497, (1977).
5. L.G. Christophorou, J.K.Olthoff, and D.S. Green, Gases for electrical insulation and arc interruption : Possible present and future alternatives to pure SF_6, *NIST technical note 1425* (Nov. 1997).
6. M.O.Pace, D.L.McCorkle, and X. Waymel , Possible high pressure nitrogen-based insulation for compressed Gas Insulated Cables, *IEEE Conference on Electrical Insulation and Dielectric Phenomena,* (1995 Annual report n°95CH35842).
7. M.O.Pace, D.L.McCorkle, and X. Waymel , N_2/SF_6 Breakdown in concentric cylinders : Consolidation of data, *IEEE Conference on Electrical Insulation and Dielectric Phenomena,* (1996)
8. T. Aschwanden, Swarm parameters in SF_6 and SF_6/N_2 mixtures determined from a time resolved discharge study, *Gaseous Dielectrics IV*, Pergamon, New York (1984).
9. X. Waymel, V.Delmon,T. Reess, A. Gibert, and P. Domens, Impulse breakdown in point-plane gaps in SF_6-N_2 mixtures, *10th Int. Symp. on High Voltage Engineering,* (Montréal, 1997).
10. A.Diessner, G.F. Luxa, W. Mosca, and A. Pigini, High Voltage of SF_6 insulated substations on site - 33-06 CIGRE session, 1986.
11. A. Sabot, D. Santos, X. Waymel, D. Feldmann, and Y. Maugain, 21/23/33-02 CIGRE session, 1998.
12. P.T. Medeiros, S.R. Naidu, and K.D. Srivastava, Lightning impulse breakdown of cylindrical coaxial gaps in nitrogen, *IEEE Intern. Symp. on Electr. Insul.*, Washington, D.C., 210-212 (1986).

DISCUSSION

B. HAMPTON: Can you say how the breakdown voltage of SF_6/N_2 mixtures at these high pressures and in the presence of free metallic particles compares with that of pure SF_6 at 0.4 MPa?

X. WAYMEL: We did make some experiments on particles. We observed that for given particles, the PF breakdown voltage is nearly the same in pure SF_6 and SF_6/N_2 mixtures at a given pressure, apparently linked to the lift-off field.

H. OKUBO: As an after-laying-test on site, will you test by AC voltage with partial discharge measurement for the detection of metallic contamination?

X. WAYMEL: Yes, AC voltage test plus PD diagnostic. But the AC level should be higher than the usual on-site test voltage (0.8 × PFWL). The on-site test voltage given by CIGRE (Working Group 33123-15, Electra No. 176, February 1998) for pure SF_6 Max (0.36 × LIWL; 0.8 x PFWL) should be modified for SF_6/N_2 mixtures to fit their natural ratio (AC breakdown voltage/LI breakdown voltage), depending on the mixture considered.

A. BULINSKI: You mentioned in your talk that you had not contemplated any gases other than SF_6/N_2 mixtures. Is this an official EDF position? What would you do if EDF is forced to significantly reduce SF_6 use?

X. WAYMEL: There is no EDF position, but a real fact. We exclusively considered pure N_2 and low-SF_6-content SF_6/N_2 mixtures. Pure nitrogen leads to unacceptable tube diameters and costs for building the GIL. If the SF_6 percentage had to be significantly reduced below the present proposed levels, EDF would certainly have to re-consider the use of GIL.

APPLICATION PROBLEMS OF SF_6/N_2 MIXTURES TO GAS INSULATED BUS

H. Hama, M. Yoshimura, K. Inami and S. Hamano

Mitsubishi Electric Corporation
Amagasaki City, Hyogo 661-8661, JAPAN

INTRODUCTION

Gas insulated bus (GIB) using SF_6 gas has widely been applied in the world, because it has the advantages in reliability and in minimizing installation space. However SF_6 is identified as a potent greenhouse gas and thus the efforts in preventing the release of SF_6 in the environment, the study on recycling SF_6 and the research for alternative gas are now being performed[1]. In designing a GIB using alternative gas, there are many application problems such as a selection of alternative gas and its rated pressure to maintain the required dielectric and heat transfer performance. The problems on recycling SF_6 are also essential in applying alternative gas using SF_6 such as SF_6/N_2 mixtures.

In this report the minimum breakdown field strength at lightning impulse and temperature rise of conductor and enclosure have been measured for SF_6/N_2 mixtures. Considering the dielectric and heat transfer properties, we clarify the application problems of the mixtures to a GIB and discuss the appropriate mixture ratio of SF_6 in designing the GIB comparable to the present dimension. And also the lowest limit of SF_6 content in a liquefied recovering method is theoretically estimated for the reference of practical SF_6 recovery from the mixtures.

EXPERIMENTAL CONDITIONS

Gas Breakdown Experiments

The experimental setup is shown in Figure 1. The gas breakdown properties of SF_6/N_2 mixtures have been measured, using the coaxial electrode system of enclosure diameter ϕ 340mm. The test electrodes of ϕ 240mm outer diameter and 100mm long are set on the conductors. The electrodes are made of aluminum alloy and the surface roughness is $\pm 3 \mu m$. The area effect of electrode in SF_6/N_2 mixtures was investigated, varying the numbers of the test electrodes. The negative breakdown voltages at standard lightning impulse were measured by the step up method to determine the minimum breakdown values.

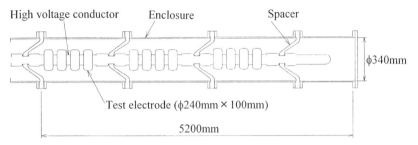

Figure 1. Experimental setup for gas breakdown measurements in SF_6/N_2 mixtures at lightning impulse

Thermal Measurements

The thermal measurements in SF_6/N_2 mixtures have been performed as shown in Figure 2. The outer diameter of conductors and the inner diameter of enclosures are ϕ220mm and ϕ590mm, respectively. The material of conductors and enclosures is aluminum alloy. Each bus unit has three gas compartments divided by two cone type spacers. The compartments of each bus unit were filled with the same constituents and pressure of SF_6/N_2 mixtures. The temperature rise at the conductors and the enclosures was measured by thermocouples. The measuring points were at the center of each unit as represented in the figure.

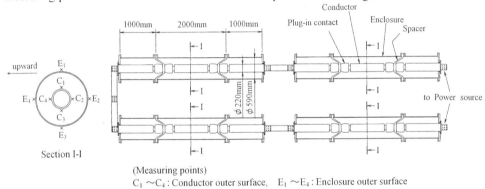

(Measuring points)
$C_1 \sim C_4$: Conductor outer surface, $E_1 \sim E_4$: Enclosure outer surface

Figure 2. Experimental setup using four bus units and measuring points for thermal measurements

EXPERIMENTAL RESULTS

Minimum Breakdown Field Strength at Lightning Impulse Voltages

Figure 3 shows the dependence of 0.1% breakdown field strength in a weibull distribution on effective electrode area. The pressure of SF_6/N_2 mixtures is 0.4MPa. The asymptotic values which correspond to the breakdown field of infinitely large electrode are normalized by the values of pure SF_6 and represented as a function of SF_6 content in Figure 4. The solid curve in the figure is the estimated value which will be mentioned later.

Temperature Rise of Conductor and Enclosure

Figures 5 (a) and (b) show the dependence of temperature rise of conductor and enclosure on the circulating current as the parameter of SF_6 content of the mixtures, respectively. The

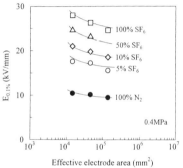
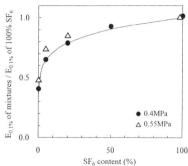

Figure 3. Minimum breakdown field strength as a function of effective electrode area in SF_6/N_2 mixtures (Negative lightning impulse)

Figure 4. Normalized minimum breakdown field strength of infinitely large electrode as a function of SF_6 content in SF_6/N_2 mixtures (Negative lightning impulse)

pressure of the mixtures is 0.6MPa. Note that the temperature rise of conductor and enclosure is the average value at the points of C_1 and C_3 and at E_1 and E_3, respectively. The curves in the figures are the analytical values which will be explained later. The dependence of conductor and enclosure temperature rise on the SF_6 content is summarized in Figures 6 (a) and (b), respectively. The circulating current is 6 kA and the pressure of the mixtures is varied between 0.3 and 0.6 MPa. The analytical values are shown again in the figures.

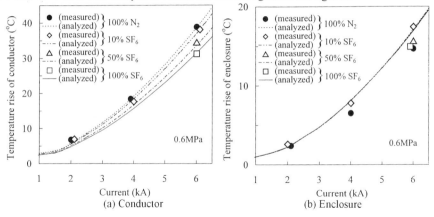

Figure 5. Dependence of temperature rise of conductor and enclosure on current in SF_6/N_2 mixtures

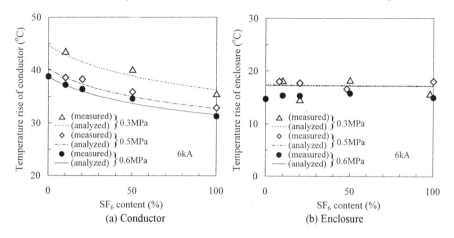

Figure 6. Dependence of conductor and enclosure temperature rise on SF_6 content in SF_6/N_2 mixtures

DISCUSSIONS

As to the dielectric design for a GIB, the following properties of SF_6/N_2 mixtures should be considered. Figure 3 shows that there is the area effect of electrode at lightning impulse in SF_6/N_2 mixtures as well as in pure SF_6 gas. Therefore the minimum breakdown field strength of infinitely large electrode in Figure 4 should be reflected on the practical design. Here the solid line in the figure corresponds to the estimated value as the following method:

$$E_{0.1\%} \text{ of mixtures} = E_{0.1\%} \text{ of pure } SF_6 \times (E_{cr} \text{ of mixtures} / E_{cr} \text{ of pure } SF_6) \quad (1)$$

where E_{cr} is the critical field strength which can be obtained from the literatures[2,3]. The experimental values are adopted for the $E_{0.1\%}$ of pure SF_6. The estimated curve in Figure 4 coincides well with the measurements at 0.4MPa. The equation (1) is effective for the practical design to evaluate the minimum breakdown field $E_{0.1\%}$ of infinitely large electrode in SF_6/N_2 mixtures at low pressures. While note that the measured values below 20% of SF_6 content at 0.55MPa are a little higher than the estimated curve. This fact suggests that area effect of electrode may be reduced in low SF_6 content at higher pressures, comparing with pure SF_6.

Secondly to discuss the heat transfer properties of SF_6/N_2 mixtures we performed the analysis on the temperature rise of conductor and enclosure. The analysis is based on the following thermal equilibrium conditions[4]:

$$P_c = W_{r1} + W_{c1}, \quad P_c + P_e = W_{r2} + W_{c2} \quad (2)$$

where P_c and P_e are power loss in conductor and enclosure, respectively. W_{r1} and W_{c1} represent heat transferred between conductor and enclosure by radiation and convection. While heat transferred from enclosure to air by radiation and convection is shown by W_{r2} and W_{c2}, respectively. The curves in Figures 5 and 6 are the analytical values by equation (2), and they agree well with the measured values.

Figures 5 and 6 show that temperature rise of conductor increases as the content of SF_6 decreases, while the temperature rise of enclosure is independent of SF_6 content. The result for enclosure can be explained by the fact that heat transfer from enclosure to air by radiation and convection does not depend on the condition of SF_6 content in the enclosure. The experimental results in Figure 5(a) are summarized in terms of the relation between the temperature rise of conductor ΔT and circulating current I as the following empirical equation:

$$\Delta T = KI^\alpha \quad (3)$$

here the values of K and α are the constants which depend on SF_6 content, and they are shown in Table 1 as the example of 0.6MPa. Figure 7 represents the constants of K and α normalized by the values K_0 and α_0 at pure SF_6, using the values in Table 1. The following properties of SF_6/N_2 mixtures are summarized from Figure 7. The constant K/K_0 indicating temperature rise of conductor increases continuously from 1.0 to 1.2, as decreasing the content of SF_6 from 100% to 0%. While the value of α/α_0 which dominates the dependence on circulating currents is independent of SF_6 content.

Thirdly considering the above dielectric and heat transfer properties of the mixtures, we estimated the relation between the required pressure and the mixture ratio of SF_6 in designing the GIB comparable to the present dimension. The conditions for the calculation are as follows. The model GIB for the calculation is the single phase enclosure type whose unit length is 14m with post type insulators in every 7m. The enclosure diameter is ϕ540mm and is assumed to be constant. For the dielectric design we considered the lightning impulse withstand voltage of 1,550kV at 0.4MPa in the case of pure SF_6 and the same dielectric performance in the mixtures.

Table 1. Dependence of constants K and α on SF_6 content of the mixtures

SF_6 content (%)	K	α
0	1.799	1.714
10	1.706	1.719
20	1.661	1.717
50	1.560	1.714
100	1.479	1.707

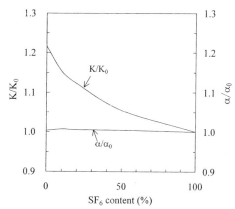

Figure 7. Constants of K and α normalized by the values K_0 and α_0 at pure SF_6 as a function of SF_6 content

The field strength $E_{0.1\%}$ including area effect of electrode obtained by the experiments are used for the calculation. As to the design of current capacity the values of maximum allowable temperature of conductor and enclosure are 105°C and 70°C, respectively, assuming the rated current is 8kA. The analytical method by equation (2) are applied for the calculation.

The solid curve in Figure 8 shows the calculated results of the relation between the minimum guaranteed pressure and the SF_6 content for the model GIB. The middle curve corresponds to the results neglecting the design of current capacity. The lowest dotted curve is derived by the simple calculation, neglecting area effect of electrode and current capacity as follows. Assuming the theoretical breakdown field for every SF_6 content of the mixtures to be constant, the relation between the pressure of the mixtures and the SF_6 content can be given by

$$P = (E/P)_{cr0} \cdot P_0/(E/P)_{cr} \qquad (4)$$

here $(E/P)_{cr}$ is the critical value at each SF_6 content of the mixtures. The value of $(E/P)_{cr0} \cdot P_0$ corresponds to the theoretical breakdown field at the pressure P_0 for pure SF_6 and is equal to 88(kV/mm · MPa)×0.4MPa in this case. Comparing the three curves, the required minimum pressure is mainly determined by the dielectric performance including area effect of electrode. It is important for the practical design of a GIB to consider both area effect of electrode and current capacity. The increase of the required minimum pressure is rapid below 10% of SF_6 content. The tendency seems to be caused by the dependence of minimum breakdown field on SF_6 content as shown in Figure 4. This result suggests that the quality control of the mixture ratio of SF_6 is difficult below 10% of SF_6 content for the practical application. Therefore in view of reducing the amount of SF_6 usage[1], the appropriate mixture ratio of SF_6 seems to be between 10 to 20% for the practical design.

Finally we discuss SF_6 recovery from SF_6/N_2 mixtures for the reference of practical recovering method. Recovering SF_6 gas has been performed since the introduction of a gas insulated equipment. Storage of SF_6 is classified into the following two methods of gaseous or liquefied recovering[5]. In the case of the gaseous recovering method the typical compressed pressure is 2.5 MPa. Gas compressed to the pressure remains mostly in a gaseous state. Larger storage volume is required for the recovering equipment. Therefore this method is not suitable for the transportable equipment. As for the liquefied recovering method the typical pressure is 3.0 MPa. SF_6 is stored in a liquid form, using compression and cooling systems. Storage capacity is larger than the gaseous recovering method. When the recover of SF_6 from the mixtures is carried out by the liquefaction method, the recovery efficiency depends on the compression and the cooling temperature of the mixtures and also on the mixture ratio of SF_6.

To estimate the theoretical limit in a liquefied recovering method we have calculated the ratio of liquid SF_6 for the various SF_6/N_2 mixtures in the cooling process based on SF_6 liquid, SF_6 vapor and N_2 gas equilibrium. Here the ratio of liquid SF_6 is defined as the mass of SF_6 liquefied to that of initial filling SF_6. The initial conditions for the calculation are represented as follows. The gas volume is a constant value of $0.1m^3$. The filling gas pressure is 3.5MPa at the temperature of 42°C, considering the critical pressure 3.77MPa of SF_6. The lowest cooling temperature for the calculation is -50°C because the melting point of SF_6 is -50.8°C. The thermodynamic parameters in the calculation are referred to the literature[6].

Figure 9 shows the calculated results which correspond to the relation between the ratio of liquid SF_6 and SF_6 content of the mixtures. The recovery efficiency of liquid SF_6 degrees as increasing the mixed N_2. Finally, when the mixed ratio of SF_6 is 7%, the liquefied ratio reaches 0% even at the temperature of -50°C. The lowest limit of the mixture ratio of SF_6 is theoretically estimated to be 7 % in the case of a liquefied recovering method. Therefore, when the SF_6 content is below 10%, the application of the recovering method is practically difficult.

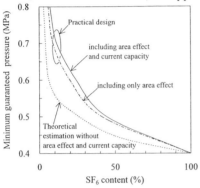

Figure 8. Relation between the minimum pressure of the mixtures and the SF_6 content for the model GIB

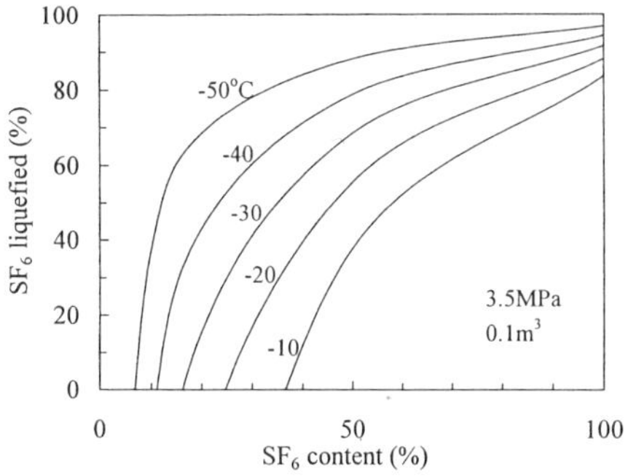

Figure 9. Relation between the ratio of liquid SF_6 and SF_6 content in the case of liquefied recovering method

CONCLUSIONS

In summary the application problems of SF_6/N_2 mixtures to a GIB were discussed from the practical point of view. It is important for the design to consider both breakdown phenomena including area effect of electrode and heat transfer properties of the mixtures. In designing the GIB comparable to the present dimension, the appropriate mixture ratio of SF_6 seems to be between 10 to 20% in view of reducing the amount of SF_6 usage. Below 10% of SF_6 content the quality control of SF_6 content filled in the GIB and the application of a liquefied recovering method are practically difficult.

REFERENCES

1. L.G.Christophorou, J.K.Olthoff and D.S.Green, "Gases for Electrical Insulation and Arc Interruption: Possible Present and Future Alternatives to Pure SF_6", NIST Technical Note 1425(1997).
2. T.Nitta and Y.Shibuya, "Electrical Breakdown of Long Gap in Sulfur Hexafluoride", IEEE Trans. PAS, Vol.PAS-90, 1065(1971).
3. S.Okabe, M.Chiba and T.Kono, "Calculation Method of Flashover Voltage of Gas Mixtures", IEE of Japan, Vol.103, 507(1983).
4. T.Kawamura, T.Nitta, N.Ito, K.Sasaki, "Problems Associated with High Service Current in GIS", CIGRE Symposium, S 06-85, 220-03(1985).
5. Recycling Guide by CIGRE 23.10 Task Force 01(1997).
6. Allied Chemical Corporation, "SF_6", Booklet of Electrical and Physical Properties(1973).

DISCUSSION

L. NIEMEYER: To which extent did you include particulate contamination into your design considerations? What is the maximum particle length you design for and under which voltage stress?

H. HAMA: In general, the particle contamination is one of the most important factors for the dielectric design. In the case of a GIS, we usually consider the allowable electric field on the inner surface of the enclosure against about a 3 mm particle under the operating voltage. However, we neglect the existence of particles in our case study of the GIB which is equipped with particle traps, considering that particles can be trapped in the particle traps.

X. WAYMEL: You are considering $E_{0.1\%}$ withstand field? How did you measure it? How many shots did you make?

H. HAMA: We measured the breakdown voltages for negative lightning impulse by means of the up-and-down method. The minimum breakdown field $E_{0.1\%}$ was determined in accordance with the Weibull distribution. The number of breakdown voltages measured for each case is about fifty.

A SEARCH FOR POSSIBLE "UNIVERSAL-APPLICATION" GAS MIXTURES

L. G. Christophorou,[1] J. K. Olthoff,[1] and D. S. Green[2]

[1]Electronics and Electrical Engineering Laboratory
[2]Chemical Science and Technology Laboratory
National Institute of Standards and Technology
Gaithersburg, Maryland, USA 20899

ABSTRACT

In an effort to respond to the recent concerns over the possible impact of SF_6 on global warming, we have searched for an SF_6 substitute gas that could be used in high voltage equipment instead of pure SF_6, with minimal changes in practice, operation, and ratings of the existing pure SF_6-insulated apparatus. Of the many unitary, binary, and tertiary gases/mixtures that have been tested to date, SF_6-N_2 mixtures seem to be the most promising and most thoroughly characterized gaseous dielectric media besides pure SF_6. Based upon research conducted world-wide over the last few decades, it appears that the optimum composition of an SF_6-N_2 mixture for use as a gaseous dielectric in place of pure SF_6 for both high voltage insulation (such as in gas-insulated transmission lines and gas-insulated transformers) and possibly also for arc and current interruption purposes may be in the range 40% to 50% SF_6 in N_2 (by volume). Most of the existing data support the use of such mixtures for gas-insulated transmission lines and gas-insulated transformers, but their use for circuit breakers is still in question requiring further exploration.

INTRODUCTION

Sulfur hexafluoride (SF_6), the electric power industry's presently preferred gaseous dielectric (besides air), has been shown to be a greenhouse gas. Concerns over its possible impact on the environment have rekindled interest in finding replacement gases. In a recent report[1] we have provided information that is useful in identifying such gases, in the event that replacement gases are deemed a reasonable approach to controlling emissions of SF_6 from high voltage electrical equipment. In that report an attempt was made to identify a gaseous mixture that could be adopted for "universal use" as an immediate replacement of pure SF_6. The large amount of available physical and laboratory data suggest that a 50%SF_6-50%N_2 mixture (percentage by volume) may exhibit dielectric characteristics suitable for use as insulation in high voltage equipment. However, it should be noted that the use of a replacement gas in *existing* systems would require equipment recertifications. This mixture may also be appropriate for arc or current interruption applications in *new* equipment designed specifically for use with SF_6-N_2 mixtures.

POSSIBLE "UNIVERSAL APPLICATION" GAS MIXTURES

The most desirable SF_6 substitute would be a gas that could be put in all existing SF_6-equipment, requiring little or no change in hardware, procedures or ratings. Such a gas we refer to as a "universal-application" gas and we define it as a gaseous medium which can be used instead of pure SF_6 in existing equipment without significant changes in practice, operation, or ratings of the existing gas-insulated apparatus. Even though it is recognized that recertification will be required, it is a useful exercise to determine if such a substitute can be identified from the existing gaseous dielectric data.

Of the many unitary, binary, and tertiary gases/mixtures that have been tested over the last three decades or so, SF_6-N_2 mixtures seem to be the most thoroughly characterized [yet not completely tested, especially at high pressures (greater than 0.5 MPa)] gaseous dielectric media besides pure SF_6 (e.g., see Refs.1-5). There is broad acceptance of the view that these mixtures may be good replacements of pure SF_6. The main reasons are[1]:
- they perform rather well for both electrical insulation applications and in arc or current interruption equipment,
- they have lower dew points and certain advantages especially under non-uniform fields over pure SF_6,
- they are much cheaper than SF_6, especially after recent increases in the price of SF_6, and
- the electric power industry and electrical equipment manufacturers have some experience with their use.

The relevant questions, therefore, are: *does an optimum mixture composition and total pressure exist that allows the use of this mixture as a "universal-application" gas, and could the industry readily use such a mixture?* While the answers are complex, it is desirable to attempt to identify, on the basis of existing knowledge, a particular mixture composition that may be best suited for consideration by the electric power industry for its needs. If such a mixture can be identified, it can perhaps be standardized in composition. Although it would be desirable to have such a standard mixture prepared and sold by chemical companies for direct use in the field, this may not be feasible, and the two gases would probably have to be mixed to the standard composition at the point of use to reduce shipping costs (see Ref. 1).

Based upon research conducted world-wide, it appears that the optimum composition of an SF_6-N_2 mixture for use by the electric power industry in place of pure SF_6 for both high voltage insulation (for gas-insulated transmission lines and gas-insulated transformers) and arc or current interruption purposes may be in the range of 40% to 50% SF_6 in N_2. Thus, possible standard mixtures that can reasonably be considered are 40%SF_6-60%N_2 or 50%SF_6-50%N_2. The annual savings of replacing pure SF_6 by a 40%SF_6-60%N_2 gas mixture are potentially large. If it is assumed that 80% of the ~ 8,000 metric tons of SF_6 produced annually is used by the international electric power industry at a price of ~\$44 / kg (~ \$20 / lb) for SF_6, the total annual savings in the cost of SF_6 would be about \$150 million.[1]

The feasibility analysis of a universal-application mixture for both insulation/transformer, and arc/current interruption purposes is based on information obtained from a number of sources.[1] Representative findings on mixtures consisting of 40% to 50% SF_6 in N_2 for *gas-insulated transmission lines and gas-insulated transformers* are listed in Table 1. Similar findings for *arc and current interruption* are listed in Table 2 (see Ref. 1).

Table 1. Representative findings on mixtures consisting of 40% to 50% SF_6 in N_2 demonstrating their usefulness for gas-insulated transmission lines and gas-insulated transformers (for a complete account see Ref. 1).

Finding/Conclusion	Testing Conditions and Reference
• A 50%SF_6-50%N_2 mixture has 88% the dielectric strength of pure SF_6 • Dielectric strength basically saturates beyond this percentage composition	Uniform and coaxial fields, many studies (see Ref. 1)
• A 50% to 60% of SF_6 in N_2 has 85% to 90% the dielectric strength of pure SF_6 • Such mixtures can have improved impulse and power-frequency breakdown strength in highly non-uniform fields (lower sensitivity to particles and surface roughness) • Mixtures can operate at higher pressures than pure SF_6-insulated apparatus, especially at lower than ambient temperatures	Review of uniform and coaxial field studies under various conditions[6]
• A 50%SF_6-50%N_2 mixture at 1000 kPa has 85% the dielectric strength of pure SF_6	Various electrode geometries including concentric cylinders. Lightning and switching impulses[7]
• A 50%SF_6-50%N_2 mixture at 540 kPa can replace pure SF_6 at 450 kPa without loss of strength (i.e., the mixture has over 83% the dielectric strength of pure SF_6 at the same pressure)	Coaxial electrode geometry. AC, lightning and switching impulse[8]
• With particle contamination, mixtures of 40-50% of SF_6 in N_2 performed well (better than 80% of the dielectric strength at about 600 kPa) compared to pure SF_6. Their performance depends on total and partial pressure and needs further study.	Coaxial electrodes; DC[9] Coaxial electrodes; AC[3]
• In the presence of spacers a 50%SF_6-50%N_2 mixture has about 90% the flashover voltage of pure SF_6	Ref. 10
• The flashover behavior of cylindrical insulators in SF_6-N_2 mixtures is similar to pure SF_6 independently of the amount of SF_6 in the mixture	AC; pressures 50 kPa to 300 kPa[11]
• The breakdown voltage of particle contaminated spacers for a 50%SF_6-50%N_2 mixture (100 kPa to 400 kPa) showed no significant difference between the mixture and pure SF_6; the mixture has only slightly lower impulse ratio than pure SF_6	Concentric cylinders; AC[12]
• The behavior of a 50%SF_6-50%N_2 mixture does not significantly differ from that of pure SF_6 under fast transient impulse conditions	Ref. 13
• Little difference is expected between the chemistry of pure SF_6 and a 50%SF_6-50%N_2 mixture. The presence of N_2 may reduce the ability of SF_6 to reform itself following an arc or a discharge. Recovery of the mixture needs investigation.	Ref. 5
• A 50%SF_6-50%N_2 mixture can be potentially useful for gas-insulated transformers in spite of less effective heat transfer and cooling by the gas. Further studies are indicated.	Refs. 1 and 14

Table 2. Representative findings on mixtures consisting of 40% to 50%SF_6 in N_2 demonstrating their usefulness for arc and current interruption (see a more complete account in Ref. 1).

Finding/Conclusion	Testing Condition and Reference
• Interrupters designed specifically for SF_6-N_2 mixtures may not require derating. They have distinct advantages over pure SF_6 in terms of higher total pressures and lower operating temperatures. *Arc interruption performance increases with total gas pressure.*	A number of studies[1,3]
• The rate of rise of recovery voltage (see Fig. 1) of a 50%SF_6-50%N_2 mixture is between 82% and 100% that of pure SF_6 depending on total pressure.	Gas-blasted arcs of various mixture compositions and total pressures 500 kPa to 700 kPa[15]
• The rate of rise of the recovery voltage of mixtures of 40% to 50% SF_6 in N_2 is more than one third greater than that of pure SF_6 at pressures of 1300 kPa to 1900 kPa. *The interruption capability increased with total gas pressure.*	Synchronous interrupter, but results may apply to other interrupter types[16]
• The voltage-current characteristic curve of the spiral arc for 50%SF_6-50%N_2 is slightly below the similar curve for pure SF_6 at 800 kPa.	DC interruption by spiral arc in SF_6-N_2 mixtures (100 kPa to 800 kPa)[17]
• The relative interruption capability of a 50%SF_6-50%N_2 mixture is only about 70% that of pure SF_6.	Evaluation of and measurements on gases for arc interruption (puffer-type, current range ~10-15 kA)[18]
• The thermodynamic properties of a 40%SF_6-60%N_2 mixture are not significantly different from those of pure SF_6.	Calculations[19]
• SF_6-N_2 mixtures may require design changes of the arc chamber to optimize arc quenching capability and these changes would depend on gas composition and pressure.	Simulations of arc dynamic behavior of gas-blasted arcs[20]

Figure 1 shows the results of Grant et al.[15] and Garzon.[16] For pressures over 600 kPa, the optimum interrupter performance, judged in terms of its voltage recovery capability, is observed to occur when the mixture composition is roughly 50%SF_6-50%N_2. Interestingly, Garzon[16] found that the rate of recovery capability of a non-synchronous circuit breaker using SF_6-N_2 mixtures was as least as good as when pure SF_6 was used.

The data in Tables 1 and 2 (and in Ref. 1), suggest that there can be considerable tolerance for variation of the percentage of SF_6 in N_2 for a proposed 50%SF_6-50%N_2 mixture without significant effect on the dielectric performance of the mixture. This is because the properties of the mixture are not generally a strong function of the SF_6 concentration at this mixture composition. Certainly a tolerance in the percentage of SF_6 content of ± 5% seems reasonable. It should also be noted that the removal of byproducts from the mixture is not expected to be much different than in pure SF_6. Furthermore, there seem to be no serious problems in making a standard gas mixture or in recovering the SF_6 from the mixture.[1]

The electric power industry seems willing to consider SF_6-N_2 mixtures for insulation, for instance, in gas-insulated transmission lines. Indeed, much work is being conducted worldwide in this area. For insulation, most such studies focus on lower concentration mixtures

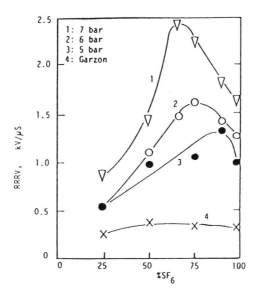

Fig. 1. Rate of rise of recovery voltage (RRRV) as a function of SF_6-N_2 mixture ratio. Curves 1-3 are the data of Grant et al.[15] and Curve 4 are the data of Garzon[16] (adapted from Fig. 2 of Ref. 15). The measurements of Garzon were made at 1700 kPa.

(10% to 15%). However, work is also being done on higher concentration SF_6-N_2 mixtures for circuit breaker use. A 40% or 50% mixture of SF_6 in N_2 performs well as an insulating medium. This "universal application" mixture has no apparent physical or chemical problems, but the fact that its dielectric performance is only 85% to 90% of that of pure SF_6 may require equipment changes and recertification or derating. This is nearly impossible for equipment already in service, and would be costly for new equipment presently certified only for pure SF_6 use. However, the application of standard gas mixtures to newly designed equipment is certainly worthy of consideration.

CONCLUSION

The physical and chemical properties of a 40% or 50% mixture of SF_6 in N_2 suggest that it may be appropriate as a "universal application" gas mixture in new equipment, particularly if designed specifically for use with SF_6-N_2 mixtures. The practical difficulties of using SF_6-N_2 mixtures in existing equipment seem to be particularly large at present.[1]

REFERENCES

1. L. G. Christophorou, J. K. Olthoff, and D. S. Green, *Gases for Electrical Insulation and Arc Interruption: Possible Present and Future Alternatives to Pure SF_6*, National Institute of Standards and Technology, NIST Technical Note 1425 (1997).
2. L. G. Christophorou (Ed.), *Gaseous Dielectrics,* Volumes I-V, Pergamon Press, New York (respectively, 1978, 1980, 1982, 1984, 1987); Volumes VI and VII, Plenum Press, New York (respectively, 1990 and 1994).
3. Electric Power Research Institute, *Gases Superior to SF_6 for Insulation and Interruption,* Report EPRI EL-2620, September (1982).

4. S. A. Boggs, F. Y. Chu, and N. Fujimoto (Eds.), *Gas-Insulated Substations,* Pergamon Press, New York (1986).
5. L. G. Christophorou and R. J. Van Brunt, "SF_6/N_2 Mixtures, Basic and HV Insulation Properties," *IEEE Trans. Dielectrics and Electrical Insulation,* 2:952 (1995).
6. N. H. Malik and A. H. Qureshi, A Review of Electrical Breakdown in Mixtures of SF_6 and Other Gases, *IEEE Trans. Electr. Insul.,* EI-14:1 (1979).
7. A. Rein and J. Kulsetås, "Impulse Breakdown of SF_6/N_2 Insulation. Influence of Electrode Covering. Polarity Effects," in: *Gaseous Dielectrics III,* L. G. Christophorou (Ed.), Pergamon Press, New York , pp. 315-321 (1982).
8. A. H. Cookson and B. O. Pedersen, "Analysis of the High Voltage Breakdown Results for Mixtures of SF_6 with CO_2, N_2, and Air," in: *Third Intern. Symposium on High Voltage Engineering,* Milan, paper 31.10 (1979).
9. M. O. Pace, J. L. Adcock, and L. G. Christophorou, "Particle Contamination in Gas-Insulated Systems: New Control Methods and Optimum SF_6/N_2 Mixtures," in: *Gaseous Dielectrics IV,* L. G. Christophorou and M. O. Pace (Eds.), Pergamon Press, New York, pp. 377-386 91984).
10. D. R. James, M. O. Pace, D. W. Bouldin, and L. G. Christophorou, Oak Ridge National Laboratory Report, ORNL/TM-9017 (1984).
11. B. Blankenburg, "Flashover Behavior of Cylindrical Insulators in SF_6, N_2, and SF_6/N_2 Mixtures," in: *Proc. Third Intern. Symp. on High Voltage Engineering,* Milan, paper No. 32.03 (1979).
12. M. Eteiba, F. A. M. Rizk, N.-G. Trinh, and C. Vincent, "Influence of a Conducting Particle Attached to an Epoxy Resin Spacer on the Breakdown Voltage of Compressed-Gas Insulation," in: *Gaseous Dielectrics II,* L. G. Christophorou (Ed.), Pergamon Press, New York, pp. 250-254 (1980).
13. W. Pfeiffer, V. Zimmer, and P. Zipfl, "Time Lags and Optical Investigations of Pre-Discharge in SF_6/N_2 Mixtures at Very Fast Transient Voltages," in: *Gaseous Dielectrics VI,* L. G. Christophorou and I. Sauers (Eds.), Plenum Press, New York, pp. 231-236 (1991).
14. *Technical Research Report I, Temperature Distribution Test in SF_6 Mixture Gas Insulated Transformer,* Electrical Engineering Department, Tsinghua University, Beijing Second Transformer Factory, Beijing, China, 1995.5 (Project: Gas-Insulated Transformer: II -Utilization, IDRC Centre File: 93-1204-04).
15. D. M. Grant, J. F. Perkins, L. C. Campbell, O. E. Ibrahim, and O. Farish, "Comparative Interruption Studies of Gas-Blasted Arcs in SF_6/N_2 and SF_6/He Mixtures," in: *Proc. 4th Intern. Conf. on Gas Discharges,* IEE Conf. Publ. No 143, pp. 48-51 (1976).
16. R. D. Garzon "The Effects of SF_6/N_2 Mixtures Upon the Recovery Voltage Capability of a Synchronous Interrupter," *IEEE Trans. Power Apparatus and Systems,* PAS-95:140 (1976).
17. H. Naganawa, H. Ohno, M. Iio, and I. Miyachi, "DC Interruption by Spiral Arc in SF_6/N_2 Mixture," in: *8th International Conference on Gas Discharges and Their Applications,* Oxford, pp. 137-140 (1985).
18. A. Lee and L. S. Frost, "Interruption Capability of Gases and Gas Mixtures in a Puffer-Type Interrupter," *IEEE Trans. Plasma Science,* PS-8:362 (1980).
19. A. Gleizes, M. Razafinimanana, and S. Vacquie, "Calculation of Thermodynamic Properties and Transport Coefficients for SF_6/N_2 Mixtures in the Temperature Range 1,000-30,000 K," *Plasma Chemistry and Plasma Processing,* 6:65 (1986).
20. H. Sasao, S. Hamano, Y. Ueda, S. Yamaji, and Y. Murai, "Dynamic Behavior of Gas-Blasted Arcs in SF_6-N_2 Mixtures," *IEEE Transactions on Power Apparatus and Systems,* PAS-101:4024 (1982).

DISCUSSION

L. NIEMEYER: Industry is using SF_6/N_2 mixtures and SF_6/CF_4 mixtures since about 20 years. In all cases the use of such mixtures implied a reduction of interruption performance. The SF_6/N_2 mixtures had to be strongly derated for thermal interruption performance and less for dielectric interruption performance. The SF_6/CF_4 mixture has to be derated more for dielectric than for thermal interruption performance. There is no gas composition for which the performance is *better* than for pure SF_6. It should be noted that, of course, any interruption performance can be obtained with any gas composition when the system is redesigned by increasing gas pressure, drive, contact separation, etc., of the circuit breaker.

L. G. CHRISTOPHOROU: As you know the issue is not *better,* but *comparable to* pure SF_6 performance. As we discussed in the NIST report (Ref. 1), there are laboratory studies which show the SF_6/N_2 mixtures perform very well or better compared to pure SF_6. Also, as you indicated, the performance of SF_6/N_2 mixtures needs to be looked at from the point of view of optimizing the entire system for using the mixture rather than pure SF_6. Indeed in Ref. 3 there is evidence that this approach can be successful. As I have mentioned in my presentation there is a need for more work in this area especially with regard to the total pressure.

A. H. MUFTI: Can you comment on the recycling of SF_6/N_2 mixtures?

L. G. CHRISTOPHOROU: In principle there should be no serious problems. However, the argument has been made that recycling of gas mixtures is more difficult and more expensive than for pure SF_6. There are others who maintain a different view. There is a poster paper by M. Pittroff et al. on recycling and you may wish to discuss the subject with the authors of that paper.

W. BOECK: The most important point is the leakage rate. It is proportional to the partial pressure of the SF_6 in a mixture and the surface of the encapsulation. In case of the Mitsubishi design 10%SF_6/90%N_2 mixtures (total pressure 0.25 MPa) in comparison to 100%SF_6 0.4 MPa this leakage rate is reduced to 20%. In the case of your proposal it would be reduced to ~ 60%. It is much easier to get this improvement by a better tightening of the encapsulation (from 1%/year down to 0.3%/year as under consideration).

L. G. CHRISTOPHOROU: Perhaps one should attempt to do both.

A. BULINSKI: Why should we try to determine or find an "universal gas mixture" if different type of equipment may require different type of mixture?

L. G. CHRISTOPHOPROU: Because, as you know, we need to do an incredible amount of work to fully understand the properties of any given medium and its use. And simply that is not possible to do for each and every gaseous dielectric medium. Also, because if a mixture is found that performs well for the main needs of the power industry, it can be standardize and its use quantified and prescribed.

SOME ASPECTS OF COMPRESSED AIR AND NITROGEN INSULATION

M. Piemontesi, F. Koenig, L. Niemeyer, C. Heitz

ABB Research Center
CH-5405 Baden, Switzerland

ABSTRACT

Air and nitrogen have been suggested as environmentally uncritical insulation media for gas insulated electrical power equipment. A thorough evaluation of such a concept requires the knowledge of the breakdown fields which control the dimensions and other design parameters of such systems. This paper contributes some preliminary discharge physical background information required to derive design fields. It reports experimental breakdown data on compressed air and nitrogen up to 950 kPa in uniform field gaps with particulate contamination. Furthermore, prebreakdown current measurements and time resolved image converter records are used to identify the relevant discharge mechanisms. In the presence of particulate contamination the breakdown of compressed air insulation is controlled by the leader mechanism; breakdown fields are in the range of 5.5 to 8 kV/cm·bar for particles in the range of 3 to 5 mm length. In compressed nitrogen streamers always cross the gap at breakdown level (at least up to 500 kPa) and the breakdown mechanism may be of the direct streamer to spark or leader type. The particle induced breakdown fields in nitrogen are in the range of 4 to 4.5 kV/cm·bar, i.e. nitrogen performs less well than air.

1. INTRODUCTION

The fact that SF_6 is a strong greenhouse gas has prompted interest in substitute gases with lower or no environmental impact. A systematic search for such gases [1] has led to the conclusion that air or nitrogen are probably the only gases that might have some potential for this purpose if it could be shown, with the help of environmental lifecycle assessment (LCA) according to the standard ISO 14040 [2], that the total environmental impact of a compressed air/N_2 insulated system is lower than that of an SF_6 insulated system of equal performance. As an LCA has to be based on equipment design, the insulation performance of compressed air/N_2 insulation has to be understood first to provide adequate design rules. The present knowledge of the discharge processes in compressed air and nitrogen under the conditions prevailing in compressed gas insulation (weakly non-uniform field gaps with small defects) is still insufficient. Only a few experimental studies [3,4] and a review containing experimental data on compressed nitrogen [5] are available. Most of the physical concepts on air breakdown have been developed in strongly non-uniform air gaps at atmospheric pressure (e.g. [6,7,8,9]) and can be used as a starting point for the exploration of the discharge processes in compressed air.
As the most critical contaminants in gas insulated systems are small conducting particles, this paper will be limited to the study of particle induced breakdown which will be studied experimentally under the simplest possible conditions, namely, in a uniform field gap with a fixed oblong anodic protrusion simulating the particle, under applied step voltage and with irradiation to minimize statistical time lag effects. On the basis of the obtained data an attempt is made to derive a preliminary understanding of the governing predischarge and breakdown processes.

2. EXPERIMENTAL

Discharge gap: The plane electrodes defining the uniform background field had diameters of 230 mm and their spacing was varied from 30 to 60 mm. The anodic protrusions were steel needle (Ogura Jewelry Ltd.) of length l (1.2, 3 and 5 mm) and tip curvature radius r (40 and 100 μm). The gas pressure was varied in the range 100 to 950 kPa. A step voltage with a rise time of about 0.1 μs and a voltage pulse with a rise time of about 1 μs and a fall time to half value of about 3 ms were applied to the gap with peak voltages up to 600 kV. In addition to the step voltage, radioactive or UV irradiation were used to minimize statistical time lags so that minimum breakdown levels could be approximately determined.

Diagnostics: The protrusion was mounted insulated on the grounded anode and its surface coated with insulating material except for its tip. The insulating coating was covered by a grounded metallic coating. This allowed one to measure the discharge current leaving the tip with a minimized displacement current component. The light emission of the discharges was recorded by two photomultipliers (focused on the needle tip and the cathode surface, respectively) and by an IMACON 790 image converter camera used in the streak and framing modes. In some cases an image intensifier was used. Two limitations of the optical diagnostics were encountered: Overexposure of the very faint predischarge phenomena by the luminosity of the breakdown and statistical time lag which rendered it difficult to trigger the image converter camera at the instant of the discharge. The latter difficulty could be overcome, to some extent, by overvolting the gap.

3. DISCHARGE PHYSICAL CONCEPTS

Breakdown in gas insulated systems develops via a sequence of processes the interdependencies of which are schematically represented in fig. 1. After the creation of an initial electron, breakdown can occur via various sequences of events. In compressed gas insulation the key initializing discharge mechanism is the streamer. The **streamer inception** criterion can be formulated in terms of the gas properties and the electric field distribution in the gap. For the special case of a small protrusion in a uniform background field, it can be expressed in terms of the undisturbed background field in the non-dimensional form [10]

$$E_{inc}/E_{cr} = f(p \cdot l, l/r) \qquad (1)$$

where p is the gas pressure and l and r are the protrusion length scale and tip radius, respectively.

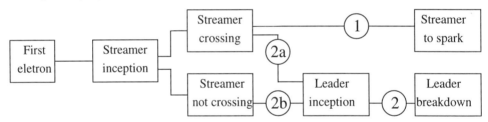

Figure 1: Sequence of discharge processes in compressed gas insulation breakdown

A streamer once started propagates into the gap. The condition for its propagation can be expressed in terms of a threshold field value which is referred to as the streamer stability field E_{stab} [11]. To a first approximation, the condition for **streamer gap crossing** can thus be expressed as average field criterion of the form

$$E_{av} = U/L > E_{stab} \qquad (2)$$

where U is the voltage applied to the gap and L the length of the streamer channel connecting the electrodes.

There are two modes by which a streamer can initiate breakdown. The first one is a **direct streamer-to-spark transition** (path 1 in fig. 1) for which streamer gap crossing is a necessary though not sufficient condition. It consists in heating the preionized streamer channel connecting the electrodes by the current which is driven through it by the applied field. The condition herefore can be approximately expressed as an average field criterion

$$E_{av} = U/L > E_{ss} \qquad (3)$$

where E_{ss} is a threshold field which will be referred to as the streamer to spark transition field. A model of the streamer to spark transition is described in [8]. The second breakdown mode is **leader**

induced breakdown (path 2 in fig.1). A leader is a discharge channel of relatively high conductivity which develops from the common stem of a branched streamer corona due to heating by the sum of the currents injected from the streamer branches. This process can be numerically modelled [6,7] and bears some resemblance to direct streamer-to spark transition except for the fact that the heating current is not controlled by the applied inter-electrode field but by the current feed from the streamer corona. Once started, the leader acts as a conducting filament and propagates into the gap, being fed from the streamer activity at its tip, until it arrives at the opposite electrode and causes breakdown. For applied voltage pulses of sufficient duration like the step voltages applied here, leader inception always leads to breakdown. For the small scale defects considered in compressed gas insulation we will tentatively introduce a background field criterion to characterize leader inception:

$$E_o > E_{ld}(p, l, r) \qquad (4)$$

where E_o is the background field at the location (and in the absence) of the defect. This criterion assumes that the leader inception process is driven by the local field in the vicinity of the defect so that major parameters controlling E_{ld} are the gas pressure p and the geometry parameters l and r of the protrusion.

4. MEASURED BREAKDOWN FIELDS

The minimum breakdown fields $E_{o,bd}$ under step voltage stress were measured for three protrusion lengths over the pressure range 100 kPa to 950 kPa for both air and nitrogen. The results are plotted in fig. 2 in dependence of the pressure, the field values $E_{o,bd}$ always being the background field.

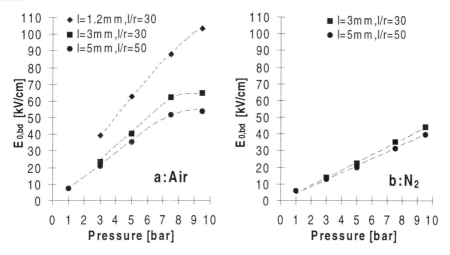

Figure 2: Particle induced minimum breakdown field $E_{o,bd}$ versus pressure p in a uniform field gap in air (a) and nitrogen (b) under positive step impulse

At low pressures the breakdown field is seen to increase proportional to the pressure and to decrease with increasing particle length l. For air, a saturation with pressure is observed above about 700 kPa. Nitrogen does not exhibit such a saturation and its breakdown levels are lower than for air.

5. OBSERVED BREAKDOWN MECHANISMS

5.1 Air

Fig. 3 shows a summary of the discharge phenomena in air at pressures up to 500 kPa. The prebreakdown phenomena **just below the breakdown level** are shown in the left column of the fig. 3 (a,b,c) by synchronized streak records and discharge currents. The applied field is given as percentage value of the minimum breakdown field $E_{o,bd}$. At 100 kPa the branched structure of the streamer corona is clearly visible and up to 300 kPa streamers cross the entire gap without causing breakdown. At 500 kPa the streamers do no more cross the gap and the corona remains confined to the vicinity of the protrusion. The current pulses associated with the pre-breakdown discharges decrease in both charge and duration with increasing pressure.

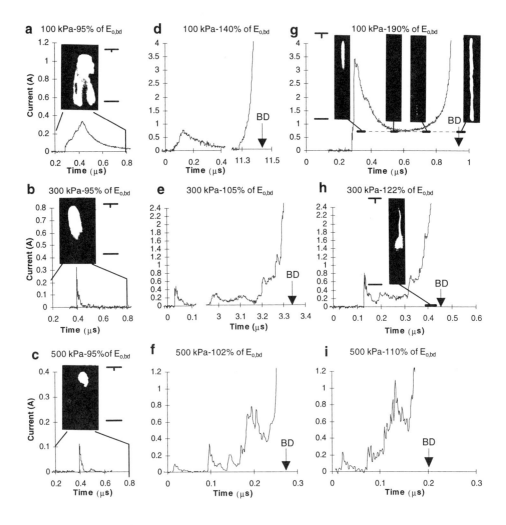

Figure 3: Prebreakdown currents and synchronized streak and framing pictures in a 40 mm gap with a 5 mm (l/r=50) protrusion in compressed air under positive step impulse

Prebreakdown phenomena **above the breakdown level** are shown in the centre column of fig. 3 at low overvoltage (d,e,f) and in the right column at high overvoltage (g,h,i). The pre-breakdown currents consist of an initial streamer corona pulse followed by low current phase and a subsequent current rise to breakdown. The instant of breakdown, i.e. the instant at which the applied voltage starts to decay, is indicated by arrows (BD). The right column shows examples of time synchronized framing pictures of the predischarge luminosity.

At 100 kPa and 90 % overvoltage (g) the first streamer pulse is associated with a straight filament crossing the gap. It is followed by a low current phase during which the filament remains faintly visible until the current re-rises and leads to breakdown. The prebreakdown current is similar to the current measured and simulated for direct streamer to spark transition in a point to plane gap under comparable average field stress [8,9]. At a pressure of 300 kPa and 22 % overvoltage (h) the image recorded during the second current rise shows a filamentary channel with a brush-like tip extending into but not crossing the gap, a structure which is similar to the leader channels propagating through long air gaps [6]. We therefore come to the tentative conclusion that the breakdown process is of the leader type in spite of the fact that the first streamer corona has crossed the gap. At a pressure of 500 kPa the prebreakdown currents (f and i) have the same structure as at 300 kPa, however, with lower and shorter initial corona pulse and faster current rise to breakdown.

Fig. 4 summarizes the quantitative aspects of the air results. It shows the pressure reduced fields for streamer inception, streamer crossing, and breakdown in dependence of the pressure for a protrusion of 5 mm length. The **streamer inception field** (broken curve) is calculated by eq. (1) on the basis of the effective ionisation coefficient given in [10]. The **streamer crossing field** E_{stab} as represented by the open points was taken from [11] up to atmospheric pressure. At higher pressures it was either determined with the help of image converter records of the kind shown in fig. 3 b or with the help of two photomultipliers (focused on the needle tip and cathode surface, respectively) using a 1,2 mm protrusion. It is seen that E_{stab} increases overproportionally with the pressure.

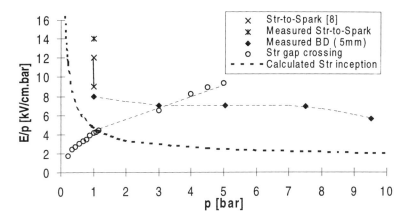

Figure 4: Pressure reduced streamer inception (broken curve), crossing (open circles) and breakdown fields (full points) versus pressure p for air in a uniform field gap with a 5 mm (l/r=50) anodic protrusion. Crosses indicate direct streamer-to-spark transition. The crosses with the bar are the simulation results of [8]

The pressure reduced **breakdown fields** (full points) are almost pressure independent between 300 and 750 kPa and decrease towards higher pressure corresponding to the saturation tendency displayed by the breakdown curves in fig. 2a. Only in overvolted gaps at atmospheric pressure direct **streamer-to-spark breakdown** is identified as the controlling mechanism (fig. 3g corresponding to the single cross in fig. 4). This is further confirmed by the fact that the breakdown field is in the same range as the one simulated in point-to-plane gaps [9], as indicated by the crosses connected by a bar. The precise limit curve for streamer-to- spark breakdown was not yet determined. At and above 300 kPa the experimental data provide strong evidence that **leader breakdown** is the controlling mechanism.

5.2 Nitrogen

The basic features of the breakdown process in compressed nitrogen were explored up to 500 kPa and are illustrated by fig. 5.

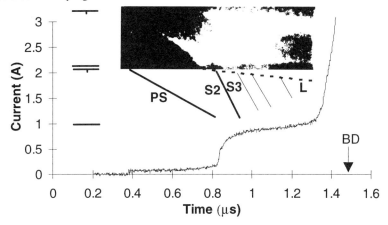

Figure 5: Prebreakdown currents and synchronized streak record in a 40 mm gap with a 5 mm (l/r=50) protrusion in nitrogen at 300 kPa under positive step impulse (165% of $E_{o,bd}$)

The gap had to be overvolted to decrease the statistical time lag so that the discharge luminosity could be recorded by the image converter. A primary streamer PS first crosses the gap associated with a relatively low current in the 100 mA range. It is followed by various further streamers (S2, S3, ...), and a much slower luminosity front L which has similarity with the streak records of leaders in long air gaps at atmospheric pressure [6] and in point to plane gaps in compressed nitrogen [4]. It is thus seen that two breakdown mechanisms are competing, namely direct streamer to spark transition and leader. Which of these two actually lead to breakdown can not be determined due to the overexposure of the streak record. Qualitatively similar prebreakdown currents are observed at lower overvoltages and over the pressure range from 100 to 500 kPa. For pressures above 500 kPa the discharge mechanisms remains to be studied.

6. CONCLUSIONS

The experimental work presented here provides some preliminary insight into the discharge mechanisms which control the insulation performance of compressed air and nitrogen in the presence of particulate contamination. The results are still restricted to a narrow field of parameters, namely, uniform background field, pressures in the range 100 to 950 kPa and step voltage stress. Within this parameter range the following is found:

In compressed air
- the streamer inception criterion can not be used to predict breakdown
- streamers crossing the gap do not necessarily lead to breakdown
- the leader mechanism is the breakdown controlling process
- for particles in the range 3 to 5 mm the pressure reduced breakdown field is in the range 5.5 to 8 kV/cm·bar.

In compressed nitrogen
- streamer always cross the gap at breakdown levels (at least up to 500 kPa)
- the actual breakdown mechanism is not yet clear, it may be of the direct streamer to spark or leader type
- for particles in the range 3 to 5 mm the pressure reduced breakdown field is in the range 4 to 4.5 kV/cm·bar, i.e. nitrogen performs less well than air.

Further work will concentrate on a generalization of the discussed discharge criteria to weakly non-uniform field geometries, an extension of the study to smaller electrode protrusions (electrode surface roughness level), a study of the role of first electron statistics and an understanding of the role of the voltage waveform for the breakdown process.

7. REFERENCES

[1] L. Niemeyer "A systematic search for insulation gases and their environmental evaluation", this conference

[2] ISO Standard 14040 Environmental Management - Life cycle assessment - principles and framework, 1st Ed. 1997 -06-15

[3] B.M.Weedy and A.E. Davis "Breakdown of Compressed Air in concentric Electrode Systems", Gaseous Dielectrics IV, Knoxville, 1984, pp. 162-167

[4] M.O. Pace, D.L. McCorkle and X. Waymel "Possible High Pressure Nitrogen-Based Insulation for Compressed Gas Insulated Cables", CEIDP, Virginia Beach, 1995, pp. 195-198

[5] L.G.Christophorou, R.J.Van Brunt, SF_6/N_2 Mixtures - Basic and HV Insulation Properties -, IEEE Transactions on Dielectrics and Electrical Insulation, Vol. 2, No.5, October 1995, pp. 952-1003

[6] I. Gallimberti, "The mechanism of the long spark formation", Journal de Physique, $\underline{40}$, Colloque C7, 1979, pp. 193 - 250

[7] L. Ullrich and I. Gallimberti, "A numerical leader inception model in air: experimental results and model predictions", IX Int. Conference on Gas Discharges and their Application, Venezia, 1988, pp. 419 - 422

[8] M. Marode "The mechanism of the spark breakdown in air at atmospheric pressure between a positive point and a plane.I. Experimental: Nature of the streamer track", J. Appl. Phys. **46**, 1975, pp. 2005-2015

[9] F. Bastien and M. Marode "Dielectric strength of electronegative gases in non uniform field: role of the vibrational exitation", Gaseous Dielectrics III, Knoxville, 1982, pp. 119-125

[10] S. K. Berger "Onset or breakdown voltage reduction by electrode surface roughness in air and SF_6", IEEE Trans on Power Apparatus and System, Vol. PAS-95 no.4, 1976, pp. 1073-1079

[11] Griffith and Phelps, "Dependence of positive corona streamer propagation on air pressure and water content", J. Appl. Phys. Vol. 47 No. 7, July 1976, pp. 2929-2934

DISCUSSION

L. G. CHRISTOPHOROU: The difference between your measurements in air and in N_2 is clearly the O_2 in air. How does this explain your measurements?

M. PIEMONTESI: The electronegativity of the oxygen molecule distinguishes the two gases. It was shown by Gallimberti [*Gaseous Dielectrics V*, L. G. Christophorou and D. W. Bouldin (Eds.), Pergamon Press, New York, 1987, p. 61] that streamer propagation is controlled (among other factors) by the voltage drop along the streamer channel, the latter being higher the more electron attaching the gas is. In this sense, we suspect that the oxygen in air restricts streamer propagation and that this might be the ultimate cause for the higher breakdown field in air. Detailed streamer simulations will eventually clear this issue.

E. MARODE: In the frame of investigating the pressure effect in studies concerning car plugs, we did some work in a Pd = 1 cm × 1 bar gap, in the range of 1 bar to 5 bar. The point radius as well as the gap length was changed with pressure keeping constant the overall *E/N* distribution in the normalized space (distances multiplied by the gas number density *N*). In such a case the comparison showed that the pressure law was approximately followed, and the transition towards spark was always streamer initiated. Would you think that increasing the distance rather than the pressure you would obtain the same result?

M. PIEMONTESI: In a plane-plane geometry with protrusion the field enhancement is confined near the protrusion tip. Increasing the electrode gap the field distribution near the protrusion tip does not change when the background field in the gap is prescribed. For longer gaps it is this background field which controls the discharge processes and not the voltage applied across the gap. This has been verified by producing the same background field with different gap distances and applied voltages.

SECTION 6: GAS DECOMPOSITION / PARTICLES

DECOMPOSITION OF SF_6 UNDER AC AND DC CORONA DISCHARGES IN HIGH-PRESSURE SF_6 AND SF_6/N_2 (10-90%) MIXTURES

Anne-Marie Casanovas,[1] Laurence Vial,[1] Isabelle Coll,[1] Magali Storer,[1] Joseph Casanovas,[1] and Régine Clavreul [2]

[1]CPAT, ESA 5002,Université Paul Sabatier
31062 Toulouse cedex 4, France
[2]EDF-DER
77250 Moret-Sur-Loing, France

INTRODUCTION

Sulphur hexafluoride (SF_6) is the gaseous dielectric the most widely used in high-voltage and very-high-voltage electrical equipment. As SF_6 is a synthetic gas that is expensive and could have a non negligible impact on environment, the current trend is to reduce its quantity in electrical equipment by mixing it with natural gases such as air or N_2. Such a mixture is envisaged for gas-insulated transmission lines, particularly for very high voltage network where a high power is necessary. Until now, transmission lines have been insulated with pure SF_6 at a pressure of 400 kPa. They could soon be insulated with a SF_6-N_2 mixture with a high proportion of nitrogen (typically 90%) at pressures between 600 and 1200 kPa.These mixtures have several advantages compared to pure SF_6 : they are cheaper and are much more environmentally friendly[1]. Apart from the purely electrical problems arising from the use of such mixtures, it is important to study their chemical stability under the electrical stresses they will have to endure and especially to compare their decomposition to that of pure SF_6 in identical experimental conditions.

The present study concerns the decomposition of high-pressure (400 kPa) SF_6 and SF_6 - N_2 (10 - 90) mixtures under DC negative polarity and 50 Hz AC coronas.

EXPERIMENTAL

The corona discharges were generated at room temperature (~23 °C) in 340 cm³ Monel 400 cells between a stainless steel point (radius of curvature 10 μm) connected to the high voltage, and a plane of AG3 aluminum (gap space of 2.3 and 3.4 mm for the pure SF_6 and the SF_6-N_2 mixtures respectively).The plane was earthed via a microammeter which gave the value of the discharge current (I_{mean}~16 μA rms for AC voltage and I_{mean}~27 μA [or DC voltage), or via a 200 MHz Philips PM 3320 oscilloscope, to visualise its temporal evolution.

Before their filling with the gas to be studied (P_{gas}=400 kPa), the cells were evacuated (without heating) down to ~1 Pa. In these conditions the residual water content was about 150 ppm$_v$. No impurities (O_2 or H_2O) were added to the gases whose initial purity was 99.995%.

At the end of each run, the gaseous byproducts were analyzed by gas-chromatography; the duration of the run depended on the transported charge required (Q) varying in the range 0 to 10 C.

With the SF_6-N_2 mixture we observed in addition to the classical byproducts seen in pure SF_6: SOF_2, SOF_4, SO_2F_2, S_2F_{10} and S_2OF_{10}, the formation of compounds like $S_2O_3F_6$, NF_3, SF_5NF_2 and $(SF_5)_2NF$. The last two were identified and assayed following the work of Castonguay and Dionne[2] and assuming that the response of our detector (TCD) was the same for these compounds and for S_2OF_{10}.

RESULTS

Before reporting the detail of the results obtained for each product, it should be noted that the small proportion of SF_6 in the SF_6-N_2 mixture gives rise to:
- an increase in the proportion of impurities (O_2, H_2O) with respect to the absolute quantity of SF_6. The impurities were desorbed from the electrode and the walls of the cell in the same amounts whatever the gas sample studied since the working pressure (400 kPa) was the same for both the mixture and the pure SF_6.
- a reduction of the ability of SF_6 to reform through recombination reactions,
- a decrease in the phenomena related to the passivation of the plane electrode by the ions produced from SF_6.[3]

(SF_4 + SOF_2): unlike in pure SF_6, these compounds did not disappear from the mixture. This confirms the role of the plane electrode and its degree of passivation on the disappearance of (SF_4+ SOF_2).[3]

(SOF_4 + SO_2F_2): it can be seen in Figure 1 that independently of the type of voltage applied, SOF_4 + SO_2F_2 was formed in similar quantities in SF_6 and in the SF_6- N_2 mixture at low transported charges (Q < 2C) and in larger quantities in the mixture for higher charges.

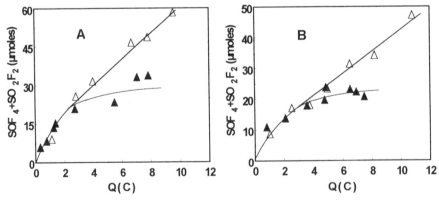

Figure 1. Measured yields of (SOF_4+SO_2F_2) versus charge transported in SF_6 : ▲ or SF_6-N_2 (10-90) mixtures: △ submitted to 50Hz AC coronas : figure 1**A** or to negative DC coronas : figure 1**B**. P = 400 kPa. Aluminum plane electrode.

S_2F_{10}: Figure 2 shows that, whatever the type of voltage, the formation of S_2F_{10} was lower in the SF_6-N_2 mixture than in pure SF_6. This effect, arising from a decrease in the number of $SF_5+SF_5 \rightarrow S_2F_{10}$ reactions, can be attributed to the small proportion of SF_6 in the mixture and can be explained by:
- the increase in the reactions between SF_5 and the dissociation fragments of water and oxygen, present in larger proportions with respect to SF_5 in the mixture, to give SOF_4 ($SF_5 + OH \rightarrow SOF_4 + HF$), ($SF_5 + O \rightarrow SOF_4 + F$),
- the generation of new compounds such as SF_5NF_2 and $(SF_5)_2 NF$ formed from the SF_5 fragment.

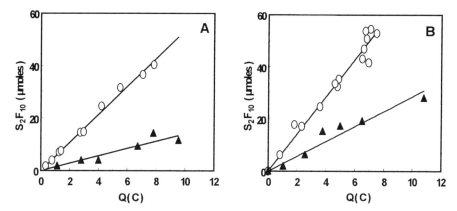

Figure 2. Measured yields of S_2F_{10} versus charge transported in SF_6 : ○ or SF_6-N_2 (10-90) mixture: ▲ submitted to 50Hz AC coronas : figure 2**A** or to negative DC coronas : figure 2**B**. P = 400 kPa. Aluminum plane electrode.

Independently of the gas studied, AC voltage enhanced the formation of (SF_4 + SOF_2) and (SOF_4 + SO_2F_2) whereas it had the opposite effect on S_2F_{10}. A similar pattern had already been noted in pure SF_6 during a previous study[4] and was correlated to the greater desorption of water from the point electrode under the AC voltage.

S_2OF_{10} **and** $S_2O_2F_{10}$: like S_2F_{10}, S_2OF_{10} was formed from the SF_5 fragment and thus, for the same reasons, in lesser quantities than in pure SF_6. The production of $S_2O_2F_{10}$ was too small to enable a comparative study of its formation in the two gas samples.

$S_2O_3F_6$: this compound only appeared in the SF_6-N_2 mixture. This could be due to its enhanced formation in the presence of water[5]: in the SF_6-N_2 mixture, the proportion of water with respect to the absolute quantity of SF_6 was higher than in pure SF_6.

NF_3, SF_5NF_2 and $(SF_5)_2NF$: the variations of the levels of these compounds versus the transported charge value are reported in Figure 3 for NF_3 and Figure 4 for SF_5NF_2 and $(SF_5)_2NF$. Overall, the quantities measured were slightly higher with DC than with AC voltage. Saturation even appeared under DC voltage for $(SF_5)_2NF$.

The first two columns in Table 1 give the quantities of the various products formed, in µmoles, for a transported charge of 7C and for the two gas samples subjected to corona discharges under 50 Hz AC voltage. The third and fourth columns give the proportion of each degradation product as a percentage of the total amount of product formed in SF_6 and in the SF_6-N_2 (10-90) mixture respectively. The last two columns express the proportion of product formed as a percentage of the initial quantity of SF_6 present in the sample (pure SF_6 and the SF_6-N_2 (10-90) mixture). At least for (SF_4+SOF_2) and ($SOF_4+SO_2F_2$) the results given in Table1 confirm those of Siddagangappa and Van Brunt[6] on the increase of the number of moles of sulphur oxyfluorides formed per mole of starting SF_6 when the proportion of SF_6 in SF_6-N_2 mixtures are low.

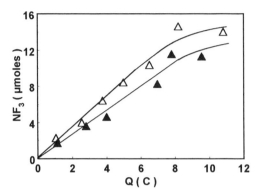

Figure 3. Measured yields of NF_3 versus charge transported in SF_6-N_2 (10-90) mixture submitted to DC negative coronas : △ or to 50Hz AC coronas : ▲ . P = 400 kPa. Aluminum plane electrode.

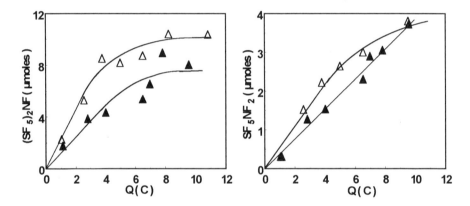

Figure 4. Measured yields of $(SF_5)_2NF$ and SF_5NF_2 versus charge transported in SF_6-N_2 (10-90) mixture submitted to DC negative coronas :△ or to 50Hz AC coronas : ▲ . P = 400 kPa. Aluminum plane electrode.

In parallel to the study of the chemical stability of SF_6 and SF_6-N_2 (10-90) mixtures, a study was also made of the electrical behaviour of the discharge under 50 Hz AC voltage. The discharge current was seen to be composed of a synchronous component in phase with the applied voltage and a pulse component in the peak synchronous current regions of each half-wave. The study of the temporal evolution of the discharge current showed that:
- at the beginning of the test, the two synchronous components had almost equivalent amplitudes and are associated to a pulse activity which was low but already higher in the positive than in the negative half-wave.
- during the test, whereas the amplitude of the negative synchronous component decreased steadily, the positive synchronous component totally disappeared for Q > about 1C. This disappearance was correlated with a considerable increase in the pulse activity on this half-wave. The discharge current in the positive half-wave was therefore no longer limited by space charge which could prove hazardous for the dielectric properties of the system.

These observations remain the same whatever the gas studied and confirm the results already obtained by several authors both in pure SF_6 [7,8] and in SF_6-CF_4 (50-50) mixtures.[8]

Table 1. Gaseous byproduct yields in SF_6 and SF_6-N_2(10-90) mixtures. 50Hz AC coronas. Q = 7C. P = 400 kPa. Aluminum plane electrode.

Byproducts	SF_6 µmoles	SF_6-N_2 µmoles	SF_6 % / Σ	SF_6-N_2 % / Σ	SF_6 % / SF_6	SF_6-N_2 % / SF_6
SOF_4	26,8	33,84	35,2	34,4	0,049	0,58
SO_2F_2	6,77	14,7	8,9	14,9	0,012	0,25
SF_4+SOF_2	0,67	4,71	0,88	4,8	~0	0,08
S_2F_{10}	40,39	14,2	53,08	14,4	0,07	0,24
S_2OF_{10}	1,46	0,4	0,02	0,41	~0	~0
$S_2O_3F_6$	0	6,8	0	6,9	0	0,12
NF_3	-	11,54	-	11,7	-	0,19
SF_5NF_2	-	3,05	-	3,1	-	0,05
$(SF_5)_2NF$	-	8,94	-	9,1	-	0,15
Sum (Σ)	76,09	98,18	100	100	0,13	1,66

CONCLUSION

Comparing the results obtained in pure SF_6 and in SF_6-N_2(10-90) mixtures, the tendencies are the following in the mixtures :
- (SF_4 + SOF_2) is formed in greater quantities, especially with high charges under 50 Hz AC voltage where it no longer disappears: passivation of the electrode is limited when SF_6 is in a small proportion,
- the production of S_2F_{10} is greatly reduced - this is an added advantage of SF_6-N_2 (10-90) mixtures,
- the nitrogenous decomposition products, such as NF_3, SF_5NF_2 and $(SF_5)_2NF$, are formed in appreciable quantities since they represent (at least in the working hypothesis proposed here - see Experimental section) almost a quarter of the total quantity of decomposition product.

The differences observed in the decomposition of SF_6 in pure SF_6 or in the SF_6-N_2(10-90) mixture can be explained by:
i) the strong dilution of SF_6 in the mixture leading to:
- an increase in the quantities of O_2 and H_2O relative to SF_6 and thus of the action of these molecules on the formation of the degradation products (for instance, the scavenging role of water is thus enhanced),
- the decrease of the effects related to the passivation of the plane electrode,
- the diminution of recombination reactions such as SF_x + (6-x)F → SF_6 and SF_5+SF_5→S_2F_{10}.
ii) the binding of fluorines to nitrogen (formation of NF_3) and the formation of compounds such as SF_5NF_2 and $(SF_5)_2$ NF which diminish the regeneration of SF_6.

Acknowledgments

Financial support from EDF is gratefully acknowledged.

REFERENCES

1. L.G. Christophorou and R.J. Van Brunt, SF_6/N_2 mixtures. Basic and HV insulation properties,*IEEE Trans.Diel. Elec. Insul.* 2:952 (1995).
2. J. Castonguay and I. Dionne, S_2F_{10} and other gaseous decomposition byproducts formed in SF_6 and SF_6-gas mixtures exposed to electrical discharges, 7th Int. Symp. on Gaseous Dielectrics, Knoxville (1994).
3. C. Pradayrol, A.M. Casanovas, A. Belarbi, and J. Casanovas, Influence de la nature et de l'état de conditionnement de l'électrode plane sur la décomposition de SF_6 soumis à des décharges couronne de polarité négative, *J. Phys. III* 5:389 (1995).
4. A. Belarbi, C. Pradayrol, J. Casanovas, and A.M. Casanovas, Influence of discharge production conditions, gas pressure, current intensity and voltage type, on SF_6 dissociation under point-plane corona discharges, *J. Appl. Phys.* 77:1398 (1995).
5. C.Pradayrol, A.M. Casanovas, C. Aventin, and J. Casanovas, Production of SO_2F_2, SOF_4, (SOF_2+SF_4), S_2F_{10}, S_2OF_{10} and $S_2O_2F_{10}$ in SF_6 and (50-50) SF_6-CF_4 mixtures exposed to negative coronas, *J. Phys. D: Appl. Phys.* 30:1356 (1997).
6. M.C. Siddagangappa and R.J. Van Brunt, Decomposition products from corona in SF_6/N_2 and SF_6/O_2 mixtures, Proc. 8th Int. Conf. on Gas Discharges and their Applications, Leeds University Press, pp 247-250, Leeds (1985).
7. M. Lalmas, H. Champain, A. Goldman, and E. Fernandez, Long-term evolution of point to plane SF_6 discharges under alternating voltage, 7th Int. Symp. on Gaseous Dielectrics, Knoxville (1994).
8. W. Vaillant, Essai de corrélation entre le spectre de lumière émis par la décharge, le courant de décharge, les quantités de produits (SOF_2, SOF_4, S_2F_{10}...) formés et le régime de la décharge lorsque du SF_6 (P=30 et 300 kPa) ou du SF_6-CF_4 (50-50; P=30kPa) sont soumis à des décharges couronne en géométrie pointe-plan sous tension alternative 50 Hz.Rapport de DEA, Toulouse (1995).

DISCUSSION

J. CASTONGUAY: It is interesting to see that your work confirms the formation of new N-containing by-products in SF_6/N_2 mixtures, as we had shown in the previous Gaseous Dielectrics Symposium in 1994. Your Table 1 data show that S_2F_{10}, for the first time, appears to be the major decomposition by-product in corona discharges (53%). How can you interpret this peculiar finding?

L. VIAL: The results shown in this table were obtained under 50 Hz AC coronas and with a transported charge of 7 C, which is quite a high charge in our case. We could explain that by the fact that during the AC test, stable negative ions are created and can make impacts on the point during the positive half-cycle. This causes a greater desorption of H_2O from the point. Previous studies at our Laboratory showed that the formation of S_2F_{10} was enhanced at high transported charges by the presence of water. This enhanced presence of water could explain, at least in part, the result shown in this table. Another reason could be the favorable effect of pressure on S_2F_{10} formation and its opposite effect on that of $SOF_4 + SO_2F_2$.

H. MORRISON: (1) How long in time was a typical experiment? One hour or three days? For very long experiments you may lose some by-products to the walls of the cell. (2) How did you confirm the identity of the compounds and calibrate the quantity measured?

L. VIAL: (1) The longest experiments were those for which we wanted to accumulate the higher charges. The duration of the application of the high voltage could be a little more than three days, it could be four days for example. But even for these longest experiments, the number of by-products lost on the walls is negligible. In some experiments, made after letting a cell charged for several days, some of the by-products were stable and some were not, but we could explain the observed differences by reactions between the by-products and not their reactions with the walls. Earlier experiments carried out at our Laboratory showed that, for example, S_2F_{10} and SOF_2 remained stable in our experimental cells over several days. Thus, reactions of these compounds on the walls are certainly negligible, at least under our experimental conditions. (2) We did not confirm, for example by mass spectroscopy, the existence of the new compounds such as $(SF_5)_2 NF$ or SF_5NF_2. We considered these compounds by using the work of Castonguay and Dionne [J. Castonguay and I. Dionne, *Gaseous Dielectrics VII*, L. G. Christophorou and D. R. James (Eds.), Plenum Press, New York, 1996, p. 449]. Thus, the $(SF_5)_2NF$ and SF_5NF_2 compounds were identified and assayed following that work and assuming that the response of our (TCD) detector was the same for these compounds and S_2OF_{10}. For the other compounds such as S_2OF_{10}, we measured the response of our detector by using mixtures of SF_6 and the compound studied under conditions as close as possible to those in the real experiment.

H. SŁOWIKOWSKA: (1) What are the parameters of the gas chromatographic technique used in your experiments? (2) What is your opinion as to the influence of oxygen on gas by-product formation?

L. VIAL: (1) To analyze the decomposition by-products we used two chromatographs: HP 5880 A (TCD detector maintained at 100 °C; column filled with Porapak N; length 3 m; diameter 1/8"), and HP 5890 II (TCD detector maintained at 110 °C- first column filled with Porapak T; length 3 m; diameter 1/8" - second column SPB1 wide bore length 55 m; internal diameter 0.53 mm, film thickness 5 μm). (2) We can assume that, as it seems to be the case with water, the effects of oxygen will be the same in pure SF_6 and in the mixture. During previous studies at our Laboratory, Catherine Pradayrol showed that for an added quantity of

O_2 between 0% and 1%, the production of $SOF_4 + SO_2F_2$ is enhanced, the formation of S_2F_{10} is inhibited, and the formation of $S_2O_2F_{10}$ and S_2OF_{10} is negligibly influenced.

L. G. CHRISTOPHOROU: Your results concerning the formation of S_2F_{10} in pure SF_6 and in SF_6/N_2 are very interesting in that they have shown that the S_2F_{10} concentration is much smaller in the mixture. But you form in the mixture other by-products. You mentioned SF_5NF_2 and $(SF_5)_2NF$. Are these other N-containing by-products toxic?

L. VIAL: The new by-products seen in SF_6/N_2 (10-90%) mixtures and that we can analyze are $S_2O_3F_6$, SF_5NF_2, and $(SF_5)_2NF$ (NF_3 is also found). I do not currently have any information on the toxicity of SF_5NF_2 and $(SF_5)_2NF$. To my knowledge no studies have been done on this subject.

INFLUENCE OF A SOLID INSULATOR ON THE SPARK DECOMPOSITION OF SF_6 AND $SF_6+50\%CF_4$ MIXTURES

Isabelle Coll, Anne-Marie Casanovas, Catherine Pradayrol, and Joseph Casanovas

CPAT, ESA 5002, Université Paul Sabatier
31062 Toulouse cedex 4, France

INTRODUCTION

It is now known that SF_6/CF_4 mixtures (50-50) make a suitable candidate to supplant pure SF_6 in high voltage circuit breakers for low temperature environment applications. When a circuit breaker is operating, the arc plasma decays in the presence of metal, carbon, oxygen and water coming from partial vaporisation and desorption of the electrodes and nozzle (composed of an organic insulator) under the stress. So, in an effort to approach conditions occurring in practice and as a continuation of previous investigations[1,2] in our laboratory, we studied the spark decomposition of SF_6 and $SF_6+50\%CF_4$ mixtures in the presence of a solid insulator struck by the electric arc.

The gaseous compounds produced during the sparks: CF_4, SO_2F_2, SOF_4, (SF_4+SOF_2), S_2F_{10}, S_2OF_{10}, $S_2O_2F_{10}$ and $S_2O_3F_6$ were quantitatively studied by varying the nature of the insulator material (Megelit, Kel'F, Teflon, polypropylene, polyethylene, nylon) and the concentration of O_2 and H_2O (0 and 0.2%) added to the gas sample.

EXPERIMENTAL

The sparks were generated by discharging a capacitor (3.59 J per spark) in a cylindrical Monel 400 cell (340cm^3) between a stainless steel point (diameter 1mm; radius of curvature 0.5 mm) and a stainless steel plane (diameter 25mm) in which a small rod of insulator (diameter 5 mm) was inserted just below the point as shown in Figure 1. The distance between the point and the insulator bar was 0.1 mm, the interelectrode gap spacing 0.9 mm.

Before filling with the gas, the cell was submitted to several heating (up to 60°C) pumping cycles and evacuated to a final pressure lower than 1Pa. The gaseous sample was used at a gas pressure of 100 kPa. At the end of each series of sparks the cell contents were assayed by gas chromatography. The analytical conditions have been described earlier in detail.[3]

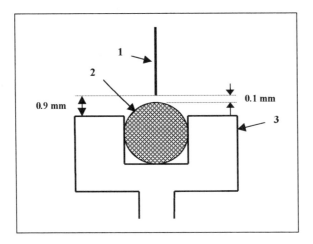

Figure 1. Schematic diagram of electrode configuration . 1: point electrode, 2: insulator bar, 3: plane electrode.

The names, the chemical formulae and the temperature of crystalline fusion of the six insulators tested are reported in Table1. Finally reproducibility tests, carried out under the experimental conditions used here, showed that the uncertainty of our data was about ∓ 10%.

Table 1. Names, formulae and temperature of crystalline fusion of the six solid organic insulators studied.

Insulator	Formula	Temperature of crystalline fusion
Megelit	$[C_9H_9O_4]_n$	—
Teflon	$[C_2F_4]_n$	330°C
Kel'F	$[C_2F_3Cl]_n$	200°C
Nylon	$[N_2C_{12}H_{22}O_2]_n$	230-260°C
Polypropylene	$[C_3H_6]_n$	160°C
Polyethylene HD	$[C_2H_4]_n$	130°C

RESULTS

The results concerning the effects of an organic insulator on the spark decomposition of SF_6 and SF_6-CF_4 mixtures at a pressure of 100 kPa are reported in figures 2, 3 and 5 for each of the SF$_6$ decomposition products studied ; the formed yields were measured with or without, addition of 2000 ppm$_v$ water or oxygen.

For the SF$_6$-CF$_4$ mixture only four insulators were tested (Megelit, Kel'F, Teflon and polypropylene) and two (Megelit and Teflon) in the presence of H$_2$O and O$_2$. The quantities of CF$_4$ produced were not reported for the mixture owing to the large amounts of CF$_4$ initially present.

As it becomes vapourised under the effect of the discharges, the insulator produces species that can trap the fluorine atoms released from SF$_6$ decomposition and enhances the formation of SF$_6$ byproducts. This explains the main results obtained we are going to present :

- In all cases the presence of the insulator increased the levels of (SF$_4$+SOF$_2$), CF$_4$, S$_2$F$_{10}$ and S$_2$O$_2$F$_{10}$ (cf. Figures 2, 3 and 5).

The strongest increase was observed for (SF$_4$+SOF$_2$) : 3 to 9 fold in pure SF$_6$ and 2.5 to 13 fold in the mixture. It should be noted that these compounds remain the major SF$_6$ decomposition products even in the presence of the insulator. The rank order of the influence of the insulator on (SF$_4$+SOF$_2$) production (Megelit < Teflon < Kel'F < nylon < polypropylene < polyethylene) is the same as that of their decreasing order of crystalline fusion (Table 1). It appears therefore that the variations in the quantities of material vapourised are largely responsible for the different effects of the insulators.

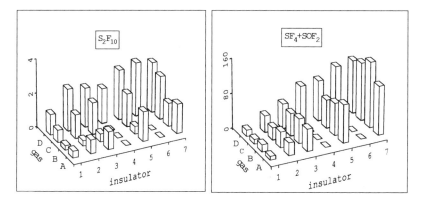

Figure 2. Production rates, in nmoles/J, of S$_2$F$_{10}$ and (SF$_4$+SOF$_2$) formed in gas **A**: SF$_6$+0.2%O$_2$, **B**: SF$_6$+0.2%H$_2$O, **C**: SF$_6$+50%CF$_4$, **D**: SF$_6$ sparked through capacitor discharge, with and without the presence of a solid insulator struck by the discharge. **1**: no insulator, **2** : Teflon, **3**: Kel'F, **4** : Nylon, **5**: Megelit, **6**: Polyethylene, **7**: Polypropylene. Gas pressure : 100kPa. Insulators 4 and 6 were not studied with the gases **A**, **B** and **C**.

The second most abundant product after (SF$_4$+SOF$_2$) formed in presence of insulator was CF$_4$ (Figure 3). The carbon and fluorine atoms were supplied by the insulator and the SF$_6$ during sparking. As reported in Figure 4 we observed that the levels of CF$_4$ formed were proportional to the increase in (SF$_4$+SOF$_2$) in the presence of the insulator. In this, the formation of CF$_4$ is at the origin of of the enhancement of (SF$_4$+SOF$_2$) by means of the fluorine atoms released from SF$_6$ and trapped not by SF$_4$ to yield new SF$_6$ but by the carbon atoms to form CF$_4$. It should be noted that the proportionality coefficient decreased with the number of fluorine atoms supplied by the insulator ; indeed insulators containing both fluorine and carbon atoms are likely to supply CF$_4$ molecules during the stress independently of SF$_6$. It can be noticed moreover that the correlation between the formation of CF_4 and (SF_4+SOF_2) remained the same whatever the water or oxygen content of the gas.

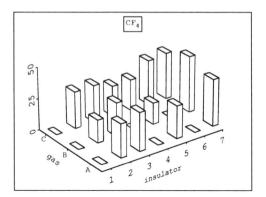

Figure 3. Production rates in nmoles/J of CF_4 in gas **A** : $SF_6+0.2\%O_2$, **B** : $SF_6+0.2\%H_2O$, **C** : SF_6 sparked through capacitor discharge, with and without the presence of a solid insulator struck by the discharge. **1** : no insulator, **2** : Teflon, **3** : Kel'F, **4** : Nylon, **5** : Megelit, **6** : Polyethylene, **7** : Polypropylene. Gas pressure : 100kPa. Insulators 4 and 6 were not studied with the gases **A** and **B**.

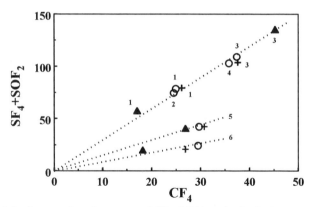

Figure 4. Correlation between the enhancement of (SF_4+SOF_2) production in the presence of an insulator and the formation rate of CF_4 measured in nmoles/J in an atmosphere of 100 kPa SF_6 under sparks from discharge capacitor. No water or oxygen added : O. 2000 ppm$_v$ water added : ▲. 2000 ppm$_v$ oxygen added : +. **1**: Megelit, **2**: Nylon, **3**: Polypropylene, **4**: Polyethylene, **5** : Kel'F, **6**: Teflon.

The enhancement of S_2F_{10} production by the insulators (Figure 2) ranged between 2 and 4 fold. This agrees with previous observations made by Sauers et al [4] for Teflon and with results on a kinetic model of an extinguishing SF_6 circuit-breaker arc in the presence of carbon [5].

- The presence of insulator enhanced the levels of S_2OF_{10} and $S_2O_3F_6$ in all cases for pure SF_6 (Figure 5). With Megelit, Kel'F and polypropylene, when 0.2% impurity was added (H_2O for the former and O_2 for the later) production of these compounds was decreased. Reactions between species arising from the stressed insulator and water or oxygen (formation of CO_n, CH, etc.) could be responsible for these particular effects . Such reactions can also explain the defavourable effect of insulator globally observed for (SOF$_4$+SO$_2$F$_2$) (Figure 5) ; indeed the formation of these products is enhanced with the water content of the gas[2].

- Finally we observed that the effect of the insulator was very similar for pure SF$_6$ and the SF$_6$-CF$_4$ mixture. On the other hand the main byproducts (SF$_4$+SOF$_2$) and S$_2$F$_{10}$ were produced in the mixture in slightly lower amounts than in pure SF$_6$.

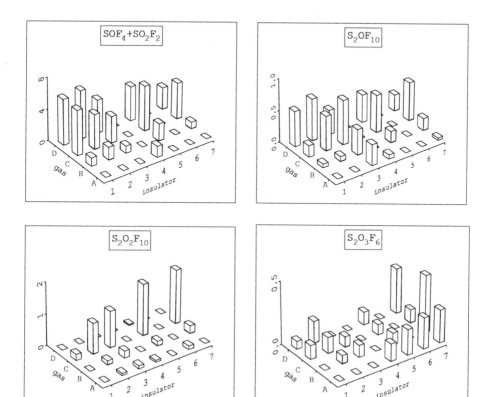

Figure 5. Production rates, in nmoles/J of (SOF$_4$+SO$_2$F$_2$), S$_2$OF$_{10}$, S$_2$O$_2$F$_{10}$ and S$_2$O$_3$F$_6$ formed in gas **A**: SF$_6$, **B**: SF$_6$+50%CF$_4$, **C**: SF$_6$+ 0.2%O$_2$, **D**: SF$_6$+0.2%H$_2$O sparked through capacitor discharge, with and without the presence of a solid insulator struck by the discharge. **1**: no insulator, **2** : Teflon, **3**: Kel'F, **4** : Nylon, **5**: Megelit, **6**: Polyethylene, **7**: Polypropylene. Gas pressure : 100kPa. Insulators 4 and 6 were not studied with the gases **B**, **C,** and **D**.

CONCLUSION

The presence of an organic insulator struck by sparks generated in a SF$_6$ medium, strongly modifies the gas composition, independently of water or oxygen (up to 0.2%) and CF$_4$ (50%) added, and leads to :
- An enhancement of (SF$_4$+SOF$_2$) , the major degradation product of such stress,
- A formation of CF$_4$ correlated to this enhancement,
- An increase in S$_2$F$_{10}$, S$_2$OF$_{10}$, S$_2$O$_2$F$_{10}$ and S$_2$O$_3$F$_6$ yields.

Acknowledgments

The authors would like to acknowlekge financial support from the GEC Alsthom Company.

REFERENCES

1. A.M. Casanovas, B. Belmadani et J. Casanovas, Produits gazeux de décomposition d'un isolant solide dans le SF_6 lors de claquages, *Revue Générale de l'Electricité*, 8 :14 (1990).
2. C. Pradayrol, A.M. Casanovas, A. Hernoune and J. Casanovas, Spark decomposition of SF_6 and SF_6+50%CF_4 mixtures, *J.Phys.D. : Appl. Phys.*, 29 : 1941 (1996).
3. C. Pradayrol, A.M. Casanovas, C. Aventin and J. Casanovas, Production of SO_2F_2, SOF_4, (SF_4+SOF_2), S_2F_{10}, S_2OF_{10} and $S_2O_2F_{10}$ in SF_6 and (50-50) SF_6-CF_4 mixtures exposed to negative coronas, *J. Phys. D : Appl. Phys.*, 30 :1356 (1997).
4. I. Sauers, S.M. Mahajan and R.A. Cacheiro, Effect of a solid insulator on the spark yield of S_2F_{10} in SF_6, *Conf. Rec.-1994 IEEE Int. Symp. on Elec. Insul.*, IEEE publ.94-CH3445-4, Pittsburg (1994).
5. A. Gleizes, E. Borge, A.M. Casanovas and B. Belmadani, Kinetic study of a decaying SF_6 arc plasma in the presence of impurities, *Trans. IEE of Japan*, 116-A : 948 (1996).

DISCUSSION

K. NAKANISHI: I am not familiar with the names of the materials such as Megelit and Kel'F. Could you please specify the materials? I am concerned whether the material of epoxy is included in your tests.

I. COLL: Samples of Megelit have been given to our Laboratory by the manufacturer. We only know the type and the number of the atoms constituting this insulator. To our knowledge epoxy is not present in Megelit. Kel'F is the trade mark for polytrifluoromonochloroethylene. Tests with epoxy resins are not planned at our Laboratory.

J. CASTONGUAY: Please check your various bar graphs since data for polyethylene seem to be wrong.

I. COLL: For the SF_6/CF_4 (50%-50%) mixtures we did not make tests with polyethylene H D because we thought that the results with this insulator would be very similar to those obtained with polypropylene as observed in pure SF_6. Concerning the graphs, we wanted to show the whole of the tests and we decided to take zero when a test has not been done.

H. MORRISON: (1) How many sparks were in each test? (2) Was there a problem with the detection limit of some compounds that determined the minimum number of sparks? In other words, would you reduce the number of sparks to obtain a cleaner system?

I. COLL: (1) In the presence of an organic insulator, the number of sparks is limited to 200 because over a certain number of sparks the insulator is no longer attacked by the arc. (2) The amounts of the decomposition products are proportional to the number of sparks. Thus, to detect these decomposition by-products it is necessary to make a minimum number of sparks. To reduce this number it would be necessary to use analytical detection methods with a better detection limit.

CHEMICAL REACTIONS AND KINETICS OF MIXTURES OF SF_6 AND FLUOROCARBON DIELECTRIC GASES IN ELECTRICAL DISCHARGES

Jacques Castonguay

Institut de recherche d'Hydro-Québec
Varennes, Québec
CANADA, J3X 1S1

ABSTRACT

Chemical reactions involving SF_6 and various dielectric gas mixtures were studied quantitatively in low-power arc discharges. The concentrations of SF_6 mixed with the following fluorocarbons: CHF_3, CF_4, C_2F_6, C_3F_6, 2-C_4F_6 and c-C_4F_8 were varied in the range of 0 to 100%. Dielectric mixtures using CHF_3 and CF_4 were found much more stable under electrical stresses than the ones containing unsaturated fluorocarbon molecules which shown the highest rates of decomposition. CF_4 and SOF_2 were the two major reaction byproducts observed. Many more sulfur-containing fluorocarbon compounds were identified in the byproducts: $C_2F_2S_2$, $C_2F_4S_2$, C_2F_6S, $C_2F_6S_2$, in addition to S_2F_2 (two different isomers found), S_3F_2 and CS_2. Finally, the dilution (at 50 and 75%) of a dielectric gas mixture with nitrogen resulted in an important reduction of the decomposition rates, larger than the dilution factor.

INTRODUCTION

The use of sulfur hexafluoride as an insulating gas in electrical equipments and substations is widely spread. Unfortunately, the unique chemical properties of the SF_6 molecules: inertness, thermal stability and electronegativity are also responsible for its accumulation into the atmosphere, over the last three decades. Consequently, a new interest has developed for dielectric gas mixtures that may help to reduce the lost of SF_6 in the air.

Indeed, works on dielectric gas mixtures were mostly concentrated a decade ago. At that time, various studies [1-3] have shown that many binary and ternary gas mixtures can exhibit breakdown and dielectric properties that often surpass those of SF_6. Mixtures of promising interest were the ones containing fluorocarbon compounds and, in particular, those prepared with perfluorocarbon molecules.

This paper presents results of the study of the chemical behavior of various SF_6-perfluorocarbon (PFC) mixtures induced by low-power arc discharges. The influence of the nature of the fluorocarbon molecule was studied over the whole range of gas composition (0-100%). In addition, some ternary gas mixtures were submitted to the same spark-like conditions in order to measure the effects of dilution with nitrogen.

EXPERIMENTAL

A simple experimental setup was used in this study. As depicted in Figure 1, it comprises a cylindrical discharge cell (290 ml) having Teflon thick wall and fitted with a point electrode (stainless steel) and a rectangular plane electrode (aluminum). These are supported by the aluminum covers, one equipped with an insulated feedthrough and the other with the gas sampling port (septum). Low-power (60W) arc discharges of short duration were sustained by means of a high voltage transformer (15 kV, 200 mA). The total arcing time was limited to 40 seconds and was cumulated in 4 distinct consecutive discharge periods (5, 5, 10 and 20s).

This arrangement was chosen to satisfy an essential experimental requirements for such kinetic studies: allow the decomposition of a sufficient quantity of the primary gases partners. The short duration also helps to minimize secondary reactions. Low-power arc experiments were done at atmospheric pressure with SF_6-PFC mixtures containing 2, 5, 10, 20, 40, 80 and 100% of the following PFCs: CF_4, CHF_3, C_2F_6, C_3F_6 C_4F_6 and c-C_4F_8.

Analysis of the gas mixtures was performed immediately after each arcing test with a gas chromatograph-mass spectrometer system using a Porapak Q packed column and manual injection with a gas syringe. This simplified analytical setup permitted the rapid quantification of the two gas mixture components and their principal gaseous decomposition byproducts: SOF_2 and CF_4.

In addition to these two major products, the following gaseous species were observed (and quantify on certain occasions) with the MS detector: $C_2F_2S_2$, $C_2F_4S_2$, C_2F_6S, $C_2F_6S_2$, S_2F_2, S_3F_2 and CS_2. These molecules seem to be members of fluorosulfane and fluorosulfide families. Since no calibration gas standards were available for the majority of these compounds, their relative MS response was taken equal to that of CS_2.

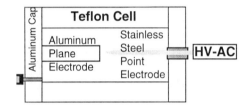

Figure 1. Experimental discharge cell.

RESULTS AND DISCUSSION

This study was principally aimed at measuring the relative chemical stability of various SF_6-fluorocarbon mixtures. A typical example of measured experimental results is shown in Figure 2. For clarity, only the data collected for the SF_6-C_3F_6 mixture are presented. These analytical results illustrate the progressive consumption of the mixture components and the increasing formation of their respective main byproducts. As shown, the evolution with time of the two primary gas concentrations fits straight lines when plotted on a semi-logarithmic graph.

All the six different SF_6-PFC mixtures studied exhibited similar types of behavior, differing only by the slope of curves obtained in identical test conditions. This kind of logarithmic time-dependent behavior of the original mixture components characterizes some sort of pseudo-unimolecular reaction mechanism. Indeed, the direct chemical interaction between the SF_6 and the PFC molecules is quite unlikely. Experiences have shown that the SF_6 disappearance is most probably due to the reduction of the recombination of its primary radicals, especially F•, which are easily caught by other radicals and reactive species, forming CF_4, for instance. Moreover, the addition of F atoms to unsaturated PFC molecules (C_3 and C_4) can certainly competes with the back recombination to SF_6.

Decomposition and formation rates

The data reported in Figure 2 clearly indicate that the SF_6 decomposition rate depends directly on the PFC concentration. In addition, The slopes of these curves can be readily used to calculate the reaction rates of the four major compounds shown. Thus, the reaction yields of all the series of experiments were computed and these values are summarized in

Figure 3. One can immediately observe that any SF_6-PFC gas mixture exhibits larger reactivities in the arcing environment than the pure SF_6.

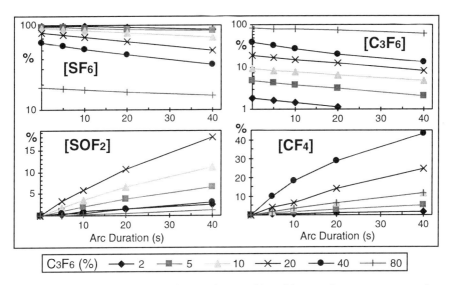

Figure 2. Arcing results for SF_6 - C_3F_6 mixtures: decomposition of the two mixture components and formation of the corresponding major byproducts.

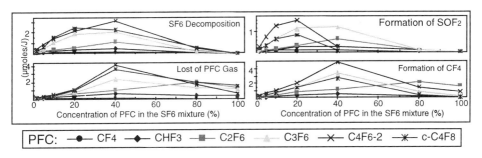

Figure 3. Variation of SF_6 and PFCs decomposition rates and of SOF_2 and CF_4 formation rates as function of gas mixture composition.

The higher the PFC molecular weight, the faster the decomposition rates of both primary mixture constituents. CF_4 is somewhat an exception, as it is found to be a more inert additive than CHF_3. However, both are seen to have the less pronounce effect on the SF_6 decomposition. Many interesting observations can be deducted from the different graphs shown in Figure 3:
- low concentrations of C_3- and C_4-PFCs added in SF_6 accelerate much more its decomposition, in comparison to the C_1 and C_2 homologues,
- except for C_2F_6, maximum consumption rates of initial components occur with 40% (or possibly slightly more) PFC mixtures in all cases,
- C_4-PFCs level higher than 20% inhibits the formation of SOF_2 from the decomposed SF_6,
- the PFC lost and the CF_4 formation curves follow almost exactly the same pattern, with the exception of the c-C_4F_8, showing a CF_4 yield ≈40% lower.

Otherwise, it interesting to note that C_2F_6 is the PFC that decompose at the fastest pace in the presence of low-power arcs. Effectively, this behavior is reported in Table 1 on a quantitative basis. Data are also tabulated to show the total decomposition rate of three different mixtures. Clearly, the binary dielectric gas composed of SF_6 and CF_4 is the most stable one under arcing. Nevertheless, some its mixtures are seen to decompose as much as 7 time faster than either pure partner.

Decomposition yield of the pure 2-C_4F_6 was rather difficult to determine, since it produced carbon particles that impair the discharge processes. Indeed, Sauers [4] has observed similar findings while also showing that carbonization was prevented when enough SF_6 was added to that PFC. In addition, he showed that maximum CF_4 was produced with ≈70%-30% SF_6-2-C_4F_6 mixtures. In the present work, mixtures of C_3- and C_4-PFCs seem to behave slightly differently, with mixture decomposition and byproduct formation maxima seen to occur at PFC contents of 40% or higher.

Table 1. Decomposition rates of pure gases and SF_6 - PFC mixtures by low-power arcs (µmoles/J) (*)

Gas mixture	SF_6	CF_4	CHF_3	C_2F_6	C_3F_6	2-C_4F_6	c-C_4F_8
Pure Gas	0.03	0.03	0.32	1.17	1.23	1.85	1.38
10% PFC/SF_6	---	0.15	0.49	0,60	1.47	1.92	1.89
40% PFC/SF_6	---	0.22	1.17	2.20	4.53	7.37	5.93
80% PFC/SF_6	---	0.07	0.52	2.34	1.53	2.21	2.46

(*) Calculated mixture rate: sum of SF_6 and PFCs decomposition rates.

Chemistry of the mixture decomposition

Over the years, the chemistry of the SF_6 decomposition in electrical discharges has became more and more complex, as new byproducts were identified with more powerful analytical techniques [5]. In the present study, the measured yield variations of both primary mixture gases and principal byproducts can shed some light on the evolution of the reaction scheme. This is best illustrated by Figure 4 which shows the variation of the reaction yield ratios of the four major compounds analyzed.

The SF_6/PFC ratios reveal the apparent stochiometry of the SF_6 decomposition by each added PFC gas. As seen, the number of SF_6 molecules consumed by low amounts of PFC addition is roughly proportional the carbon content of the PFC, except for CHF_3 which seems slightly more efficient in that process. On the other side and in the same conditions, the relative formation of CF_4 appears also roughly related to the number of carbon in the PFC molecules.

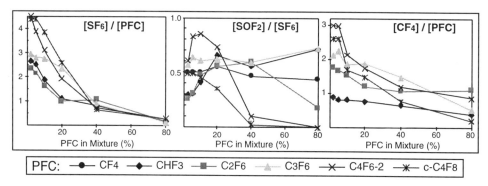

Figure 4. Variation of decomposition and formation yield ratios with mixture composition.

As the mixtures are enriched in PFC, all the SF_6/PFC ratios are reduced and converge to almost the same values. For 40% PFC mixtures, the main partners are consumed at about the same rate in all cases and the corresponding CF_4/PFC ratios are also found around 1:1.

For CF_4, these same ratios are observed with the 80% PFC mixtures. This means that the majority of the PFC molecule fragments react through paths.

On the other hand, the SOF_2/SF_6 ratios average only about 60%. Furthermore, those obtained with the 40% and up C_4-PFC mixtures are much lower, reaching even zero. This also means that other reaction paths are getting more important and that the reaction mechanism is becoming more complex. Indeed, many other by C-, F-, and S-containing byproducts were identified, as it will be discussed later.

The data just discussed indicate that a larger portion of the arc-decomposed molecules are missing from the analyzed byproducts, as the PFC content is increased in the mixtures. This is well illustrated in Figure 5 which reports the evolution of the atomic mass balances calculated for the three atoms present in the primary gases. Except as noted, these corresponds to the combined atom content of SOF_2 and CF_4. The amounts of fluorine found in these two byproducts are about 50% of those lost by the decomposed SF_6, with the exception of the c-C_4F_8 which was decomposing into lower fluorocarbons and was also probably lost through polymerization.

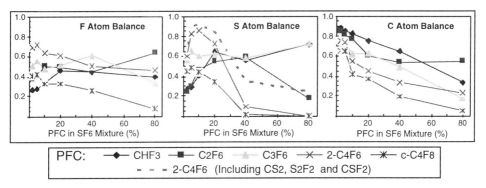

Figure 5. Ratio of atomic mass balance: byproducts formed / (SF_6 + PFC) consumed.

The same seems to happen with the other C_4 and the C_3-PFCs, as the quantity of lost carbon atoms (found as CF_4) is progressively increasing with the PFC content in the mixtures. Surely, polymerization and sulfurization-vulcanization of unsaturated PFC molecules can be invoked to explain the diminution of the recovered atoms, and especially the S atoms. In addition, more important quantities of some other gaseous byproducts (S_2F_2, S_3F_2 and CS_2) were analysed during the 40% and higher 2-C_4F_6 tests (to a lesser extend in those using c-C_4F_8), proving that a significant fraction of the SF_6 and PFC molecules are fragmented to their elements by the low-power arcs. The S-atom graph in Figure 5 shows that these three new products accounted for one third of the S• missing in those conditions.

Simple reaction mechanism

Despite the crude and violent source of chemical excitation chosen for these tests, the experimental data presented earlier provide sufficient evidence about the predictable reactivity of the SF_6-PFC mixtures. The measured evolution with time of all the mixture compositions was found to obey fairly well the following simple kinetic relations:

$$- d[SF_6] / dt = k' [SF_6] [PFC]$$
$$- d[PFC] / dt = k'' [SF_6] [PFC]$$

Obviously, the k's in these equations are not really constant but are function of the mixture composition. However, for PFC levels of 10% or less, the k' value certainly represents the relative propensity of one PFC to *virtually* accelerate the arc decomposition of an SF_6 molecule (Table 2). Indeed, the dominant initial chemical process is most likely the thermal dissociation occurring immediately around the arc channel.

The increase SF_6 consumption observed in the presence of PFC molecules is resulting from the trapping of its primary radicals, and especially the F•, by the fragments of the arc-decomposed PFC. Consequently, the two kinetic equations do not represent the absolute reaction reality but express the overall apparent processes, since there is no direct chemical interaction between the two gas mixture constituents.

Table 2. Relative decomposition rates of SF_6 in the presence of PFC (20% or less) (μmoles / J / %PFC)

PFC	Pure SF_6	CF_4	CHF_3	C_2F_6	C_3F_6	2-C_4F_6	c-C_4F_8
SF_6 absolute destruction rate	0.03	0.03	0.31	1.15	1.25	1.85	1.38
Rate relative to pure SF_6	1.0	1.05	9.9	35.0	36.9	55.3	41.4

Dilution effect of N_2 (ternary mixtures)

Further investigations were carried on to evaluate the behavior of diluted SF_6-PFC mixtures under the same low-power arcing conditions. Only one kind of mixture was tested, the one that showed previously the least chemical stability towards the low-power arcs: 60% - 40% SF_6-2-C_4F_6 previously

It is worth noting that the general reaction mechanism was not significantly modify by the dilution with N_2. Indeed, data in Table 3 indicate that the relative SOF_2 formation was still greatly reduced even with the highest diluted gaseous solution and where its quantity was also too low to be analyzed.

Table 3. Effect of dilution with N_2 on the SF_6 - 2-C_4F_6 mixture decomposition rate (*)

Composition (dilution ratio)	Decomposition - Formation Rate (μmoles/J)				Ratio of rates, relative to undiluted mixture		
(Mixture / N_2 (%))	SF_6	2-C_4F_6	SOF_2	CF_4	SF_6	2-C_4F_6	CF_4
1 : 1 (*pure mixture*)	5.74	5.70	0.32	7.33	1	1	1
1 : 2 (50% - 50%)	1.58	1.51	0.07	2.53	3.6	3.8	2.9
1 : 4 (25% - 75%)	0.26	0.23	- - -	0.48	22	24.8	16.3

(*) Mixture composition used: SF_6 - 2-C_4F_6 : 60% - 40%

Interestingly, the lowering of the total mixture is seen to be proportional to the square of the dilution factor. This is an interesting feature which play in favor of the usage of such ternary gaseous dielectrics, if other electrical characteristics are measured to satisfy the usual criteria.

CONCLUSIONS

This paper has discussed results of the decomposition and kinetics of six SF_6-fluorocarbon gas mixtures exposed to low-power arcing. In all the experimental tests performed, the thermal decomposition within the arc sheath was the dominant initial chemical process and all gaseous compounds added to SF_6 tend to accelerate its apparent rate. Fluorocarbons of higher molecular weight and those of cyclic or unsaturated structure were seen to form less stable mixtures with SF_6, when exposed to the electrical discharges.

SOF_2 and CF_4 were the major byproducts analyzed. These correspond to the principal gaseous decomposition species formed, respectively, by each of the two parent components of all the mixtures. According to the experimental data, the maximum decomposition rate all SF_6-PFC gas mixtures seem to occur at 40% PFC content or higher. In low concentrations (<20%), the C_3- and C_4-PFCs are those who increase most the arc-decomposition of SF_6. However, higher content of the C_4-PFCs were seen to inhibit the formation of SOF_2 from the decomposed SF_6. Polymerization reactions and other reaction paths are invoked to explain this curious finding that does not take place with the other unsaturated PFC tested (C_3F_6). In fact, higher PFC concentration mixtures were found to produce larger quantities of CSF-containing compounds ($C_2F_2S_2$, $C_2F_4S_2$, C_2F_6S, $C_2F_6S_2$, S_2F_2, S_3F_2 and CS_2). As such, the atomic mass balances computed from the amount of usual byproducts analyzed were always declining with the increase of PFC level in the SF_6 mixtures.

Lastly, the dilution of SF_6-PFC mixtures with nitrogen to form ternary gaseous dielectrics was found to reduce the overall mixture decomposition rate as the square of the dilution factor. This finding had support for the use or such new dielectric gas mixtures which may otherwise show synergetic improvements of their breakdown characteristics.

REFERENCES

1. L.G. Christophorou, D.R. James, I. Sauers, M.O. Pace, R.Y. Pai and A. Fatheddin, "Ternary Gas Dielectrics", Gaseous Dielectrics III: Proceedings of the Third International Symposium on Gaseous Dielectrics, Knoxville Tn, Pergamon Press, pp. 151-163, March 7-11, (1982).

2. L.G. Christophorou, I. Sauers, D.R. James, H. Rodrigo, M.O. Pace, J.G. Carter, and S.R. Hunter, "Recent Advances in Gaseous Dielectrics at Oak Ridge National Laboratory", IEEE Trans. on Electrical Insulation, Vol. EI-19 No. 6, pp. 550-566, (1984).

3. D.W. Bouldin, D.R. James, M.O. Pace and L.G. Christophorou, "A Current Assessment of the Potential of Dielectric Gas Mixtures for Industrial Applications", Gaseous Dielectrics IV Proceedings of the Fourth International Symposium on Gaseous Dielectrics, Knoxville Tn, Pergamon Press, pp. 204-211, April 29-March 3, (1984).

4. I. Sauers, T.J. Havens and L.G. Christophorou, "Carbon Inhibition in Sparked Perfluorocarbon-SF_6 Mixtures", J. Phys. D: Appl. Phys., Vol 13, pp. 1283-1290, (1980).

5. J. Castonguay and I. Dionne, "S_2F_{10} and Other Heavy Gaseous Decomposition Byproducts Formed in SF_6 and SF_6-Gas Mixtures Exposed to Electrical Discharges", Gaseous Dielectrics VII: Proceedings of the Seventh International Symposium on Gaseous Dielectrics, Knoxville Tn, April 24-28, Plenum Press, pp. 449-462, (1994).

DISCUSSION

S. MATSUMOTO: SF_6 and PFCs are greenhouse gases. Why do you choose such gases as the gas mixtures?

J. CASTONGUAY: The PFCs have certainly a "relative" greenhouse effect. However, the relative effect of PFCs that may eventually be added to the atmospheric burden will certainly be small due to typical utility losses. Even if the other actual PFC sources are stopped, the eventual PFC usage in electrical equipment will certainly be limited to "special cases" and should not be of significant concern in comparison to more important ones. For example, the SF_6 actual contribution to global warming is only ~ 0.08 % and even continuing at actual atmospheric rate of increase, it will still be less than ~ 2.5 % in 50 years.

E. MARODE: What kind of discharge regime have you as a plasma source? What is the current value? Is your plasma in an LTE (local thermodynamic equilibrium) state or would you think that you have a non-thermal plasma? If you increase the current and decrease correspondingly the exposure time would you have the same result?

J. CASTONGUAY: We called our discharge conditions as "sparks," but they could be considered as "low-power arcs." The oscilloscope observations showed a square voltage form of slightly fluctuating amplitude with mean value of ~ 240 V to 250 V. The current was ~ 250 mA. Thus, the spark-arc filament was very thin and serves as a thermal source to decompose the molecular constituent of the mixture. I think that at P = 1 atm, we had a thermal plasma and LTE. If we increase the current, we expect an increase of the decomposition rate from a few μmoles /J to what are obtained with high-power arc, i.e., 100 to 500 μmoles/J. The purpose of using this specific sparking setup is to get a simple, reproducible electrical discharge system for comparative chemical kinetic studies.

M. S. NAIDU: Generally, C_2F_6, C_3F_6, etc., are chemically very active. When mixed with SF_6, which is chemically very stable, what is the chemical stability of these mixtures?

J. CASTONGUAY: The stability of C_2F_6 or C_3F_6 - SF_6 mixtures is certainly less than for pure SF_6 and it also depends on the mixture composition. Also, C_2F_6 - SF_6 mixtures proved to be about twice as stable as C_3F_6 - SF_6 mixtures, due to the saturated nature of the former. In general, our present work has shown that C_2F_6 - SF_6 mixtures (percentage of SF_6 >20%) decompose 25 to 55 times faster than pure SF_6 under the sparking (low-power arcing) conditions used in those tests.

M. FORYŚ: What is the influence of ionic processes in the decomposition process?

J. CASTONGUAY: We do not think that ion-molecule reactions are of significance in our spark decomposition studies.

DIELECTRIC COATINGS AND PARTICLE MOVEMENT

IN GIS/GITL SYSTEMS

M.M. Morcos,[1] K.D. Srivastava,[2] M. Holmberg,[3] and S. Gubanski[3]

[1]Kansas State University, Manhattan, KS 66506, USA
[2]University of British Columbia, Vancouver, BC, V6T 1Z4, Canada
[3]Chalmers University of Technology, Gothenburg, Sweden

ABSTRACT

Gas Insulated Switchgear (GIS) and Gas Insulated Transmission Line (GITL) technology is now well established, with sulfur hexafluoride (SF_6) as the principal insulation medium. In field installations it has been observed that metallic particle contamination is often present in GIS/GITL and such contamination adversely affects the insulation integrity. Power industry has proposed, and in some cases utilizes, several methods to control and minimize the effect of particle contamination. One such technique is to apply a dielectric (high resistance) coating to the inside surface of the outer GIS/GITL enclosure.

Dielectric coatings help to improve the insulation performance in several ways. For example, it is known that dielectric coated electrodes in compressed gas give a somewhat higher breakdown voltage. It has been suggested that such coatings have the effect of "smoothing" the surface and reducing the pre-breakdown current in the gas gap. Also, in the presence of metallic particle contamination, it has been shown that the electrostatic particle charging is impeded and hence the maximum particle excursion in a coaxial GIS/GITL is significantly reduced, for a given applied 60 Hz voltage. A simple particle charging model, however, has some significant shortcomings, for example, the charge exchange mechanism between the particle and the electrodes is poorly understood and very complex to model. Some suggestions for improving the dynamic particle excursion model are discussed.

KEYWORDS

Compressed-gas insulation; particle contamination; particle movement; dielectric coating.

INTRODUCTION

Development of compressed gas insulated switchgear (GIS) and compressed gas insulated transmission line (GITL) equipment has progressed rapidly (Kopejtkova and co-workers, 1992). However, all practical GIS/GITL systems operate well below their theoretical insulation limit due to the deleterious effects of conductor surface roughness, metallic particle contaminants and the presence of support insulators. The presence of particle contaminants is by far the most significant factor for the deterioration of insulation integrity. Free conducting particles may lower the corona onset and breakdown voltage of a GIS/GITL system considerably. Contamination control in GIS can be achieved either by designing the insulation conservatively, or by providing designated low field areas inside as particle traps where the particles can be safely trapped.

Conductors in a GIS/GITL system may be coated with a dielectric material to restore some of the dielectric strength. Coating reduces the degree of surface roughness on conductors (Chee-Hing and Srivastava, 1975; Honda and co-workers, 1987; Morgan and Abdellah, 1988; Parekh and co-workers, 1979). Also, the high resistance of the coating impedes the development of predischarges in the gas, thus increasing the breakdown voltage (Chee-Hing and Srivastava, 1975). Coating thickness has been varied from a few microns to several millimeters, and the influence of coated electrodes on the insulation performance has been studied under dc, 60 Hz ac and lightning impulse voltages (Cookson, 1970; MacAlpine and Cookson, 1970; Skipper and McNeall, 1965).

PARTICLE MOTION

A metallic particle resting on a coated enclosure surface may acquire free charge through several physical processes, such as charging from an already charged dielectric surface, conduction through the coating and through a partial discharge (PD) between the particle and coating. There is experimental evidence to suggest that the PD mechanism dominates at lower gas pressures and thin coatings. The electric field necessary to lift a particle resting on the bottom of a GIS enclosure is much increased due to the coating (Parekh and co-workers, 1979). Once a particle begins to move in the gas gap under the applied voltage, it may collide with either conductor. If the conductors are coated, the particle will acquire a smaller charge, thus the risk of a breakdown initiated by a discharge is reduced significantly.

Particle Charging

The main purpose of the coating is to decrease the net charge on the particle, thus making the particle less influenced by the electric field. Charging a metallic particle on the surface of a dielectric coating is considered to be based on two different charging mechanisms;

   i) conduction through the dielectric coating (higher gas pressures)
   ii) microdischarges between the particle and the coating (lower gas pressures and higher electric fields).

Moving particles can also acquire net charge by contacting an already charged dielectric surface if the surface resistivity of the material is high enough to enable charges to be trapped on the surface.

Charging by Conduction. The charge mechanism based on conduction through the dielectric coating can be represented by a circuit model as shown in Figure 1. C_c and G represent the capacitance and conductance between the particle and the grounded electrode formed by the channel where the current flows through the coating. C_g represents the capacitance in the gap between the particle and the upper electrode. In a parallel plane configuration, the charge on a particle resting on a coated electrode can be expressed as (Anis and Srivastava, 1989; Hara and Akazaki, 1977; Parekh and co-workers, 1979),

$$Q(t) = \frac{V/\omega}{\left[\frac{1}{G^2}\left(1 + \frac{C_c}{C_g}\right)^2 + \frac{1}{\omega^2 C_g^2}\right]^{0.5}} \cdot \left[\cos\phi - \cos(\omega t + \phi)\right] \quad (1)$$

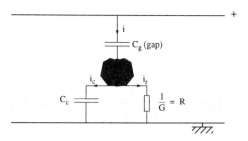

Figure 1. Circuit model of charging a particle situated on a dielectric coating through conduction.

where V is the applied voltage, ω the angular frequency and ϕ the phase angle between the applied voltage and the charging current. The particle lifts off when the maximum electrostatic force equals the gravitational force. The applied field required to lift the particle depends on the resistivity (ρ_c) and the thickness (T) of the coating material, according to the expression (Holmberg, 1995),

$$E_{LO} \simeq \left(\frac{\rho_c T}{S} \right)^{0.5} \qquad (2)$$

where S is the surface area of the particle in contact with the coating. The lift-off field is approximately proportional to the square root of the thickness and the resistivity of the dielectric coating, respectively.

Charging by Microdischarges. Another possible mechanism through which charges can be transferred to a contaminating particle from a coated conductor in a GITL is microdischarges (Parekh and co-workers, 1979). If the field near the particle is high enough discharges may occur between the particle and the dielectric coating, thus charging the particle. When the particle is in contact with the dielectric coating and the field is sufficiently high, electron avalanches will be initiated at a point on the particle surface and propagate along a field line towards the coating layer. The point on the particle surface at which avalanches will develop is that which ensures maximum avalanche size which is a function of both field and space.

In general, if the coating thickness is increased the electric field near the particle surface will be reduced, and consequently, a higher applied field will be necessary to lift the particle. In order to study the influence of the coating thickness on the magnitude of the electric field along the surface of a wire particle, finite element calculations were performed for a standing particle. A vertical wire particle was chosen since it is possible to fully account for its shape in the finite elements program.

Dielectric Coated Enclosure

If it is assumed that the free charge on a particle resting on a dielectric coating is due to a PD, it is possible to estimate the field necessary for lift off against gravity (Anis and Srivastava, 1989). Also, some experimental data exists (Morcos and Srivastava, 1996). Once levitated and before it returns to the enclosure, the particle movement can be described by a simple model. The charge acquired upon impact with the enclosure will depend upon the instantaneous voltage magnitude and polarity. It was assumed that if the instantaneous ac voltage is lower than that necessary for lifting, the particle would rebound without losing its charge and with a reduced velocity if the coefficient of restitution is < 1.0 (Prakash and co-workers, 1997). If the instantaneous voltage, on the other hand, is higher and of the opposite polarity, there could be a charge neutralization and the particle would acquire opposite polarity charge just sufficient to lift against gravity.

After the particle hits the outer enclosure, the field is computed from the applied voltage. Figure 2 shows the calculated effect of the applied voltage level on the maximum height reached by the particle for an uncoated and a 25 μm thick anodized enclosure for a coefficient of restitution of 0.95 and a total time of 100 s. Whenever the particle hits the outer enclosure, it rebounds and reaches a new larger height with random movement if the rebound velocity is substantial (Prakash and co-workers, 1997). Calculations were done for a system that corresponds very closely to the dimensions of a 138 kV GITL system. The aluminum wire particle has a diameter and length of 0.45 and 6.4 mm, respectively. This particle size was chosen since considerable data exists for this specific size from a previous study (Wootton, 1979). The inner and outer diameters of the coaxial system are 70 and 190 mm, respectively. For modeling, a wire particle may be represented by a longitudinal cylinder, hemispherically terminated at both ends or a semi-ellipsoid. The effect of drag is neglected since the particle velocities are small. The computational algorithm was developed to study particle motion with the applied 60 Hz ac sinusoidal voltage. In the case of uncoated electrode the coefficient of restitution is assumed to be 0.95.

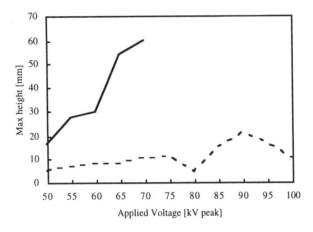

Figure 2. Effect of applied voltage on maximum height reached by an aluminum wire particle (0.45 mm diam/6.4 mm long) in a 70/190 mm GIS/GITL system (— uncoated,--- coated) for a coefficient of restitution of 0.95 (Prakash and co-workers, 1997).

For an uncoated enclosure, the maximum height reached by the particle generally increases with the applied voltage. For a coated enclosure, as the applied voltage increases, the maximum height reached by the particle does not necessarily increase; sometimes it decreases. This is because the maximum height reached depends on the velocity and the point-on-wave of the applied voltage at which the particle hits the outer enclosure.

Pseudo Resonance

It has been reported that, after a certain time, every time the particle hits the outer enclosure, the rebound velocity may be below a certain limit (Prakash and co-workers, 1997); the particle does not reach greater heights and gets captured very close to the outer enclosure. This leads to a 'pseudo-resonance' condition (Wootton, 1979). Pseudo-resonance occurs when the particle impacts the enclosure with a speed and at a phase angle of the electric field which is nearly same as on a previous bounce. The bounce sequence is more or less repeated several times. This pseudo-resonance may continue indefinitely and occurs most readily with bounces of small amplitude. The requirement for pseudo-resonance that the phase of the impact with the electrode be repeated almost exactly, often is not satisfied for bounces of large amplitude where particle may be in flight for several periods of the 60 Hz ac voltage.

PARTICLE LIFT-OFF MEASUREMENTS

Experiments were conducted in the High Voltage Laboratory at Chalmers University of Technology. Tests were performed with maximum applied voltage level of 250 kV, which corresponds to 16 kV/cm (rms) at the enclosure surface. Wire-like aluminum particles were used in the lift-off measurements. Particles were cut into desired lengths and no attempt was made to smooth the wire ends. The lift-off voltage was determined by applying a linearly rising ac ramp voltage (8-9 kV/s) and observing the particle when it lifted from its resting position on the enclosure. After the particle had been lifted, the voltage was lowered to a value low enough to make it rest on the surface in a horizontal position. The procedure was repeated approximately 40 times.

Figure 3 shows the lift-off voltage, and the calculated corresponding lift-off field at the enclosure surface for aluminum particles of 0.27 mm diameter and 6 mm long (Holmberg, 1997). Every lift-off value is marked by a dot. Both the measured mean and minimum lift-off values are displayed by continuous lines. The mean values are calculated using approximately 40 individual lift-off values. A zero coating thickness correspons to a bare aluminum sheet. The SF_6 gas pressure was 0.45 MPa. Often the first lift-off value in a series of measurements was above 250 kV. This voltage level was considered to be the upper permissible limit in order to avoid a flashover in the system since a flashover may damage the coating surface by creating a small crater in the dielectric material.

The variation of individual lift-off values from the mean value is sometimes more than 100%. The reasons for the spread are that the lift-off value depends on the surface of the particle in contact with the coating, on the microscopic angle between the coating and the particle, and on whether or not the particle starts its lift-off from an already charged surface area on the coating. The minimum lift-off levels increased between 30 and 100% from zero-coating thickness to 187 μm thickness, depending on the particle used. The lift-off levels for particles on the bare electrode (0 μm) are approximately 60 kV (4 kV/cm) and practically independent of the particle lengths.

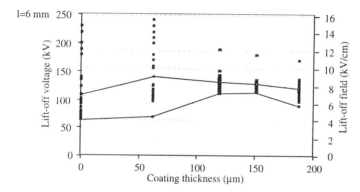

Figure 3. Calculated and measured values of lift-off voltage (field) for aluminum wire particle (0.27 mm diam/6 mm long) as a function of the coating thickness.

CONCLUSIONS

A metallic particle resting on a dielectric coated electrode will acquire less charge than a particle on a bare electrode subjected to an equivalent electric field (Anis and Srivastava, 1989; CIGRE, 1994; Parekh and co-workers, 1979). Depending on the properties of the coating, areas of surface charges may be formed due to the low conductivity of the materials that are generally used. These areas of surface charge on the coating can either adhere an oppositely charged particle or repel a particle with similar polarity. This physical process is random and very difficult to model. Moreover, some coating may also have mechanical adhesive properties. This is also difficult to model. It may, however, be noted that except for electrostatic repulsion from an already charged surface, the other two processes will tend to reduce the particle motion. That is, the simple movement model will provide a conservative prediction of the maximum excursion.

The work presented focuses on the lift-off and motion characteristics of an elongated metallic particle in a one phase GIS bus. Calculated and experimental results are presented with the aim of improving mathematical models of the particle motion. Simple models can usually be utilized for calculating the motion of a particle when the applied electric field is low. For higher fields, the influence of such parameters as carried charge, particle shape, and discharge activity around the particle become important. The influence of these parameters is often neglected in particle motion control. The maximum height reached by the particle for a dielectric coated enclosure is much less than for an uncoated enclosure, thus minimizing the deleterious effects of metallic particle contamination. For a smaller coefficient of restitution, the particle bounce sequence is repeated, leading to pseudo-resonance. More modeling and experimental work is in progress to evaluate further the usefulness of such coatings.

Acknowledgments

This research is supported in part from an NSF grant (No. ECS-9706297) and from the Natural Sciences and Engineering Research Council of Canada. Financial and technical support from ABB Sweden, and ABB Switzerland for the experimental part is greatly appreciated.

REFERENCES

Anis, H., and Srivastava, K.D., 1989, Breakdown characteristics of dielectric coated electrodes in sulphur hexafluoride gas with particle contamination, *Sixth International Symposium on High Voltage Engineering*, New Orleans, LA, Paper No. 32-06.

Chee-Hing, D., and Srivastava, K.D., 1975, Insulation performance of dielectric-coated electrodes in sulphur hexafluoride gas, *IEEE Trans.*, Vol. EI-10, pp. 119-124.

CIGRE WG 15.03, 1994, Effects of particles on GIS insulation and the evaluation of relevant diagnostic tools, *CIGRE*, Paper No. 15-103.

Cookson, A.H., 1970, Electrical breakdown for uniform fields in compressed gases, *Proc. IEE*, Vol. 117, pp. 269-280.

Hara, M., and Akazaki, M., 1977, A method for prediction of gaseous discharge threshold voltage in the presence of a conducting particle, *Journal of Electrostatics*, Vol. 2, pp. 223-239.

Holmberg, M., 1995, *Metallic Particles in GIS, Identification and Particle Lift-off/Fall-down on Coated Electrodes*, Technical Report No. 195L, Chalmers University of Technology, Gothenburg, Sweden.

Holmberg, M., 1997, *Motion of Metallic Particles in Gas Insulated Systems*, Ph.D. Thesis, Chalmers University of Technology, Gothenburg, Sweden.

Honda, M., Okubo, H., Aoyagi, H., and Inui, A., 1987, Impulse breakdown characteristics of coated electrodes in SF_6 gas, *IEEE Trans.*, Vol. PWRD-2, pp. 699-708.

Kopejtkova, D., Molony, T., Kobayashi, S., and Welch, I.M., 1992, A twenty-five year review of experience with SF_6 gas insulated substations (GIS), *CIGRE*, Paper No. 23-101.

MacAlpine, J.M.K., and Cookson, A.H., 1970, Impulse breakdown of compressed gases between dielectric covered electrodes, *Proc. IEE*, Vol. 117, pp. 646-652.

Morcos, M.M., and Srivastava, K.D., 1996, Control of metallic particle contamination in compressed SF_6 insulated switchgear by conductor coating, *Nordic Insulation Symposium*, NORD-IS 96, Bergen, Norway, pp. 371-378.

Morgan, J.D., and Abdellah, M., 1988, Impulse breakdown of covered cylinders in SF_6 - gas mixtures, *IEEE Trans.*, Vol. EI-23, pp. 467-472.

Parekh, H., Srivastava, K.D., and van Heeswijk, R.G., 1979, Lifting field of free conducting particles in compressed SF_6 with dielectric coated electrodes, *IEEE Trans.*, Vol. PAS-98, pp. 748-755.

Prakash, K.S., Srivastava, K.D., and Morcos, M.M., 1997, Movement of wire particles in compressed SF_6 GIS upon dielectric coating of conductors, *IEEE Trans.*, Vol. DEI-4, pp. 344-347.

Skipper, D.J., and McNeall, B.I., 1965, Impulse strength measurement of compressed gas insulation for extra high voltage power cables, *Proc. IEE*, Vol. 112, pp. 103-108.

Wootton, R.E., 1979, *Investigation of High Voltage Particle Initiated Breakdown in Gas-insulated Systems*, Electric Power Research Institute, Report No. IL-1007.

DISCUSSION

K. NAKANISHI: I presented the same kind of paper at the last meeting which studied the charging mechanisms of the particle placed on the dielectric coating electrode. In the case of particle charging the electrical conduction, the conductivity of the dielectric material should be high at application of AC and impulse applications. It is because the charge quantity for the particle lift-off is not obtained in a short time. So, if we use materials such as epoxy and polyethylene, the mechanism of particle charging should be through microdischarges.

M. MORCOS: We agree with Dr. Nakanishi that microdischarges between the particle and the coating are going to be the dominant mechanism. Also, there is experimental evidence, contained in references Rarekh et al. (1979) and Morcos and Srivastava (1996) of our paper showing the pressure dependence of the particle lifting field, thus strongly supporting the microdischarge mechanism.

H. HAMA: What kind of coating materials do you recommend for practical design?

M. MORCOS: Taking into account the manufacturing convenience, the coating material should be suitable for spraying. Therefore, epoxy coatings may prove to be most suitable.

BREAKDOWN CHARACTERISTICS OF A SHORT AIRGAP WITH CONDUCTING PARTICLE UNDER COMPOSITE VOLTAGES

R. Sarathi and M. Krishnamurthi

Department of Electrical Engineering
Indian Institute of Technology
Madras-36, India.

INTRODUCTION

The idealistic approach to insulation coordination involves selecting electrode clearance such that the largest surges can be safely withstood. The more practical approach to insulation coordination thus involves minimising electrode clearance while at the same time ensuring that number of flashover resulting from surges on the high voltage network is kept to acceptably small values. The insulation coordination is universally based on the standard lightning impulse voltage characteristics of the insulation system. This internationally agreed upon voltage waveshape is designated as 1.2/50 μs wave defining the time to crest as 1.2 μs and the time to subsequent decay to half value of the crest as 50 μs. This specific waveshape has helped in standardising the insulation levels of power apparatus and systems. In practice, however, a dielectric system is stressed by transient voltages of wide varieties of nonstandard waveshapes which are oscillatory in nature caused by lightning, line faults and switching operation. These non-standard waveshape are complex in nature identified by its frequency, decay rate etc. These surges are either unidirectional, sine/cos bi-directional. Oscillatory frequency produced by lightning is the order of several hundred Hertz and that produced by line faults and switching operations is one to 50 kHz [1].

With the increase of power transmission capacity, it has become necessary to design and develop a compact, highly reliable and cost effective systems and now the design safety margin are diminishing and the need to reliably predict the performance of a dielectric system under non-standard lightning impulse voltage has become very much necessary. In addition it is equally important to know about the breakdown characteristics of airgap with floating conducting objects, the reason being that open air has various floating particle such as insects, rain drops, leaves and other objects [2]. Kubuki et al.,[2] have studied the breakdown characteristics of airgaps with floating metal particle under the DC voltages. Hara et al.,[3] studied the particle initiated impulse flashover characteristics in airgap to understand the influence of floating metal particle on the breakdown characteristics. F.A.M. Rizk [4] analysed the influence of floating metal objects in long airgap under switching impulse voltages utilising mathematical physical problems.

In addition, in a DC power transmission system when a lightning strikes a tower or an earth wire, the airgap will be subjected to transient voltages, which adds the DC voltage across the airgap. Hence to understand the characteristics it necessitates to know the influence of the composite voltages formed by superposition of transient voltages, which will give a very clear understanding to the insulation engineer to provide proper insulation coordination.

A review of past research showed a lack of experimental data on the breakdown characteristics of airgaps with conducting particle under OIV. Also the influence of DC bias voltage on the breakdown characteristics of the airgap with the conducting metal particle with the composite voltages, the results available are scanty. Having this in mind in the present work, the influence of floating metal particle on the breakdown characteristics of airgap subjected to Unidirectional, Sine/Cos bi-directional OIV of different frequencies and dampings and with composite voltages formed by superposition of transient voltages especially under DC+LI, DC+SI and DC+OIV of different frequency and dampings have been studied for an airgap of 3cm length in a sphere-sphere configuration with floating conducting particle at different positions in the gap. For the purpose of comparison the results under AC, DC, LI and SI voltages are also presented.

EXPERIMENTAL

Fig.1 shows the schematic representation of the electrode system with the floating conducting object. The sphere-Sphere gap consist of two spheres of diameter 10 cm mounted at a height of 1m above the ground horizontally. A capacitor divider and a tektronics TDS-220 model digital oscilloscope was used for measuring the applied impulse voltages. The DC voltage was determined by measuring the current through 140MΩ resistor connected across the output of the DC voltage generator.

Fig. 2 shows the circuit used for the generation of UOIV and SBOIV and the CBOIV generated through Fig.3. Fig. 4 shows the circuit used to generate the standard LI and SI voltages. The required frequency and dampings[5] were obtained by suitably choosing the values of series inductance (L_s), shunt inductance (L_c), load capacitance (C_l), shunt resistance (R_s) and tail resistance(R_t). In the present context the decay rate is defined as the ratio of the magnitude of the first peak to the immediate next peak of same polarity.

Fig.5 Shows the circuit used to generate the composite voltages. It has a DC generator and a impulse generator circuit. In the case of combined voltages, both the voltage generators are uncoupled using the appropriate coupled capacitors and blocking resistance to obtain reaction free superposition of voltages.

In the present work, the 50 % breakdown voltage of air obtained by up and down method by applying 30 pulses. Each pulse applied at the gap of one minute to reduce the effect of preceding pulse on subsequent breakdowns. No humidity correction factor were applied, as the humidity correction factor recommended for lightning and switching impulse voltages, may not be applicable for the non-standard waveshapes[6]. The difference in breakdown voltage of oscillatory impulse voltage due to humidity variation must be less, as the humidity was between 70-80 % through out the series of the experiments.

The breakdown voltage under AC and DC voltages were determined in the following manner. First the voltage was raised at the rate of 3kV/sec up to 70% of the expected breakdown voltage and then at the rate of 1 kv/sec until breakdown occurred. The summation of 20 breakdowns is taken as the breakdown voltage for the configuration. The floating needle with a tip radius of 50 µm has been used. In the present work the floating metal particle were placed at a position by an insulating support aligning with the main gap axis.

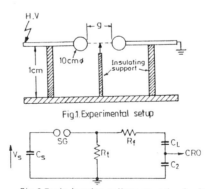

Fig.1 Experimental setup

Fig.2 Basic impulse voltage generator circuit

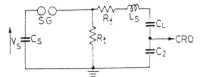

Fig.3. Uni- and sin-bidirectional OIV circuit

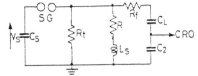

Fig.4. Cos-bidirectional OIV generator circuit

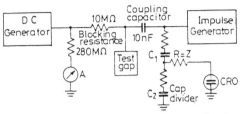

Fig.5. Composite voltage generator circuit

RESULTS AND DISCUSSION

The required fundamental information concerns the dielectric strength under slow front overvoltages which are more more crucial to EHV and UHV systems. In general the important results expected is to understand the critical impulse shape leading to lowest voltage breakdown for the examined configuration. The important aspect is the corona inception plays a major role in the breakdown characteristics of the gap. Hence perturbed field is developed and examined the influence of breakdown characteristics of airgap under different voltage profiles.

Influence of Frequency and damping of OIVs

Fig. 6 shows the variation in breakdown voltage as a function of location of the floating metal particle in the sphere-sphere gap under different frequencies and dampings of uni-directional, sine/cos bi-directional oscillatory impulse voltages. It is very clear that the breakdown voltage is much less when the floating metal particle is close to the positive electrode in the gap. The cause may be due to inception of positive corona discharge, reducing the breakdown voltage of the gap. Also it is observed at a particular location in the gap the breakdown voltage is the maximum and almost equal to the breakdown voltage under the clean airgap. This phenomenon is predominant under BOIV and under the higher frequencies of the UOIV. Also the breakdown voltage with floating metal particle in the sphere-sphere configuration reduced much especially under the UOIV compared to the BOIV. It is very clear, the influence of damping much observed in the case of UOIV whereas, under SBOIV it shows a slight increase in the breakdown voltage with increase in damping and a reverse tendency under CBOIV.

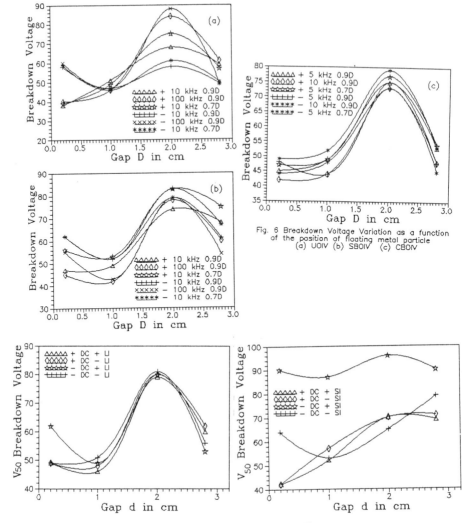

Fig. 6 Breakdown Voltage Variation as a function of the position of floating metal particle
(a) UOIV (b) SBOIV (c) CBOIV

Fig. 7 Breakdown Voltage Variation as a function of Floating Metal Particle Position Under Composite Voltage Profile.

Influence of Composite voltages

Fig.7 show the breakdown voltage variation as a function of position of the floating metal particle under the composite voltages formed under DC superposed with LI and DC superposed with SI voltages. The switching impulse voltages shows much of the polarity dependency. Fig.8 shows the variation in breakdown voltage of sphere-sphere configuration with floating metal particle at different positions in the gap under different levels of DC voltages superposed with UOIV, SBOIV and CBOIVs of different frequencies. The breakdown voltage increases with increase in the DC bias voltage especially when the DC bias voltage and the superposed voltages are of opposite voltages. This characteristics was observed especially under UOIV, whereas under BOIV, increase in DC bias voltage shows reduction in the breakdown voltage. Fig. 9 shows the variation in breakdown voltage as a function of damping of the OIVS. It is observed with increase in the damping of the OIvs shows increase in the breakdown voltage.

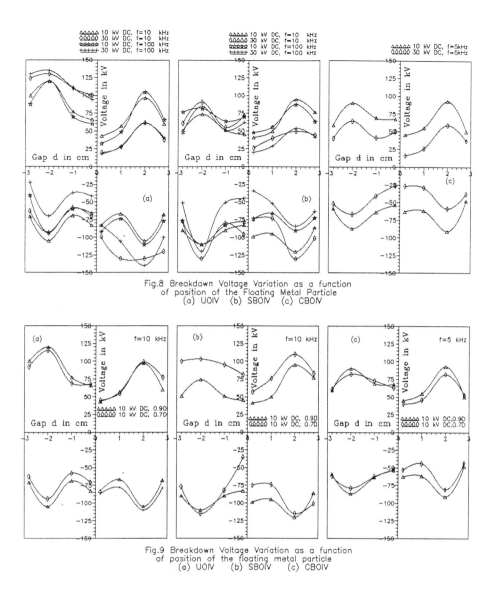

Fig.8 Breakdown Voltage Variation as a function of position of the Floating Metal Particle
(a) UOIV (b) SBOIV (c) CBOIV

Fig.9 Breakdown Voltage Variation as a function of position of the floating metal particle
(a) UOIV (b) SBOIV (c) CBOIV

Comparison between standard voltages and composite voltage profiles

Fig. 10 shows the breakdown characteristics of short airgap with floating metal particles at different positions under AC, DC, LI and SI voltages. It is observed the AC voltage have lower breakdown voltage. The composite breakdown voltage is less compared to the AC voltages when high DC bias voltage superposed with UOIVs especially both of positive polarity. Also the breakdown voltage of DC superposed with OIVs are much higher than the composite voltages formed under standard LI and SI voltages.

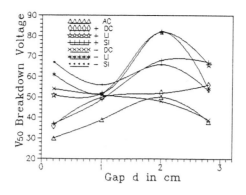

Fig 10 Breakdown Voltage Variation as a function of position of the floating metal particle Under Different Voltage Profiles

CONCLUSIONS

The important conclusions obtained based on the present work are the following
- The presence of floating metal particle reduces the breakdown voltage of airgap predominantly and the effect is more effective when the floating metal particle is much near to the positive electrode of the gap.
- At a particular position of the floating electrode in the gap, the influence on reduction in breakdown not much observed.
- Under the composite voltages formed the breakdown voltage is much less especially with the positive polarity voltages and at high positive DC voltages.
- Influence of damping of the OIV's is observed effective under the bi-directional oscillatory impulse voltages. Similar observation has-been observed under the composite voltages.

REFERENCES

1. R.C.Degeneff, IEEE Trans. on Power Apparatus and Systems, PP1457-59, (1982).
2. Kubuki, R. Yoshimoto, K. Tanowe and M. Hara, IEEE Trans. on Dielectrics and Electrical Insulation, Vol-2, PP155-165, (1995).
3. M. Hara and M. Akazaki, J. of IEE of Japan, Vol-91, PP 557-566, (1971).
4. F.A.M Rizk, IEEE Trans. on Power Delivery, Vol-10, PP1360-1370, (1994).
5. I.D Couper, A.Mejia and M.B.Abraham, IEE Fifth Int. Conf. on Dielectric Materials, Measurements and Applications, Kent, June, (1988).
6. Cigre Task Force 07-03 of study committee-33, Humidity influence on non-uniform field breakdown in air, Electra, No. 134, PP62-90, (1991).

SF$_6$ HANDLING AND MAINTENANCE PROCESSES OFFERED BY QUADRUPOLE MASS SPECTROMETRY

Constantine T. Dervos,[1] and Panayota Vassiliou [2]

[1] Department of Electrical and Computer Engineering
[2] Department of Chemical Engineering
National Technical University of Athens
9, Iroon Polytechniou Str
Zografou 157 73, Athens, Greece

INTRODUCTION

High Voltage (HV) insulation relying on the use of sulfur hexafluoride SF$_6$ has passed the required qualification tests to become the most commonly used insulating gas in electrical systems to date[1]. Gas Insulated Switchgear (GIS) power substations have all of their components interconnected and insulated via compressed SF$_6$, i.e. circuit breakers, dissconnectors, power switches, bushbars, current transformers, power transformers, cable insulation etc[2,3]. Due to their compactness and steel shielding, they offer significant savings in land use, are aesthetically acceptable, have relatively low radio and audible noise emission levels, and enable substation installation in cities very close to the loads.

While SF$_6$ insulated HV equipment adequately fulfill functional properties, unanswered scientific questions are currently the object of various research activities on SF$_6$ or related mixtures,[4] in the hope that the results will help to improve either the short or the long term material properties, the design and cost of GIS substations, installations maintenance and safety aspects. In situ monitoring and analysis of the gaseous insulant in operating GIS power plants[5] may aid effectively *(a)* equipment *reliability (b)* personnel *health* assurance, and *(c)* elimination of global *environmental effects*.

Systematic monitoring of the SF$_6$ byproducts[6] could be practically utilized to assure proper functionality of the GIS installation and maintenance requirements. For example, detection of metal fluorides might reveal fretting of field smoothing devices in which case, when the problem persists the employed epoxy resin insulators would develop enhanced surface conductivity, resulting to temperature fatigue and insulator degradation.

Decomposition byproduct yields are also expected to provide useful information concerning safe gas handling approaches during periodic maintenance process[7]. Specifically, concentration levels of S_2F_{10}[8] and oxyfluorides[9] in a SF$_6$ gas matrix could be used to predetermine the overall gas toxicity. Thus, personnel health may be ensured and accidents caused by acute exposure to SF$_6$ degradation products[10] would be prevented, provided that cytotoxicities of the related byproducts are kept well below the recommended threshold limit values (TLV)[11].

Finally, SF_6 leaking quantities towards the air of the working environment could also be suppressed by immediate detection, especially during the maintenance periods. Since SF_6 is an anthropogenically produced long lived compound that will accumulate in the atmosphere and persist for centuries or millennia[12] such steps are expected to have a significant effect on overall global warming potential and atmospheric "greenhouse" effect[13]. SF_6 recycling or reduction to environmentally compatible products has not yet been globally adopted, and as a natural consequence, its concentration levels in the lower stratosphere and upper troposphere increased over the last decades with annual rates of the order of 8.7%[14]. Its overall emission into the atmosphere is currently estimated as 5.9 ±0.2 Gg SF_6/yr[15] corresponding to a global warming contribution of the order of 0.01%.[16,17]

The purpose of this work is to demonstrate some of the direct advantages offered by continuous byproduct concentration monitoring of the gaseous insulant in a GIS plant. Gas monitoring has been accomplished in pilot test sites by quadrupole mass spectrometry[18].

EXPERIMENTAL

The reliability of the insulating gas of six 150 kV GIS power substations operated continuously over a period of thirty years has been investigated. As shown in Figure 1, the design of the substations incorporated single phase double pressure dead tank switches[19] commonly produced in late 60's. During all these years periodic inspection and maintenance (i.e. 13X filter substitution, gas dehydration and gas room cleaning) had been primarily based upon water dew point measurements. Recently, quality control and assurance of the insulating gas has been exploited by in situ mass spectrometry and byproduct production yield determination. Such measurements can be performed while the substation is in normal operation, thus offering operational safety and exact maintenance prognosis.

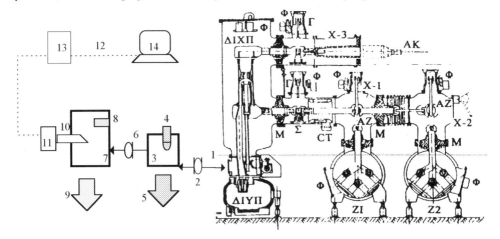

Figure 1. Typical view of the examined GIS plant and gas analysis measuring system.

Z1	Bus 1	AZ	Disconnecting switch	5	high vacuum pumping unit
Z2	Bus 2	Σ	Support insulator	6	needle valve
X-1	Room 1	M	Conical insulators	7	ultra high vacuum chamber
X-2	Room 2	ΔIXΠ	Power switch low pressure	8	cold cathode ion gauge
X-3	Room 3	ΔIYΠ	Power switch high pressure	9	turbo-drug pumping unit
Φ	Filter position	1	Flexible stainless steel	10	quadrupole mass analyzer
Γ	Earth switch		interconnect with pressure gauge	11	ion current amplifier (SEM)
CT	Current transformer	2	gas dozing valves at either end	12	data bus
AK	Cable bushing	3	prechamber	13	quadrupole electronics
		4	pressure gauge (2000 -10^{-3} mbar)	14	result storage and data bank PC

Gas samples could be obtained from all rooms of the GIS plant, Figure 1, according to the following two different modes: In the *pulsed sampling mode*, a series of pneumatically operated gas dosing stainless steel valves located along a flexible stainless steel interconnect enabled small gas quantities (30 cm^3 at a pressure of 1000 mbar) to be trapped in discrete steps and investigate mass flow phenomena. Alternatively, *continuous* gas circulation could be established via the prechamber and back to the GIS tank. This sampling mode could be used for gas analysis without any mass discrimination. The gas dosing system was heated to avoid gas condensation during mass flow towards the ultra high vacuum chamber under pressure variation. The interconnect network between the GIS rooms and the prechamber could be cleaned by prolonged pumping. Its state of cleanliness was monitored by analyzing the degassing spectrum. The pumping unit of the prechamber consisted of a 300 l/s diffstack high vacuum pump offering ultimate total pressure 10^{-6} mbar.

The gas to be analyzed was injected towards the ultra high vacuum (UHV) chamber via a bleed-in valve of the pumping station. The main chamber included the ion source, the quadrupole mass filter, the ion current detection unit, and the total pressure monitoring gauge. The UHV chamber was pumped via a Balzers 500 l/s turbo drug molecular pump backed by a 8m^3/h double stage rotary pump. The ultimate total pressure, measured by a cold cathode ionization gauge, was less than 1.10^{-9} mbar and it was increased by two orders of magnitude during gas sampling and analysis. Partial pressure measurements were performed by Quadrupole Mass Spectrometry (QMS). The employed analyzer head (Balzers 125) consisted of *(a)* an axial ion source where molecules could be ionized by 100 eV electron beam, thus producing positive ions, *(b)* a quadrupole mass filter unit, operating in the atomic mass range between $1 \leq$ amu/q ≤ 200, and *(c)* a Faraday collector offering minimum ion current detection 10^{-14} A. Lower current intensities could be monitored by incorporating the 90° off-axis Secondary Electron Multiplier (SEM) unit, which increased further the overall amplification factor, by an electron cascade effect, up to 10^6. The quadrupole electronics were interfaced to the data bus of a PC platform offering result storage, and further analysis.

Exact ion concentrations were determined by the corresponding ion current values in the obtained spectra, according to the following equation:

$$\text{ppm}_{(ion)} = I_{(ion)} \cdot 10^6 / P_{(mbar)} \cdot E_{(amu/q)}$$

where ppm$_{(ion)}$ is the ion concentration in ppm, $I_{(ion)}$ the detected ion current (A), $P_{(mbar)}$ the total pressure -measured in mbar- in the analyzing system, and $E_{(amu/q)}$ the quadrupole sensitivity (A/mbar) at the examined atomic mass unit over charge (amu/q) position.

The characteristic peaks of singly ionized fragments of pure dry SF$_6$ are encountered at the following amu/q positions: $F^+ = 19$, $^{32}S^+ = 32$, $^{32}SF^+ = 51$, $^{32}SF_2^+ = 70$, $^{32}SF_3^+ = 89$, $^{32}SF_4^+ = 108$, and $^{32}SF_5^+ = 127$. Doubly ionized SF$_6$ fragments may be detected at the following amu/q positions: $^{32}SF^{++} = 25.5$, $^{32}SF_2^{++} = 35$, $^{32}SF_3^{++} = 44.5$, $^{32}SF_4^{++} = 54$, and $^{32}SF_5^{++} = 63.5$. Singly ionized fragments formed by the ^{34}S isotope, having natural abundance 4.2%, are shifted by two mass units from the corresponding ^{32}S ions. For the doubly ionized fragments this difference turns out to be only one mass unit. $^{32}SF_3^+$ and $^{32}SF_5^+$ provide large characteristic peaks and therefore, suitable mass tracking sites for SF$_6$ detection in air. For the employed analyzing system, the minimum detection capability was of the order of 10 ppb.

For electrically stressed SF$_6$ gas matrix in a GIS environment, the most frequently mentioned molecular ion byproducts (^{32}S isotope) are detected at the following amu/q positions: $H_2O^+ = 18$, $N_2^+ = 28$, $HF^+ = 20$, $H_2S^+ = 34$, $CO_2^+ = 44$, $SOF^+ = 67$, $SOF_2^+ = 86$, $SOF_4^+ = 124$, $SO_2F^+ = 83$, $SO_2F_2^+ = 102$, $SO_2F_4^+ = 140$, and $S_2F_{10}^+ = 254$.

RESULTS AND DISCUSSION

Figure 2 provides typical successive spectra of the gas flow concentrations when a synthetic mixture between 90% SF$_6$ and 10% air, flows through a tubular (8mm diameter) longitudinal crossection simulating pressure flow parameters for the double pressure GIS power switch construction. It can be clearly noticed that the air constituents (i.e. mainly N_2 and O_2 molecules) first flow across the lower pressure end of the tube, followed by the heavier SF_6 molecules. This mass discrimination effect dominates the differential pressure gas flow of SF_6 mixtures with air constituents, and may be practically met during all tank refilling processes. On line gas flow monitoring will aid workers undertaking service tasks towards appropriate chamber gas refilling following gas dehydration and installation cleaning during maintenance. However, mass stratification has also been encountered in the pressurized volumes of the GIS plants. In this case, the included air and water quantities stagnate towards the highest sites of the GIS gas chambers, where filter positions should normally be implemented. Implications induced by stratification and mass discrimination effects may have to be systematically considered when innovative gas mixtures are introduced instead of a single insulating pressurized gas. In the event of binary or ternary gas systems such mass transfer behavior might also affect the overall arc extinction capabilities, especially in the older GIS installations. In the present case, gas flow from the high pressure (13 atm) tank towards the low pressure compartment incorporating the power switch (3 atm), had to be established to allow for the arc extinction. Alternatively, the same result can be attained by mechanical motion of the later introduced puffer designs[19].

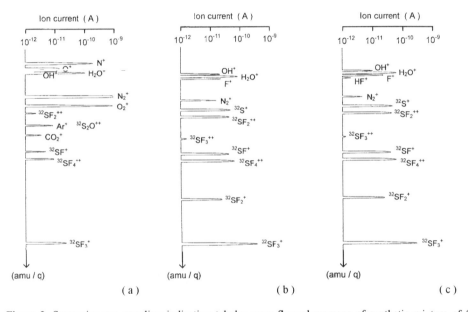

Figure 2. Successive gas sampling indicating tubular mass flow phenomena of synthetic mixture of SF_6 incorporating 10% air constituents. The air mass initially flows (**a**), followed by the heavier SF_6 molecules (**b**). Finally, as the air concentration is further reduced, (**c**) chemical reactions towards byproduct formation could be encountered i.e. OH^+ concentration reduction and introduction of HF^+ are noticed.

Information concerning the maintenance prognosis requirements can be practically attained by the CO_2 concentration levels in the SF_6 gas matrix. The CO_2 is gradually introduced by the interactions occurring between the SOF_2 species, always present in a large quantity of electrically stressed SF_6, and the molecular sieve type 13X,[20] which is the most popular sorbent in use in HV devices. This sorbent typically appears as superimposed SiO_2 and Al_2O_3 groups which together form a well-defined three dimensional structure with a 10 Å pore size. This property allows contaminants to diffuse and interact at all adsorption sites available on the surface, i.e. Si^{+4}, Al^{+3} and H^+. Sorbent exposure to oxyfluorides releases CO_2 species through preferential oxyfluoride adsorption process. Although it is well known that the sorption of CO_2 on the aluminosilicates is of complex nature, recently it has been experimentally pointed out that the small CO_2 amounts released during the early stages of the sorbent exposure to SOF_2 originates from the free species present in the pores[20]. The additional cummulative release is attributed to the displacement of initially sorbed molecules on the media, usually carbonate, the most favored formula for 298K applications being $Na_x^+(CO_3)^{-x}$ [21]. In practice, the aforesaid filter saturation process induced by prolonged sorbent exposure to high oxyfluoride levels can be exploited by the CO_2 concentration levels in the SF_6 gas matrix. Figure 3 presents typical spectra from GIS Room 1 (of Figure 1) monitored during plant operation before and after the filter material substitution. The reduction of the CO_2 concentration from 5 ppm to sub ppm levels (100 ppb) as a result of sorbent substitution is evident. It should be noted that for CO_2 detection in SF_6 matrix the characteristic peaks between CO_2 and SF_3^{++} differ only by 0.5 in amu/q scale.

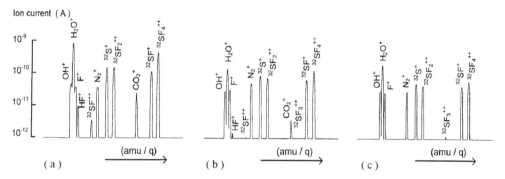

Figure 3. The effect of filter substitution on CO_2 concentration levels as a function of time in GIS Room 1 incorporating the current transformer. **(a)** before filter substitution; water dew point -23°C, **(b)** two days following filter substitution; water dew point -37.8°C and **(c)** 60 days following filter substitution; water dew point -38°C. Concentrations of SF_6 fragments are slightly reduced under identical measuring conditions.

Finally, the oxyfluoride levels may be directly detected at higher current sensitivities, as shown in Figure 4. The displayed results were selected from the low pressure power switch room, following a switch operation. Greater concentrations of detected oxyfluorides were recorded for the following amu/q positions: $SOF^+ = 67$, $SOF_2^+ = 86$, $SOF_3^+ = 105$, $SOF_4^+ = 124$, $SO_2F^+ = 83$ and $SO_2^+ = 64$.

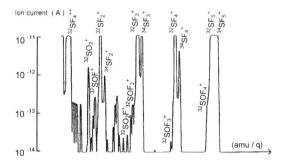

Figure 4. Oxyfluoride detection levels in the volume of the switch during a switch operation.

CONCLUSIONS

In situ sampling and analysis, by the QMS, of the pressurized insulating gas in a GIS power installation has been proved in practice to be a powerful tool concerning assurance and reliability. Precise prognosis of maintenance requirements may be based upon the SF_6 byproduct production yields. Evacuation and re-filling processes of the GIS rooms during periodic maintenance or implementation of innovative gas mixtures may be greatly assisted in practice by the mass flow detection. The oxyfluoride and disulfur decafluoride levels could be used to predetermine the overall gas cytotoxicity prior maintenance to allow for safe gas handling and good housekeeping approaches. Finally, leaks into the atmosphere can be detected to sub ppm levels at amu/q positions 89 or 127.

AKNOWLEDGMENTS

The authors would like to acknowledge for the collaboration with the Public Power Corporation, Greece, and specifically Director Dr. D. Stavropoulos for project implementation and Mr. A. Fraefel, Balzers, Liechtenstein, for offering technical support.

REFERENCES

1. W.T.Shugg. *Handbook of Electrical and Electronic Insulating Materials*, IEEE Press, New York, (1995).
2. D. Kopejkova, T. Molony, S. Kobayashi, and I.M. Welch. A twenty five year review of experience with SF_6 Gas-Insulated Substations, *CIGRE* paper 23-1091 (1992).
3. A. Elliker, H. Karrenbauer, and A. Lerondeau. Recent evolution of SF_6 circuit breakers and the impact on driving mechanism technology, *CIGRE* paper 13-303 (1990).
4. L.G. Christophorou, and R.J. Van Brunt, SF_6 / N_2 mixtures: basic and HV insulation properties, *IEEE Trans. Diel. Elec. Insul.* 2:952 (1995).
5. C.T. Dervos, I.C. Papaioannou, and J.M. Michaelides, Detecting aging byproducts of SF_6 using quadrupole mass spectrometry, 1997 Annual Report CEIDP, (Cat. No. 97CH36047):571 (1997).
6. I. Sauers, H.W. Ellis and L.G. Christophorou, Neutral decomposition products in spark breakdown of SF_6, *IEEE Trans. Elec. Insul.* 21:111 (1986).
7. G Mauthe, K. Pettersson, D. Gleeson, D. Konig, J. Lewis, T. Molony, P.O'Connell, A. Porter, and L. Niemeyer, Handling of SF_6 and its decomposition products in gas-insulated-switchgear (GIS), *Electra*, 136: 69 and 137: 81 (1991).
8. D.R. James, I. Sauers, G.D. Griffin, R.J. Van Brunt, J.K. Olthoff, K.L. Stricklett, F.Y. Chu, J.R. Robins, and H.D. Morrison, Investigation of S_2F_{10} production and mitigation in compressed SF_6-insulated power systems, *IEEE Electr. Insul. Mag.* 9:29 (1993).
9. R.J Van Brunt, Production rates for oxyfluorides SOF_2, SO_2F_2, and SOF_4 in SF_6 corona discharges, *J. Res. Nat. Bur. Stand.* 90: 229 (1985).
10. A. Kraut, and R. Lilis, Pulmonary effects of acute exposure to degradation products of sulfur hexafluoride during electrical cable work, *Brit. J. Ind. Med.*, 47: 829 (1990).
11. G. D. Griffin, K. Kurka, M.G. Nolan, M.D. Morris, I. Sauers, and P.C. Votaw, Cytotoxic activity of disulfur decafluoride (S_2F_{10}), a decomposition product of electrically stressed SF_6, *In Vitro*, 25: 673 (1989).
12. A.R. Ravishankara, S. Solomon, A.A. Turnipseed, and R.F. Warren, Atmospheric lifetimes of long-lived halogenated species, *Science*, 259: 194 (1993).
13. L.G. Christophorou and R.J. Van Brunt, SF_6 insulation: possible greenhouse problems and solutions, *National Institute of Standards and Technology*, Report 5685, (1995).
14. C.P. Rinsland, E. Mahieu, R. Zander, M.R. Gunson, et.al., Trends of OCS, HCN, SF_6, $CHClF_2$ (HCFC-22) in the lower stratosphere from 1985 and 1994 atmospheric trace molecular spectroscopy experiment measurements near 30 degrees N latitude, *Geophys. Res. Lett.*, 23 : 2349, (1996).
15. L.S. Geller, J.W. Elkins, J.M. Lobert, A.D. Clarke et.al. Tropospheric SF_6: observed latitudinal distribution and trends, derived emissions and interhemispheric exchange time, *Geoph. Res. Lett.*, 24: 675, (1997).
16. L.G. Christophorou J.K. Olthoff, and D.S. Green, Gases for electrical insulation and arc interruption: possible present and future alternatives to pure SF_6 *National Institute of Standards and Technology*, Technical Note 1425, (1997).
17. R.J. Van Brunt, Present and future environmental "problems" with SF_6 and other fluorinated gases, in 1996 Conference Record IEEE ICPS, (Cat. No. 96CH35939) :145 (1996).
18. P. Dawson. *Quadrupole Mass Spectrometry*, Elsevier, Amsterdam (1976).
19. J.S.T. Looms *Insulators for High Voltages*, Peter Peregrinus Ltd, London (1988).
20. T. Lussier, M.F. Fréchette, and R.Y. Larocque, Interactions of SOF_2 with molecular sieve 13X , CEIDP 1997 Annual Report, (Cat. No. 97CH36047):616 (1997).
21. H. Knöyinger, Specific poisoning and characterization of catalytically active oxide surfaces for n-pentane isomerization *Ad. Catal. Relat. Subj.* 2:497 (1970).

SECTION 7: ENVIRONMENTAL ASPECTS OF GASEOUS DIELECTRICS / RECYCLING

THE UNITED STATES ENVIRONMENTAL PROTECTION AGENCY'S SF$_6$ EMISSIONS REDUCTION PARTNERSHIP FOR ELECTRIC POWER SYSTEMS: AN OPPORTUNITY FOR INDUSTRY

Eric Jay Dolin[1]

[1]Program Manager,
Atmospheric Pollution Prevention Division
United States Environmental Protection Agency
Washington, D.C 20460

INTRODUCTION

Sulfur hexafluoride (SF$_6$) is an extremely important gaseous dielectric used by the electric power industry in circuit breakers, gas-insulated substations, and switchgear. SF$_6$ is also a very potent and persistent greenhouse gas that contributes to global climate change. Due to this second characteristic, the United States Environmental Protection Agency (EPA) is launching a voluntary SF$_6$ Emissions Reduction Partnership for Electric Power Systems. The partnership will work with the electric power industry to pursue technologically and economically feasible actions aimed at minimizing SF$_6$ emissions, thereby reducing the threat of global climate change. EPA plans to have the partnership operating by the Fall of 1998.

GREENHOUSE GASES AND GLOBAL CLIMATE CHANGE

EPA's concern about global climate change is rooted in science. Over the past century, the average temperature of the planet has increased roughly 1 degree Fahrenheit, and the eleven warmest years on record have all occurred since 1980. The primary cause of this rise in global temperature is the increase in the anthropogenic emissions of greenhouse gases, e.g., carbon dioxide (CO$_2$), methane (CH$_4$), perfluorocarbons (PFCs), hydrofluorocarbons (HFCs), nitrous oxide (N$_2$O), and SF$_6$. As these gases accumulate in the atmosphere, they act like a blanket, warming the planet by trapping heat that otherwise would be radiated from the earth back to space. The burning of fossil fuels (coal, oil and gas) by power plants for energy and fuel used for transportation is the primary source of these emissions; other industrial sources, changing land-use patterns through agriculture, and deforestation also contribute a significant share.

The need for society to take action to counteract global climate change became more urgent in 1995, when the Intergovernmental Panel on Climate Change (IPCC), a group of more than 2,500 of the world's most distinguished scientists with expertise in the physical, social, and economic sciences, concluded that "the balance of evidence suggests that there is a discernible human influence on the global climate."[1] For the first time ever, scientists agreed that the world's changing climatic conditions are due to more than the natural variability of the weather. They concluded that human beings are altering the Earth's natural climate system. Furthermore, even under the best scenario put forth by the IPCC, the warming of the planet's atmosphere is likely to accelerate at a rate "greater than any seen in the last 10,000 years.[2]

Concerns about global warming stem from what it may do to people and the environment. Warmer temperatures could lead to the increased frequency of extreme weather events, such as heat waves and droughts. Changes in precipitation and increased evaporation from higher temperatures could affect water supplies and water quality, posing threats to hydropower, irrigation, fisheries and drinking water. Agriculture and forest ecosystems may be hard hit by shifts in precipitation patterns and growing zones. Sea level could rise, placing heavily populated, coastal regions in jeopardy. And human health might be harmed through an increase in the rate of heat-related mortality and in the potential for the spread of diseases such as malaria, yellow fever and encephalitis. The United States and other countries throughout the world would have to invest large amounts of money to cope with these and other negative impacts.

SF_6 AND THE EVOLUTION OF NATIONAL AND INTERNATIONAL CONCERN

The United States government has focused considerable attention on reducing greenhouse gases since the 1992 Earth Summit, in Rio de Janeiro, when the U.S. and 172 other countries hammered out the Framework Convention on Climate Change, the goal of which was to "stabilize greenhouse gas concentrations in the atmosphere at a level that would prevent dangerous anthropogenic interference with the climate system."[3] As part of its commitment to the Rio Convention, the U.S. moved forward with a wide variety of programs designed to reduce its greenhouse gas emissions to 1990 levels by the year 2000. In the ensuing years, although the programs instituted by the U.S. government were successful in reducing emissions, they did not go far enough; it soon became clear that the 2000 goal set out in Rio would not be met. This was not just a problem for the U.S.. The vast majority of countries that signed the convention fell short of the convention's goals.

Between 1992 and 1997, the parties to the convention met three times, culminating in the December, 1997 meeting in Kyoto, Japan. At Kyoto, the United States and more than 150 other nations reached an historic agreement to join together to reduce greenhouse gas emissions by channeling the forces of the global marketplace to protect the environment. The agreement establishes strong, binding targets for reducing greenhouse gases, but allows each country the flexibility to determine the most cost-effective policies for meeting the targets. The U.S. negotiators agreed to reduce U.S. greenhouse gas emissions 7% below 1990 levels by the period of 2008-2012. Before the Kyoto agreement is ratified and these reduction targets become national goals, the agreement will require the advice and consent of the U.S. Senate.

The Kyoto agreement significantly raised the profile of SF_6, domestically and internationally. It is one of the six greenhouse gases that the agreement specifically targets for emissions reductions – the others are CO_2, CH_4, N_2O, HFCs, and PFCs. Estimates place annual U.S. emissions at roughly 8 million metric tons of carbon equivalent (MMTCE).[4] Worldwide production of SF_6 is on the order of 6,500 to 7,500 metric tons. Whereas a decade ago, the atmospheric concentration of SF_6 was barely perceptible, today it stands at roughly 3.2 pptv. Measurements indicate that the concentration of SF_6 is increasing at 7-9% per year, and could reach 10 pptv by 2010, depending on the assumed release rate.[5]

EPA's FOCUS ON ELECTRIC POWER SYSTEMS

The increased emphasis on SF_6 combined with the U.S. government's longstanding concern about gases with high GWPs is the driving force behind EPA's current efforts to establish a voluntary SF_6 Emissions Reduction Partnership for Electric Power Systems (utilities and cogenerating facilities). EPA is focusing on these systems because of the numbers.

According to one estimate, the electric power industry uses 80% of all the SF_6 produced worldwide.*[6] Ideally, none of this gas would be emitted to the atmosphere, either directly from operating SF_6 equipment or as a result of losses due to maintenance and/or recycling activities. In practice, however, there are emissions from these sources. Leakage rates from the equipment are largely a function of the quality of the sealants and enclosure porosity.[7] These factors vary widely depending on the age and make of the equipment. Some older, dual-pressure circuit breakers, for example, can leak at rates greater than that of more recent equipment. Newer SF_6 insulated electrical equipment is designed much tighter, with national and international standards calling for leakage rates on the order of 1% or less.[8]

If all SF_6 equipment leaked at 1% or less, annually, those emissions would still be a cause for concern for two reasons. First, SF_6 is an extremely potent greenhouse gas, with a global warming (GWP) potential of 23,900, over a 100-year time horizon.[9] That means that on a per-unit basis, and over a 100-year period, SF_6 is 23,900 times more effective at trapping infrared radiation than an equivalent amount of CO_2. A relatively small amount of SF_6 can have a significant impact on global climate change. Second, SF_6 is very persistent, with an atmospheric lifetime of 3,200 years.[10] As the gas is emitted it accumulates in an essentially undegraded state for many centuries. Therefore, a 1% annual leak rate for a piece of equipment in service thirty years is roughly equivalent, in terms of global warming potential, to the instantaneous release of 30% of the equipment's gas.

SF_6 is not a major greenhouse gas in terms of its impact on global climate change. For example, in 1995 CO_2 emissions in the U.S. were estimated at 1,305 MMTCE, 163 times as large as the emissions of SF_6 during that same year (roughly 8 MMTCE).[11] Nevertheless, SF_6's potency and persistence combine to make it a significant and increasingly important contributor to global climate change that must be addressed in any comprehensive approach to greenhouse gas reduction. From a broader perspective, SF_6 is much like other greenhouse gases, such as N_2O, PFCs, and HFCs, that have a comparatively small impact on climate change, but which we cannot afford to neglect because they are nonetheless critical to solving this environmental problem.

THE SF_6 EMISSIONS REDUCTION PARTNERSHIP FOR ELECTRIC POWER SYSTEMS

The primary goal of EPA's proposed voluntary SF_6 Emissions Reduction Partnership for Electric Power Systems is clearly indicated in its name. As with many of the other voluntary programs administered by EPA, this new program will ask partners to sign a Memorandum of Understanding (MOU) which details the roles of both EPA and the partner (in this case an electric power company) for furthering the goal of emissions reductions.

---

* The gas is also used as a cover gas in magnesium casting, a reactive gas in aluminum recycling operations, for thermal and sound insulation, in airplane tires, in scuba diving voice communication, and for selected consumer products.

As of this writing the MOU is still in draft form, but when finalized it will certainly include, at least, the following elements. For its part under the terms of the MOU, EPA will serve as a clearinghouse for technical information on successful strategies for reducing SF_6 emissions and will endeavor to fund additional research on such strategies. EPA will also provide partners with recognition for their achievements in reducing SF_6 emissions, serve as a credible repository for data on the emissions reduction achievements of the partners, and work to obtain commitments from all electric power system operators to join the partnership. The MOU will ask partners, in turn, to determine SF_6 emissions during the designated baseline, annually inventory and assess leaks of SF_6, submit annual progress reports, institute an SF_6 maintenance training program, mandate SF_6 recycling and establish an official corporate policy requiring the proper handling of SF_6. Although most of the MOU will be the identical for each partner, the section that details specific emissions reduction actions will be individually tailored to reflect the needs and abilities of the partner.

SF_6 Partnership Poised For Success

There are a many reasons why a voluntary program for SF_6 is likely to succeed. The first comes from EPA's past experience with other voluntary efforts. The agency implements eleven other voluntary programs, all of which are designed to reduce emissions of greenhouse gases.[12] The best known of these is the ENERGY STAR® Buildings and Green Lights® Program, which works with all types of organizations (e.g., companies, schools, non-profits, government) to improve their energy efficiency by installing energy-efficient technologies, such as lighting, heating, and air conditioning systems. As of February 1998, this program has developed MOUs with more than 2,600 partners, who have cumulatively upgraded the lighting in nearly 3 billion square feet of building space, saved more than $1 billion on their energy bills, and eliminated almost 20 billion pounds of CO_2 emissions to the air.

Another of EPA's programs is the Voluntary Aluminum Industrial Partnership (VAIP), an innovative pollution prevention program developed jointly by EPA and the primary aluminum industry, with the assistance of the Aluminum Association (U.S).[13] Participating companies (partners) work with EPA to improve aluminum production efficiency while reducing PFC emissions, potent greenhouse gases that, like SF_6, have high GWPs and are very persistent in the atmosphere. Since the VAIP program was launched in 1995, membership has grown to include 12 of the nation's 13 primary aluminum producers, representing 22 smelters and 94 percent of U.S. production capacity. While reduction goals vary according to site-specific conditions at each smelter, the partners have cumulatively committed to reduce PFC emissions, through various operational changes, 45 percent from the 1990 levels by the year 2000 -- roughly equivalent to 2.2 MMTCE. As of this writing, VAIP is already well over halfway to achieving this goal.

EPA's experience with these and other voluntary programs gives the agency confidence that it will be able to do as well with the new SF_6 partnership. Future success for one program, however, cannot be solely predicated on the past success of other programs, no matter how similar they are. Fortunately, with respect to the SF_6 Emission Reduction Partnership for Electric Power Systems there are other, much more specific reasons why EPA expects to be able to successfully work with the electric industry to reduce SF_6 emissions. First, in light of recent sharp increases in the cost of SF_6 there is now a strong economic incentive for industrial users to reduce the loss of gas to the atmosphere. Second, many in the industry have already taken steps to reduce emissions. EPA's new voluntary program is intended to compliment and build upon those actions, creating a valuable opportunity for industry and EPA to work together to achieve a common goal. Finally, voluntary programs provide a unique opportunity for industry to demonstrate proactive behavior outside of the typical environmental regulatory scheme.

As one of the gases included in the Kyoto agreement, SF_6 will be under mounting scrutiny in the years to come, by the U.S. as well as other parties to the framework convention. That makes it all the more important for the electric power industry to prove that it is part of the solution to the problem of global climate change. Assuming that the SF_6 Emissions Reduction Partnership for Electric Power Systems establishes a credible tracking and reporting scheme, the industry will be in an excellent position to reliably demonstrate its proactive behavior.

EPA's SF_6 partnership will not only result in emissions reductions, it will improve our understanding of the magnitude of the SF_6 problem. The numbers quoted earlier in this paper with respect to annual SF_6 emissions are our best estimate given current information. Unfortunately, information on emissions is limited. EPA's ultimate goal for the partnership is to sign up every utility and cogeneration company in the country. Since one of the pre-requisites for partnership is the provision of accurate emissions data, the more comprehensive EPA's program, the better the national database for SF_6 emissions and cumulative greenhouse gas emissions becomes.

CONCLUSION

The voluntary SF_6 Emissions Reduction Partnership for Electric Power Systems provides an exciting opportunity for industry to work with the EPA towards the common goal of reducing SF_6 emissions. EPA's intention is to make the partnership comprehensive, ultimately signing on all of the utilities and cogenerating companies in the U.S.. A comprehensive program will greatly improve the estimates of annual U.S. SF_6 emissions, significantly reduce such emissions, and contribute to ameliorating the problem of global climate change.

REFERENCES

[1] J.T. Houghton, et al., editors, *Climate Change 1995, The Science of Climate Change,* Cambridge University Press, Cambridge (1996): 4.

[2] Ibid, 6.

[3] Office of Global Change, U.S. Department of State. *Climate Action Report – 1997 Submission of the United States of America Under the United Nations Framework Convention on Climate Change,* U.S. Department of State, Washington, D.C. (July 1997): 1.

[4] Ibid, 56.

[5] L.G. Christophorou, J.K. Olthoff, and D.S. Green. *NIST Technical Note 1425, Gases for Electrical Insulation and Arc Interruption: Possible Present and Future Alternatives to Pure SF_6,* National Institute of Standards and Technology, Gaithersburg (November 1997): 3-4.

[6] Ibid, 3.

[7] CIGRE Task Force 23.10.01. *SF_6 Recycling Guide, Re-Use of SF_6 Gas in Electrical Power Equipment and Final Disposal,* CIGRE, Paris (August 1997): 3.

[8] Ibid.

[9] J.T. Houghton, et al., editors, *Climate Change 1995, The Science of Climate Change,* 22.

[10] Ibid.

[11] Office of Global Change, U.S. Department of State. *Climate Action Report – 1997 Submission of the United States of America Under the United Nations Framework Convention on Climate Change,* 56.

[12] Eric Jay Dolin. EPA's atmospheric P2 division succeeds with voluntary programs, *Pollution Prevention Review.* 7:75-86 (Autumn 1997).

[13] Eric Jay Dolin. U.S. on target for 40% PFC gas reduction by 2000, *Aluminium Today.* 9:54, 56 (August/September 1997).

DISCUSSION

H. MORRISON: I have two questions. (1) You spoke of a "basket of gases" in the Kyoto Protocol. Would you clarify how a country could meet its obligations under the protocol with respect to the gases in the "basket?" (2) Would you clarify the issue of selecting a base year for calculating emissions?

E. DOLIN: (1) The emissions target is referred to as a basket because it represents the sum of the six gases (converted to CO_2 equivalent terms using 100-year global warming potentials), rather than separate targets for each gas. Thus, cost-effective tradeoffs in emissions reductions among the gases are feasible, so long as the overall comprehensive target is met. A country could meet its obligations under the protocol through a number of means, including domestic emissions reductions, domestic activities to absorb carbon (e.g., planting trees), or international emissions trading. The United States led the effort in Kyoto to reject proposals that require all countries with targets to impose specific mandatory measures, such as energy taxes.

(2) The Kyoto Protocol sets a baseline year of 1990 for carbon dioxide, methane, and nitrous oxide. Parties have the option of either 1990 or 1995 for HFCs, PFCs, and SF_6. Use of these latter three gases in the U.S. and other countries has grown since 1990, so that permitting a 1995 baseline allows for a higher overall baseline than a uniform 1990 baseline.

B. DAMSKY: What fraction of the greenhouse gas effect is now "man-made" as opposed to "natural?"

E. DOLIN: Variations in both natural and human activities can affect climate change. Although most of the greenhouse gases occur naturally (except for HFCs, PFCs, and SF_6), the emissions and concentrations of all of these gases are changing as a result of human activities. Indeed, it is important to note that atmospheric concentrations of naturally occurring greenhouse gases were relatively stable since the end of the last ice age, 10,000 years ago, until relatively recently, when human inputs to atmosphere began having a significant impact on those concentrations (added to this, of course, are the impacts of the human-made greenhouse gases). To quote from the 1997 U.S. Climate Action Report (Department of State Publication 10496 – July 1997, page 4): "The atmospheric concentration of carbon dioxide has risen about 30 percent since the 1700s - an increase responsible for more than half of the enhancement of the trapping of infrared radiation due to human activities. In addition to their steady rise, many of these greenhouse gases have long atmospheric residence times (several decades to centuries), which means that atmospheric levels of these gases will return to pre-industrial levels only if emissions are sharply reduced, and even then only after a long time. Internationally accepted science indicates that increasing concentrations of greenhouse gases will raise atmospheric and oceanic temperatures and could alter associated weather and circulation patterns." Similarly, the Intergovernmental Panel on Climate Changes, which included more that 2,500 of the world's most distinguished scientists with expertise in physical, social, and economic sciences, concluded that "the balance of evidence suggests that there is a discernible human influence on the global climate" (Climate Change 1995, The Science of Climate Change, Cambridge University Press, 1996, page 5). Thus, the important point is that while greenhouse gases come from both natural and anthropogenic sources, it is the anthropogenic activities that have, in a relatively short time, led to the increase in greenhouse gas concentrations that has contributed to the increase in the average temperature of the planet, over the past 100 years, of roughly $1°$ F.

SF$_6$ RECYCLING IN ELECTRIC POWER EQUIPMENT

Lutz Niemeyer

ABB Research Center
CH–5405, Baden, Switzerland

Abstract

SF$_6$ has become a key insulation material in electric power transmission and distribution equipment. It allows to realize highly compact insulation and switching equipment which, due to its functional superiority, has replaced older insulation and switching technologies to a large extent. SF$_6$ has, however, recently been identified to be one of the strongest greenhouse gases. Unlike other substances with high environmental impact, SF$_6$ in electrical power equipment can not be substituted by an equivalent gas with better environmental performance. Although the SF$_6$ released hitherto does not contribute significantly to the greenhouse effect, concern remains about its long-term accumulation in the atmosphere as it has a long atmospheric lifetime. This makes it imperative to limit the application of SF$_6$ to closed systems and to reuse and recycle the gas systematically. In this contribution the basic concepts of handling and recycling the SF$_6$ in electric power equipment will be outlined based on a recently published committee document, the CIGRE SF$_6$ Recycling Guide. Major topics treated are the terminology used, the major SF$_6$ recycling fluxes, main features and performance characteristics of SF$_6$ reclaimers, quality control and categorization of used SF$_6$, basics features of handling SF$_6$ mixtures and environmentally correct final disposal of SF$_6$.

1. INTRODUCTION

Due to an exceptional combination of physico-chemical properties, SF$_6$ has become an indispensable insulation material for electric power transmission and distribution equipment. Due to its functional superiority it has replaced older insulation and switching technologies to a large extent. Particularly, high voltage transmission circuit breakers and gas insulated substations (GIS) are almost uniquely based on SF$_6$ as switching and insulation medium.

SF$_6$ has, however, recently been identified to be one of the strongest man-made greenhouse gases with a global warming potential about 25 000 times higher than CO$_2$ and an atmospheric lifetime of the order of 3000 years [1]. Extensive research to find better substitute gases for SF$_6$ (e.g. [2]) has been without success and it has been shown that it is difficult to substitute SF$_6$ by another gas without increasing the total environmental impact of the equipment [3].

Quantitatively, the contribution of the hitherto released SF$_6$ to the greenhouse effect is still negligible due to the small quantity in absolute terms [4]. Nevertheless, concern remains for the long-term future as the released gas, though released at a low rate, may accumulate in the atmosphere due to its long atmospheric lifetime. This makes it imperative to

(1) limit the application of SF_6 to closed systems
(2) minimize leakage from these systems
(3) minimize gas handling losses
(4) systematic reuse and recycle the gas
(5) develop processes for an environmentally correct final disposal of the gas

In electric power transmission and distribution equipment SF_6 is, by its very function, contained (point (1)) and minimal leakage (point (2)) is an imperative feature to satisfy reliability and availablity requirements. The present paper will therefore concentrate on points (3) and (4) and present a short outlook on point (5). Additionally, some basic considerations will be presented on the handling of gas mixtures containing SF_6 which are being used or considered to be used for special applications in electric power transmission and distribution.

2. BASIC CONSIDERATIONS

The scale of the SF_6 handling, reuse and recycling issue is characterized by the quantity of SF_6 that is presently stored in operating electrical equipment and by the rate at which SF_6 will be installed in new equipment in the forseeable future. From the estimates given in [5] the following figures can be extrapolated:

- about 35 000 metric tons of SF_6 are presently (1998) stored in operating electrical equipment
- 1500 to 2000 metric tons/year are presently installed in new equipment. This rate has been approximately constant over the last 5 years and is expected to stay constant in the foreseeable future.

The basic features of SF_6 handling in electric power equipment have recently been described in an SF_6 Recycling Guide prepared by a CIGRE committee [6]. This document will be referred to for details throughout the text.

3. DEFINITION OF TERMS

The technical terms used in the CIGRE Recycling Guide and in this text are summarized in **table 1**.

Gas handling operations include the evacuation of the equipment prior to filling it with SF_6, the inverse process of recovering the gas from the equipment for service or repair purposes and the purification of the gas and its intermediate storage until it can be reused on-site or off-site. These functions are provided by mobile gas handling devices which is referred to as **SF_6 reclaimers**.

Prior to handling, the gas has to be assigned, one of three **used gas categories** which define the handling modalities and personnel protection measures:

Non-arced gas is gas from an equipment compartment of which it is known that no arcing has occurred in it. In such gas, reactive decomposition products are absent or only present in small concentrations, if partial discharges have occurred. Personnel safety measures are not required.

Normally arced gas is gas that has been exposed to normal arcing in switchgear. Such gas usually contains relatively low concentrations of gaseous decomposition products as the switchgear is equipped with adsorbers designed to remove such products. However, substantial quantities of solid decomposition products ("switching dust") may have been generated so that personnel which opens the equipment has to be protected against this dust by gloves, security goggles and dust filter masks.

Heavily arced gas, i.e. gas which is known to have experienced heavy arcing by insulation failure ("internal arcing") or switchgear failure. Under these (rarely occuring) circumstances, large quantities of gaseous and solid decomposition products have to be expected requiring improved protection for personnel opening the equipment (protective overalls, gas masks).

Table 1. Technical terms used in SF_6 recycling

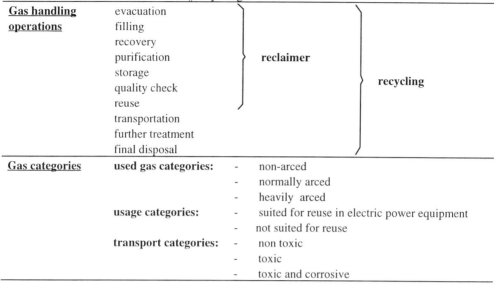

After having been processed by the reclaimer the gas has to undergo a **quality check** as a result of which it can be assigned one of the two **reuse categories** "suited for reuse in electric power equipment" or "not suited for reuse". In the latter case the gas requires further treatment which either renders it reusable or requires its final disposal. Further treatment and final disposal are normally carried out off-site by enterprises specialized in gas recycling or by SF_6 producers which offer such services. The totality of the above described operations will be referred to as SF_6 recycling. Gas which requires transportation has to be assigned a transport category according to its level of toxic and corrosive contamination. The three main **transport categories** are "non-toxic", "toxic" and "toxic and corrosive". The major **contaminants in used** SF_6, their origins and the concentrations above which they start to degrade the equipment function are summarized in **table 2**. For details see the CIGRE SF_6 Recycling Guide.

Table 2: Major contaminants in SF_6, origins and deteriorating effects on equipment function

contaminant	main origin	deteriorating effect	critical concentration
Non-reactive gases		reduction of insulation and switching performance	
- air	handling		5 % vol
- CF_4	arc erosion of polymers		
Reactive gases			
- WF_6, SF_4,	arc erosion	toxicity	>1000 ppmv
- SOF_2, SO_2, HF	hydrolysis with H_2O	toxicity and corrosion	>1000 ppmv
- SOF_4, SO_2F_2	partial discharges	toxicity	>1000 ppmv
Humidity H_2O	desorption from polymers	condensation on insulators → flashover	partial pressure of H_2O > 610 Pa
"Switching dust" CuF_2, WOF_4; WO_2F_2	arc erosion of contacts	adsorbed toxic gases	

4. MAJOR SF$_6$ RECYCLING LOOPS

The major recycling loops to which the SF$_6$ is subject in the course of its usage in electric power equipment are schematically represented in **fig.1**. The presently most important recycling **loop (1)** is the reuse of SF$_6$ on-site on the occasion of equipment maintenance or repair. In most cases the gas can be purified in a mobile reclaimer and immediately reused. A similarly important SF$_6$ flux is **loop (2)**, i.e. the installation of newly produced SF$_6$ in newly commisioned equipment. It is presently still important because most of the operating SF$_6$ equipment has not reached its end of life, so that there is not yet much gas available from the decommissioning of old equipment (**loop (3)**). This loop will, however, grow in the future. When used SF$_6$ is recovered from equipment, a very small fraction of it (estimated < 1 %) can not be purified for reuse on-site and requires further treatment off-site (**loop (4)**). Most of this gas can be rendered reusable and refed into equipment (**loop (5)**) and only a very small fraction of it has to be finally deposited (**loop (6)**). Only in the long term future when SF$_6$ might eventually become obsolete, loop (6) might increase in importance.

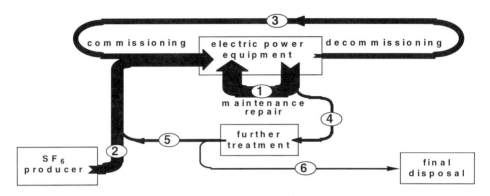

Fig.1: Major fluxes and recycling loops of the SF$_6$ in electric power equipment

In order to monitor the fluxes in fig. 1 it is important that SF$_6$ inventories and fluxes be systematically recorded in all areas of usage. Inventories of the SF$_6$ used in electric power equipment can be implemented with relative ease, as this equipment is not a consumer product but an investment good which is handled professionally. SF$_6$ inventories can be implemented in standard accounting procedures.

5. SF$_6$ RECLAIMERS

The devices serving for handling SF$_6$ on-site are called reclaimers. They have been developed through the last three decades and are commercially available for all application ranges in T&D equipment. Their design is essentially determined by the gas quantity to be handled and their operation mode has to consider the used gas category as defined in section 4. **Fig. 2** shows the **basic structure** of an SF$_6$ reclaimer. Its main elements are a vacuum pump which serves to evacuate the equipment prior to filling with SF$_6$, a compressor or compressore set which recovers the SF$_6$ from the equipment and compresses it into a storage tank, a dust filter which removes solid particles from the recovered gas, a gas/humidity filter which removes reactive gaseous decomposition products and humidity and a storage tank for the purified gas. An additional prefilter is inserted at the input of the reclaimer when strongly contaminated ("heavily arced gas") gas is expected.

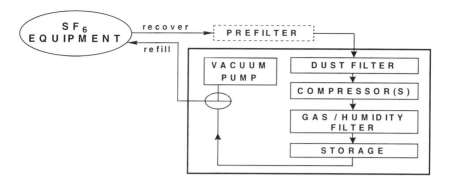

Fig. 2: Basic structure of SF_6 reclaimer

The major **performance characteristics** and features of reclaimers are summarized in table 3. A first characteristic is the **residual evacuation pressure**, i.e. the pressure p_{ev} down to which the air can be removed from the equipment before filling in SF_6. This pressure controls the air contamination introduced per handling cycle. The residual air which is not evacuated adds to the filled-in SF_6 as contaminant. The added air concentration per fill-in operation is

$$\Delta c_{air} = p_{ev} / p_f \tag{1}$$

As an example, with p_{ev} = 5 mbar and p_f = 5 bar the SF_6, 0.1 % air contamination is added per fill-in cycle and would accumulate to 1 % after 10 reuse cycles. The gas can thus be reused the more frequently the lower the residual recovery pressure p_{ev} is kept. A second performance characteristic, which is of environmental importance, is the **residual recovery pressure** p_{rr}, i.e. the pressure down to which the compressor can recover the SF_6 from the equipment, or, otherwise stated, the residual SF_6 pressure which has to be left in the equipment and is released into the atmosphere upon opening. Together with the filling pressure p_f, p_{rr} determines the SF_6 release rate λ per gas recovery cycle:

$$\lambda = SF_6 \text{ released} / SF_6 \text{ used} = p_{rr} / p_f \tag{2}$$

Modern reclaimers with two stage compressor systems allow to reach residual recovery pressures of typically p_{rr} = 50 mbar which, with a typical filling pressure of p_f = 5 bar yields

$$\lambda = 1 \% \quad \text{loss rate per recovery cycle for pure } SF_6 \tag{2a}$$

A third performance characteristic is the **storage capability** of the reclaimer, usually specified in terms of kg of SF_6. It has to be chosen such as to match the size of the power equipment for which it is used. Typical storage capacity requirements range from a few kg for small distribution equipment over several tens of kg for outdoor high voltage circuit breakers up to hundreds of kg for large GIS and GIL (gas insulated lines). The most frequently used storage containers are standard high pressure cylinders in which the SF_6 can be stored compactly in liquefied form at pressures of several tens of bars. This is why 50 bar compressors are normally used in reclaimers. An alternative storage concept is gaseous storage in large containers at pressures down to atmospheric. Also, cryogenic low temperature/low pressure liquid storage devices are available.

Table 3: Major characteristics of SF_6 reclaimers

Residual evacuation pressure	1 mbar[1]	1 mbar[1]
Residual recovery pressure	50 mbar[1]	50 mbar[1]
Recover-refill cycle time[2]	a few hours	
Storage capability	as defined by	equipment size

Recommended features:
- failsafe operation control
- connection to standard gas cylinders for storage
- built-in balance to determine cylinder weight
- cylinder heater for fast refilling
- oil-free compressors, pumps, sealings etc.
- gas hoses of 10 and 20 mm ø and 10 m lenght

[1] recommended by the CIGRE SF_6 Recycling Guide
[2] to be specified for given gas quantity to be processed and given gas hose (see fig. 3 below).

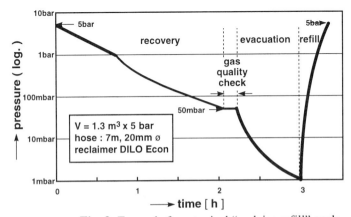

Fig. 3: Example for a typical "reclaim-refill" cycle

A further important reclaimer characteristic is the **"reclaim-refill" cycle time** which may be critical for factory testing, equipment erection and on-site testing and which determines the equipment outage time during maintenance and repair work. This cycle time depends on the reclaimer design and on the quantity of SF_6 to be processed. **Fig. 3** gives an example of a measured reclaim-refill cycle showing the time history of the gas pressure in the equipment. It is seen that in this case the total cycle time is about 3 hours half of which is required for gas recovery. The further entries in table 2 are self-explaining.

6. DETERMINATION OF GAS CATEGORIES

The SF_6 categories defined in section 2 can be defined with the help of a set of criteria and corresponding quality check measurements. The criteria comprise reuse and transport criteria and are summarized in **table 4**. They are expressed in terms of limit contaminant concentrations, expressed in terms of volume concentrations. The **reuse criteria** have been proposed in the CIGRE SF_6 Recycling Guide and are now in the process of being implemented in the international standard IEC 480 [7]. Note that the humidity levels are equal to those of new SF_6: SF_6 for reuse is required to be as dry as new SF_6. The limit concentrations of non-reactive gases are higher than in new SF_6 because the SF_6 performance as switching and insulation medium is insensitive against such contaminants. The limit concentrations of reactive decomposition products are determined by toxic risk considerations (see the CIGRE SF_6 Recycling Guide). It has to be noted that the reuse criteria discussed here are restricted to reuse in electrical power equipment.

Transport criteria were derived from existing international transportation regulations as shown in the CIGRE SF_6 Recycling Guide and can be specified in terms of the total concentrations of reactive decomposition products. The data given in table 4 are a simplified version of the CIGRE proposal which covers the vast majority of practically occuring cases. In practice, most of the SF_6 recovered from electric power equipment can be purified suffiently on-site to fall into the non-toxic category so that it can be transported like new SF_6.

Table 4: Reuse and transport criteria for SF_6

	Contaminant	Limit concentration		Gas category
Reuse criteria	Non-reactive gases (air, N_2, CF_4)	< 2% vol in switchgear < 5% vol in insulation		
	Reactive gases (SF_4, WF_6, SOF_2, SO_2, HF, SOF_2, SO_2F_2)	< 50 ppmv total or (eqivalently) < 12 ppmv SO_2+SOF_2 [1]		suited for reuse in electric power equipment
	Humidity H_2O	< 120 ppmv (liqufied SF_6) < 320 ppmv (at 500 kPa) < 1600 ppmv (at 100kPa)		
Transport criteria	Reactive gases (mainly HF, SOF_2, SO_2)	< 200 ppmv total	→	non-toxic
		> 200 ppmv total	→	toxic
		> 1000 ppmv HF and > 1000 ppmv H_2O	→	toxic and corrosive [2]

[1] SO_2 + SOF_2 are easily detectable indicator gases   [2] Corrosion-proof container required

The **quality check** of reclaimed gas consist in measuring the concentration of non-reactive gases (normally measured as SF_6 content), the humidity and the sum concentration of all reactive decomposition products. The devices for these measurements only have to function as level detectors, i.e. they have to indicate if the critical levels are exceeded or not. This may help to simplify these devices to keep their cost low. **Fig. 4** shows the optimal sequence in which the quality check measurements should be carried out to minimize working time. The figure also indicates the gas category assignment as result of the measurement.

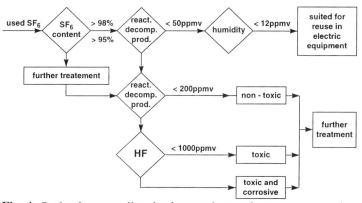

Fig. 4: Optimal gas quality check procedure and gas category assignment

7. HANDLING OF SF_6 MIXTURES
7.1 General considerations

Admixtures of low boiling point gases to SF_6 are being used in equipment exposed to low ambient temperatures in arctic regions to avoid SF_6 liquefaction. Such a reduction of the SF_6 partial pressure is compensated by "filling up" with a non-liquefying admixture gas such as nitrogen or CF_4. The SF_6 concentration is typically of the order of 50 % in these applications.

Another concept presently under investigation is the use of diluted SF_6 mixtures with nitrogen as background gas. This concept exploits the strong insulation synergy between these two gases due to which the SF_6 enters the insulation performance of the mixture overproportionally [8]. The advantage of such a concept is that the absolute quantity of SF_6 required to fulfill an insulation function is reduced. As an example, with a 10 % SF_6 - 90 % N_2 mixture, about 65 % of the insulation capabilty of pure SF_6 is reached. At a given gas pressure, this requires that the insulation distances in the system be increased by a factor of 1: 0.6 = 1.7 which, for a coaxial system, corresponds to a gas volume increase by a factor of $1.7^2 = 2.9$. Thus, about two thirds of the SF_6 that would be required with pure SF_6 insulation could be saved. From an environmental point of view it is, however, important to note that it is not the quantity of SF_6 *used* which is critical but the quantity *released*. This means that also the losses associated with the handling of the gas mixture must be lower than when using pure SF_6. In order to quantify this issue, we use, as benchmark value for pure SF_6, the handling loss rate of 1% per cycle according to eq. (2a). A loss rate of 3 % would then be the breakeven figure for the 10 % SF_6 mixture as the latter only requires one third of the SF_6. Hence, to reach an environmental advantage, the SF_6 release rate when handling a 10 % mixture should be

$$\lambda \ll 3\% \quad \text{per handling cycle for 10 \% } SF_6 \text{ mixtures} \tag{3}$$

The presently available SF_6 reclaimers are not suited for handling SF_6 mixtures, particularly diluted mixtures, as they are mostly based on compression liquefaction and as their compressors are insufficient to liquefy the SF_6 in the mixture. Reclaimers for handling low percentage mixtures are presently under development at various reclaimer producers. This paper will therefore be limited to discussing some basic aspects of gas mixture handling. A key quantity in this issue is the vapor pressure curve of SF_6 which is shown in **fig. 4** [9]. It gives, on the left side scale, the partial pressure p_v of SF_6 at which condensation begins in dependence of the temperature T_c to which the gas is cooled. Note that the condensate is only liquid down to the triple point temperature of SF_6 which is $T_m = -51\ °C$ and solid below this value. The two right side scales show the total pressure of the mixture above which SF_6 condensation sets on. The scales refer to mixtures with 10 and 50 % vol of SF_6, respectively. As an example, at ambient temperature $+20\ °C$, pure SF_6 liquefies above 2 MPa, a 50% SF_6 mixture above 4 MPa, and a 10% mixture above 20 MPa.

The basic concepts for handling and recycling SF_6 mixtures can be subdivided in three main groups which are characterized in **table 5**. The major evaluation criteria for choosing among these concepts are feasibility, cost and the SF_6 release rate as the key environmental characteristic.

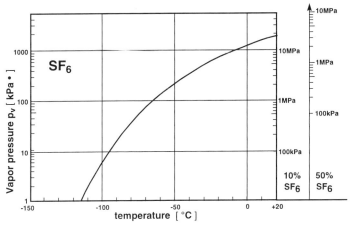

Fig. 5: Vapor pressure curve of SF_6. Pressure scale on the left side refers to pure SF_6 or to the SF_6 partial pressure. Pressure scales on the right refer to total mixture pressures with 50% and 10 % vol of SF_6.

Table 5: Basic concepts for handling SF_6 mixtures

Basic concept	method	main problems
Recycling of unseparated mixture	handling in gaseous state at ambient temperature	high storage volume required
SF_6 separation	a) condensation of SF_6 by compression and release of admixed gas	high compression and low temperatures required, handling of solid condensate
	b) sorption/diffusion process - selective sorption - membrane separation - thermodiffusion - gas centrifuges	high release rate process time ?
Chemical SF_6 destruction in mixture	a) decomposition at hot surfaces b) discharge decomposition c) chemisorption	process time ? thermal activation required

7.2 Handling and recycling of unseparated mixtures

The simplest concept consists in handling the unseparated gas mixture. This implies that the gas has to be stored as mixture in the gaseous state which requires a high storage pressure times volume (pV) product but has the advantage that the cost and problems of separation devices are avoided. In order to illustrate the storage requirements, we take liquid storage of pure SF_6 in standard gas cylinders as a reference. They allow to store about 40 kg of liquid SF_6 in a volume of V = 50 liters at a pressure of p = 2 MPa at + 20°C, hence with a pV product of 100 MPa liter. The same quantity of SF_6 admixed to 90 %vol nitrogen requires a storage pV product of 6 700 MPa liter which is almost two orders of magnitude higher.

7.3 SF_6 separation

Separation of the SF_6 from the mixture has the advantages of compact SF_6 storage in liquefied form and reuse for arbitrary purposes but has disadvantage of requiring additional cost for a separation device. The separation device can be based on two types of physical processes, namely, separation by SF_6 condensation and sorption/diffusion processes.

Separation by SF_6 condensation can be realized by compressing the mixture to above the SF_6 condensation pressure (fig.4, right side scales), withdrawing the liquid or solid SF_6 and releasing the cushion of admixed gas remaining above the condensated SF_6. The feasibility and efficiency of this process are determined by the compression pressure p_c and the temperature T_c to which the gas is cooled. The release rate associated with this separation process is determined by the residual SF_6 partial pressure $p_v(T_c)$ in the released gas cushion and is

$$\lambda_{mix} = SF_6 \text{ in the gas cushion / total } SF_6 \text{ in mixture} = p_v(T_c)/(n\, p_c) \qquad (4)$$

where n is the SF6 concentration in the mixture. The release rate λ_{mix} thus decreases with decreasing cooling temperature T_c and increasing compression pressure (i.e compressor performance). **Fig. 6** shows a plot of the loss rate λ_{mix} in dependence of the cooling temperature

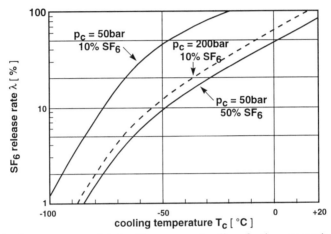

Fig. 6: SF_6 release rate associated with compression liquefaction separation process

T_c for the typical SF_6 concentrations of 10 % and 50 % (solid curves). The broken curve is valid for 10 % SF_6 and a compression pressure of p_c =200 bar. The benchmark value for the release rate of 3 % defined in eq. (3) can only be reached with cooling temperatures in the -100 °C range, i.e. by a cryo-technology which would also have to handle a solid SF_6 condensate.

Separation by sorption and diffusion processes such as selective sorption, membrane separation, thermo-diffusion and gas centrifuge technologies still have to be evaluated.

7.4 Chemical removal of the SF_6 from the mixture

In this concept, the gas mixture is brought in contact with a chemical reactant which transforms the environmentally critical components of the mixture (SF_6, CF_4) in substances with much lower or no environmental impact. Although a detailed concept for SF_6 has not been published, to the knowledge of the author, some preliminary information can be obtained from the chemical literature. Thermodynamically (i.e. based on the Gibbs enthalpy), SF_6 can react with a number of substances when the reaction is appropriately activated. **Thermally activated SF_6 reactions** with metals and metal oxides were studied in [10] where it was shown that activation temperatures above 500 °C are required. They transform the SF_6 in environmentally compatible fluorides, sulphides and sulphates. Decomposition at **hot surfaces** which was observed at temperatures below 300°C [11] suggest that catalytic decomposition processes might be eyploited at lower temperatures. Decomposition by **discharge activation** in a microwave discharge was demonstrated to be efficient for fluorinated compounds [12] and might also be applicable to SF_6.

8. FINAL DISPOSAL OF SF_6

If SF_6 can not be efficiently recycled or should not be required any more it can be re-transformed into natural substances by various chemical processes. They include the thermally activated processes mentioned above. A typical SF_6 conversion reaction is

$$SF_6 + 4\,CaO \rightarrow 2\,CaF_2 + CaSO_4 \qquad (5)$$

where CaO is lime, CaF_2 is fluorspar (used in toothpaste) and $CaSO_4$ is gypsum. At present, the only industrially developed SF_6 destruction process is burning in high temperature kilns [13], a processes which is also used for the destruction of chlorinated fluorocarbons (CFCs). The injection of SF_6 as co-flow in domestic waste incinerators is presently under study.

9. CONCLUSIONS

(1) To avoid long-term accumulation of SF_6 in the atmosphere, this gas must not be released.
(2) The SF_6 used in electric power equipment is, by its very function, enclosed. It can and must be systematically reused and recycled.
(3) Mobile SF_6 reclaimers for the correct and efficient handling of the gas are fully developed and commercially available. They allow to reuse most of the SF_6 on-site.
(4) Procedures and quality criteria for handling, reuse and transportation of used SF_6 from electric power equipment have been established and are being included in international standards.
(5) SF_6 which can not be reused on-site can still be rendered reusable, in most cases, by further treatment off-site. The corresponding technology is available at industrial level.
(6) If necessary, SF_6 can be re-transformed into natural materials by various chemical processes one of which is already industrially available.
(7) Mixtures of SF_6 with other gases can not be handled with presently available reclaimers. Various processes are under evaluation and equipment is being developed.

10. REFERENCES

1. "Radiative Forcing of Climate Change", the 1994 report of the scientific assessment working group of IPCC, P.28.
2. L.G. Christophorou, I. Sauers et.al., " Recent Advances in Gaseous Dielectrics at Oak Ridge National Laboratory", IEEE Trans. Electr. Insul.,19 (6) (1984), 550-566"
3. L. Niemeyer, "A systematic search for insulation gases and their environmental evaluation", 8th Internat. Sympos. on Gaseous Dielectrics, Virginia Beach, June 1998
4. CIGRE WG 23-10 " SF_6 and the global atmosphere", ELECTRA 164 (1996), 121-131
5. UNIPEDE/CAPIEL (EU associations of manufacturers and utilities): "Position paper on the influence of SF_6 switchgear on the greenhouse effect", March 1997
6. CIGRE WG 23-10, " SF_6 Recycling Guide", ELECTRA 173 (1997) 43-69
7. IEC (International Electrotechnical Commission) Publication 480, 1974
8. L. G. Christophorou and R. J. Van Brunt, "SF_6 - N_6 mixtures", IEEE Trans. Dielectr. Electr. Insul., 2 (5) 1995, 952-1003
9. I. M. Bortnik, "SF_6 thermophysical properties: engineering approach", Proc. of the 3rd Internat. Sympos. on Gaseous Dielectrics, Knoxville 1982, Pergamon, 533-543
10. A.A. Opalovsky and E. U. Lobkov, "Interaction of sulphur hexafluoride with metals and oxides", J. Fluorine Chem., 2, (1972/73), 349-359
11. K. Hirooka and M Shirai, "Thermal characteristics of SF_6", J. Chem. Soc. Japan, 2 (1980), 165-169
12. G. A. Askar'yan et. al., "Efficien chlorofluorocarbons (CFCs) destruction in a microwave discharge", 11th Intern. Conf. on Gas Discharges and their Applications, Tokyo, 1995, II-438
13. The recycling company SAKAB (Gustavsberg, Sweden) offers an SF_6 destruction sevice using a thermal process.

DISCUSSION

B. DAMSKY: What standards exist, if any, concerning the allowable heating of SF_6 cylinders to extract all gas?

L. NIEMEYER: I have no specific knowledge about this issue, but I would expect that the national standards for pressure containers should cover the formal aspects. The heating of the cylinders has merely the function to keep the liquid SF_6 from being cooled below the freezing point (−51 °C). It is to provide heat and not a high temperature.

H. MORRISON: (1) The slide you showed of the cycle time to proceed from a full compartment, evacuated and refilled seems to leave out a couple of steps. After the SF_6 is removed to ≤ 50 mbar then the equipment must be filled with air and opened for maintenance. After maintenance the compartment is closed and evacuated to ~ 1 mbar to remove air and humidity. The compartment must then be left at 1 mbar for a while, perhaps an hour or several hours, to check the vacuum integrity before finally refilling with SF_6. This time is all separate from the maintenance time. (2) The problem of cryogenic separation has been solved for many years for other gas systems, such as hydrogen or carbon monoxide for isotope separation. Known as cryogenic distillation, there are many "plates" of separation in series to achieve an arbitrary separation of the components. You have shown in your slide only one "plate" of separation. What should be done is to investigate the effectiveness and size of a cryogenic still needed to separate SF_6 from N_2. Other gas mixtures would require similar study. For example, cryogenic stills for hydrogen separation can be a few meters tall, while for carbon monoxide separation some systems are on the order of 100 meters tall.

L. NIEMEYER: (1) You are right. The cycle-time diagram contains only the items characterizing the SF_6 handling process. Other process steps like repair or maintenance and leakage checks which do not relate to the SF_6 handling process, have to be accounted for separately. (2) I am not sufficiently familiar with fractional distillation separation techniques to give a precise answer, but it seems to me that there must be a range of state parameters (pressure, temperature) in which both SF_6 and N_2 are liquids. Such a range does not exist as the critical temperature of N_2 (~ −160 °C) is much smaller than the triple point temperature of SF_6 (~ −51 °C). Thus, the only coexistence of condensed SF_6 and condensed N_2 can be SF_6 "snow" solid in liquid N_2 below ~ −160 °C. I feel that it might be difficult to fractionally separate such a "snow-mud" system, at least practically.

B. DAMSKY: Does your suggestion of removing SF_6 from a N_2/SF_6 mixture by hot chemical reaction mean that the user would lose the SF_6 and so suffer an economic loss?

L. NIEMEYER: Yes, the user would lose the SF_6 but I think that the economic loss is to be weighted against the ecological advantage of not releasing the SF_6. The global warming potential of SF_6 is by far the dominant property of SF_6 in the frame of environmental impact analysis according to ISO 14040. The economic loss is probably negligible with respect to the total cost of the insulation system.

STUDY OF THE DECOMPOSITION OF SF_6 IN THE LOWER ATMOSPHERE: THE EXPERIMENTAL APPROACH

Jean-Marc Gauthier[*] and Jacques Castonguay

Institut de recherche d'Hydro-Québec
Varennes, Québec
CANADA, J3X 1S1

ABSTRACT

Sulfur hexafluoride is considered a very long-live (800-3000 years) gas species that has accumulate into the atmosphere due to leaks and releases during the last three decades. About 80 kT of SF_6 is now dispersed uniformly, mainly in the troposphere, where there seems to be no known sinks for it. An experimental program was started to specifically study the chemical stability and reaction(s) of the 'inert' SF_6 molecule in the conditions of the lower atmosphere (<10 km). It was focused on the almost exclusive decomposition path resulting from the dissociative capture of an electron. Chemical behavior of SF_6/air mixtures, with concentrations ranging from 1000 to 1 ppmv, was studied in specially designed cells and setups permitting the interaction of SF_6 and low-energy electrons. These were generated in situ by photo-emission from the inner aluminum walls of the cells exposed to UV irradiation. Indeed, the SF_6 decomposition was observed but proceeding at very low rates, with sulfuryl fluoride (SO_2F_2) being the only byproduct found. Its quantitative analysis at the ppbv level had needed the development of a sensitive analytical method. Finally, a simple reaction mechanism is proposed to describe the SF_6 chemical fate in the above experimental conditions.

INTRODUCTION

Sulfur hexafluoride is used as an insulating gas in electrical equipments and substations for the last three decades. During these years, slow continuous leaks on the sealed equipments, in addition to maintenance operations, had released important amounts of the contained SF_6 into the atmosphere. Moreover, many open uses of SF_6 were developed and have directly added to the SF_6 atmospheric burden which approaches 80 kT today.

The unique chemical inertness and thermal stability of this gas has permitted its continuous accumulation into the atmosphere during all those years [1-3], at a rate of ≈8% per year. Indeed, the actual ground-level concentration of SF_6 reaches about 4 parts per trillion (4 pptv) [1-3]. This gas does not participate to the ozone depletion, like the freons, but it demonstrates the largest global warming potential of all the anthropogenic gaseous impurities found in the atmosphere (25 000 to 40 000 time that of CO_2) [4].

---

[*] PhD Student, Dept. of Chemistry, University of Montreal.

Moreover, simulation models have revealed that the SF_6 life time in the atmosphere exceeds a millennium (3200, 800 and 1950 years) [5-6], since the only sinks known for this molecule are the decomposition by electron capture in the mesosphere (at altitude > 50 km) and by Lyman-α radiations.

Indeed, the SF_6 molecule exhibits a large electronegativity (electron affinity = 1.05 eV) [8] and can capture relatively easily thermal electrons of few tens of eV [9]. In addition, the SF_6 negative ion may readily dissociate into SF_5^- and F radical [10]. In the conditions of the troposphere, these two species would certainly react very rapidly with the surrounding gases.

This paper presents the experimental approaches developed to study the chemical behavior SF_6 in the lower atmosphere. Indeed, as reported in Table 1, the major portion of the atmospheric SF_6 is present in the lower altitudes. In these circumstances, the SF_6 molecules may not capture electrons directly but charge transfer by O_2^- is certainly an feasible alternative [11]. The results of such tests will be discussed in parallel with the experimental setups used to obtain them.

Table 1. Distribution of SF_6 in the atmosphere (by absolute weight)

Altitude (km)	0 - 5	5 - 10	10 - 15	15 - 20	20 - 25	25 - 30	> 30
Altitude (%) distribution	55.7	24.1	10.3	5.5	2.8	1.3	0.3
Cumulated %	55.7	79.8	90.1	95.6	98.4	99.2	100

EXPERIMENTAL

Various experimental setups were developed to achieve the fundamental goal of this study. These were so chosen to favor conditions allowing the dissociative capture of electrons by SF_6 molecules, in a situation that approach that of the atmosphere. They will be described in the next section along with the discussion of their respective performance. All experiments were conducted at atmospheric pressure with air mixtures of SF_6 in concentrations of 1000 ppmv or lower.

Before and after tests, the analysis of the SF_6/air mixtures was performed on a gas chromatograph-mass spectrometer system using capillary columns, automatic loop-injection ($\approx$100ml) and cryogenic oven temperature programming. This analytical setup is very similar to the one developed for the analysis of S_2F_{10} and described elsewhere [12]. In particular, it permits the quantitative measurement of trace quantities of SF_6 gaseous byproducts at the level of few tens of ppbv and up, an essential need for this kind on study.

In summary, the quantification of the various gas mixture components is done by integrating the signal of their major characteristic ions in the corresponding single ion chromatograms. For SO_2F_2, the signals from three ions were summed: m/z 48, 83 and 102.

It should be noted that the hardware and the software of the Saturn II (Varian) mass spectrometer were upgraded to permit operating in SIS (Selected Ion Storage) mode. In this mode, one can specify the mass ranges (up to 5) of ions to be kept in the ion trap during the pre-scan preparation time. Ions excluded from these ranges are simply ejected from the ion trap prior to each full scan recording. This new mode is particularly useful in reducing the high 'ion background noise' in the tail of large peaks (air, SF_6) and thus facilitate the quantification of the SO_2F_2. Typical examples are shown in Figure 1.

EXPERIMENTAL RESULTS AND DISCUSSION

This section presents the evolution of the different experimental setups developed to promote electron capture and/or transfer to SF_6 molecules and facilitate the quantitative measurement of the resulting decomposition byproducts. As such, the test cells were designed to have the minimal practical volume: 200 - 500 ml. On another practical aspect, it was not experimentally feasible to work with the real SF_6/air mixture because of its to low SF_6 content. Instead, synthetic SF_6-dry air mixtures of 1000 ppmv and less were

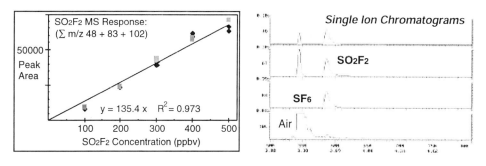

Figure 1. Calibration of the MS response to SO_2F_2 at ppb level, obtained from single ion chromatograms.

prepared especially for this study. Also, it should be noted that all series of experiments were conducted in duplicata (or more replicates), simultaneously in identical cells.

Field emission of electrons

The first experiments were realized in quite simple test cells described in Figure 2. They consists of cylindrical metallic (Al) enclosures equipped with a thin chromel wire (#26) supported in the centers of the Teflon caps. These cells of coaxial geometry were used to generated electrons by field emission, with the axial wire maintained at a high negative potential. To prevent the development of partial discharges, the voltage was adjusted at 80-85% of the PD inception point.

Interestingly, these conditions did induce some decomposition of the SF_6 molecules. However, as it is shown in the right part of Figure 2, the byproducts analyzed were not those expected. Indeed, the only degradation products found were COS and CS_2, with no traces of the usual sulfur oxyfluorides. This confirms that no PD were occurring at the HV wire. On the other hand, These results are an indication that the field-emitted electrons have sufficient energy to split the SF_6 molecules down to their elements.

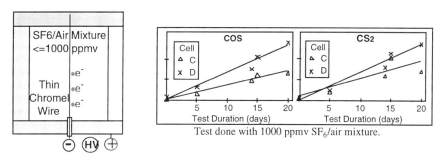

Figure 2. Experimental cell with HV(-) wire and results of decomposition byproducts analysis.

The quantities of carbonyl sulfide and carbon disulfide were not quantify but estimated to correspond to few ppm at the end of the tests. Nevertheless, it also indicates that rather slow reaction rates were prevalent during that kind of experiments. It was indeed interesting to observe the decomposition of SF_6 in the absence of discharge. However, the field emission approach was abandoned as a non suitable source of electrons for this study.

Electrons generated by UV photodetachment

UV irradiation is another well known technique for generating electrons at metallic surfaces. It is mostly employed to reduce the electron emission time lag during breakdown probability measurements. For this study, a low-power UV source was chosen since it has to be inserted into rather small experimental cells. Figure 3 reports the electrical and UV emission characteristics of the low-pressure mercury lamp selected. Its major UV radiation at 254nm (4.5 eV) is of particular interest on aluminum, since the work function of this metal is 4.3 eV. Consequently, low-energy electrons (<= 0.2 eV) will be photodetached from the metallic surfaces in this setup.

The first version of the new test cells was not exactly built as illustrated in Figure 3. Rather, the previous design was modified to accommodate the insertion of a small UV lamp, while keeping the HV thin wire in place. However, the polarity of the wire voltage was changed, so the electric field can now favor the extraction of the detached electrons from the irradiated inner walls. The UV lamp was positionned directly into the cell in these experiments which conducted to the formation of SO_2F_2.

This compound was the only gaseous SF_6 byproduct analyzed. Further, its presence was not due to any kind of electrical discharges, as other experiments done without applying any electric field have also given rise to the same amount SO_2F_2. The thin chromel wire and the field-extraction approach were never used again. Unfortunately, the wire-free setup was not satisfactory, still showing the following flaws:
- the 195 nm radiation was producing lot of ozone in the reaction volume,
- the non-UV energy (4W+) emitted by the lamp was heating significantly the whole setup during the relatively long lasting tests (15 - 60h).

Figure 3 shows how the problem of insulating the lamp from the SF_6-air gas-phase was solved. A tube made of quartz or Vycor was fitted across the whole cell, passing through the two modified covers. It offers a simple way of cooling the lamp, an important need to satisfy in order to stabilize its emission power.

Low-Power UV Lamp Output
(ORIEL #6035, 5W)
(UV power ≈ 700 mW)

UV Radiation (nm)	Relative Intensity (%)
366	15
254	80
195	5

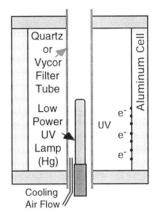

Figure 3. Experimental cell with UV lamp and filter tube and UV lamp output characteristics.

Effects of filtering the UV radiations. The improved model in Figure 3 represents the final version in the evolution of the experimental cylindrical cell setup. It offers one interesting feature: the filtering the UV radiation at 195nm, by the use of a Vycor tube. This kept the generated ozone out of the reaction volume. However, the presence of O_3 was not found to influence measurably the electron-induced decomposition of SF_6, as it is shown in Figure 4. Within the analytical error margin, the two upper SO_2F_2 formation curves are almost identical. However, it was preferred to pursue the rest of the study using Vycor filter tubes, to limit the formation of CO_2. Indeed, abundant amounts of carbon dioxide were especially found when O_3 was allowed to form in the cell volume.

A series of tests was done in similar conditions, using Pyrex tubes as UV filter. This type of glass completely block the 254nm radiations. No SO_2F_2 was ever detected in these cases. This is a clear indication of the implication of the 254nm photons in the electron photoemission from the aluminum walls, inducing the SF_6 degradation. In addition, as shown in Figure 4, the SF_6 consumption rate is clearly dependent on its concentration in the air mixture.

Again, one must note that the SF_6 reaction rate is quite slow. Moreover, only a very small fraction of that gas was decomposed in these experiments. Finally, there is an evident diminishing SO_2F_2 yield with time. This aspect, one can relate to the passivation of the walls by the growing aluminum oxide layer, will be treated later.

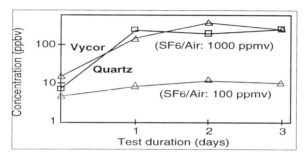

Figure 4. Formation of SO_2F_2 with and without the presence of O_3 generated in the SF_6-air mixture.

Influence of the material of the cell walls. Aluminum is a metal that exhibits a relatively good photoemission efficiency but this characteristic is sensitive to the oxide thickness on the irradiated surface. Stainless steel does not suffer such problem. However, as data reported in Table 2 show, it is not a good metal candidate for increasing the already slow reaction rate. Teflon cells were also tried with even less success.

Table 2. SO_2F_2 formation rate in cells made of different material. (*) (SF_6 / Air = 1000 ppmv)

	Aluminum Cell				Stainless Steel Cell	
Cell Id.	E	F	G	H	I	J
SO_2F_2 formation (pmoles/h)	85	62	80	113	2.1	2.2

(*) SO_2F_2 found in tests done with Teflon cells was almost not detectable after 90 days (calculated yield ≈ 0.1 - 0.4 pmole/h).

Passivation effects of the aluminum walls. As said earlier, some progressive reduction of the yield of formation of SO_2F_2 was observed over time, as cells were reused in many consecutive test runs. This has also contributed to the relatively large dispersion (few hundreds %) of the calculated reaction rates, measured either as inter-cell variations than as time-changing results within the same cell. The culprit was thought to be the increasing passivation of the inner aluminum walls as the oxide layer thickness [13] kept growing during each relatively long irradiation runs.

This kind of behavior is well illustrated in Figure 5. These results were recorded during a series of tests done in three identical aluminum cells with freshly polished inner walls. It is worth noting that the initial SO_2F_2 formation rates then calculated were almost two orders of magnitude higher than the ones measured in the previous conditions. Unfortunately, these yields are found to diminish very rapidly during the first hours of UV exposure. Nevertheless, this behavior seems to be well behave and make possible the calculation of correction factors for experiments lasting few hours.

In reality, these interesting experimental results convince us to limit to one or two hours the duration of future UV exposures, This option was also supported by the improved operating software and hardware of the GC-MS analytical instrument. Furthermore, the use of such shorten test duration contributed to improve by almost a factor of ten the reproducibility of inter-cell results and the overall experimental error margin to only about ± 30%.

Influence of UV radiation intensity. A last series of tests was performed to validate the experimental approach taken in this study of the reactivity of trace concentrations of SF_6 in air. A new reaction cell had to be developed in order to verify the influence of the UV light intensity on the in situ electron generation rate. Figure 6 describes the aluminum flat cell built for this purpose. It can accommodate up to three small UV lamps on its Vycor window (Ø 12mm). For the tests done at the lower radiation intensities, two such cells were put sideways and face to face, with the UV lamps sandwiched in between.

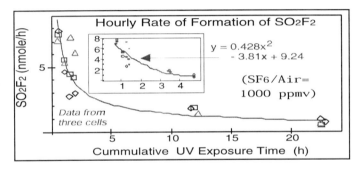

Figure 5. Reduction of the formation rate of SO_2F_2 with cumulative irradiation time.

The higher intensity tests were realized with a totally different lamp and setup arrangement. A medium-pressure Hg lamp (emitting most of its radiations in the near UV and visible) was used. Its output had to be filtered by a water-filled window, 1-inch thick, in order to absorb the intense heat emitted and prevent the warming of the test cell. For all the tests done in both setups, the 254nm radiation intensity was measured at the equivalent surface position of the cell, with a radiometer (Cole-Parmer model 97503-00) equipped with a 254 nm specific head.

The results shown in Figure 6 confirm that the SF_6 decomposition rate is directly proportional to the 254nm light intensity and thus, to the electron photodetachment at the Al surfaces. Finally, it is worth mentioning that all these experiments were done with 100ppm SF_6/air mixtures. The final experimental conditions found and cell preparation will now permit testing to be realized with lower SF_6/air mixture concentrations.

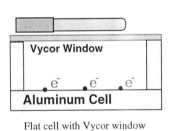
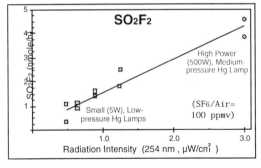

Figure 6. Design of new flat cells and variation of the SO_2F_2 formation rate with the UV light intensity reaching the walls of that type of cells.

CONCLUSIONS

This paper has described the development of an experimental approach aimed at the study of the decomposition reactions of trace concentrations (<1000 ppmv) of SF_6 in tropospheric air. The decomposition reactions were induced by the capture of low-energy electrons, generated on the inner walls of experimental cells by UV photoemission. Cells made of aluminum have shown much larger reaction yield in comparison to those made of stainless steel.

At atmospheric pressure and in very low concentration conditions used, the electron-capture by SF_6 probably never occurs directly and the charge transfer by O_2^- seems to be the favored initiation step of the reaction path. Indeed, SF_6 was observed to decomposed slowly during these tests. The sensitive analytical technique developed in our laboratory for such traces analysis has permitted the quantitative detection of the only SF_6 byproduct found: SO_2F_2. This finding is compatible with the fact that the SF_6 negative ion requires only 0.1-0.2 eV of activation energy to decomposed into SF_5^- and F radical.

Finally, these unique experiments have confirmed the inertness of the SF_6 towards ozone and UV radiations in the conditions of the lower atmosphere. Further studies are still needed to measure the SF_6 degradation rate and evaluate the influence of important atmospheric parameters: SF_6 level, pressure, temperature and humidity content.

REFERENCES

1. M. Maiss, L.P. Steele, R.J. Francey, P.J. Fraser, R.L. Langenfelds, N.B.A. Trivett and I. Levin, "Sulfur Hexafluoride - a Powerful New Atmospheric Tracer", Atmospheric Environment., Vol 30, pp. 1621-1629, (1996).
2. L.S. Geller, J.W. Elkins, J.M. Lobert, A.D. Clarke, D.F. Hurst, J.H. Butler and R.C. Myers "Tropospheric SF_6: Observed Latitudinal Distribution and Trends, Derived Emission and Interhemispheric Exchange time", Geophys. Research Letters, Vol. 24, pp. 675-678, (1997).
3. C.P. Rinsland, E. Mathieu, R. Zander, M.R. Gunson, R.J. Salawicth, A.Y. Chang, A. Goldman, M.C. Abrams, M.M. Abbas, M.J. Newchurch and F.W. Irion, "Trends of OCS, HCN, SF_6, $CHClF_2$ (HCFC-22) in the Lower Stratosphere from 1985 and 1994 Atmospheric Trace Molecule Spectroscopy Experiment Measurements near 30°N Latitude", Geophys. Research Letters, Vol. 23,, pp. 2349-2352, (1996).
4. E. Dutrow, Proceedings of the EPA "Electrical Transmission and Distribution Systems, Sulfur Hexafluoride and Atmospheric Effects of Greenhouse Gas Emissions" Conference, Washington DC, August 9-10, 1995.
5. A.R. Ravishankara, S. Solomon, A.A. Turnipseed, and R.F. Warren, "Atmospheric Lifetimes of Long-lived Species", Science, Vol. 259, pp. 194-199, (1993).
6. R.A. Morris, T.M. Miller, A.A. Viggiano, J.F. Paulson, S. Solomon and G. Reid, "Effects of Electron and Ion Reactions on Atmospheric Lifetimes of Fully Fluorinated Compounds,, J. Geophysical. Research., Vol. 100, 1287-1294, 1995.
7. P.K. Patra, S. Lal and B.H. Subbaraya, "Ovserved Vertical Profile of Sulfur Hexafluoride (SF_6) and its Atmospheric Applications", J. Geophysical Research, Vol. 102, pp. 8855-8859, (1997).
8. G.E. Streit, "Negative Ion Chemistry and Electron Affinity of SF_6", J. Chem. Phys., Vol. 77, pp. 826-833, (1982).
9. P.J. Hay, "The Relative Energies of SF_6^- and SF_6 as a Function of Geometry", J. Chem Phys., Vol. 76, pp. 502-504, (1982).
10. D. Smith, P. Spanel, S. Matejcik, A. Stamatovic, T.D. Märk, T. Jaffke and E Illenberger, "Formation of SF_5^- in Electron Attachment to SF_6; Swarm and Beam Results Reconciled", Chem. Phys. Letters, Vol. 240, pp. 481-488, (1995).
11. A.A. Viggiano, R.A. Morris and J.F. Paulson, "Effects of O_2^- and SF_6 Vibrational Energy on the Rate Constant for Charge Transfer between O_2^- and SF_6", Int, J. Mass Spectrom., Vol. 135, pp. 31-37, (1994).
12. J. Castonguay and I. Dionne, "Analysis of S_2F_{10} in Decomposed SF_6 by Gas Chromatography: Various Aspects", Gaseous Dielectrics VII: Proceedings of the Seventh International Symposium on Gaseous Dielectrics, Knoxville Tn, April 24-28, Plenum Press, pp. 465-473, (1994).
13. A. Buzulutskov, A. Breskin, R. Chechik and J. Va'vra, "Study of Photocathode Protection with Thin Dielectric Films", Nuclear Instrum. and Meth. in Phys. Research A, Vol. 371, pp. 147-150, (1996).

DISCUSSION

L. NIEMEYER: Did you account for diffusion and sorption of SF_6 in the experiments where teflon was used in the enclosure? Teflon has a high diffusivity and sorption capability for SF_6.

J. CASTONGUAY: We did not experience any sign of SF_6 absorption or diffusion during tests done with the teflon cell or with cells housing teflon caps. In all experimental tests, the concentration of SF_6 in the air mixture was analyzed and showed no sign of variation with time. Is should be remembered that our SF_6-air mixture contained a low concentration of SF_6 gas (1000 ppmv or less).

A. GOLDMAN: Does dissociative adsorption of SF_6 on the aluminum walls of your cell, which may lead to fluorination of their surface, have an influence in your experiments?

J. CASTONGUAY: Letting low concentrations of SF_6-air mixtures stand in our last generation of experimental cell did not produce any measurable SO_2F_2 or other S-containing by-products over the detection limit of our analytical instrument (~ 20 ppbv). Nor did this standing without irradiation produce significant passivation of the aluminum walls. Thus, this seems to indicate that the "covering" of the Al surface with fluorine by exposure to SF_6 does not interfere with our experiment.

P. HAALAND: What is the rate of reaction between OH and SF_6?

J. CASTONGUAY: I do not know, but the OH reaction with SF_6 was not considered in the model used to calculate the lifetime of SF_6. One can then think that this reaction is unimportant. The fact that OH is the "atmospheric cleaner species" and SF_6 was always considered inert to low atmospheric oxidation processes, would support the low probability of the SF_6-OH reaction.

EXTREMELY LOW FREQUENCY ELECTRIC AND MAGNETIC FIELD MEASUREMENT METHODS

Martin Misakian

National Institute of Standards and Technology[†]
Electricity Division
Gaithersburg, MD 20899

INTRODUCTION

During the early 1970's, reports originating in the Soviet Union described a variety of ill effects experienced by personnel working in 500 kV and 750 kV switchyards [1,2]. The effects were attributed to the presence of the ac electric fields as well as to the occurrence of spark discharges between the workers and ground. At about the same time questions were raised regarding the possible environmental impact of high voltage transmission lines in the U.S. by the American author L.B.Young [3]. In response to the concerns generated by these and similar reports, numerous bioeffect studies with extremely low frequency (ELF) electric and magnetic fields were initiated in the U.S. by the Department of Energy [4], the Electric Power Research Institute, and during the early 1980's, the New York State Department of Health. The results of these and many newer studies which have focussed on magnetic field effects are now readily found in the technical literature. However, after more than 20 years of research, questions related to possible health effects from exposure to power frequency and other ELF fields remained unresolved. Evidence for the unresolved situation was given by Congress when it initiated an expanded 5-year research program which began in the early 1990's [5].

This paper focuses on the characterization of ELF magnetic and electric fields. Reliable characterization of these fields in the work place, residences, transportation systems, and in laboratory apparatus designed to simulate the fields (during biological studies) is essential if risk assessments are eventually to be performed. Because the mechanisms for effects reported in the literature are not yet understood, the question of what characteristics of the field constitutes the "dose" during exposure is also unresolved. One consequence of this uncertainty during field characterizations is the need to measure more than one parameter associated with the field. For example, in addition to measuring the magnitude of the field at a given location, measurements might also be performed to determine the polarization, temporal variation, frequency content, or time-weighted-average [6].

Following a brief description of ELF electric and magnetic fields, the types of field meters and their principles of operation will be surveyed. Consideration will be given to features necessary in the design of instrumentation to adequately characterize electric and magnetic fields with harmonics and fields which are highly nonuniform. Calibration techniques for magnetic field meters which are used to characterize fields over the dynamic range of a few nanotesla to about a millitesla will be noted. Measurement techniques in various environments as well as their limitations will also be briefly discussed and examples of measurement results in different environments will be presented.

Gaseous Dielectrics VIII, Edited by Christophorou
and Olthoff, Plenum Press, New York, 1998

EXTREMELY LOW FREQUENCY ELECTRIC AND MAGNETIC FIELDS

While concepts of terms such as magnitude and frequency, as they are used to describe electric and magnetic fields are familiar to many readers, the concept of field polarization is perhaps less familiar. Single sources of magnetic fields consisting of currents in straight wires or loops of wire (one or more turns) can be represented at a point in space by a vector that oscillates in magnitude and direction along a straight line. Such fields are produced by some grounding systems and electrical appliances and are said to be linearly polarized. Multiple sources of magnetic fields, in which the currents are out of phase with respect to one another, can also be represented at a point in space with a vector. In contrast to the case of linear polarization, however, the vector rotates and in general traces an ellipse. For some conditions, the trace can be approximately a circle. Such fields are said to be elliptically or circularly polarized. Both types of fields have been used during *in vivo* and *in vitro* bioeffects studies.

Electric fields also can be linearly, elliptically or circularly polarized. The largest most likely encountered electric fields occur near ground level in the vicinity of power lines. Because the electric fields near ground level are approximately linearly polarized, such electric fields normally have been used for exposure purposes during bioeffects experiments.

The electric and magnetic fields of a three-phase transmission line have been calculated and illustrated by Deno [7] and a somewhat simplified sketch of the electric fields in the vicinity of a single-circuit three-phase transmission line is shown in Figure 1. Further discussion of ELF field polarization is provided in the IEEE standard for ELF measurement instrumentation, IEEE Std 1308-1994 [8].

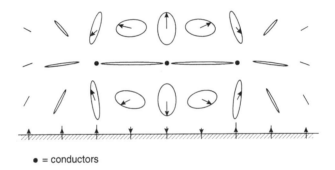

• = conductors

Figure 1. Electric field ellipses in the vicinity of a three-phase transmission line after Deno [7]. The phase of the power frequency voltage applied to each conductor differs by 120 degrees with respect to the other conductors. The electric field vector at a point in space rotates and traces an ellipse in the plane perpendicular to the conductors.

INSTRUMENTATION FOR MEASURING ELECTRIC AND MAGNETIC FIELDS

Electric Field Meters

Three types of electric field meters have been used to characterize power frequency and other ELF electric fields: free-body meters, ground reference meters, and electro-optic meters. Free-body meters are usually battery operated, electrically isolated from ground potential, and are supported in the field at the end of a long insulating rod. The magnitude of the alternating electric field is determined by measuring the induced current oscillating between two halves of an electrically isolated conductive body which makes up the probe or sensor. Figure 2 show geometries that have been used for commercial electric field strength meters. The free-body meter is suitable for survey-type measurements because it is portable, allows measurements above the ground plane, and does not require a ground reference.

Ground referenced meters are normally used with the probe or sensor located on grounded surfaces[1]. Ground reference meters determine the magnitude of the field by measuring the induced current to ground.

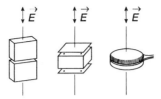

Figure 2. Geometries used for commercial free-body electric field meters.

Because the currents induced into the probes of free-body and ground referenced meters are proportional to the time-derivative of the electric field, the waveform of the current no longer reflects that of the electric field when harmonics are present in the field. That is, the contributions of the harmonics to the current will be weighted by the harmonic numbers due to the differentiation operation by the probe. Consequently, to avoid measurement errors, a stage of integration is often incorporated into the signal processing circuitry to recover the waveform of the field.

The underlying physics for performing measurements with electro-optic meters (typically Pockel's effect devices) differs from that of free-body meters, but both types of meters are used in a similar fashion. Further discussions of the various types of electric field meters and their principles of operation are found in reference [8].

Magnetic Field Meters

Magnetic field meters are available with single- and three-axis coil probes or sensors. Fluxgate magnetometers, capable of measuring the static and ELF magnetic fields, are also available with single- and three-axis probes. A schematic view of a simple magnetic field meter consisting of a single coil (or single-axis) probe and voltmeter detector is shown in Figure 3. The principle of operation is based on Faraday's law which predicts that a voltage proportional to the time-derivative of the magnetic field will be produced by the coil. As for free-body and ground referenced electric field meters, a stage of integration should be added to the signal processing circuit to avoid measurement errors when the field contains more than one frequency. While the simple device described in Figure 3 shows a single coil probe, there is increasing use of probes consisting of three coils which have their axes orthogonally oriented.

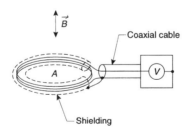

Figure 3. Design of simple magnetic field meter.

Although the devices described are useful for survey type measurements, for long term and more comprehensive measurement applications there is available three-axis instrumentation which periodically records the field level. The recorded values are normally downloaded to a personal

---

[1] A notable exception occurs when electric fields from video display terminals are measured in accord with requirements in an IEEE standard [9].

computer for analysis. Yet more sophisticated instrumentation which can simultaneously record periodically the waveform of the field in three orthogonal directions for later analysis is also available [10].

CALIBRATION METHODS

Electric Field Meters

Several standards exist which provide guidance for the measurement of power frequency and other ELF electric fields [8,11,12]. These standards describe the use of a parallel-plate apparatus for generating a known electric field for purposes of calibration. A nearly uniform electric field can be produced with parallel plates provided that the side dimensions of the plates are more than twice the plate spacing. Further details of the calibration process as well as verification of calibrations are discussed in the standards cited above.

Magnetic Field Meters

Calibration of magnetic field meters is normally done by introducing the probe into a nearly uniform magnetic field of known magnitude and direction. Helmholtz coils have frequently been employed to generate such fields, but the more simply constructed single loop of many turns of wire with a square geometry has also been used [8,11]. The simplicity in construction is at the expense of reduced field uniformity, but sufficient accuracy is readily obtained.

Establishing a known magnetic field for calibrating the more sensitive scales of a magnetic field meter (e.g., 0.2 µT range) is usually complicated by the presence of ambient fields that are of the order 0.1 µT. This problem can be overcome in some cases by using an alternative calibration technique known as voltage injection. With this approach, voltages corresponding to signals that are produced by small magnetic fields are injected into the detector circuit. Further details of this approach can be found in References [8] and [13].

SOURCES OF MEASUREMENT UNCERTAINTY

Electric Field Measurements

Effects of handle leakage, humidity, temperature, harmonic content in the electric field, observer proximity, and reading an analog display from a distance may all contribute to errors in measurements when electric fields are being characterized [8]. The effect of the observer's proximity to the field meter probe was not fully appreciated during some early measurements of electric field strength. Figure 4 shows the perturbation, in percent, of the electric field strength reading as a function of distance between the observer and field meter, and the height of the field meter above ground. The

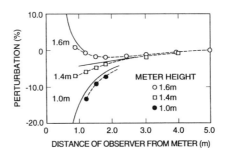

Figure 4. Perturbation due to a 1.8 m tall grounded observer as function of meter/observer distance and meter height above ground. Theoretical values are indicated with solid lines; dashed lines pass through measured values.

data points, which were obtained beneath a 500 kV transmission line, represent perturbations by a 1.8 m tall observer at ground potential and the solid curves are corresponding theoretical predictions

[14]. The perturbations depend markedly on both observer distance and field meter height. Other parameters which can introduce measurement errors are discussed in Reference [8].

Magnetic Field Measurements

The sources of error during measurements of magnetic fields are fewer than for the electric field case. Because there are no significant observer proximity effects, the observer can hold the field meter and thereby reduce errors due to reading an analog display from a distance (as during electric field measurements). However, consideration must be given to choosing the correct probe size when measuring nonuniform magnetic fields near their sources. Significant errors can occur when using circular coil probes because of averaging effects over the cross sectional area of the probe. Recently, calculations have been made of the error probability distributions when three-axis coil probes (with a common central point) and single-axis coil probes are used to measure dipole-like magnetic fields produced by some electrical appliances [15,16]. Figure 5 shows the probability distribution of measurement errors for a three-axis coil probe when the probe is positioned three probe-radii away from an appliance that can be approximated as a dipole (e.g., an electric shaver or hair clipper). The results can explain why two measurements made at the same point in space can differ by more than 30% if the orientations of the probes are different [15].

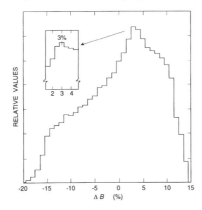

Figure 5. Probability distribution of measurement errors (ΔB) when three-axis probe is three probe-radii away from dipole magnetic field source. The inset shows an expanded view of the distribution near the most probable error.

MEASUREMENT UNCERTAINTY AND VARIABILITY

During measurements of magnetic fields, it is useful to distinguish between measurement uncertainties associated with calibration and instrument design, and uncertainties due to spatial and temporal variations. The uncertainties in the first category are normally associated with measurement accuracy and can be made small (e.g., <5%) by careful instrument design and calibration procedures. There is less control over the second category of uncertainty because the magnetic fields can have, for example in a residence, unknown spatial and temporal variations. The second category of uncertainty may be better referred to as measurement variability, distinct from measurement accuracy. Thus, while a spot measurement at some location may be performed with good accuracy, it will not be possible to specify with confidence what the variability will be without further measurements. Figure 6 shows 24-hour histories of the resultant magnetic field (i.e., root-mean-square of three spatial components) at the center of a living room on two days during which the load currents varied significantly because of weather conditions. The data were obtained with a three-axis meter that recorded the field at a height of 1 m above the floor. Figure 6 (top) shows measurements during a hot and humid July day in the metropolitan Washington area, when air conditioners were presumably in great use. The data were recorded every 15 seconds and the short-term variations, which could last as long as several minutes, could not be attributed to any known sources in the residence. Field measurements at the same location during a cooler, less humid day in September [Fig. 6 (bottom)], reveal a significantly different range of values with an average field of about one-half as large as that during the July observations. The anecdotal data shown in Figure 6 demonstrates that the temporal variability can exceed by far the uncertainties associated with the calibration process and field meter design.

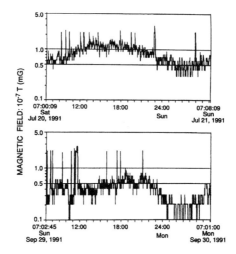

Figure 6. Twenty-four hour measurements of magnetic field at center of living room during hot and humid weather (upper record) and during cool dry weather (lower record).

†U.S. Department of Commerce, Technology Administration. Supported by the Office of Energy Management, U.S. Department of Energy.

REFERENCES

1. V.P. Korobkova, Yu. A. Morozov, M.D. Stolarov and Yu. A. Yakub, Influence of the electric field in 500 and 750 kV switchyards on maintenance staff and means for its protection, *CIGRE Paper* 23-06, (1972).
2. *Study in the USSR of Medical Effects of Electric Fields on Electric Power Systems*, translated from Russian by G.G. Knickerbocker, IEEE Special Publication No. 10, 78 CH01020-7-PWR, IEEE Power Engineering Society, The Institute of Electrical and Electronics Engineers, Inc., New York (1975).
3. L.B. Young, *Power Over People*, Oxford University Press, New York (1975).
4. A.O. Bulawka, W.G. Wisecup, L.A. Rosen, W.E. Feero, and F.M. Dietrich, The U.S. department of energy, 60-Hz electric fields bioeffects research, *IEEE Trans. Power Appar. Syst.*, PAS-101: 4432 (1982).
5. *U.S. House Congressional Record*, Sec. 2118, Electric and Magnetic Field Research and Public Information Dissemination Program, October 5, (1992).
6. IEEE Std 1460-1996, *IEEE Guide for the Measurement of Quasi-Static Magnetic and Electric Fields*, The Institute of Electrical and Electronics Engineers, Inc., New York (1996).
7. D.W. Deno, Transmission line fields, *IEEE Trans. Power Appar. Syst.*, PAS-95: 1600 (1976).
8. IEEE Std 1308-1994, *IEEE Recommended Practice for Instrumentation: Specifications for Magnetic Flux Density and Electric Field Strength Meters-10 Hz to 3 kHz*, The Institute of Electrical and Electronics Engineers, Inc., New York (1994).
9. IEEE Std 1140-1994, *IEEE Standard Procedures for the Measurement of Electric and Magnetic Fields from Video Display Terminals (VDTs) from 5 Hz to 400 kHz*, The Institute of Electrical and Electronics Engineers, Inc., New York (1994).
10. R.M. Sicree, G.B. Rauch, and F.M. Dietrich, Comparison of magnetic flux density meter responses over a database of residential measurements, *IEEE Trans. Power Delivery*, 8: 607 (1993).
11. ANSI/IEEE Std 644-1994 (Revision of IEEE Std 644-1987), *IEEE Standard Procedures for Measurement of Electric and Magnetic Fields from AC Power Lines*, The Institute of Electrical and Electronics Engineers, Inc., New York (1987).
12. IEC Publication 833-1987, *Measurement of Power Frequency Electric Fields*, International Electrotechnical Commission, Geneva, Switzerland (1987).
13. P.M. Fulcomer, *NBS Ambient Magnetic Field Meter for Measurement and Analysis of Low-level Power Frequency Magnetic Fields in Air*, National Institute of Standards and Technology Report NBSIR 86-3330, prepared for Department of Energy (1985).
14. J. DiPlacido, C.H. Shih, and B.J. Ware, Analysis of the proximity effects in electric field measurements, *IEEE Trans. Power Appar. and Systems*, PAS-97: 2167 (1978).
15. M. Misakian and C. Fenimore, Distributions of measurement error for three-axis magnetic field meters during measurements near appliances, *IEEE Trans. Instrum. and Meas.*, 45: 244 (1996).
16. M. Misakian and C. Fenimore, Distributions of measurement errors for single-axis magnetic field meters during measurements near appliances, *Bioelectromagnetics*, 18: 273 (1997).

DISCUSSION

H. OKUBO: In the case of a power transmission line for the electric field there is no problem because the transmission-line voltage is almost constant. However, in the case of the magnetic field, the current flowing condition changes instantaneously like unbalance component of 3-ϕ current and so on. Could you please comment on this point?

A. BULINSKI: A general kind of impression is that the higher the voltage of the transmission line the more danger it is, i.e., the higher is the magnetic field it produces. However, a higher voltage transmission line is taller than a lower voltage line and it could be that the lower voltage rated line produces higher magnetic fields than the higher transmission line. Do you have any data on that? What about fields from underground lines?

M. MISAKIAN: The answers to the questions from Dr. Okubo and Dr. Bulinski overlap sufficiently that I shall give a single response.

Excellent discussions of magnetic fields from transmission lines and distribution lines (overhead and underground) are given in a paper prepared by a Task Force of the AC Fields Working Group in the IEEE Power Engineering Society. The reference is, "Magnetic Fields from Electric Power Lines - Theory and Comparison to Measurements," IEEE Transactions on Power Delivery, Vol. 3, pp. 2127-2136 (1988). Power frequency magnetic fields can be calculated using the law of Biot-Savart which predicts that the field decreases with distance from the source. As implied in the question from Dr. Bulinski, although distribution lines carry less current than transmission lines, the location of the conductors may be closer to where there is human activity, resulting in greater exposure. However, as will be evident from reading the above article, there are many other considerations when estimating exposure to magnetic fields, e.g., the magnitude and location of the return current, buried pipes, and the geometry of the energized conductors. Because of the temporal variation of magnetic fields in transmission and distribution lines, a statistical description of the field levels may be more informative as discussed in the Transaction paper. Local sources in residences such as electrical appliances and currents in metallic water pipes (used as connections to ground) can lead to higher background magnetic fields.

Depending on the measurement location, the magnetic field from buried conductors may be greater than from overhead lines carrying the same current. The shielding effectiveness of metal pipes used in underground power lines is described in a conference paper presented at the 1993 U.S. Department of Energy Conference on Electric and Magnetic Fields in Savannah, Georgia. The paper, "Reduction of Power Frequency Magnetic Field Emission Through the Use of Metallic Conduit," was presented by M. Major and D. March from Montana State University.

T. KAWAMURA: Do you have any experience and experimental data concerning the magnetic field environmental condition in the GIS transmission line and substation?

M. MISAKIAN: NIST does not have experience in characterizing magnetic fields from gas-insulated power lines or substations. A journal that would likely have articles related to this question is the IEEE Transactions on Power Delivery. For example, the article "Measurements and Computations of Electromagnetic Fields in Electric Power Substations," IEEE Transactions on Power Delivery, Vol. 9, pp. 324-332 and the references cited therein, would be of interest. Another recent article, entitled "Five Years of Magnetic Field Measurement," [IEEE Transactions on Power Delivery, Vol. 10, pp. 219-228 (1995)], may also be of interest as it includes measurements of magnetic fields from a variety of sources found in the electric power industry.

A SYSTEMATIC SEARCH FOR INSULATION GASES AND THEIR ENVIRONMENTAL EVALUATION

Lutz Niemeyer

ABB Reasearch Center
CH-5405, Baden, Switzerland

Abstract

The high global warming potential of SF_6 makes it essential to search for substitutes with lower environmental impact. This paper presents a systematic approach to searching and evaluating gaseous SF_6 substitutes for insulation and switching applications in electrical power equipment. The procedure consists of three steps. Firstly, a structured set of criteria is defined which accounts for the functional requirements, environmental effects and safety aspects. Secondly, a comprehensive search strategy for candidate gases is defined which combines combinatorics and chemical systematics. The third and final step an evaluation of the integral environmental impact of a functional unit of equipment redesigned for a substitute gas. It is found that only air/nitrogen fulfill all relevant criteria both for insulation and switching but that they exhibit a much lower performance than SF_6.

1. INTRODUCTION

SF_6 has been found to be a strong greenhouse gas with an atmospheric lifetime of about 3000 years. The gas is a key material in modern electrical transmission and distribution equipment where it is used as both switching and insulation medium. As many environmentally critical substances such as the CFC (chlorinated fluorocarbons) are now being phased out and replaced by environmentally more compatible substances the question arises, if also SF_6 could be substituted, in its electrotechnical applications.

About two decades of intense research have been devoted to finding gases with *better* insulation and switching performance than SF_6 without particular consideration of all functional and environmental aspects (e.g.[1]). Another search is therefore required which focusses on environmental characteristics and also accounts for all important functional requirements. The procedure described in this paper consists of three levels:

(1) The definition of a set of functional and environmental gas evaluation criteria
(2) A comprehensive search strategy for candidate gases
(3) A final evaluation of gases fulfilling the criteria (1) by an integral environmental impact evaluation of equipment redesigned to be used with these gases.

These three steps will be discussed with restriction to pure gases. They will, however, also have to be applied to gas mixtures.

2. GAS EVALUATION CRITERIA

Three sets of criteria will be distinguished, namely, primary selection criteria which have to be fulfilled irrespective of the insulating or switching performance of the gas, insulation specific criteria and additional switching specific criteria. These criteria will be numbered in boldface for later reference.

Primary selection criteria. A gas for application in electric power equipment has to fulfil a set of basic functional requirements independent of its insulation or switching performance. The first of these criteria is **non-toxicity (1)**. Toxic gases will be considered non-acceptable for reasons of personnel safety.

The second criterion is **non-liquefaction (2)** down to the minimal operating temperature T_{min} of the equipment at the chosen operating pressure p. The liquefaction temperature T_{liq} depends on the gas pressure p and can be derived from the boiling temperature T_b (at atmospheric pressure) by an approximate expression of the form $T_{liq}(p) = T_b + 27 \ln p$ where T_{liq} and T_b are in °C and p in bar. This relation can be derived from the Clausius-Clapeyron equation for the vapor pressure curve and allows to express the non-liquefaction criterion in the form of an upper limit to the boiling point T_b:

$$\mathbf{T_b < T_{min} - 27 \ln p \text{ [bar]} < -20 \text{ °C}} \text{ (assuming p >1.2 bar and } T_{min} < -20 \text{ °C)} \tag{1}$$

Chemical stability (3). Chemical stability can be expressed in the form of two criteria: The **decomposition temperature** T_{dec} of the gas must be higher than the maximal temperature which may occur in the equipment (typically around 200 °C at hot spots): $\mathbf{T_{dec} > 200}$ **°C**. The gas must not be ignited into a decomposition reaction by partial discharges of spark type (in which temperatures above 10000 K can occur). A safe criterion herefore is that the gas be an exothermic compound, i.e. that its standard enthalpy of formation ΔH_f be negative: $\mathbf{\Delta H_f < 0}$.

Environmental criteria. The most critical environmental impact indicators are the ODP (ozone depletion potential), the GWP (global warming potential) and the atmospheric lifetime. In view of the Montreal Protocol which prescribes the phasing out of ozone depleting substances, **zero ODP (4)** is considered to be imperative. This entails the a priori exclusion of chlorinated compounds. The magnitude of the global warming potential **GWP (5)** will be used as a quantitative environmental ranking criterion as well as the **atmospheric lifetime (6)** which characterizes the long-term accumulation of a gas in the atmosphere. The GWP will be referred to CO_2.

Insulation specific criteria. The prime insulation performance criterion is the **critical field** E_{cr} **(7)** which measures the insulation capability and will be expressed with reference to the value for SF_6. The second insulation criterion is that **conducting decomposition products must not be generated (8)** by partial discharges because these deposit on solid insulators where they create surface conductivity and surface flashover.

Switching specific criteria. Besides high critical field E_{cr}, switching gases have to meet four additional criteria. Of these, **arcing stability (9)** is the most important: A switching gas has to recombine to its original molecular structure after having been decomposed by the arc in the course of the switching process. Arcing stability can be assessed by a thermodynamic equilibrium calculation which minimizes the Gibbs free enthalpy. The second criterion is the **specific thermal interruption performance (10)**, i.e. the capability of the gas to interrupt the current flow at ac current zero. This property can be empirically characterized by laboratory reference experiments in which, for fixed arcing configuration, gas pressure and current, the limit rate of rise of the recovery voltage after current zero is determined, as a relative performance measure [2]. A further important criterion is that the reaction of the (principally unavoidable) **arc erosion products** from contacts and nozzles **must not form conducting deposits (11)** such as carbon, metal or semiconductor layers because these would cause surface conductivity and lead to surface flashover. Finally, an important criterion is related to the generation of the gas flow required for arc extinction. The higher the velocity of sound c of the gas is the faster it escapes through the flow generating nozzle and the more mechanical energy is required to keep up the gas flow over the necessary arcing time. **Low velocity of sound (12)** is thus is an essential switching gas criterion.

3. SEARCH STRATEGY

A *comprehensive* scan of all existing gases for evaluation as SF_6 substitutes requires a systematic search strategy which makes sure that no gas is overlooked. The strategy proposed here uses, in a hierarchical sequence, the periodic system of elements, molecular combinatorics and the chemical systematics of compounds. At each of these three levels the above primary selection criteria (1) to (3) and the environmental criterion (4) are used for preliminary screening.

The uppermost level of the search procedure is a bottom-up **scan of the periodic system** by elements. For each element, the gaseous compounds are identified according to usual chemical systematics according to textbooks on anorganic and organic chemistry. The result of this scan is that up to and including the 4th row all elements are metals or semiconductors and their gaseous compounds, mainly hydrides, carbonyls and some fluorides, are all chemically unstable and toxic. The only exceptions are the noble gases Ar and Kr. In the third row, the metals Na, Mg and Al do not form stable gaseous compounds and silicon only forms the toxic gaseous compounds SiH_4 and SiF_4. Phosphorus P has to be excluded because all its gaseous compounds like phosphane PH_4 are extremely toxic. Chlorine was already excluded above by requiring zero ODP (criterion (4)). The only surviving atoms in the third row are thus the noble gas Ne and sulphur S. In the second row the metallic or metal-like elements Li, Be and B have to be excluded because all their gaseous compounds are unstable and toxic. The surviving atoms in this row are thus the noble gas helium He, carbon C, nitrogen N, oxygen O and fluorine F. Finally, hydrogen H remains in the first row. Apart from the noble gases which do not form stable compounds the only remaining elements to form compounds are **H, C, N, O, F**, and **S**.

These elements are used, at the second level of the search, to construct compounds by a **combinatoric procedure** which is structured as follows: The above 6 elements are subdivided in two categories, the bi- to hexavalent elements **O, N, C, and S** which will be referred to as **backbone elements** and the monovalent elements **H and F** which will be referred to as **member elements.** Each compound formed from these elements can then be represented, without loss of generality, as a backbone structure (boldface) with attached member elements M:

```
    M   M                F   F F             H    H
M  backbone  M         F  C - S  F         H - C - O - C - H
    M   M                F   F F             H    H
```

This concept is illustrated by the two examples CF_3SF_5 and C_2OH_6. Each combination of different backbone elements (irrespective of the number with which they are present) defines a compound **group**. Complexer backbone structures can be generated by considering the backbone as a combination of atoms and/or functional groups of atoms in the sense of chemical systematics. Examples for functional groups are the saturated hydrocarbon backbone chain C-C, the ether bridge C-O-C and the cyano group -C=N-. From each combinatorically generated backbone structure, compound **families** can be derived by attaching the member elements H or F or H and F. These families correspond to usual chemical classification schemes and their physico-chemical data are readily available from chemistry text- and reference books. The combinatoric procedure is illustrated by **table 1** which shows the combinatoric set for backbone structures with two different atoms. Not all possible compounds are given in the scheme and the compounds and families which fulfill the primary selection criteria (1) to (3) are highlighted by boldface.

Table 1: Compound groups and families with two different backbone elements

backbone	member elements			
	none	H	F	H and F
C N		HCN		
C O	CO	ethers (E)	**PF-ethers (PFE)**	**HF-ethers (HFE)**
	CO_2		COF_2	
C S	CS	thioethers		
			CF_3SF_5	
N O	N_2O	HNO	NOF	
	NO_x	HNO_x	NOF_3	
N S	SN			
	SN_2			
S O	S_2O		SOF_2	
	SO		SO_2F_2	
	SO_2		SOF_4	

4. RESULTS OF THE SEARCH

The above search procedure has been applied comprehensively up to backbone structures with two different atoms. Compounds with 3 different atoms have not yet been explored comprehensively because of the enormous combinatoric complexity of the task.

Among the compounds which only consist of member atoms (i.e. **zero backbone elements**) hydrogen H_2 is the only one. For compounds with only **one** kind of **backbone element** one finds the "memberless" molecues N_2 and O_2. For a C-backbone the compound families are hydrocarbons (**HC**) (with H as member elements), perflourocarbons (**PFC**) (with F as member elements) and hydrated fluorocarbons (**HFC**) (with H and F as member elements). Sulphur as only backbone element yields SF_6 only. Among the compounds with **two backbone elements** the CO group contains CO_2 as only memberless compound and the first members CF_3-O-CF_3 and CF_3-O-CF_2H of the fluorinated ether families PFE and HFE, respectively. In the CS group only the SF_6-CF_4 hybrid CF_3SF_5 exists. All gases in the CN, NO, NS and SO groups are toxic.

The selected gases are collected in **table 2** together with their most important physico-chemical, environmental and functional properties. From the HC, PFE and HFC families only the members with the highest dietric performance have been included. These data illustrate the following general **trends and correlations:**

(1) Increasing molecular complexity favors the dielectric performance but tends to increase the GWP. The two main desired and undesired properties are thus in a physically based conflict: Insulation performance requires molecular complexity which entails a large number of oscillatory and rotational degrees of freedom leading to high IR absorption and, hence, greenhouse efficiency.

(2) Perfluorination is a necessary (although not sufficient) prerequisite for high dielectric and switching performance but also gives rise to atmospheric stability. Partial hydration of fluorinated compounds makes the molecule vulnerable to atmospheric decomposition but also reduces the insulation performance.

(4) None of the environmentally acceptable gases (the noble gases, H_2, N_2, O_2, CO_2, CH_4, C_2H_6) has a significantly higher insulation performance than air.

Table 2: Selected candidate gases

family	Gas	Environmental		functional criteria			switching specific criteria		
		GWP (100y) [CO_2]	Atmos. lifetime [years]	Boiling point [°C]	Sound velocity [m/s]	Dielectric strenght [SF_6]	arcing stability	Thermal interrupt [SF_6]	conduct. arcing products
	SF_6	24900	3 200	- 51	140	1	yes	1	none
	CF_3SF_5	> SF_6	< SF_6	- 24.5	115	1.3	no		none
PFC	CF_4	6 300	50 000	- 128	168	0.41	yes	0.6 [1]	carbon
	C_2F_6	12	10 000	- 78	156	0.8	no	0.45 [1]	
	C_3F_8	500		- 38	124	0.94	no		
PFE	C_2OF_6			- 59	137	0.86	no		
HFC	$C_2F_4H_2$	1300	14	- 27	168	0.46	no		
	$C_3F_4H_3$	4400	55	- 111	185	0.47	no		
noble gases	He	-	irrelev.	- 269	980	0.02	yes	0.033 [1]	metals
	Ar	-	irrelev.	-168	308	0.07	yes	0.007 [1]	metals
HC	CH_4	24.5	14.5	-164	418	0.47	yes	60 [2]	metals
other F- free gases	H_2	-	irrelev.	-253	1200	0.18	yes	120 [2]	metals
	N_2	-	irrelev.	- 196	336	0.30	yes	0.02 [1]	metals
	air	-	irrelev.		330	0.33	yes	0.04 [1]	none [3]
	CO_2	1	irrelev.	- 78	254	0.3	yes	0.11 [1]	metals

[1] In terms of limit RRRV at 24 kA_{RMS} / 60 Hz and a blast pressure of 2 MPa [2], [2] at 50 kA_{RMS} / 60 Hz and 3.8 MPa [4], [3] when carbon arcing contacts are used

(5) The only gases with insulation performance comparable to SF_6 are CF_3OCF_3 (14 % worse) CF_3SF_5 (30 % better). Both are very stable and must thus be expected to have a high GWP and a long atmospheric lifetime.

These results lead to the following conclusions:
- **For insulation applications only air or nitrogen can be considered as SF_6 substitutes with, however, only about one third of the SF_6 performance.**
- **For switching applications air is the only choice. Used with carbon arcing contacts the oxygen suppresses the formation of carbon deposits. Its thermal interruption performance is, however, much lower than for SF_6.**

H_2 and CH_4, although excellent thermal interruption gases, have only poor dielectric performance and much too high velocities of sound.

6 ENVIRONMENTAL LIFECYCLE ASSESSMENT (LCA)

The above evaluation procedure only considers the gas as a material but does not yet account for the re-design of the equipment which would be necessary to account for its lower performance with respect to SF_6. Therefore, the third and final evaluation step has to consider the *integral* environmental impact of the whole equipment when redesigned (for equal performance) for the substitute gas in comparison to SF_6 equipment). The procedural aspects of this evaluation step, which is referred to as environmental lifecycle assessment (LCA), is defined in a series of ISO standards [3] and consists of the following four steps:

(1) **A functional unit** of the equipment has to be identified for which the assessment is carried out. For electrical power equipment the natural functional unit is a three-phase switching bay.

(2) **Life cycle inventories (LCI)** have to be established. They include materials and energies used and emissions caused over the full lifecycle of the unit, starting from mining raw materials over the production of the materials employed, transportation, manufacturing of the equipment, commissioning and operating life till decommissioning and scrapping. For SF_6 equipment, these inventories have to include, in particular, the SF_6 losses due to testing, leakage in operation, maintenance and repair, decommissioning and

gas recycling. From this it becomes clear that an LCI can not be carried out generically but only on a concrete equipment design.

(3) The inventory items have to be assigned **environmental indicators** which characterize their specific environmental impacts such as health damage, resource depletion, atmospheric pollution etc.. For SF_6, the most important indicator is the global warming potential (GWP) whereas for the other materials used in the equipment a mix of various indicators is relevant.

(4) The final step consists in applying an **environmental value system**, which assigns each indicator a quantitative weighting factor. The choice of an environmental value system is difficult because the various impacts are normally not commensurable: Health risk by smog and acid rain can not be directly quantified against resource depletion and global warming. Presently, a globally accepted value system is still in the phase of negotiation and only national systems are available which still differ in several respects. For the following reasoning we will use the swedish EPS (Environmental Priority Strategy) which weights the GWP impact and thus SF_6 most severely.

LCA has been carried out for a presently manufactured 300 kV GIS assuming 20 % SF_6 lifetime losses. This loss rate can be achieved with present day GIS sealing and gas handling technology. The EPS value system then yields that about half of the environmental impact of the GIS is due to SF_6 losses and the other half mainly to containment materials. In order to evaluate the consequences of substituting SF_6 by air a redesigned GIS would be required in order to carry out a comparative assertion in the sense of the LCA ISO standard 14040. As such a design is presently not yet available, the following rough estimate will be made for preliminary orientation: A GIS redesigned for air would have to operate with at least 2 times higher gas pressure and/or insulation distance to compensate for the lower insulation performance. The quantity of material required for the increased enclosure pressure rating or size is determined by pressure container design rules and scales roughly with a power of 1.5 with pressure and dimensions. This corresponds to an increase of required materials by a factor of $2^{1.5} \approx 2.8$. Referred to the SF_6 GIS impact values (0.5 impact from materials, 0.5 from SF_6) the impact of the air enclosure material would result $0.5 \times 2.8 = 1.4$ (plus zero impact of the gas). The total environmental impact of the equipment would thus be higher than for SF_6. This example allows the following conclusions:

- **Gas properties alone do not allow to decide if it is environmentally rational to substitute SF_6.**
- **The design changes called upon by the substitute gas have to be incorporated into the evaluation. The procedural tool for this is LCA.**
- **An LCA can only be carried out properly on the basis of a complete equipment (re-)design.**

8. REFERENCES

1. L.G. Christophorou, I. Sauers et.al., " Recent Advances in Gaseous Dielectrics at Oak Ridge National Laboratory", IEEE Trans. Electr. Insul.,19 (6) 191984, 550-566"
2. G. Frind et al: " EPRI Report EL-284, Jan. 1977
3. ISO Standard 14040 Environmental Management - Life cycle assessment - principles and framework, 1st Ed. 1997 -06-15
4. H.O. Noeske, "Arc thrmal recovery speed in different gases and gas mixtures", IEEE-PAS-100,1981,4612-4620

SF_6 ReUse Concept and SF_6 New Applications

Michael Pittroff[1], Andreas Schütte[2], Andreas Meier[3]

Solvay Fluor und Derivate GmbH, Germany
[1]Technical Service, Hans-Boeckler-Allee 20, 30173 Hannover, Germany
[2]Siemens AG EV HGIL, Paul-Gossen-Str. 100, 91050 Erlangen, Germany
Project Manager Gas insulated Transmission Lines
[3]Marketing, Solvay Fluorides, Inc.; 41 West Putnam Avenue, Greenwich, CT 06830,USA

Introduction

Solvay is a worldwide supplier of SF_6 since the late nineteen sixties and started to develop a ReUse concept for used SF_6 at the begining of 1990. The ReUse concept was developed for SF_6 coming from electrical equipment, where most of the worldwide SF_6 production is used. In these applications SF_6 has proven to be essential due to its combination of favourable properties such as chemical inertness, high thermal stability, non-flammability and non toxicity.

A less favourable property of SF_6 is its comparatively high greenhouse warming potential due to its persistance in the atmosphere. As a consequence this requires SF_6 to be kept in closed cycles as far as possible in order to minimize emissions and by utilizing recycling and reuse potential.

For this reason Solvay developed a ReUse concept which includes environmental consulting, analytical services of used SF_6, packaging and transport of used SF_6 and the reclaiming of used SF_6. The reclaiming of used SF_6 is divided into three cases: The Normal Case (on site), the Special Case (return of used SF_6 to the manufacturer) and the Exceptional Case (disposal company).

As a new application of SF_6, Gas Insulated Transmission Lines (GIL) can be a suitable solution to replace overhead lines in urban areas. In the GIL SF_6 is filled with N_2 as the insulating medium.

The ReUse Concept from Solvay

1. Normal Case (on site)

The technology for SF_6 on-site treatment is available today and exists at most utilities. This requires a corresponding guideline or instruction of the user, including the support of his management. Essential to this "no release" concept is that all SF_6 insulated switch gears are equipped with a safe gas connection.

The first step for the on-site treatment has to be the determination of the gas quality, in order to decide if it is reclaimable on site or has to be sent back to the SF_6 manufacturer. The interesting analytical data are for SO_2, SOF_2, inert gas, oil content, HF and moisture content to select the corresponding method of regeneration.

Before determining the overall SF_6 purity (via SF_6 volume percentage measuring devices) the acidity should be measured, to avoid potential corrosion on the measuring device. Mineral oils, SO_2, dew point and total acidity as hydrogen fluoride are easily measured on site with detection tubes having detection limit down to 1 ppm.

The transfilling and storage of used SF_6, should be done with the conventional devices from maintenance suppliers. Polluted SF_6 should not be filled in cylinders for new SF_6. There are several reasons to avoid a mixing of the cylinders for new and used SF_6:

- health risk for workers (using new SF_6 cylinders filled with used SF_6)
- cylinders for new SF_6 have no permission for used SF_6, linked to the acidity and corrosion effects
- avoid additional cleaning of new cylinders
- no guaranty and no overview about the SF_6 gas management

For clear and good gas handling practices, used SF_6 should be stored in cylinders with higher corrosion resistance and labeled with the orange label (UN 3308).

Most of the impurities could be purified with adequate filters from the maintenance supplier. Changing filters should not require disassembly of any fittings, tubing, or any other connection, eliminating the additional possibility of leakage. It would be useful if these filters have indicators when they are saturated with impurities.
Some examples of possible impurities and the corresponding reclaiming method are:

contamination	moisture (water vapour)	gaseous decomposition products	solid decomposition products	mineral oil	mixed with gases other than SF_6 (e.g. air)
reclaiming method	adsorption with molecular sieve	adsorption with activated alumina	retaining with solid filters	activated charcoal filter	SF_6 gas separation device

Table: 1 Different reclaiming procedures

For both safety and warranty reasons, the purified SF_6 gas should be analysed. Ideally the quality of reclaimed SF_6 should approach the quality of new gas. This determination can only be made with more costly analytical, laboratory methods. Upon request Solvay Fluor und Derivate will carry out an analysis of the used/reclaimed SF_6 gas.

Refilling of electrical equipment with reclaimed SF_6, should be done with devices operating oil free, gas tight, safe filling to pre-set pressures and transportable.

SF_6 gas reclaimed in this way does **not** fulfill all the quality requirements of DIN IEC 376 or ASTM D 2472.

2. Special case

In case of a non on-site reclaimable SF_6 gas quality, the user should contact Solvay. If the impurity is not known, Solvay sends sampling cylinder with sampling guidelines to the SF_6 user. The SF_6 user performs the sampling procedure and sends the sampling cylinder back to Solvay. The technical department carries out the analysis and decides if the gas quality is within the maximum tolerable values for reclaimable SF_6 gas at Solvay. Up to now all SF_6 gas qualities have been found to be in line with the maximum tolerable values of Solvay.

The gas must fulfill the following conditions:

- The total amount of degradation products may not exceed 5 % by weight
- The inert gas content (e.g. air, N_2, O_2 etc.) may not exceed 30% by volume
- The CF_4 content may not exceed 5% by volume
- The water content may not exceed 1000 ppm by weight
- The HF content may not exceed 1000 ppm by weight
- The material must be pure in regards to grade (no mixture), oil-free (< 0,1 %) and non radioactive
- It must be specified from which area of application the contaminated SF_6 originates

The SF_6 user decides, if the polluted gas will be sent to Solvay. The used and polluted SF_6 is filled in special cylinders with high corrosion resistance. These containers comply with the requirements for hydrogen chloride and used SF_6. Therefore the connections and valves are different from the valves of new SF_6. These cylinders have an orange collar to easily identify the used gas quality. The user is responsible for the transfilling of the contaminated SF_6 from the switching unit into appropriate special containers. Normally these containers are at the customer's disposal from Solvay Fluor und Derivate GmbH on a temporary loan basis.

The used SF_6 gas has to be prefilted and afterwards introduced in the production process. Within the production process the impurities are destroyed and reformed to SF_6.

3. Exceptional case

In case of a non reusable SF_6 quality, the gas mixtures are destroyed through incineration.

SF$_6$ New Applications

Today power transmission is ensured almost completely by overhead lines. These overhead lines allow high power to be transmitted from power generating sites to the substations. High population in city centers and environmental aspects create more and more problems to install overhead lines. In the future a suitable solution can be the Gas Insulated Transmission Lines (GIL) where mixtures of SF$_6$/N$_2$ will function as the insulating medium.

Solvay and Siemens (as manufacturer of GIL) have invested in research activities in order to find a procedure to prepare SF$_6$/N$_2$ - mixtures with the requested SF$_6$/N$_2$ ratio for GIL-projects. It was a requirement that the procedure is workable "on-site" and "on an industrial scale".

1. Filling procedure

Electrical arc discharge tests have been performed on a GIL test sample. The total amount of the filling volume was 13 m^3 at 7 bar. The required mixture was 20 vol. % SF$_6$ and 80 vol.% N$_2$ with a tolerance of maximum + 1 vol.% SF$_6$.

SF6/N2 Mixing station

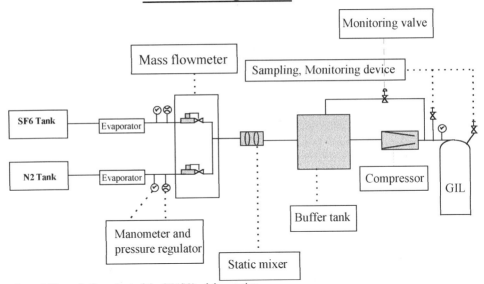

Figure 1: Shematic flow chart of the SF6/N2 mixing station.

Solvay's mixing station has a maximum flow rate of the mixture 1-40 vol. % SF$_6$ and 60-100 vol. % N$_2$ up to 250 m^3/h. With the mixing station a maximum filling pressure of 10 bar can be achieved.

During the filling process with the above mentioned plant Solvay monitored the mixtures with a gas chromatograph. In case of mixtures outside of the required specification, the mixing ratio was verified manually. The filled GIL (Gas Insulated Transmission Line) was once again proofed via gas chromatograph and cross checked via a DILO SF$_6$ percentage monitoring device.

2. Analysis of the decomposition products

- Electrical arc discharge test 1 (50 kA, 330 ms)

There were no gaseous decomposition products above 50 vppm (detection limit) observed (GC chromatogram below). The formed solid decomposition product was AlF_3, aluminium trifluoride.

Further analysis of the gas sample identified a SO_2 concentration of 48 ppm. S_2F_{10} could not be detected (detection limit 10 - 15 ppm). No NF_3 was detected via gas chromatograph (GC) and IR(Infra Red) spectroscopy. No $S_2F_{10}O$ (detection limit 1 - 2 ppm) and no SO_2F_2 was measured.

- Electric arc discharge test 2 (50 kA, 330 ms)

The mainly formed solid decomposition product was AlF_3, aluminium trifluoride. There were also some traces of other inorganic fluoride salts coming from the metallic surfaces. As gaseous decomposition product, SOF_2 was formed in the range of 1%. NF_3 as decomposition product could not be measured in a quantity above 50 vppm (detection limit).

Further analysis of the gas sample identified a SO_2 concentration of 165 ppm. S_2F_{10} could not be detected (detection limit 10 - 15 ppm). No NF_3 via GC and IR (Infra Red) spectroscopy (detection limit 50 ppm) was detected. No $S_2F_{10}O$ (detection limit 1 - 2 ppm) and no SO_2F_2 was measured. The SOF_2 concentration was 0,3 %. The detector was a DID (Discharge Ionisation Detector).

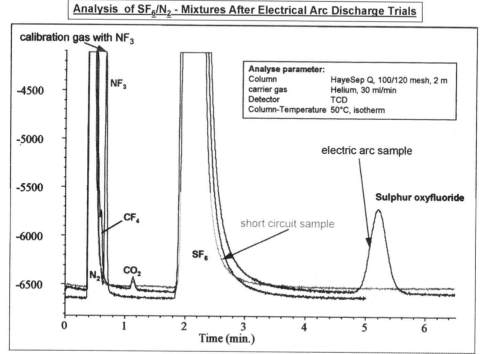

Figure 2: The chromatogram of the calibration gas and the decomposition samples:

The SF_6 destruction during the electrical arcing was between 1 - 2 %.

Final conclusions

For an efficient and controlled GIL (with gas mixture) filling process a gas mixing station is necessary. If the filling process is done without a mixing station (e.g. filling by weight), it is a time consuming and uncontrolled process, because the emptying of the cylinders, specially with counter pressure in GIL, is time consuming.

The analysis of the decomposition products assured, that even under non normal working conditions like short circuit or electrical arc there is no formation of NF_3 above 50 vppm, no S_2F_{10} above 10 ppm and no SO_2F_2. The other gaseous decomposition products are sulphur oxyfluorides (SO_2 and SOF_2) which are formed as well from pure SF_6 applications. The CO_2 and CF_4 amounts are coming from organic insulating material. The solid decomposition products mainly depend on the metal enclosure of the GIL.

SECTION 8: SURFACE DISCHARGES / DESIGN ENGINEERING

THE SURFACE FLASHOVER CHARACTERISTICS OF SPACER FOR GIS IN SF6 GAS AND MIXTURES

Y. Hoshina, J. Sato, H. Murase, H. Aoyagi, M. Hanai and E. Kaneko

TOSHIBA CORPORATION
Power and Industrial Systems Technology Development Center
Kawasaki, Japan

INTRODUCTION

The SF_6/N_2 gas mixture is probably the most promising dielectric gas among SF_6 substitutes, because nitrogen is cheap and harmless for an environment. Because the boiling point of SF_6/N_2 gas mixture is lower than that of pure SF_6, this mixture will be used in apparatus designed for cold areas.

There have been many investigations of insulating properties with the SF_6/N_2 gas mixture, but the most cover fundamental properties. We have studied SF_6/N_2 gas mixture insulation for engineering applications of substation equipments (gas-insulated switchgears, circuit breakers and gas-insulated transformers).

It is well known that the SF_6/N_2 gas mixture shows desirable synergisms. We investigated surface flashover characteristics in SF_6 gas and mixtures using a spacer model. The surface flashover characteristics were compared with the breakdown characteristics in a uniform field gap.

We also investigated the partial discharge and flashover characteristics from metallic particle attached to the surface of an insulator in SF_6 gas and mixtures. Different synergisms were shown between partial discharge characteristics and flashover characteristics.

EXPERIMENTAL MODEL AND METHODS

Figure 1(a) shows the experimental model. This model is composed from a spacer model and shielding electrodes. The spacer model is an epoxy resin of 80mm in diameter and 100mm in height, and this has aluminum electrodes embedded its inside. The discharge pattern of this model shows that the flashover starts from an insulating surface, not from a shielding electrode[1]. The flashover electric field strength over the insulator surface is lower than that on the electrode surface. Figure 1(b) shows potential distribution on an epoxy molding of this model.

We investigated the surface flashover characteristics of epoxy molding in pure SF_6,

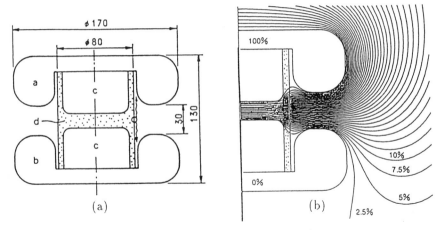

Figure 1: Experimental model. (a) Sectional view (a:high-voltage electrode, b:grounding electrode, c:embedded electrode, d:epoxy molding) (b) Potential distribution

pure nitrogen and SF_6/N_2 gas mixtures (mixing SF_6 2%, SF_6 5%, SF_6 10%, SF_6 15% into nitrogen). The surface condition of the spacer was clean without metallic particle attached. The surface flashover voltages under a negative lightning impulse were measured in the gas pressure range of 0.1 to 0.3MPa. Maximum voltage was 350kV. The gases used are commercial ones.

Next, we investigated the partial discharge and flashover characteristics from metallic particle attached to the surface of spacer. For metallic particles, aluminum wires of 0.20mm in diameter and 5mm in length were used. They were attached to the center of the spacer in parallel with the direction of the electric field. Gas pressure was kept at 0.5MPa. Partial discharge inception voltages and flashover voltages under AC 50Hz voltage were measured. The applied voltage was increased by 5kV step for a certain period of the voltage applying time (one minute) until flashover.

Partial discharge behavior was measured and analyzed using a tuning-type (400± 40kHz tuning) PD detector "CD-5" of Nihon-Keisokuki, which has a detection level below 0.1pC.

EXPERIMENTAL RESULTS AND DISCUSSION

Surface Flashover Characteristics under Clean Conditions

The flashover voltage depends on gas pressure as shown in Figure 2. The data almost have a linear correlation. The dielectric strength of pure SF_6 is about 2.5-3.0 times higher than that of pure nitrogen.

When mixing 15% SF_6 into nitrogen, flashover properties are closer to that in pure SF_6 than in pure nitrogen. Flashover voltages in $SF_6 15\%/N_2 85\%$ mixtures are about two times higher than that in pure nitrogen.

Figure 3 shows the flashover voltage as a function of SF_6 content in gas mixtures. As shown in Figure 3, in the range where SF_6 content was small, flashover voltage increased greatly with the increase of SF_6 content. The flashover voltages in gas mix-

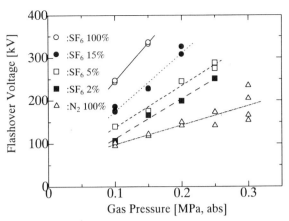

Figure 2: Gas pressure dependence of surface flashover voltages for various gas mixtures under clean conditions.

tures were higher than the sum of the partial-pressure-weighted flashover voltages in pure SF_6 and nitrogen. The surface flashover characteristics in SF_6/N_2 gas mixture show desirable synergisms. The surface flashover characteristics are similar to breakdown characteristics in a uniform field gap, which have been reported by Qiu et al. [2, 3]. The breakdown voltages are proportional to the SF_6 content raised to the power of 0.18. The flashover voltage in the SF_6/N_2 gas mixture is expressed by the following equation.

$$V_{BD} = V_0 \cdot X^{0.18} \quad (1)$$

Where V_0 is the flashover voltage in pure SF_6 and X is SF_6 content in gas mixture.

The curve in Figure 3 is obtained by calculating eq (1). It shows a good agreement between the measured data and the calculating curve.

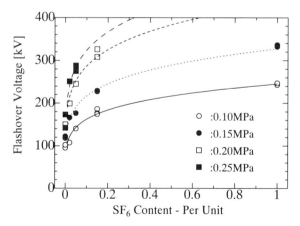

Figure 3: SF6 gas content dependence of surface flashover voltages. (Marks : measured data, Curves : calculated results).

Partial Discharge and Flashover Characteristics with Particle Attached

We investigated the partial discharge and flashover characteristics from metallic particle attached to the surface of an insulator in SF_6 gas and mixtures.

The surface flashover voltage in SF_6/N_2 gas mixtures as a function of SF_6 content is shown in Figure 4. As shown in Figure 4, in the range where SF_6 content was small, flashover voltage increased more with the increase of SF_6 content. The flashover voltage in SF_6 15%/N_2 85% mixture was highest in these sample gases, and was higher than that in pure SF_6. In SF_6/N_2 gas mixtures, the surface flashover characteristic with metallic particle attached (in a non-uniform field) shows more desirable synergisms than in a uniform field.

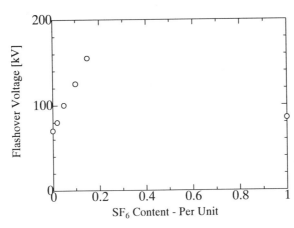

Figure 4: SF6 gas content dependence of surface flashover voltages with particle attached.

The partial discharge inception voltages in SF_6/N_2 gas mixtures as a function of SF_6 content are shown in Figure 5. As shown in Figure 5, the partial discharge inception voltage was lowest in pure nitrogen, and was highest in pure SF_6. When mixing 2% SF_6 into nitrogen, the partial discharge inception voltage (PDIV) increased slightly. But, in the range where the SF_6 content was more than 2%, the PDIV hardly increased. The PDIVs in gas mixtures were lower than the sum of the partial-pressure-weighted PDIVs in pure SF_6 and nitrogen. In SF_6/N_2 gas mixtures, the partial discharge characteristic with metallic particle attached shows negative synergisms.

Figures 6 and 7 show variations of partial discharge magnitude (maximum value) versus applied voltage. In the case of AC 35kV, the partial discharge magnitude was about 100pC in pure nitrogen, and was about 2pC in pure SF_6. Although the discharge magnitude in pure nitrogen was about 30-100 times larger than that in pure SF_6, the discharge magnitude when mixing 2% SF_6 into nitrogen was about two to four times larger. The discharge magnitude in 15% SF_6 into nitrogen was as large as that in pure SF_6. Although the repetition rate of the partial discharge in pure nitrogen was large, that in SF_6/N_2 gas mixtures and pure SF_6 was smaller than in pure nitrogen.

The partial discharge inception voltage in SF_6/N_2 gas mixtures resembles that in nitrogen, and is strongly affected by nitrogen. But the partial discharge behavior in

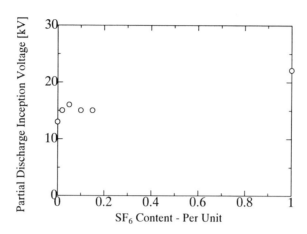

Figure 5: SF6 gas content dependence of partial discharge inception voltages in gas mixtures with particle attached.

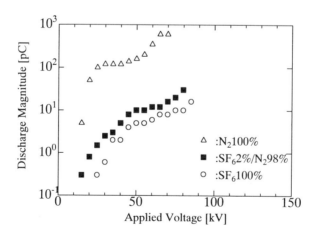

Figure 6: Variations of partial discharge magnitude versus applied voltage in pure N_2, pure SF_6 and $SF_6 2\%/N_2 98\%$.

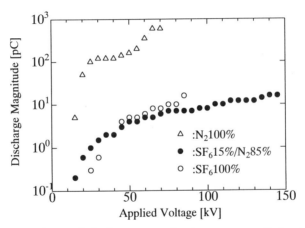

Figure 7: Variations of partial discharge magnitude versus applied voltage in pure N_2, pure SF_6 and $SF_6 15\%/N_2 85\%$.

SF_6/N_2 gas mixtures resembles that in SF_6, and is strongly affected by SF_6.

CONCLUSION

We investigated the surface flashover characteristics in SF_6/N_2 gas mixtures using a spacer model. The surface flashover characteristics under clean conditions are similar to the breakdown characteristics in a uniform field gap. The breakdown voltages are proportional to the SF_6 content raised to the power of 0.18.

The surface flashover characteristics with metallic particle attached (in non-uniform field) show more desirable synergisms than in a uniform field. The partial discharge inception voltage with metallic particle attached shows negative synergisms. However, the partial discharge behavior in SF_6/N_2 gas mixtures resembles the behavior of SF_6, and is strongly affected by SF_6.

References

[1] H. Aoyagi et al, "Study of Flashover Characteristics over the Surface of an Insulator in SF_6 Gas", JEE B, Vol.117, No.3, pp395-401, 1997 (in Japanese)

[2] Y. Qiu and Y. P. Feng, "Calculation of Dielectric Strength of the SF_6/N_2 Gas Mixture in Macroscopically and Microscopically Non-uniform Fields", 4th International Conference on Properties and Applications of Dielectric Materials, Vol.1, pp87-90, 1994

[3] Y. Qiu and D. M. Xiao, "Dielectric Strength of the SF_6/CO_2 Gas Mixture in Different electric Fields", 9th ISH, 2255, 1995

DISCUSSION

W. PFEIFFER: It is obvious that Eq. (1) cannot be used for $x = 0$. Could you comment about the validity range of Eq. (1)?

Y. HOSHINA: Equation (1) is valid except for $x = 0$. Some justification may be necessary in order to apply it generally.

K. D. SRIVASTAVA: There is some evidence to suggest that in the case of a particle attached to the spacer, the minimum flashover voltage may occur at a location of attachment closer to one of the electrodes. Have you varied the position of the particle on the spacer?

Y. HOSHINA: Yes, we have some data. As shown in Fig. 1, in this model the position of maximum electric field is the center of the spacer. The minimum breakdown voltage is given when a particle is attached at this point, in this particular model.

A. H. MUFTI: In this paper and in a previous one as well you mentioned the 15%SF_6/85%N_2 mixture. Can you comment as to why this mixture?

Y. HOSHINA: We are now investigating the suitable mixing ratio in the range of less than 20%SF_6.

H. OKUBO: Did you detect partial discharge under impulse voltage application, especially with metallic particle condition? If there would be a partial discharge on the solid insulator, the charging would influence the next shot of impulse voltage application. Could you please comment on this point?

Y. HOSHINA: Our experiments were carried out using a single impulse shot. Thus, we think that there is no influence of previous voltage application. However, detection of partial discharge is important. We will measure partial discharge under impulse voltage application.

B. HAMPTON: The AC flashover voltage of the insulator with a particle on the surface is much higher with a mixture than with pure SF_6. Is this due to the higher level of PD which has been measured? If so, since the PD will itself degrade the surface and eventually lead to flashover, should the mixture be regarded as giving only a short-term benefit?

Y. HOSHINA: Your comment is well-taken. We need more long-term test data to answer your question.

GENERATION AND INVESTIGATION OF PLANAR SURFACE DISCHARGES

S J MacGregor, R A Fouracre and S M Turnbull
Department of Electronic and Electrical Engineering
University of Strathclyde
Glasgow G1 1XW, UK

INTRODUCTION

In the present work the term surface discharge refers to an electrical discharge occurring in a gas in the region of a solid dielectric surface. The discharge is closely coupled to the underlying solid surface and can result in considerable degradation of the solid which can subsequently lead to electrical failure. Although surface discharges have been the subject of numerous investigations[1], many of the basic physical processes affecting the characteristics are still not fully understood.

EXPERIMENTAL EQUIPMENT

In order to investigate the properties of surface discharges, two sets of apparatus have been used. The first experimental arrangement, Figure 1, was employed to measure the propagation velocity of the discharge front[2]. The surface discharge was initiated on a 350 μm thick, 30cm wide polyester sheet by a voltage impulse applied to a profiled aluminium electrode mounted on the top surface of the polyester. An earthed 50mm wide aluminium strip, attached to the back of the polymer sheet acted as a ground plane. Above the top surface, an array of corona electrodes provided the source to charge the polymer surface. Charge would be trapped as a result of Coulombic image forces. The discharge was initiated after the polymer surface had been charged with a polarity opposite to that of the applied impulse voltage. The discharge propagated across the surface but at no time did it make a low impedance electrical connection between the upper and lower surface electrodes. The impulse generator used was a 50kV, 320nF, single-stage, inverting Marx generator.

The second apparatus was used to record the light from the discharge. The discharge was generated between two copper sheet electrodes attached to the upper surface of the mylar dielectric. To define the position of the discharge, the electrodes tapered to a

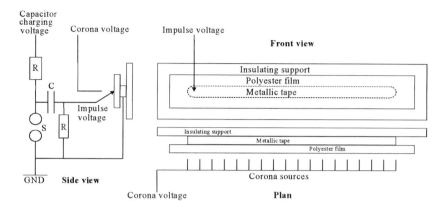

Figure 1. A schematic diagram of the surface discharge generating apparatus.

point and these were separated by a distance of ~30mm. The under surface of the polymer was attached to a third copper electrode which was connected to earth and acted as a guide for the surface discharge. An Ocean Optics, four channel spectrometer (Type SQ2000, resolution ~1.3nm) was used to detect the discharge light emission. Input to the spectrometer was through optical fibres which could be positioned close to the discharging surface. Corona electrodes were mounted over the region of the free polymer surface. A single stage coaxial cable PFN impulse generator, used to initiate and sustain the discharge, produced a 20kV impulse of 90ns duration. The system was contained in a perspex enclosure which could be evacuated to rotary pump pressures and filled with gas to the required pressure.

SURFACE DISCHARGE MECHANISMS

Previous investigations into the mechanism of surface discharge propagation have indicated that the process is significantly affected by the so called specific capacitance of the substrate. However, the definition of this capacitance is not clear. In discussion, it should be remembered that capacitance relates voltage to charge associated with two equipotential surfaces. An electrostatic model describing the propagation of the surface discharge has been developed. After discharge initiation, a capacitance will be formed between the advancing plasma sheet, one equipotential, and the underlying earthed strip, the other equipotential. This capacitance increases as the discharge advances and therefore the energy stored in the dielectric increases. Since the charge in the system remains constant, in the absence of surface charge, the potential difference between the plasma and the ground electrode must decrease. It is straightforward to calculate the energy stored in the dielectric and the total energy stored in the system as a function of the length of discharge. It is also possible to allow for the surface charge. Figures 2 and 3 show the energy stored in the system and in the dielectric as a function of the discharge length for a charged surface. The model shows that the energy stored in the dielectric, whilst the discharge is propagating, increases to a peak value before starting to decrease.

Electrostatic field analysis (Electro³) has been used to determine the potential distribution in the region around the advancing discharge front. Figures 4 and 5 show the potential distribution associated with an advancing discharge in the case of an uncharged and

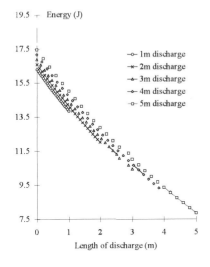 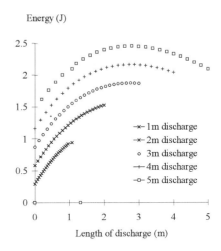

Figure 2. The energy stored in the system. Figure 3. The energy stored in the dielectric.

charged dielectric. For an uncharged surface, the results show that the resultant high electric field between the discharge and the substrate forces the discharge close to the underlying surface. For a pre-charged surface, the results indicate that the normal component of the field in the region around the discharge head is reduced. These results indicate that the presence of charge in the surface reduce the forces which couple the discharge to that surface. The magnitude of these forces is strongly dependent upon the ratio of the dielectric constant of the solid to the gas.

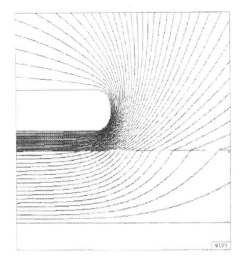 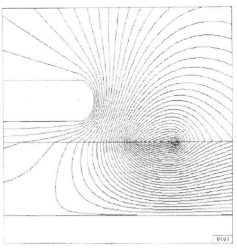

Figure 4. Potential distribution with an uncharged dielectric.

Figure 5. Potential distribution with a charged dielectric.

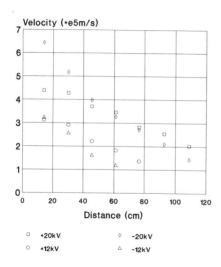
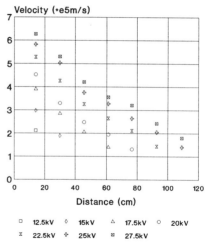

Figure 6. Discharge propagation velocities- effects of applied impulse voltage.

Figure 7. Discharge propagation velocities - effects of corona voltage.

ELECTRICAL MEASUREMENTS

A measurement system was developed in order to investigate the characteristics of surface discharges during propagation [4]. In this system the earthed strip present on the under surface of the polymer was broken to form six separate probe regions situated at positions along the earthed backing strip. As the discharge advanced over each probe, the electric flux changed as the surface charge was altered. The change in electric flux results in an induced current which could be measured using an appropriately located viewing resistor. The onset of the induced current is used to indicate the arrival of the discharge front in the vicinity of the probe. This effect could be used to accurately measure the discharge velocity at a number of different positions along the surface.

The effect of varying the impulse voltage polarity on the discharge propagation velocity was measured and the results for applied voltages of 12kV and 20kV are shown in Figure 6. These measurements were undertaken for a corona voltage of 20kV with a polarity opposite to that of the impulse voltage. The results have shown that the initial velocity associated with the negative applied impulse is significantly affected by the magnitude of the applied voltage to an extent greater than that which occurred for a positive impulse. The effect of changing the corona voltage when the applied impulse voltage is held constant is shown in Figure 7. As the corona voltage is increased, and hence the surface charge density, the discharge propagation velocity also increases. Measurements have shown that the surface charge increased monotonically with the corona supply voltage.

OPTICAL MEASUREMENTS

A study has been carried out to determine the spectral content of the light emitted from the surface discharge experiment using the spectrometer. This was used to measure the spectral content (200-500nm) of the light emitted from a surface discharge initiated along a polyester surface by a 20kV voltage pulse of 90ns duration. The results of Figure 8 show the discharge spectra obtained from discharges in air at a pressure of 0.1 and 0.6 bar, with no surface charge present. The emission consists of a continuum spectrum with a discrete line spectrum superimposed. The results show that the continuum spectrum increases by a factor of ~2 when the pressure is increased from 0.1 bar to 0.6 bar. The relative magnitudes of the line spectra for the two pressures is also different, particularly in the range 220-260, 320-350, 400-425and 470-510nm. The results shown in Figure 9, illustrate the effect of pre-charging the surface at 0.6 bar. It can be seen that the presence of the surface charge increases the magnitude of the continuum and alters the relative intensity of the line spectrum.

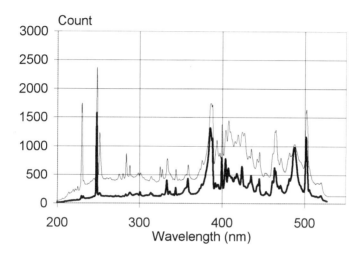

Figure 8. Discharge spectra in air at pressures of 0.1bar (bold) and 0.6bar (light).

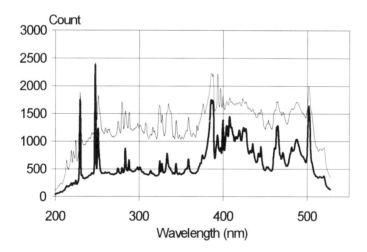

Figure 9. Discharge spectra in air at 0.6bar, for an uncharged surface (bold) and a charged surface (light).

CONCLUSION

Electrostatic modelling has shown that the deposition of surface charge on a dielectric substrate results in a decrease in the normal component of electric field surrounding an advancing surface discharge. This finding indicates that for a pre-charged surface, the propagating discharge may become less coupled to the underlying surface.

Discharge velocity measurements have shown that the initial velocity of the surface discharge is significantly affected by the polarity of the applied impulse voltage. An increase in the impulse voltage from 12kV to 20KV has been found to increase the initial velocity of the discharge by ~100% in the case of a negative impulse voltage and by only 30% in the case of a positive impulse voltage.

A spectroscopic study has shown that light emission from a surface discharge consists of a continuum spectrum with discrete lines superimposed. Initial spectral measurements in air have shown that the continuum in the 200-500nm spectral range increases by ~100% for an increase in the pressure from 0.1 bar to 0.6 bar. Due to the presence of surface charge, the continuum has been found to increase in magnitude and the relative line intensities have been altered. The reasons for this are believed to be associated with the reduced coupling which takes place between the discharge and the dielectric in the presence of surface charge.

ACKNOWLEDGMENTS

The authors would like to thank F A Tuema and E Bryce for their assistance in gathering some of the data for this paper. The financial support of the Centre of Electrical Power Engineering and the Faculty of Engineering is acknowledged with thanks.

REFERENCES

1. R E Beverly, Light emission from high current surface spark discharge, Progress in Optics XVI, Edited by E Wolf, pp359-411, 1978.
2. S J MacGregor, R A Fouracre, M J Given and S M Turnbull, The propagation of a 4 metre guided surface discharge across a dielectric surface, IEE Int. Conf. On Dielec. Mats. Meas. And Appl., Bath, pp184-189, 1996.
3. ELECTRO-Boundary Element Method, Integrated Engineering Software, Winnipeg, Canada.
4. E Bryce, R A Fouracre, M J Given, S J MacGregor and S M Turnbull, An investigation into the characteristics of large scale surface discharges, IEE Colloquium Pulsed Power '97, Savoy Place, London, 1997.

DISCUSSION

P. HAALAND: What is the source of continuum emission which is particularly strong with SF_6? Are there hot particles in the system?

R. A. FOURACRE: Infrared emission is very strong. All but the U.V. is compatible with hot particle production. However, we have not looked for residual particles in the chamber.

A. BULINSKI: Was there any change in the detected spectra along the surface where the discharge took place?

R. A. FOURACRE: To date we have no measured lateral effects. I would expect decreasing interaction between the plasma and the substrate along the discharge due to voltage drop along the length resulting in a less energetic plasma front.

D. SCHWEICKART: Did you measure the amount of charge (i.e., charge density) deposited on the surface by the corona source? If so, what was the relationship between the deposited charge density and the velocity of propagation of the pulsed surface discharge?

R. A. FOURACRE: Yes, we measured the surface charge density. There is an approximately linear relationship between corona voltage and surface charge density.

M. GOLDMAN: What proportion of the deposited charge disappears in the surface discharge and how far from the surface discharge axis?

R. A. FOURACRE: We have not yet measured (i) the surface charge distribution relative and transverse to the discharge axis prior to discharge, and (ii) the remaining charge post discharge.

ELECTRICAL SURFACE DISCHARGES ON WET POLYMER SURFACES

Hans-Joachim Kloes, Dieter Koenig and Michael G. Danikas[1]

Darmstadt University of Technology
High Voltage Laboratory
Landgraf-Georg-Str. 4
64283 Darmstadt
Germany
[1]Democritus University of Thrace
Dept. of Electrical and Computer Engineering
67100 Xanthi
Greece

ABSTRACT

This paper tackles the problem of the influence of water droplets on the surface of HV insulators on the electric field distribution. During the lifetime of such a high voltage component, the surface of the solid insulation is very often stressed with layers of water droplets. These droplets tend to deform the electric field. Local field enhancements ensue. Field forces causing a deformation of the droplets may be produced as well as surface discharges starting from these droplets. Chemical deterioration of the insulator surface may follow.

The present paper offers an overview on the formation of such surface discharges and of their influence on the deterioration of the polymer surface. The sort of the described phenomena can be observed in both indoor and outdoor insulation. The importance of different parameters, e.g. discharge intensity and repetition rate, of the surface discharges on the deterioration of the polymer surface is also emphasized. Additionally it is shown that field inhomogeneities, caused by droplets on a polymer surface, can be estimated by using a numerical calculation program.

INTRODUCTION

Interfaces between solid and gaseous and between liquid and gaseous insulating materials are considered to be "weak" zones in insulating systems. The study of the interfaces becomes thus of paramount importance in enabling us to design medium and high voltage equipment capable to have a lifetime of several decades.

Former studies have dealt with the aging performance of cylindrical samples under the simultaneous stressing of wet conditions and high electric field[1,2,3]. Aging depends on the type of resin and filler material as well as the processing method[4]. During the so-called "early" aging stage a loss of hydrophobicity occurs. High voltage and humidity on the insulator surface generate surface discharges in the droplet layer[2]. These discharges generate in turn ozone and nitric oxides. The surface of the insulator is attacked by a combination of physico-chemical stresses resulting in the so-called "electrolytic partial discharge erosion"[5]. Increasing aging time results into an increase of the number and of the apparent charge of surface discharges[6]. No matter whether the HV insulator is wet, lightly contaminated and highly electrochemically stressed or whether is subjected to more severe environmental stresses, it may be said that the aging process is driven to a great extent in both cases by surface discharges which start at the droplets. Consequently, an investigation of the performance of single droplets under the influence of electric field is important in the effort to understand some of the factors contributing to the mechanism of aging of HV polymer insulators.

SURFACE CONTAMINATION AND SURFACE AGING MECHANISMS

Under dry and clean conditions the electric field distribution on the surface of a solid insulating material can be simulated with the help of a chain of n capacitors (Fig. 1a). In case of wet conditions, however, there is an increased surface conductivity and the electric field can be simulated with a chain of n capacitors and n resistors in parallel, representing the superposition of relevant ohmic and capacitive components of the electric field (Fig. 1b)[2]. The presence of droplets on the surface of a solid insulation changes the electric field distribution and it may lead to an overstressing of the gaseous domain, initiating thus surface discharges. Since the presence of humidity is unavoidable for both indoor and outdoor insulators, it can be claimed that the presence of surface discharges due to the droplets is a common denominator underlying the aging in the aforementioned technologies.

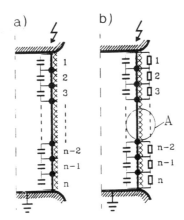

Fig. 1 Modelling of an insulator surface : a) dry and clean conditions,
b) wet and /or contaminated conditions

As has been remarked above, a combination of high voltage and humidity generate surface discharges which may lead to the formation of byproducts. Erosion of the surface ensues and this leads to an increase of surface roughness and a loss of hydrophobicity of the insulator surface. The consequence of these phenomena is double: the surface absorbs more water and builds in turn an increasing thickness of the moisture layer which contributes to an increasing layer conductivity. The creation of solvable nitrates and relevant dissociation processes increase the volume conductivity of the surface layer which also contributes to an increasing layer conductivity[2]. Such a mechanism of aging can well be common to both indoor and outdoor insulators. Both types of insulators are prone to be subjected to the effects of humidity (of course with different degrees of severity) and to natural and industrial pollution (again with different degrees of severity) although their creepage distances vary significantly (outdoor insulators have larger creepage distances than indoor ones)[2]. The implication of all the above is that for both indoor and outdoor insulation aging progresses in stages and that an "early" stage of aging as well as a "late" stage of aging exist. Various researchers payed great attention to the "late" stage of aging, the "early" , however, stage of aging has been rather overlooked. The onset of dry band zones and the change of the nature of surface discharges from low energy discharges to "mini"-arcs indicates mostly the transition from the "early" stage of aging to the "late" stage.

Since the "early" stage of aging is particularly interesting from both a scientific and an industrial point of view, the present paper concentrates on the study of the performance of droplets on the surface of an HV insulator under the influence of an electric field.

INFLUENCE OF THE SHAPE OF WATER DROPLETS

Condensation droplets on the surface of an HV insulator can come about from droplet germs. In Fig. 2, the forces which are exercized on the droplet, in case that no electric field is applied, are shown. They are the surface tension of the liquid (τ_L), of the solid (τ_S) and the interface tension between liquid and solid (δ_{SL}). In case of the presence of an electric field, the droplet will deform because of an additional force. The state of the surface of the solid insulation should not be underestimated. Aging provokes a change of the state of the surface and a loss of hydrophobicity[7].

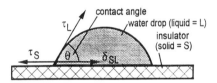

Fig. 2 Force balance at the interface solid/liquid at a water droplet on a solid surface

The tangential electric field on the surface of the insulation creates a force on the surface of the droplet. The droplet is deformed. The deformation of the droplet influences in turn the field distribution and strength. Local field enhancements may follow, which cause micro-discharges between the droplets. This is the beginning of the electrochemical deterioration of the insulator surface. Hydrophobicity is lost, eventhough at a local level. The droplet layer becomes even more inhomogeneous. This has as a result the decrease of the voltage difference across the droplet with decreased resistivity[8] and consequently the decrease of the inception voltage of the micro-discharges. When the electrochemical deterioration sets in, solvable nitrates are created which result in a higher conductivity of the water droplets. The continuation of this process results in the creation of dry zones. The latter are the indication of the beginning of the "late" stage of aging.

EXPERIMENTAL PROCEDURE AND THEORETICAL CONSIDERATIONS

To experimentally observe the behavior of a water droplet on a solid insulation surface, we had to concentrate firstly not on a cylindrical insulator or on an insulator with sheds (the whole system would be too complex) but on a plane test specimen with single droplets on its surface. The uniform electrode arrangement together with the plane specimen and the initial droplet distribution are shown in Fig. 3a. The applied field was 2 kV/cm. The application of electric field deforms immediately the droplets. A few minutes later, a complete water path connects both electrodes (zone P in Fig. 3b). The time duration needed for the change from Fig. 3a to Fig. 3b depends on a number of parameters: electrical field strength, droplet resistivity, surface conductivity and locally stored energy[7, 8].

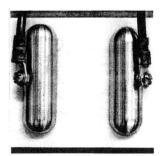 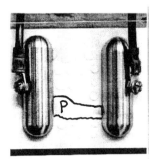

Fig. 3 Performance of a droplet layer at the surface of a plane insulaton in an electric field :
a) droplet distibution without electric field, b) water path after the field was applied for some minutes

The above described geometry was simulated with the field calculation program MAFIA. Fig. 4 shows a cross-sectional view of the calculated field distribution with a train of three droplets between the electrodes. It has to be noted that electric field maxima occur at the points better known as "triple points", i.e. at the common points where air, solid insulation and metallic electrode meet each other. The importance of "triple points" for the dramatic increase of electric field as well as for the initiation of surface discharges is shown. Fig. 5 shows a cross-section of the whole arrangement together with the "triple points".

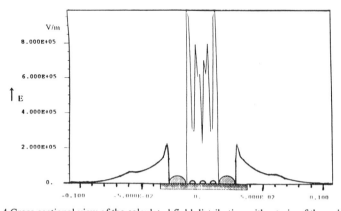

Fig. 4 Cross-sectional view of the calculated field distribution with a train of three droplets

DISCUSSION

The experimental results and the resulting field plot show that the presence of water droplets on the surface of a polymer may have serious effects. Destabilization of the droplets under the influence of the applied electric field may result. Such a behavior is due to the difference in the permittivities of the materials involved. Elongation of droplets and the appearance of discharge was also observed by others[9], although with a different electrode arrangement. The aging of the polymer surface may lead into a possible increase of the conductivity of the droplet with the simultaneous decrease in voltage across the droplet. In actual fact, as was noted in the past[8], the voltage along the droplet will be lower the higher the conductivity. Surface discharges will follow which accelerate the aging process. The droplet performance greatly depends on the conductivity of both the surface and of itself. The duration of the "early" stage of aging can be extended in case of a reduction of the conductivity of the water droplets[10]. Increase of the volume conductivity of the surface layer reduces the duration of the "early" stage of aging, too[2]. Subsequently, even "light" contamination can be an important stress factor. Self-contamination generated by low energy surface discharges can also be important for the further aging of the insulator. Since self-contamination can not be avoided - even to a low degree - also in outdoor insulation (despite the self-cleaning effects of rain), it is evident that some parallel exists between indoor and outdoor insulation aging, especially if we take into account the "early" stage aging phenomena.

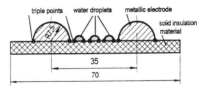

Fig. 5 Geometry of the specimen and critical zones with droplets

A more complete picture of the above phenomena should emerge if one takes into account not only the influence of the applied electric field on the shape of the droplet but also the influence of the resulting shape of the droplet on the field distribution. There are still a number of parameters that are not as yet well defined, such as the conductivity of the droplets and the non-linear behavior of the droplets. However, such parameters are important because they have an influence on the applied field. The discharges taking place between the droplets are indeed discharges in air and a modelling of such phenomena is possible, particularly if ones bears in mind that for a first rough approximation the droplets can be considered as metallic. As a later step, the simulation of the droplet performance taking into account the continuous variation of the droplet conductivity can also be carried out. Furthermore, a comparison between experimental and simulation data with electrode arrangements and insulator samples closer to practice than that in Fig. 3a is needed.

CONCLUSIONS

Some experimental results on water droplet performance under electric field infuence have been presented in this paper together with simulation results. Surface discharges play a significant role in setting an "early" stage of aging for both indoor and outdoor insulation. This stage has been often overlooked although it can be of vital importance to diagnose correctly and early enough the state of an insulation. The performance of the droplet depends on various parameters, such as electrical field strength, droplet conductivity and surface hydrophobicity.

REFERENCES

1. I. Quint, Investigations according to the influence of surface contamination of low conductivity on the surface ageing performance of cylindrical test specimen under a.c. stress, (in German), Ph. D. Thesis, Darmstadt University of Technology, 1993
2. D. Koenig, Surface and aging phenomena on organic insulation under the condition of light contamination and high electric stress, Proc. of Nord. Insul. Symp. NORD-IS 94, June 13-15, 1994, Vaasa, Finnland, paper 01, pp. 17-35
3. O. Claus, Characterization of the state of the surface of cylindrical epoxy resin samples before and after the stressing with wet polluted layers and 50-Hz a.c. voltage, (in German), Ph. D. Thesis, Darmstadt University of Technology, 1995
4. H.-J. Kloes and D. Koenig, Thin solid silicone layers on the surface of model insulators made of resin and their influence on wet surface performance under high electrical stress, Ann. Rep. Conf. Elec. Insul. Diel. Phen., October 19-22, 1997, Minneapolis, USA, pp. 121-124
5. D. Koenig and B. Müller, Contribution on the specification of an aging test for the insulation of enclosed air-insulated switchgear, *IEEE Trans. Electr. Insul.*, vol. 22, no. 6, pp. 769-774, 1987
6. O. Claus, D. Koenig and W.-K. Park, Partial discharges on wet and lightly contaminated insulating surfaces and relevant diagnostics, *Gaseous Dielectrics VII*, L. G. Christophorou and D. R. James Ed., Plenum Press, New York (1994), pp. 625-633
7. H.-J. Kloes and D. Koenig, Basic investigations of the performance of droplets on electrically stressed polymer surfaces, Ann. Rep. Conf. Elec. Insul. Diel. Phen., October 19-22, 1997, Minneapolis, USA, pp. 374-377
8. T. Schütte, Water drop dynamics and flashover mechanisms on hydrophobic surfaces, Proc. of Nord. Insul. Symp. NORD-IS 92, June 15-17, 1992, Vasteras, Sweden, paper 8.1, pp. 1-14
9. D. Windmar, L. Niemeyer, V. Scuka and W. Lampe, Discharge mechanisms at wetted hydrophobic insulator surfaces, Proc. of Nord. Insul. Symp. NORD-IS 92, June 15-17, 1992, Vasteras, Sweden, paper 8.2, pp. 1-13
10. D. Koenig and I. Quint, Multifactor aging mechanisms on the surface of model post insulators, Ann. Rep. Conf. Electr. Insul. Diel. Phen., October 18-21, 1992, Victoria, B.C., Canada, pp. 664-670

DISCUSSION

K. NAKANISHI: Thank you for your interesting presentation. My question is concerned with Fig. 1 of your paper. Figure 1 shows a simple circuit which accounts for the surface discharge with droplets. I would like to know whether the shape and the distribution of the droplets are considered in the circuit or not?

D. KOENIG: You are right that Fig. 1 presents a simplified circuit which aims only to explain the basis of a change from dry to a wet surface. In order to do field calculations one has to take into account the shape and distribution of the droplets and microdry zones and their relevant material data. However, this is not a trivial problem and support from field calculation experts is required.

M. GOLDMAN: A problem arising from your talk is that of the initial stages. At the beginning your polymer surface is hydrophobic. Fixation of water on it implies the presence of dipoles. How do you see the phenomenon initialization?

D. KOENIG: We did not take into account nor have observed the phenomena you are reporting. However, we are interested to learn from your experience and experiments on the role of dipoles you suppose to be present, if water droplets are present on the insulating surface. I agree that my presentation aims at simplifying the very complex conditions on an insulating surface, and at concentrating on those phenomena which we can identify and quantify with our diagnostic tools and which we assume to be the main and dominant ones.

M. S. NAIDU: Similar problems do occur with line insulators. Can you comment please?

D. KOENIG: You are right that the same phenomena as reported do occur on line insulators and are of special interest for modern composite insulators under the aspect of ageing. However, in the outdoor insulation community the phenomena in the early stage of ageing based on micropartial discharges have mainly been overlooked, while phenomena in the late phase of ageing have been in the center of interest. But this situation is changing now and work dealing with these early basic phenomena has been started at several laboratories.

A. BULINSKI: Have you tried epoxy resin coated with silicon rubber?

D. KOENIG: We did not yet, but we are preparing a relevant experiment aiming at finding out details on the (supposedly limited) amount of compounds in the silicon rubber bulk which are able to restore hydrophobicity. It is supposed that their "repair" effect is dependent on the bulk volume, i.e., the thickness of the silicon rubber coating, and tends to disappear if the stored compounds are more-or-less spent, i.e., the bulk is "exhausted."

TERMINATION OF CREEPING DISCHARGES ON A COVERED CONDUCTOR

Lars Walfridsson, Udo Fromm, Anna Kron*, Li Ming, Rongsheng Liu,
Thorsten Schütte, Dan Windmar and Mats Sjöberg†

ABB Corporate Research, S-721 78 Västerås, Sweden
* Casco Nobel AB, 85013 Sundsvall, Sweden
† Chalmers Tekniska Högskola, Gothenburg, Sweden

ABSTRACT

Under laboratory conditions, a polymer covering of electrodes has resulted in a reduction of the insulating distances, under impulse stress, by a factor of two. One problem is creeping discharges on the surface of the covering.

This paper presents a new approach to stop the propagation of a creeping discharge on a covered conductor. This was achieved by inserting a stress cone of the type used for distributing stress in high voltage cables, into the discharge path.

The experiments were performed with 1.2/50 µs lightning impulses which showed a considerable increase of the breakdown voltage, in arrangements comprising covered electrodes by using cable stress cones to stop the propagation of creeping discharges. In addition, the breakdown voltage of the arrangement became virtually polarity independent.

INTRODUCTION

For urban areas, real estate prices will continue to increase due to land shortages. There is a demand for substations designed to occupy a small land area. This demand has led to the usage of electronegative gases such as SF_6 in switchgears. However, a number of factors make it desirable to replace SF_6 insulated systems. Firstly, air insulated switchgears are avoiding the intrinsic complications of pressurized gas filled equipment. Furthermore, SF_6 is one of the six gases included in the Kyoto protocol, due to its possible greenhouse effects, and may thus in future be subject to restrictions.

Optimized air-polymer insulation can become an alternative to enclosed gas insulated switchgears, (GIS). The insulation behavior of air/polymer insulation can be improved by utilizing insulating barriers [1-3], foam [4] or electrodes with insulating cover [5,6]. Under laboratory conditions, polymer covering the electrodes has resulted in the reduction of the insulating distances by a factor of two [5,6]. One problem with covered conductors is, however, creeping discharges. The propagation mechanism for these discharges is described in standard textbooks, eg [7]. The propagation of creeping discharges is favoured by conductive contaminations and by a large specific capacitance of the covering material. (The thinner the insulating layer and the higher its permittivity, the higher the specific capacitance of the covered conductor surface). A creeping discharge may propagate along the insulating layer until it meets a conducting surface. This limits the ability of covered conductors to raise the breakdown voltage of covered voltage electrode arrangements [6]. One possible counter measure is to insert one electrical insulating shed into the discharge part and another is to terminate the covered conductor by an electric field steering element, (eg a stress cone as used in HV cable termination).

The aim of this work was to study this new approach to stop the propagation of a creeping discharge on a covered conductor before it reaches the other electrode and thus causes a flashover.

EXPERIMENTAL METHOD

The test sample was a XLPE insulated high voltage cable, arranged as shown in Figure 1. All layers outside the XLPE insulation were removed, (semiconductor layer, metallic screen etc.). The diameter of the aluminum core was 14 mm and the thickness of the XLPE insulation was 6 mm. Tests were performed both with and without cable stress cones of the type ABB DON APIT 6. The cable stress cones were made of EPDM and had a semiconductive layer making the electric field distribution more even, see Figure 2. In the middle of the cable length there was a 10 mm wide and 0.1 mm thick copper foil which was connected to earth. At both cable ends there were toroidal shaped shields to distribute the electric field more evenly. A lightning impulse voltage, 1.2/50 µs was applied at one end of the cable. The XLPE insulation was thick enough to withstand the test voltages without puncture. All flashovers took place along the surface of the cable and of the stress cones, between the copper foil and the toroidal shields.

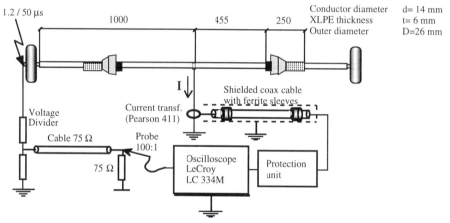

Figure 1. XLPE cable with stress cones. (All measures in mm)

The current measurement was done as shown in Figure 1. A current transformer was placed around the ground connection and the signal was recorded by a digital oscilloscope, LeCroy 884M, through a protection unit, Protec Devices CX12LC. The impulse voltage was measured simultaneously with the same oscilloscope by means of a voltage divider and a probe.

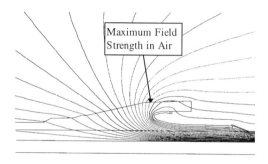

Figure 2. Equipotential lines around the stress cone when a creeping discharge connects the semiconductive layer of the stress cone with earth.

Lightning impulse tests on the test object were performed according to the IEC 60-1 "up and down" method. Each test sequence consisted of 25 consecutive impulses applied when an approximate break down voltage level had been reached. Between each impulse the test object was discharged and cleaned with ethanol, to ensure that the results were reproducable. The impulse tests were performed with both positive and negative polarity. The test results were corrected for temperature and air pressure.

RESULTS AND DISCUSSION

The rapid oscillations found approximately at the time of triggering the oscilloscope (0), as shown in Figures 3-10, were caused by the trigger circuit of the impulse generator.

Voltage and current oscillograms for a covered conductor without stress cones are presented in Figures 3-6. Two typical observations may be seen both in Figures 3 and 5 where no breakdown occurred and in Figures 4 and 6 where breakdowns occurred. The current during a positive impulse has a higher amplitude than during a negative impulse. On the other hand, the duration of the current is longer for a negative impulse than for a positive impulse. The breakdown occurs shortly after crest time for positive impulses, as shown in Figure 4, while breakdown for negative impulses occurs later, typically close to 8 µs after voltage application, see Figure 6.

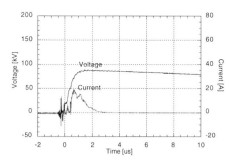

Figure 3. Voltage and current oscillogram for a voltage level close to breakdown. Crest voltage = +89.2 kV on a test sample without stress cones. No breakdown occurred.

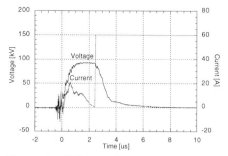

Figure 4. Voltage and current oscillogram for a voltage level close to breakdown. Crest voltage = +93.4 kV on a test sample without stress cones. Breakdown occurred.

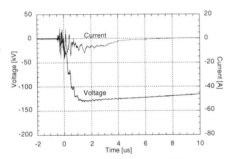

Figure 5. Voltage and current oscillogram for a voltage level close to breakdown. Crest voltage = -130.8 kV on a test sample without stress cones. No breakdown occurred.

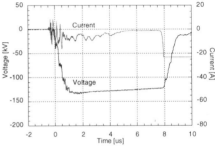

Figure. 6 Voltage and current oscillogram for a voltage level close to breakdown. Crest voltage = -134.2 kV on a test sample without stress cones. Breakdown occurred.

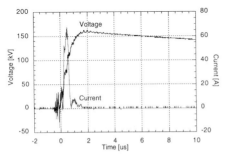

Figure 7. Voltage and current oscillogram for a voltage level close to breakdown. Crest voltage = +159.2 kV on a test sample with stress cones. No breakdown occurred.

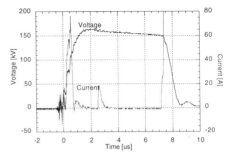

Figure 8. Voltage and current oscillogram for a voltage level close to breakdown. Crest voltage = +163.2 kV on a test sample with stress cones. Breakdown occurred.

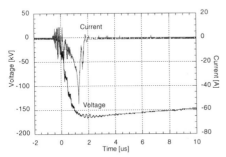

Figure 9. Voltage and current oscillogram for a voltage level close to breakdown. Crest voltage = -161.6 kV on a test sample with stress cones. No breakdown occurred.

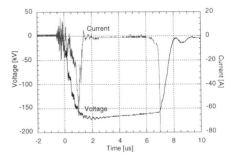

Figure 10. Voltage and current oscillogram for a voltage level close to breakdown. Crest voltage = -170.1 kV on a test sample with stress cones. Breakdown occurred.

Voltage and current oscillograms for a covered conductor with stress cones are presented in Figures 7-10. Typical observations shown in these Figures are that the current during a positive and negative impulse has almost the same amplitude. The duration of the current is however longer for a negative impulse than for a positive impulse. Other observations are that the current obtained at positive impulses has a very short rise time, reaching its maximum before voltage crest and rapidly dropping back to zero. The current obtained at negative impulses, however, increases more slowly, with a final fast rise, (probably when the discharge reaches the semiconductive layer of the stress cone), at a time close to voltage crest, before it drops back to zero. The shape at the front of the current at breakdown, shows similarities to the front seen in the first large current pulse during voltage rise. A few of the current oscillograms obtained at voltages close to breakdown contained one or more smaller pulses, as shown at time 2.5 μs in Figure 8.

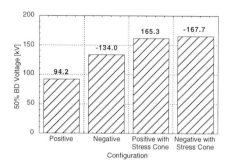

Figure 11. 50% Breakdown voltage vs. Configuration

The 50% breakdown voltages for the four different tests are shown in Figure 11. The experiments presented in this report show a considerable increase of the breakdown voltage in arrangements comprising covered electrodes by using cable stress cones to stop propagation of surface discharges. As an additional effect, the breakdown voltage of the arrangement became virtually polarity independent. For a covered conductor without a stress cone, positive polarity of impulse (negative creeping discharge) results in a lower breakdown voltage than for negative polarity. This is consistent with the result reported by Takahiko Yamashita et al. [8]. Figure 2 shows the equipotential lines around the stress cone, in the situation where the discharge connects the semiconductive layer of the stress cone with ground. The field strength calculated at 50% breakdown level, indicates that the maximum electrical field strength in air will be about 3.6 kV/mm. This field strength is above the withstand level for air, resulting in a complete breakdown of the test sample. The breakdown field for air, normally 3.0 kV/mm, is exceeded because the voltage drop in the discharge channel, ≤1 kV/mm, is not considered.

Figure 12 shows a picture of a discharge that was stopped by the stress cone. In the case of a flashover, as shown in Figure 13, the flashover always occurred on the outside of the stress cone. The stress cone was never punctured. The corresponding voltage and current oscillograms are shown in Figures 7 and 8.

Figure 12. A discharge, stopping at the stress cone. Crest voltage = +159.2 kV

Figure 13. A flashover. Crest voltage = +163.2 kV.

The current measurements and visual observations show clearly that the discharges are stopped by the stress cones. The high value of the current, found in Figures 3-10, is believed to be due to the large capacitive current produced by a fast growing surface discharge in combination with a short voltage impulse rise time. In the case of measurements where no breakdown occurred, (Figures 7 and 9), it was confirmed visually that the streamer reached the stress cone. Therefore, it is possible to approximate the magnitude of the current. The speed of the discharge is assumed to be constant and the current is calculated for the moment when the streamer reaches the semiconductive layer. The capacitor consists of the conductor, the insulating XLPE layer and the fast propagating discharge being the counter electrode (values from Figure 1).

$$dC \approx C = 2\pi \ \varepsilon_0 \varepsilon_r \frac{l}{\ln(D/d)} = 2\pi \ \varepsilon_0 2.3 \frac{1 \text{ m}}{\ln(26 \text{ mm}/14 \text{ mm})} = 0.21 \text{ nF} \qquad (1)$$

During the propagation of the discharge a current I is caused (voltage and time taken from Figure 7), where

$$I = C\frac{dU}{dt} + U\frac{dC}{dt} = 0.21 \text{ nF} \frac{80 \text{ kV}}{0.5 \ \mu s} + 100 \text{ kV} \frac{0.21 \text{ nF}}{0.5 \ \mu s} = 74 \text{ A} \qquad (2)$$

This value is a first order approximation, because the voltage drop along the discharge and a possible change of the speed of the discharge are not taken into account, whereas it is assumed, that the discharge covers 100% of the surface.

CONCLUSION

1. The use of covered electrodes is restricted due to creeping discharges.
2. Propagation of creeping discharges can be stopped by cable stress cones.

Further work is required on variations of the streamer stopping devices, e.g. by changing the size of the stress cones or replacing them with simple sheds. In addition, other geometrical arrangements of covered conductors could be tested.

REFERENCES

1. H. Roser, Schirme zur Erhöhung der Durchschlagsspannung in Luft, *E.T.Z.*, Vol. B17, pp. 411-412 (1932)
2. Li Ming, M. Leijon, T. Bengtsson, M. Darveniza, Barrier Effect in a Rod/Rod Air Gap under DC Voltage, *7th Inter. Symp. on Gaseous Dielectrics*, Knoxville, USA, April 24-28 (1994).
3. A. Beroual, A. Boubakeur, Influence of Barrier on Lightning and Switching Impulse Strength of Mean Air Gaps in Point/Plane Arrangements, *IEEE Trans. on El. Ins.* Vol. 26, pp 1130-1139 (1991).
4. U. Fromm, Li Ming, T. Schütte, L. Walfridsson, A. Kron, Rongsheng Liu, D. Windmar, Electric Field Control by Polymer Foam, *ICSD*, Västerås, Sweden (1998)
5. P.O. Geszti, J. Patk, Overhead Lines with Insulated Phase Conductors, *Acta Technica Academiae Scientarum Hungaricae*, Vol 73 (3-4), pp. 265-277 (1972)
6. Li Ming, U. Fromm, M. Leijon, D. Windmar, L. Walfridsson, A. Vlastos, M. Darveniza, J. Kucera, Insulation Performance of Covered Rod/Plane Air Gap Under Lightning Impulse Voltage, *10th ISH*, Montreal, Canada, August 25-29 (1997)
7. G. Hilgarth, *Hochspannungstechnik*, B.G. Teubner Verlag Stuttgart, pp 80-81 (1981)
8. T. Yamashita, G. Satoh, T. Fujishima, H. Matsuo, T.Shibata, Propagation Characteristics of a Surface Discharge on Covered Conductors and Cylindrical Dielectric Materials, *10th ISH*, Montreal, Canada, August 25-29 (1997)

DISCUSSION

S. MATSUMOTO: What is the difference of design concept between impulse and AC voltage?

L. WALFRIDSSON: The concept is like this: It is the impulse voltage that determines the distance between conductors and ground or conductor-conductor. If the conductors are covered, they can be located closer to each other and to ground. If there is a lightning impulse overvoltage, discharges can cross the air gaps but the following creeping discharges will be stopped by the stress cones. The covered conductors can however not be located too close, allowing PD discharges to occur at the surface of the cover under normal and abnormal AC-voltage stress.

K. NAKANISHI: In your test the gas flows perpendicularly to the surface of the insulating board. Why didn't you make the gas flow along the surface of the board? In the transformer the gas flows along the surface of the coils absorbing the heat from the surface.

L. WALFRIDSSON: The electrification due to the parallel flow case is similar to that of the vertical flowing to the board.

STATIC ELECTRIFICATION PHENOMENA ON DIELECTRIC MATERIALS OF SF6 GAS-INSULATED TRANSFORMER

M. Ishikawa,[1] T. Kobayashi,[1]
T. Inoue,[2] T. Goda,[2] A. Inui,[2] T. Kobayashi,[2] T. Teranishi,[2] and M. Meguro[2]

[1]The Tokyo Electric Power Company, Inc.
Tokyo, Japan
[2]Toshiba Corporation
Kawasaki, Japan

1. INTRODUCTION

In Japan, large-capacity gas-insulated transformers (275kV, 300MVA) designed to be cooled by flowing SF6 gas have been developed and are operating in underground substations.[1] They use insulating plastics with excellent heat resistance and insulation strength, in addition to cellulose materials such as insulating paper and pressboard used in conventional oil-immersed transformers. Such insulating plastics have a high surface resistance and can be electrified through contact with each other, delamination, or friction. Again, the resistivity of SF6 gas is so high that once the insulation material surface is electrified, charges are slow to leak out.

In the gas-insulated transformer designed to be cooled by flowing SF6 gas, the SF6 gas flows at high speed over or against the insulation material surfaces. Thus, to ensure the long-term reliability of gas-insulated transformers, it is important to investigate the static electrification of insulation materials due to a high-speed flow of SF6 gas. In addition, although gas-insulated transformers are manufactured under carefully controlled conditions to prevent intrusions of foreign matter, insulating dust can be generated or induced. Thus, it is also important to investigate electrification due to SF6 gas mixed with insulating dust. This paper describes the results of investigations of electrification caused by SF6 gas flowing over insulation surfaces.

2. EXPERIMENTS

2-1 Static Electrification Due to the Flowing of Pure SF6 Gas

Fig. 1 shows the schematic diagram of experimental setup. It consisted of a dielectric

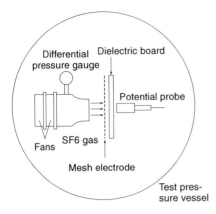

Fig. 1 Schematic diagram of experimental set-up for static electrification due to SF6 gas flow

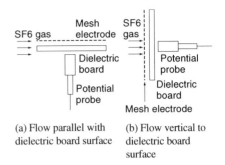

(a) Flow parallel with dielectric board surface

(b) Flow vertical to dielectric board surface

Fig. 2 SF6 gas flow approach angles to dielectric board

board, fans for blowing SF6 gas over the board surface and a probe to measure the surface potential of the dielectric board. The test vessel was vacuumized to below 13Pa, then filled with SF6 gas to 0.5MPa.

With the help of two fans arrayed in series, SF6 gas was blown out of a nozzle with a diameter of 50mm and onto the surface of the dielectric board. Two flow directions of SF6 gas were chosen: parallel and vertical to the dielectric board as shown in Fig. 2 (a) and (b). The flow rates of SF6 were obtained by measuring pressure differences between the fan inlet and outlet with a differential pressure gauge housed in the tank.

In the test, the flow rates of SF6 gas ranged from 5 to 25m/s. Static potential on the dielectric board surface was measured with a rotating-sector surface electrometer which was previously calibrated in SF6 gas. As shown in Fig. 1, 10mm off the fan side of the dielectric board was a metal mesh earth electrode (mesh: 1.0mm, mesh wire: 0.3mm) set to keep the capacitance between the dielectric board and the mesh electrode constant when measuring the surface potential of the dielectric.

For the dielectric board, three typical types used in gas-insulated transformers were employed. Their properties are shown in Table 1. The mixed board was prepared by making paper from a mixture of polyethylene terephthalate fiber and aromatic polyamide pulp and pressing it into the board under a high pressure at a high temperature. Before the test, the dielectric boards were dried at 110°C for more than 48h, and their surfaces were deelectrified by cleaning them with ethyl alcohol.

2-2 Electrification Due to Flow of SF6 Gas Mixed with Dielectric Particles

An investigation was carried out on static electrification due to the flow of SF6 gas mixed with dielectric particles. Table 2 shows the dielectric particles used in the test. They were dried at 110°C for more than 48h. They were prepared by crushing dielectric boards

Table 1 Properties of dielectric materials

Material	Dielectric constant	Surface resistivity (Ω)	Density (g/cm^2)
PET film	3.2	5.0×10^{15}	1.40
Mixed board	2.7	4.7×10^{15}	1.24
Pressboard	3.4	2.8×10^{13}	1.15

Table 2 Grain sizes of three kinds of dielectric particle

Particles	Mean grain size (μm)	Amount (g)
From PET	36	5
From mixed board	42,200	5
From pressboard	40,150	0.25, 1, 5, 10

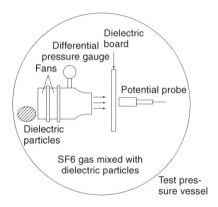

Fig. 3 Schematic diagram of experimental set-up for static electrification by SF6 gas flow mixed with dielectric particles

Table 3 Surface potential of dielectrics due to flow of pure SF6 gas

Time	Surface potential (V)		
	PET film	Mixed board	Press-board
When installed in tank	+900	+1100	+80
When tank filled with SF6 after vacuumization	0	0	0
After receiving vertical SF6 gas flow (25m/s, 10h)	0	0	0

and classifying particles by sifting. As shown in Fig. 3, they were placed near the suction port of the fans. The test vessel was vacuumized to below 13Pa, then after more than 1h, it was filled with SF6 gas to 0.5MPa. By running the fans, particles were blown into the SF6 gas to prepare SF6 gas mixed with particles. This gas was blown vertically against the dielectric board (200×200×2mm), which was the same one used in 2-1. In this test, no mesh electrode was set between the dielectric board and the fans to prevent it from obstructing the flow of particles.

The SF6 gas mixed with particles flowing against the dielectric board was observed through a window provided in the shell of the test tank with the help of a high-speed video (400 pictures/s). It was found that directly after starting the fans, many of the positioned particles were sucked up and blown against the dielectric board, but within about 1 min after the fans reached the specified speed the particle quantity against the board became constant.

Tests were carried out with several combinations of dielectric boards and particles, as well as different sizes and amounts of particles. The SF6 gas pressure in the tests was 0.5MPa and the SF6 gas speed ranged from 5 to 25m/s.

3. TEST RESULTS

3-1 Static Electrification by Pure SF6 Gas

While varying its flow rate from 5 to 25m/s, the pure SF6 gas was blown parallel or vertical to the surface of the dielectric board for 10h each. Some measured values of potentials on the surface of the dielectric board are shown in Table 3. Subsequently, in the test, SF6 gas was blown at a rate of 25m/s to the surface of the dielectric board for 10h, which caused no changes in the potential and no electrification of the surface. In addition, when there was no mesh electrode between the dielectric board and the fans, the potential on the surface of the dielectric board under a SF6 gas flow remained at zero Volt.

3-2 Electrification by SF6 Gas Mixed with Dielectric Particles

Figs. 4 and 5 show some of the test results for the relation between the surface potential of the dielectric board and the flow rate of the SF6 gas mixed with dielectric particles. The values of surface potential here are those of 1h after the gas flow was started.

With SF6 gas mixed with dielectric particles blown against the surface of the dielectric

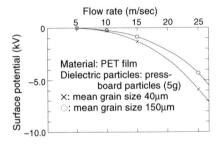

Fig. 4 Surface potential dependence of dielectric film on SF6 gas flow rates (at 1h from start)

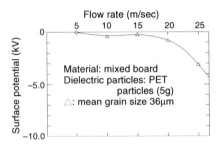

Fig. 5 Surface potential dependence of dielectric board on SF6 gas flow rates (at 1h from start)

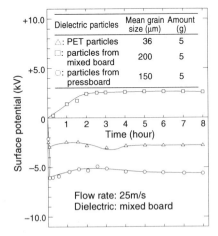

Fig. 6 Time dependency of surface potential of dielectrics

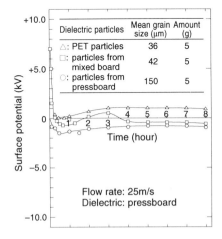

Fig. 7 Time dependency of surface potential of dielectrics

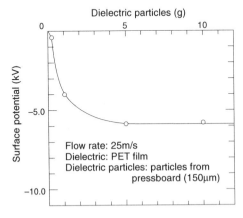

Fig. 8 Surface potential dependence of dielectric film on amount of dielectric particles

board, no electrification was detected in gas flow rate ranges below 5m/s with any combination of dielectric board and particles in the SF6 gas. As the gas flow rate increased above 5m/s, the surface of the dielectric board was electrified and its potential also increased almost in proportion to the square or cube of the flow rate.

Figs. 6 and 7 show changes in the surface potential of the dielectric board with time under a flow of 25m/s for SF6 gas mixed with dielectric particles.

The surface potential and the polarity of the dielectric board are determined from combinations of dielectric boards and dielectric particles in the SF6 gas. With any combination, the surface potential saturates within 1 or 2h. As shown in Fig. 6, with the mixed board employed as the dielectric board, the potential was highest in case with the pressboard particles used as dielectric particles. Even if the material of the particles in the SF6 gas is the same kind as the dielectric board, electrification was also detected. With the pressboard as the dielectric board, the potential was lower than with the mixed board.

An example of the results about effects of the amount of particles included in the SF6 gas on the electrification is shown in Fig. 8. As the content of the particles increased, the potential increased, ultimately toward saturation.

4. DISCUSSION

4-1 Static Electrification Due to Flow of Pure SF6 Gas

SF6 gas is composed of neutral vapor molecules. It is generally recognized that when flowing at a high speed, such molecules are themselves not electrified.[2-4] The present study experimentally proved that a flow of pure SF6 gas, at different speeds ranging up to 25m/s, over the surface of the dielectric board for 10h caused no electrification.

Using the same test apparatus, the authors proved that a flow of ionized SF6 gas over the surface of dielectrics caused the dielectrics to be electrified to a certain level. Ordinarily, SF6 gas is composed of neutral vapor molecules that are so stable they are unlikely to be ionized simply by contact, friction, or collision with dielectric solids. The energy needed to ionize a neutral molecule by collision is estimated to be above several eV.[5] Even if all the kinetic energy of SF6 gas molecules flowing at a speed of 25m/s is converted into collision energy, the kinetic energy of one SF6 molecule is as small as below 1eV. Therefore, a flow of pure SF6 gas, composed of neutral molecules, cannot generate flow electrification.

4-2 Static Electrification Due to the Flow of SF6 Gas Mixed with Dielectric Particles

In flow rate ranges below 5m/s of the SF6 gas mixed with dielectric particles, with any combination of dielectric board and dielectric particles included in the SF6 gas, no electrification was caused by the gas flow. This suggests that electrification due to the flow of SF6 gas mixed with dielectric particles is caused by friction between the dielectrics and the dielectric particles, subject to a threshold of gas flow rates. Electrification probably arises from kinetic energy due to collisions of dielectric particles combined with other factors such as work function and surface condition of the dielectric board. The kinetic energy of particles at the time of collision is proportional to the square of the flow rate. On the other hand, there may be repulsion and neutralization between the electrified dielectric board and particles. With these factors added, the charges probably increased in proportion to the square or cube of the flow rates.

The dust content in the gas-insulated transformers manufactured under carefully controlled conditions to prevent dust inclusion is estimated to be much lower than the content of dielectric particles in the SF6 gas used in the present test. Again, electrification is generated only in flow rate ranges above 5m/s, which are higher than the flow rates in actual gas-insulated transformers. From these facts, it is presumed that in gas-insulated transformers produced under carefully controlled conditions, electrification cannot be caused by the flow of SF6 gas.

Six banks of the 275kV-300MVA gas-insulated transformer have been installed in substations of Tokyo Electric Power Company and have been operated trouble-free for over three years.

This shows that this type of gas-insulated transformer is free from electrification problems due to gas flow.

5. CONCLUSION

Basic tests were carried out on electrification due to the flow of SF6 gas in an assumed gas cooled gas-insulated transformer, yielding the following results:
1. A flow of pure SF6 gas at a rate below 25m/s over the dielectric board surface causes no electrification.
2. A flow of SF6 gas mixed with dielectric particles can cause the electrification on the surface of dielectric board if the flow rate is high, and its potential increases in proportion to the square or cube of flow rate.
3. If electrification is caused by a flow of SF6 gas mixed with dielectric particles, surface potential of dielectric board gradually rises, ultimately reaching saturation within several hours. In relation to the content of particles, it also saturates.

REFERENCES

1. E. Takahashi, K. Tanaka, K. Toda, M. Ikeda, T. Teranishi, M. Inaba, and S. Menju, Development of large-capacity gas-insulated transformer, *T. IEE Japan*, Vol. 115-B, No. 4 (1995) (Japanese).
2. Ph. Pothmann, Zur Selbstentzundung Ausstromen den der Wasserstoffes, *Zeitschrift des Vereines Deutscher Ingenieure*, 66,39,938 (1922).
3. H. Geitel, Zur Frage nach dem Ursprunge der Niederschlagselektritat, *Physik. Zeitschr.*, Vol. 17455 (1916).
4. W.A.D. Rudge, On the electrification given to the air by a steam jet, *Proc. Cambridge Phil. Soc.*, 18,127 (1915).
5. H.S.W. Massey, *Atomic and Molecular Collisions*, Taylor & Francis Ltd. (1979).

PREBREAKDOWN PHENOMENA IN DRY AIR ALONG PTFE SPACERS

Henk H.R. Gaxiola, Joseph M. Wetzer

High Voltage and EMC Group, Eindhoven University of Technology
PO Box 513, 5600 MB Eindhoven, The Netherlands
phone +31 40 2473993   fax +31 40 2450735
e-mail: e.h.r.gaxiola@ele.tue.nl   j.m.wetzer@ele.tue.nl

INTRODUCTION

In the design of electrical gas insulated systems data on breakdown fieldstrengths is extensively used. Breakdown voltages for gases are usually given by Paschen curves, which serve as an engineering tool in high voltage engineering. This work is a continuation of earlier work[1,2] and deals with a study of avalanche and streamer formation in air along teflon (PTFE) insulators.

A fast experimental setup is used to measure the discharge current and the optical discharge activity with and without a spacer present. Sophisticated simulation programs describe the physical discharge processes in the gap (so far without a spacer present). This model is used to obtain a better insight in the relevant mechanisms and processes by a comparison of measurement and simulation data.

MODELLING OF PREBREAKDOWN PHENOMENA

Particle density distributions enable insight in the formation phases leading to breakdown. These are compared with data from optical measurements of discharge activity. From this we can determine relevant processes and parameters, and thereby strip the complex physical 2-D hydrodynamic model developed initially to a "simple" model only incorporating the dominant species.

To obtain a better understanding of the processes fundamental to gas-breakdown we have developed a 2-D model to describe the prebreakdown phenomena and thereby reproduce and get an understanding of the externally measurable quantities. The modelling techniques are presented in detail by Kennedy[1]. The 2-D model describes the spatio-temporal development of the charged species. The incorporated processes are drift, diffusion, ionization, recombination, attachment, detachment, charge exchange, gas-phase photo-ionization and secondary photo-electron emission. The model does not (yet) involve the influence of a spacer.

EXPERIMENTS

An experimental setup with a *250 MHz* bandwidth is used to measure the discharge current (Fig.1). An Intensified Charge Coupled Device (ICCD) camera and two photomultipliers with U.V./I.R. filters record the optical discharge activity. A more detailed description of the experimental setup is given by Gaxiola[2].
The type of gas under study is dry air; pressure = *1,02*10^5 Pa*; voltage = *29 to 30.5 kV*; electrode gap = *1 cm*.

DIAGRAM OF THE EXPERIMENTAL SETUP

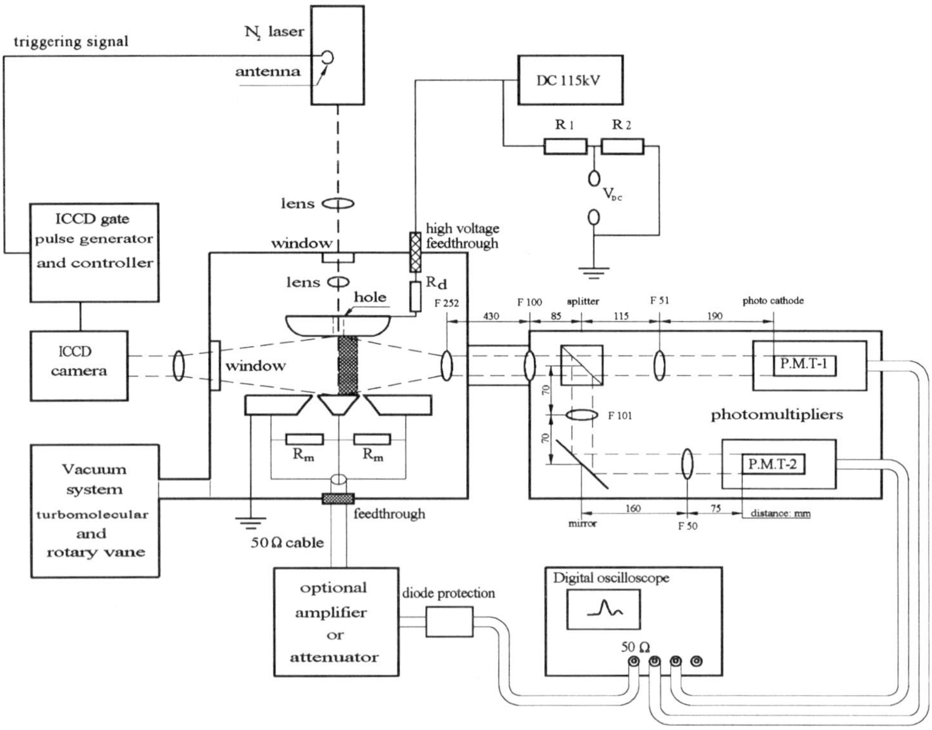

Fig.1 Experimental setup.

RESULTS AND DISCUSSION

In Fig.2 the measured prebreakdown currents, the photomultiplier output with "no filter", "filter 89B (*>700 nm*)" or "filter 03FCG163 (*300nm< <700nm*)" are shown during breakdown for an oblate conical PTFE spacer geometry. On the horizontal axis the time is normalized to the electron transit time in a uniform field T_e. The photomultiplier output signal (PM2) shows the start of the laser-induced discharge activity. Filter 89B filters this dominant N_2-laserpulsline (*337nm*) out completely. At breakdown and overexponential current growth also optical activity in the I.R. (*>700nm*) is generated. There is no difference in photomultiplier signal (PM2) with or without filter.

Figure 3 shows the measured prebreakdown currents, the photomultiplier output representing space-integrated light activity in the gap, and a sequence of optical ICCD images at various time moments during breakdown for different insulator geometries.

The situation with no spacer is also shown; we find a time-to-breakdown of $t_{bd}=110ns$ and a peak current at $T_e=78ns$ of 9 mA (the vertical scale is divided by 5).

Oblate conical spacer I: Electric field attenuation at the cathode side (and field enhancement ($2*E_0$) at the anode along the spacer surface) leads to an increase in time-to-breakdown ($t_{bd}=121ns$) relative to the situation with no spacer present. The small number of initial electrons $N_0=2*10^6$ released results in a relatively small peak current $i(T_e)$. For times beyond T_e the current increases. Due to the increased growth at these times near the anode region the time-to-breakdown t_{bd} is shortened. Image a shows the primary avalanche's arrival at the anode. Images b and c show the streamer formation, whereas images d through f show the pinching of the formed streamer.

Cylindrical spacer: the time-to-breakdown ($t_{bd}=160ns$) increases; electrons trapped at (and later detrapped from) the surface delay breakdown formation.

Oblate conical spacer II: small electric field enhancement at the cathode side and attenuation at the anode side leads to a lower breakdown voltage (29.9kV). We observe a decrease in the time-to-breakdown ($t_{bd}=89ns$). Current growth deviates considerably

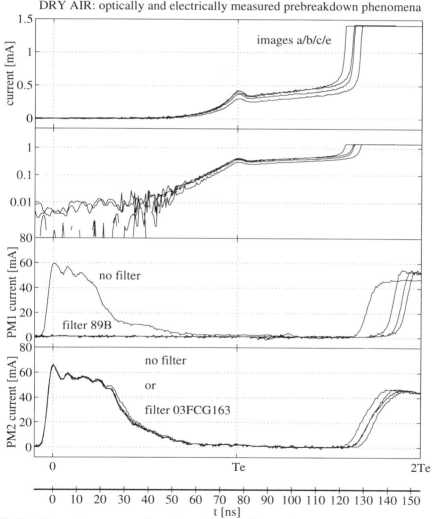

Fig.2 Measured current waveforms (on a linear- and log-scale) and photomultiplier outputsignals for prebreakdown along oblate conical PTFE spacer geometry I.

from exponential avalanch growth. Due to the surface charging by trapped electrons a considerable radial space charge field is build up enabling shorter streamer formation times. The electron transit time T_e slightly decreases. ICCD images j and k show the streamer formation, whereas in images l and m the streamer and subsequent pinching are seen.

Conical spacer geometry: the time-to-breakdown ($t_{bd}=172ns$) increases compared to the case with no spacer present. Like in the situation of oblate conical spacer I only a very small number of initial electrons is released resulting in a relatively small peak current $i(T_e)$. Image n shows the primary avalanche's arrival at the anode, image o streamer formation and images p through r the pinching of the streamer.

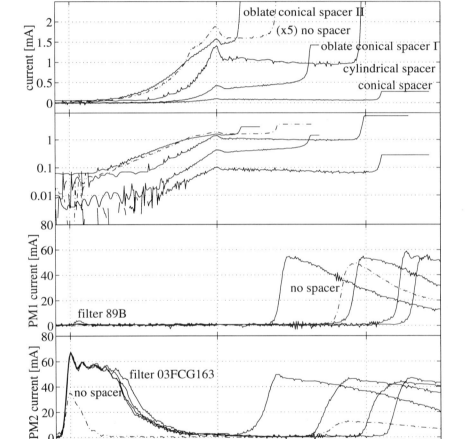

Fig.3 Measured current waveforms, photomultiplier output and sequence of ICCD camera images describing a laser-induced avalanche and it's transition to a streamer. The corresponding times on the measured current waveforms are indicated. The cathode (bottom) and anode (top) surfaces are located at 0 and 1 cm respectively on the vertical axis. The horizontal scaling in centimetres is also indicated. The initial electrons are released from the cathode (bottom) at t=0s indicated by the region around x=0 (-0.05<x<0.05). The maximum photon count per pixel is denoted in the upper right corner of each image. 5 ns ICCD camera gate-shutter time.

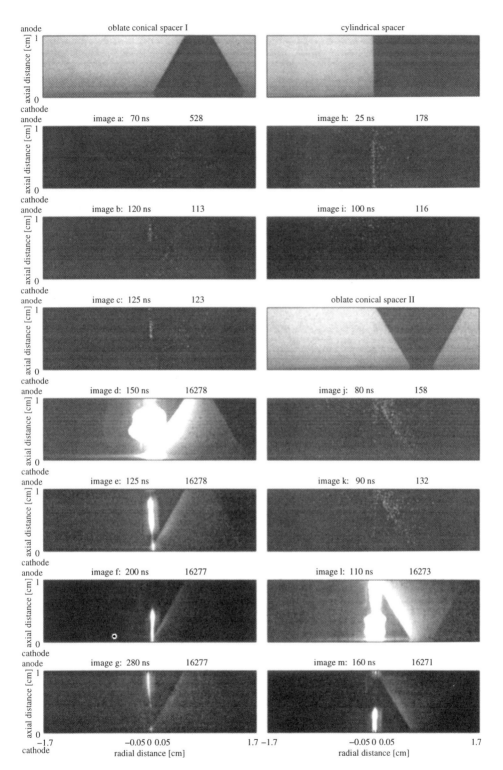

Fig.3 (continued) ICCD camera images of breakdown along PTFE spacer geometries.

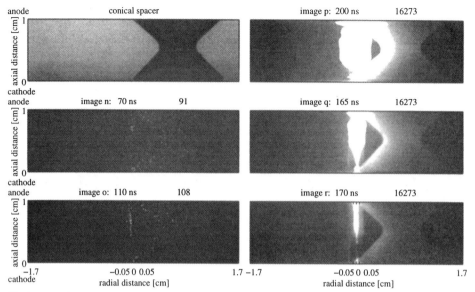

Fig.3 (continued) ICCD camera images of breakdown along a conical PTFE spacer geometry.

Use of photomultipliers helps to depict the laser-induced start of activity, and the time occurence of actual breakdown. The wavelength range for optical activity before breakdown is found to be below *700 nm*. The ICCD camera used is however far more sensitive to optical activity.

From the above we conclude that the modification in electric field by a spacer being present is dominating the discharge resulting in an increase/decrease of the breakdown fieldstrength and the time-to-breakdown. This is ascribed to the field distortion and (in case of a cylindrical spacer) to surface trapping or detrapping processes. For other than cylindrical shapes the insulator surface itself seems to be playing a minor role on the prebreakdown behaviour.

CONCLUSIONS

Earlier investigations without spacers showed that:
* A strong radial expansion is observed during the primary avalanche's gap transition.
* The 2-D model predicts the externally measured current. It correctly describes the streamer formation leading to breakdown.

From the present work the following can be concluded:
* Differences in time-to-breakdown curves for dry air with a spacer indicate a different discharge behaviour. The large radial expansion is not present. This is consistent with the proposed physical explanation of the occuring phenomena.
* No breakdown along the surface occurs! (except for a cylindrical spacer)

REFERENCES

1. J.T. Kennedy, "Study of the Avalanche to Streamer Transition in Insulating Gases". Ph.D. thesis, Eindhoven Univ. of Techn., The Netherlands (1995).
2. E.H.R. Gaxiola, J.M. Wetzer, "A study of streamer formation in nitrogen". 12th Int. Conf. on Gas Discharges and Their Applications, Greifswald, Germany, Vol.1, pp.232-235 (1997).

SECTION 9: GAS-INSULATED EQUIPMENT I

DEVELOPMENT OF THE SF_6/N_2 GAS MIXTURE INSULATED TRANSFORMER IN CHINA[*]

Y. Qiu[1] and E. Kuffel[2]

[1]Xi'an Jiaotong University
Xi'an, China
[2]4-K Engineering Inc.
(also University of Manitoba)
Winnipeg, Canada

INTRODUCTION

SF_6 gas found its use in China as an insulating medium in 1971 when Xi'an High Voltage Switchgear Research Institute developed the first Chinese GIS rated 110kV for a hydroelectric power station. In 1975 Xi'an Electric Power Capacitor Factory and Xi'an Jiaotong University jointly developed the first 500kV SF_6 gas-insulated standard capacitor in China. And as an extension of that joint project, a 1.1MV SF_6 gas-insulated standard capacitor with some floating potential electrodes for improving the electrical stress distribution[1] was developed in 1980. However, the development of SF_6 gas insulation technology in China was rather slow in 1970's.

With the rapid growth of the Chinese electric power industry starting in early 1980's, development of the novel reliable apparatus based on advanced concepts and modern technology became an urgent need, which stimulated research of SF_6 gas insulation in China. In view of this background, the International Development Research Centre of Canada (IDRC) awarded a research project to Xi'an Jiaotong University in co-operation with the University of Manitoba in 1985. The three-year project "Gas Insulated System (China)" laid emphasis on investigation of SF_6 gas mixtures, which led to a second phase of the IDRC funded project "Gas Insulated

---

[*] An IDRC funded project for which E. Kuffel was the team leader and Y. Qiu the co-ordinator of four Chinese participating groups: Xi'an Jiaotong University (Y. Qiu and Y. Feng), Tsinghua University (J. Zhang, Q. Qi and D. Zhu), Beijing Second Transformer Factory (J. Liu), and Hanzhong Transformer Factory (J. Zhou).

Transformer: II-Utilization" starting in September 1994. As a result, two prototype SF_6 gas mixture insulated apparatuses were developed in early 1996, i.e. the SF_6/N_2 gas mixture insulated power distribution transformer, and the SF_6/air gas mixture insulated cubicle type GIS[2]. This presentation deals only with the gas mixture insulated transformer.

REASON FOR USING SF_6 GAS MIXTURE

In China development of 10 kV SF_6 gas-insulated power distribution transformers started in early 1985 with two teams working separately, i.e. Changzhou Transformer Factory in co-operation with Xi'an Jiaotong University for 200 kVA and 1000 kVA prototypes, and Beijing Second Transformer Factory in co-operation with Tsinghua University for a 500kVA prototype. Both teams completed their development work successfully in 1987, and as a further step towards developing higher voltage transformers, a 150kV/30kVA SF_6 gas-insulated testing transformer was developed by Xi'an Jiaotong University and Hanzhong Transformer Factory in early 1991.

Based on the above-mentioned research for developing the gas insulated transformer (GIT), it has been found that SF_6 gas mixtures have some important advantages over pure SF_6 gas in the case of transformers other than the well-known arguments for switching to SF_6 gas mixtures[3]. This is because highly non-uniform fields at the winding edge and voids in the film insulation are inevitable in the transformer, and the following paragraphs will show that SF_6 gas mixtures can be superior to SF_6 gas in both cases.

HIGHLY NON-UNIFORM FIELD BREAKDOWN

It is well-known that the breakdown strengths of SF_6/N_2 and SF_6/CO_2 are slightly lower than that of SF_6 gas in uniform and quasi-uniform fields. For engineering application purposes, the relative electric strength (*RES*) of SF_6/N_2 and SF_6/CO_2 gas mixtures can be expressed by the following expressions[4] when the fractional concentration of SF_6 (*F*) is greater than 0.1.

$$(RES)_{SF_6/N_2} = F^{0.18} \qquad (1)$$

$$(RES)_{SF_6/CO_2} = F^{0.32} \qquad (2)$$

But in highly non-uniform fields the breakdown strength of SF_6 gas mixtures can be higher than that of SF_6 probably because of the better corona stabilization effect on breakdown in the gas mixtures. For instance Fig.1 compares SF_6, SF_6/N_2 (0.8/0.2) and SF_6/CO_2 (0.8/0.2) in a 15mm point-plane gap[5], which indicates that the positive dc breakdown voltage of both gas mixtures is 43% higher than that of SF_6 gas at the transformer gas pressure of 0.22MPa, and the low probability positive impulse breakdown voltages of SF_6/N_2 and SF_6/CO_2 are 25% and 60% higher than that of SF_6 gas, respectively.

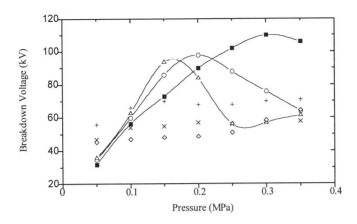

Fig. 1   Breakdown voltages of SF_6, SF_6/N_2 (0.8/0.2) and SF_6/CO_2 (0.8/0.2) in a 15mm point-plane gap
dc breakdown : △ SF_6 ; o SF_6/N_2 ; ■ SF_6/CO_2
low probability impulse breakdown : ◊ SF_6 ; × SF_6/N_2 ; + SF_6/CO_2

It can be seen in Fig.1 that the impulse breakdown voltages of SF_6 and both gas mixtures are much lower than their corresponding dc breakdown voltages in the corona stabilization region, which means that the insulation dimensions of the GIT are determined by the specified basic impulse level, and therefore from this point of view SF_6/CO_2 is superior to SF_6/N_2 and both gas mixtures are better than SF_6 gas.

DISCHARGE IN THE GAS-FILLED VOID

The major problem of gas-impregnated film insulation used in the GIT is its limited ability to withstand partial discharges, and therefore partial discharges (PD) should be avoided in the transformer insulation system under the operating voltage. As the PD extinguishing voltage can be as low as 85% of the PD inception voltage[6], the following expression can be written.

$$E_w \leq 0.85 E_i \tag{3}$$

where E_w is the working electrical stress under the maximum operating voltage, and E_i the electric field corresponding to partial discharge inception.

Figure 2 shows an experimental specimen modeling the gas/film insulation, of which the PD inception voltage (V_i) can be expressed in terms of the breakdown voltage of the gas-filled void (V_g) in the GIT[6].

$$V_i = V_g \left[1 + (d - d_g)/\varepsilon d_g \right] \tag{4}$$

$$\varepsilon = \varepsilon_f \Big/ \left[\varepsilon_f - k(\varepsilon_f - 1) \right] \tag{5}$$

where ε_f is the relative dielectric permittivity of the plastic film, d_g the dimension of the void along the electric field which in the worst case can be as large as the film thickness, d the separation of the electrodes, and k the filling factor of the film insulation.

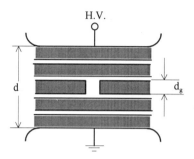

Fig. 2   Gas / film insulation with a mid-dielectric void

Since V_g can be obtained from the Paschen curve for SF_6 or the gas mixture in question, the PD inception of the gas/film insulation can thus be predicted quite accurately[6] as shown in Fig.3.

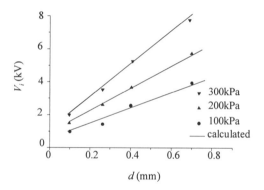

Fig. 3   PD inception voltages of SF_6/PET film insulation shown in Fig. 2
(PET film thickness : 0.04 mm)

Expression (4) shows that V_i is directly proportional to V_g, which can be expressed as[7]

$$V_g = (E/p)_o \left[pd_g + M \right] \quad (6)$$

where $(E/p)_0$ is the theoretical breakdown threshold, and M the figure-of-merit defined by A. Pedersen[8]. Table 1 gives the measured $(E/p)_0$ and M values for SF_6/N_2 and SF_6/CO_2 reported by the authors previously[7].

Calculations showed that when $p = 0.22$ MPa and $d_g = 0.025$ mm, V_g for SF_6/N_2

(0.5/0.5) and SF_6/CO_2 (0.5/0.5) is 2.4% and 18.6% higher than that for SF_6, respectively. This means SF_6/CO_2 is superior to SF_6/N_2 and both mixtures are better than SF_6 gas, which is the same conclusion as in the previous paragraph.

Table 1. $(E/p)_0$ and M values measured in SF_6/N_2 and SF_6/CO_2

fractional concentration of SF_6	buffer gas	$(E/p)_0$ (V/mm·kPa)	M (mm·kPa)
1	/	88.6	3.88
0.73	N_2	85.1	4.17
0.5	N_2	79.0	5.27
0.7	CO_2	79.7	5.94
0.5	CO_2	70.3	8.52

Validity of Exp.(6) was verified by the experimental result shown in Fig.4, where the Paschen curves for SF_6/N_2 (0.5/0.5) and SF_6/CO_2 (0.5/0.5) intersect each other at the pd_g value of ~20 kPa·mm, which is in agreement with the calculated value of 21 kPa·mm given by Exp(6).

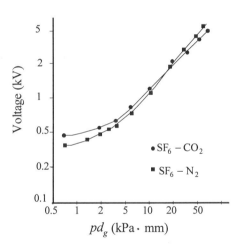

Fig. 4   Breakdown voltages of SF_6/N_2 (0.5/0.5) and SF_6/CO_2 (0.5/0.5) at low pd_g values

GAS DECOMPOSITION IN SF_6/N_2 AND SF_6/CO_2 MIXTURES

SF_6/N_2 and SF_6/CO_2 gas mixtures were also compared in respect of gas decomposition. Figure 5 shows the gas chromatograph analysis of the concentration of SO_2F_2 and SO_2 in SF_6, SF_6/N_2, and SF_6/CO_2 after 2000 impulse breakdowns in a 15mm point-plane gap[9]. It can be seen that SF_6/N_2 is as good as pure SF_6 while SF_6/CO_2 is apparently worse than SF_6/N_2 and SF_6 gas in this regard.

The sparked SF_6/N_2 gas mixture was further examined using the mass spectrum analysis, which indicated that no more toxic species was found in the mixture than those in the sparked SF_6 gas. And therefore the SF_6/N_2 gas mixture was chosen as the gaseous dielectric to be on the safe side.

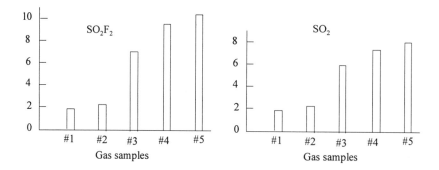

Fig. 5   Comparison of two gas decomposition products in sparked SF_6, SF_6/N_2 and SF_6/CO_2 using gas chromatograph analysis (the ordinates show the peak area in an arbitrary unit system)

gas samples : #1—SF_6, #2—SF_6/N_2 (0.5/0.5), #3—SF_6/CO_2 (0.67/0.33)
#4—SF_6/CO_2 (0.5/0.5), #5—SF_6/CO_2 (0.4/0.6)

HEAT TRANSFER PERFORMANCE

It is well known that the heat-transfer performance of the SF_6/N_2 gas mixture is not as good as that of pure SF_6, which is actually the decisive factor for limiting the nitrogen content in the gas mixture.

The temperature rise (ΔT) of the gas mixture and the H.V. and L.V. windings was both measured and calculated[10]. Table 2 shows good agreement between the measured and calculated values of the gas temperature rise.

Table 2.   Calculated and measured gas temperature rises[10]

concentration of N_2	ΔT of top gas (K)		ΔT of bottom gas (K)	
	calculated	measured	calculated	measured
0	40.2	39.6	10.8	9.5
0.15	41.5	40.6	9.7	10.0
0.30	43.0	42.5	8.5	9.3
0.45	45.2	44.8	7.2	9.0

Tables 3 and 4 show that the hottest place is the top part of the L.V. winding, where ΔT increased by 20% when SF_6 gas was substituted with SF_6/N_2 (0.55/0.45). For the transformer tank with the radiators originally designed for the SF_6-insulated transformer the highest nitrogen concentration should be limited to 0.45 with the average temperature rise of the L.V. winding being 70K.

Table 3. Calculated ΔT (K) of the H.V. winding in comparison with measured values

N_2 concentration	top	bottom	average	measured
0	64.6	20.5	45.6	44.0
0.15	67.2	20.0	46.7	46.0
0.3	70.3	19.5	48.2	48.0
0.45	74.7	19.0	50.5	50.0

Table 4. Calculated ΔT (K) of the L.V. winding in comparison with measured values

N_2 concentration	top	bottom	average	measured
0	84.8	24.6	60.0	58.3
0.15	89.7	24.3	62.7	61.3
0.3	94.7	24.2	65.6	65.8
0.45	102.0	24.4	70.0	70.5

TRANSFORMER INSULATION PERFORMANCE

To facilitate the development work, a transformer tank with the radiators designed for SF_6 gas-insulated transformers was adopted for the 500kVA/10kV gas mixture insulated transformer. The high voltage winding was of the conventional disk type structure, and the low voltage winding was of dual-spiral structure. Electrical field computations[11] were carried out for different parts of the transformer insulation, and the transient voltage distribution was also calculated for the H.V. winding to check the safety margin of the disk-to-disk and turn-to-turn insulation[12]. It is quite obvious that the electric field and transient voltage distributions were not affected by the composition of the gaseous medium. The prototype transformer passed the specified factory tests with all four gas mixing ratios shown in Tables 2, 3 and 4. It is quite interesting to note that the partial discharge measured in the transformer[13] was around 150 pC, which did not change with the nitrogen concentration ranging from 0 to 0.45.

CONCLUSIONS AND RECOMMENDATIONS

The prototype SF_6/N_2 gas mixture insulated transformer passed all the specified factory tests, which showed the feasibility of the gas mixture used as an insulating and cooling medium in GIT. With about the same heat transfer performance, the SF_6/CO_2 gas mixture may have better electrical insulation properties in GIT than the SF_6/N_2 gas mixture, and therefore the gas decomposition problem of the SF_6/CO_2 gas mixture should be further investigated in more detail to see whether it can be used in GIT. To cut down the transformer loss and thus to reduce the temperature rise, it is recommended to use amorphous steel core in the gas-mixture insulated transformer. It is also recommended to use the aluminum sheet winding to simplify the manufacture technology and to improve the dielectric and mechanical properties of the transformer.

ACKNOWLEDGMENT

The financial support of the International Development Research Centre of Canada is greatly appreciated.

REFERENCES

1. Y. Qiu and Z. Zhu, A 1.1 MV SF_6 gas filled standard capacitor with new design features, in: Proc. Int. Seminar on Capacitors, II-27 (1986).

2. Y. Qiu and Y.P. Feng, Investigation of SF_6-N_2, SF_6-CO_2 and SF_6-air as substitutes for SF_6 insulation, in: Conf. Record of the 1996 IEEE Int. Symp. on Electrical Insulation, 776 (1996).

3. L.G. Christophorou and R.J. Van Brunt, SF_6/N_2 mixtures, basic and HV insulation properties, IEEE Trans. Dielectrics and Electrical Insulation, Vol.2, No.5, 952 (1995).

4. Y. Qiu and D.M. Xiao, Dielectric strength of the SF_6/CO_2 gas mixture in different electric fields, in: Proc. 9th Int. Symp. on High Voltage Engineering, Vol.S2, paper 2255 (1995).

5. W. Gu, Low probability impulse breakdown characteristics of SF_6 in non-uniform field gaps, Doctor Thesis (in Chinese), Xi'an Jiaotong University (1998).

6. Y. Qiu, A. Sun and Z. Zhang, Factors affecting partial discharges in SF_6 gas impregnated polymer film insulation, in: Proc. 2-nd Int. Conf. on Properties and Applications of Dielectric Materials, 594 (1988).

7. Y. Qiu, S.Y. Chen, Y.F. Liu and E. Kuffel, Comparison of SF_6/N_2 and SF_6/CO_2 gas mixtures based on the figure-of-merit concept, in: Annual Report Conference on Electrical Insulation and Dielectric Phenomena, 299 (1988).

8. A. Pedersen, On the assessment of new gaseous dielectrics for GIS, IEEE Trans. Power Apparatus and Systems, Vol.104, 2233 (1985).

9. H. Ma, Y. Qiu, Y. Meng and H. Sun, Gas chromatograph analysis of spark decomposition products in SF_6/CO_2 gas mixtures, High Voltage Apparatus (in Chinese), Vol.31, No.3, 16 (1995).

10. Tsinghua University and Beijing Second Transformer Factory, Temperature rise test and calculation method, Technical Research Reports I and II for the IDRC project (1995).

11. Tsinghua University and Beijing Second Transformer Factory, Electric field distribution in gas insulated transformer, Technical Research Report IV for the IDRC project (1995).

12. Tsinghua University and Beijing Second Transformer Factory, Transient voltage distribution in gas insulated transformer, Technical Research Report III for the IDRC project (1995).

13. Tsinghua University and Beijing Second Transformer Factory, The insulation structure of SF_6 mixtures GIT and its test, Technical Research Report V for the IDRC project (1995).

DISCUSSION

K. NAKANISHI: I would like to ask a question from a rather practical point-of-view. The gas mixture of SF_6/N_2 is inferior in both insulating and thermal properties. So, what is the size and the cost of the gas-mixture-insulated transformer as compared with the pure-SF_6-insulated transformer? If the size is the same, does it mean reduction in reliability?

Y. QIU: Yes, you are quite right about that. In our case we used all the parts including the tank designed for the SF_6-insulated transformer manufactured by the Beijing Second Transformer Factory, which means that the size and cost of the SF_6/N_2-insulated transformer are the same as for the SF_6-insulated transformer with the nitrogen content being limited to up to 45% in the mixture. The safety margin for the mixture-insulated transformer is actually less than that for the SF_6-insulated transformer.

E. H. R. GAXIOLA: For the inception voltage you make/propose the use of the Paschen curve for SF_6, which actually applies to breakdown between (metal) electrodes. In the case of voids you have insulating surfaces which also can charge. Isn't this approach too crude? Do you not need something more sophisticated for the description of the PD's in voids, like a critical streamer (length) criterion?

Y. QIU: Yes, the Paschen curve in the literature was obtained by using plane-parallel electrodes, but our experiments have proved that the Paschen curve we obtained by using the model shown in this paper is in agreement with the published data, and is also in very good agreement with the calculation using the Townsend streamer breakdown criterion (Eq. 6).

K. OKUBO: Figure 1 shows the BDV characteristics under impulse voltage application. I think the PDIV curve would be much lower than the BDV. For the insulation design of gas-mixture-insulated transformer, do you use PDIV or BDV characteristics for impulse voltage application? In the case of BDV, what do you think of the charging problem on the film surface?

Y. QIU: Yes, the PD inception voltage is much lower than the impulse breakdown voltage, but on the other hand the maximum operating voltage is also much lower than the BIL (Basic Impulse Level) of the transformer. Therefore, in transformer design one has to take into consideration both factors, which is different from GIS design. In GIT internal breakdown is not allowed, so it is not necessary to consider film charging problems in case of breakdown.

W. BOECK: Corona stabilization is necessary to get the necessary withstand level in the case of GIT. Is this the case for usual operation stress or only in the case of overvoltage? Are there further decomposition products besides the two presented?

Y. QIU: It is not for the case of normal operation. It is only for the case of overvoltages. There are other decomposition products in addition to the two products mentioned in the paper, but the conclusion given is based on all decomposition products.

H. SŁOWIKOWSKA: In the case of SF_6/CO_2 mixtures the presence of low-energy discharges can be the source of carbon particles in the insulating system. What is your opinion about this negative factor?

Y. QIU: We have never observed phenomena of carbon particle deposition in sparked SF_6/CO_2 mixtures, but we will try to do more work in this regard if SF_6/CO_2 is to be used in a prototype transformer.

POSSIBILITY STUDIES ON APPLICATION OF SF_6/N_2 GAS MIXTURES TO A CORE-TYPE GAS INSULATED TRANSFORMER

K.Tsuji,[1] M.Yoshimura,[1] T.Hoshino,[2] T.Yoshikawa,[2]
H.Fujii[1] and K.Nakanishi[1]

Advanced Technology R&D Center[1]
Transmission & Distribution, Transportation Systems Center[2]
Mitsubishi Electric Corporation
8-1-1, Tsukaguchi-Honmachi, Amagasaki, Japan

INTRODUCTION

Lots of concern have been given to the potent global warming gases (greenhouse gases) which reflect back infrared radiation to the Earth's surface and work to increase the average temperature of the Earth's atmosphere. Last December, the protocol which regulates the emission of six greenhouse gases to the atmosphere was adopted at the Kyoto meeting of COP 3. Three man-made gases of HFCs, PFCs and SF_6 were designated as the greenhouse gases together with the three natural gases of CO_2, CH_4 and N_2O in the Kyoto protocol.

As recognized well, SF_6 gas has been used as the insulating and arc-quenching gas for high voltage and high power apparatus and contributed to realization of the compact and highly reliable gas insulated switchgears and gas insulated transformers. Since the gas superior to SF_6 gas has not been found and the modern gas insulated apparatus are considered to be more efficient in the life cycle analysis than the conventional ones, the gas insulated apparatus are indispensable untill the alternative method will be developed. Therefore, we need to make much efforts in reducing the emission of SF_6 gas into air. The various actions concerning SF_6 gas such as the studies on quantities of holdings, the gas leaked from the apparatus and the technologies of recovery and reuse are conducted in the world.[1,2] To reduce the consumption and emission of SF_6 gas to air, application of 50%/50% or 40%/60% gas mixtures of SF_6/N_2 to the apparatus without any major change of the structure was proposed at the NIST meeting.[3]

In the paper, we studied the possibility of application of SF_6/N_2 gas mixtures to a core-type gas insulated transformer. The dimension and the rating of the transformer are mainly determined by the insulating and thermal properties of the gas. Selecting an actual gas insulated transformer for a case study, we estimated the dielectric strength and the temperature rise of the coil in the transformer at application of the gas mixtures, changing the conditions of the ratio of gas mixture and the gas pressure. Since the gap spacing between the high and low (H-L) voltage coils occupies a large part of the transformer, the dielectric strength at the H-L gap spacing was estimated, based on the test results obtained from a major insulation model in the laboratory. Concerning the thermal simulation, the basic physical properties of gas mixture which were necessary for thermal simulation is firstly shown. The principle of natural gas cooling and the method to obtain the flow velocity of the gas are described. The temperature rise at cooling tubes, gas and coils is mentioned. The temperature rise at the coils which showed the maximum in the transformer is calculated, changing the conditions of the ratio of gas mixture, the pressure and the load current. Finally, we discuss the possibility of application of the gas mixture to the transformer from the results obtained in the studies.

TRANSFORMER SELECTED FOR CASE STUDIES

We chose a core-type gas insulated transformer with natural cooling as a model for the case studies. The rating of the transformer is 66kV/3.3kV, 10MVA. The high and low voltage coils are disc-type as shown in Fig.1. The gas pressure is 0.24MPa at 20°C which is exempted from the regulation of the second class pressurized tank in Japan. The three cases we studied are, (1) the ratio of SF_6/N_2 gas mixture we can use on the condition that the structure of the transformer and the gas pressure are not changed, (2) the gas pressure in the transformer we have to increase at the gas mixture of $SF_6 50\%/N_2 50\%$ on the condition that we have to satisfy the rating of the transformer, (3) the rating of the transformer we have to lower at the gas mixture of $SF_6 50\%/N_2 50\%$ if we do not increase the gas pressure.

INSULATION DESIGN FOR GAS MIXTURES OF SF_6/N_2

Experiments Using a Major Insulation Model

Since there have not been available data of dielectric strengths of the gas mixtures at the actual coil structure for estimating the dimension of the transformer, we conducted the experiments using a major insulation model shown in Fig.2. We used the three pieces of coils as the high voltage coil which face the grounded plane electrode simulating the low voltage coil. The coil with the cross-section of $2.4 \times 10 mm^2$ wrapped by the fourfold PPS (Poly-Phenylene Sulfide) film (thickness:50 μm) was used. The gap spacing between the coils and

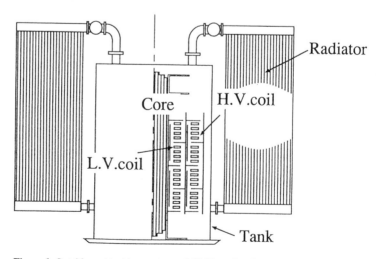

Figure 1. Outside and inside structure of 66kV gas insulated transformer

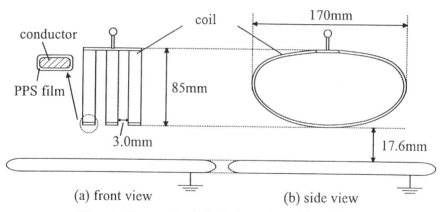

Figure 2. Structure of major insulation model used in the laboratory test

plane was set to 17.6mm. The tests were performed in the following four cases of SF_6/N_2 gas mixtures : 100/0, 50/50, 20/80 and 0/100. The standard lightning impulse voltages were applied between the coils and the grounded plane at the pressure of 0.22MPa, the alarm pressure where the insulation design is conducted. For reference, the breakdown voltages using the bare coils - grounded electrode were also obtained at the selected gas mixtures.

The experimental results are shown in Fig.3. The coils wrapped by the PPS showed the higher breakdown voltages than the bare coils, except for the case of the ratio of 50/50. We consider wrapping the bare coils with the PPS film covered the irregularities on the coil surface and relaxed the electric stress on the coil. However, we could not find the wrapping effect for the case of 50/50. The breakdown properties at the gas mixture of 50/50 also showed large dispersion, compared with other cases. The solid curves in Fig.3 show the theoretical value estimated using the Boltzmann equation.[4] The breakdown voltages except for those of 50/50 agreed with the theory. Although we conducted the breakdown tests, changing the procedure of filling N_2 gas and SF_6 gas to the tank, we obtained the same results. We cannot give a satisfactory explanation for the reason of the decrease from the expected value at the gas mixture of 50/50.

Figure 4 shows the gas pressure dependences of the breakdown voltages for SF_6 50%/N_2 50% gas mixture and pure SF_6. The breakdown voltage 270kV at the gas pressure of 0.22MPa of pure SF_6 was attained at the pressure of 0.5MPa for the gas mixture of 50/50.

Estimation of Dielectric Strength at Actual Coil Structure with Gas Mixture of 50/50

The insulation design is performed so that the actual breakdown voltage at the H-L coil spacing of the transformer is higher with a certain margin than the lightning impulse withstand level (LIWL). As shown in Fig.3, theoretical and experimental breakdown voltages for the gas mixture of 50/50 are about 10 % and 26 % (at mean value) lower than that of pure SF_6, respectively. In the estimation, we assume that the electrode area effect of breakdown voltages observed in pure SF_6 gas acts in the same manner to the gas mixture and the breakdown voltages increase linearly with the gap spacing. The estimation indicates that the gas mixture of 50/50 may be applied though the same reliability with the pure SF_6 cannot be obtained. While, if the decrease from the theoretical value observed in the experiment takes place in the actual transformer, the breakdown voltages at the H-L coil would become close to the LIWL for the transformer. In order to maintain the same breakdown voltage with pure SF_6 at 0.22MPa, we need to increase the gas pressure to 0.5MPa which may impair the advantages of the gas insulated transformer.

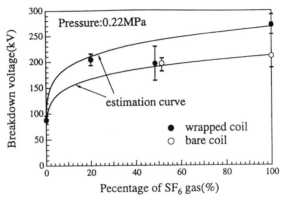

Figure 3. Breakdown voltages of SF_6/N_2 gas mixtures for wrapped and bare coils and theoretical curve calculated using Boltzmann equation

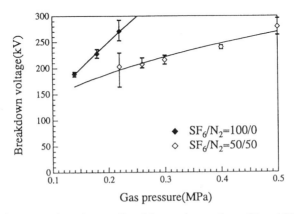

Figure 4. Pressure dependences of breakdown voltages of pure SF_6 and SF_6/N_2 gas mixture

COOLING CAPACITY OF GAS MIXTURES

Estimation of basic physical properties of SF_6/N_2 Gas Mixtures

We needed the physical properties such as density, viscosity, kinematic viscosity, thermal conductivity, specific heat capacity at constant pressure, Prandtl number and Sutherland number for the gas mixtures to perform the simulation of cooling capability for gas insulated transformer. Since there are not their data available for SF_6/N_2 gas mixtures, we calculated them using the estimation methods described in details in the literatures.[5-12] Since the space in the paper is limited, we show the calculation results of density, viscosity, kinematic viscosity, thermal conductivity, specific heat capacity and Prandtl number in Fig.5 (a)-(f), which were estimated on the condition of the gas temperature of 50 °C.

Natural Cooling of Gas Insulated Transformer

The heat generated in the core and coils is absorbed by the gas which flows along the surface of the core and coils. The warmed gas is lead to the cooling tubes in the radiator where the heat is taken away from the gas by transfer to the air. Temperature distribution of the gas in the transformer is shown in Fig.6.

The gas whose temperature is θ_1 at the bottom of the transformer goes up along the coils

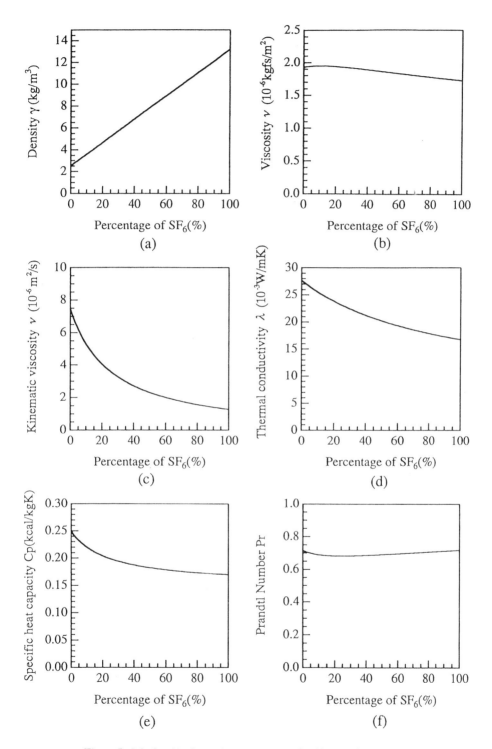

Figure 5. Calculated basic physical properties of SF_6/N_2 gas mixture
(a)density (b)viscosity (c)kinematic viscosity (d)thermal conductivity
(e)specific heat capacity (e)Prandtl number

and absorbs the heat from the surface of the coils and reaches to the temperature θ_2 at the top of the coils. The gas moves to the entrance of the cooling tubes in the radiator, keeping the temperature θ_2. In the cooling tubes, the gas is cooled down by flowing down and once again returns to the original temperature θ_1. The gas with the temperature θ_1 continues to circulate through the coils and the cooling tubes.

As shown in Fig.6, the gas density varies with the temperature in the transformer, according to the Boyle-Charles law. The weight per unit area of the gas in the coils is lighter than that in the cooling tubes in the radiator according to the temperature distribution. The temperature distribution of the gas generates the thermal buoyancy. The quantity of gas which circulates by natural force through the coils and the cooling tubes is determined by equalizing the thermal buoyancy with the flow resistance of circulation paths. When H_0 is much smaller than $(H_R - H_C)$, the thermal buoyancy ΔP_1 and the flow resistance ΔP_2 are expressed by equations (1) and (2), respectively.

$$\Delta P_1 = \beta \cdot \gamma \cdot (\theta_2 - \theta_1) \cdot (H_R - H_C)/2 \tag{1}$$

$$\Delta P_2 = (\gamma/2g) \cdot \Sigma \xi_i \cdot (Q/S_i)^2 \tag{2}$$

where β: thermal expansion coefficient of gas ($\beta = 1/(273.15+T)$), θ_1: gas temperature at the bottom, θ_2: gas temperature at the top, γ: gas density, H_0: difference in height between the bottom end of cooling tubes and that of coils, H_R: height of cooling tube, H_C: height of coils, ξ_i: coefficient of gas flow resistance, g: gravity, Q: quantity of gas flow, S_i: cross-sectional area at various part. The temperature difference $(\theta_2 - \theta_1)$ is obtained from equations (1) and (2). Pressure drop along the gas circulation is caused by friction resistance through ducts and form resistance through fittings such as square elbow, sudden enlargement and contraction. Since the total radiation of heat loss W is cooled down by the quantity of gas flow Q, Q can be written by equation (3), using specific heat capacity of the gas C_p.

$$Q = W / \{C_p \cdot \gamma \cdot (\theta_2 - \theta_1)\} \tag{3}$$

By inserting the temperature difference into equation (3), the quantity of gas flow is given by equation (4).

$$Q = \left[\frac{\beta \cdot g \cdot W \cdot (H_R - H_C)}{\gamma \cdot C_p \cdot \Sigma \frac{\xi_i}{S_i^2}} \right]^{\frac{1}{3}} \tag{4}$$

The velocity V_{ct} and Reynolds number R_{ect} at the cooling tube are Q/S_{ct} and $V_{ct} \cdot d_{ct}/\nu$, respectively. The cross-sectional area of cooling tube, equivalent diameter at cooling tube and kinematic viscosity are given by S_{ct}, d_{ct} and ν, respectively.

Calculation of Temperature Rise

The average temperature rise at the coils from ambient temperature $\Delta \theta_c$ is given by summation of temperature rise at cooling tubes $\Delta \theta_{ct}$, temperature difference between gas and cooling tubes $\Delta \theta_{g-ct}$ and temperature difference between coils and gas $\Delta \theta_{c-g}$ as shown in Fig.7.

$$\Delta \theta_c = \Delta \theta_{ct} + \Delta \theta_{g-ct} + \Delta \theta_{c-g} \tag{5}$$

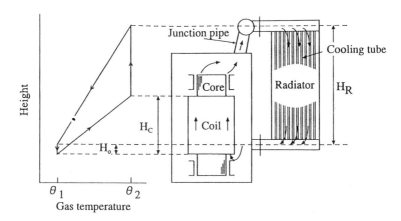

Figure 6. Natural circulation and temperature distribution of gas in the transformer

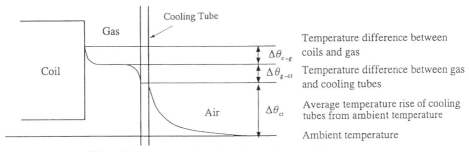

Figure 7. Temperature rise of transformer from ambient temperature

The average temperature rise of cooling tubes from ambient temperature can be expressed by the equation (6)[10] which does not depend on the kind of the gas in the transformer.

$$\Delta \theta_{ct} = K_1 \cdot (W/S_T)^{0.8} \tag{6}$$

where K_1: constant, S_T: total area of cooling, W: total heat loss. The temperature difference between gas and cooling tubes $\Delta \theta_{g-ct}$ is given using the heat transfer coefficient between gas and cooling tubes α_{gas} as follows,

$$\Delta \theta_{g-ct} = W/(\alpha_{gas} \cdot S_T) \tag{7}$$

$$\alpha_{gas} = \lambda \cdot N_u / d_{ct} \tag{8}$$

where λ: thermal conductivity, N_u: Nusselt number. The heat transfer coefficient α_{gas} is expressed differently, depending on the flow conditions such as laminar flow, turbulent flow and transition region between them.

The temperature difference between coils and gas $\Delta \theta_{c-g}$ can be expressed by equation (9) using the heat flux of the coils q, heat transfer coefficient between coil and gas α_{coil}. The thermal flux is given by equation (10).

$$\Delta \theta_{c-g} = q/\alpha_{coil} \tag{9}$$

$$q = i^2 \cdot R + q_e \tag{10}$$

where i: current which flows in the coils, R: total resistance of the coils, q_e: eddy current loss.

Simulation Results for Cooling Capability of SF$_6$/N$_2$ Gas Mixtures for the Transformer

Calculation of Cooling Properties in the Transformer

Case study (1) for the Ratio of Gas Mixture The calculation results show that the gas flow at the coils is laminar along radial duct and turbulent along axis duct. Since the coils are disc-type as shown in Fig.1, the area which effectively absorbs the heat from the coils is the coil surface along radial duct. The gas flow along axial duct hardly contributes to cooling. Since the heat transfer coefficient of the radial duct at the coils is rather constant, independent of the ratio of gas mixture, the temperature difference between coil and gas is not influenced by the ratio of gas mixture.

While, the flow condition in the cooling tubes is influenced by the ratio of gas mixture. Figure 8 shows the temperature rise at the coils, gas and cooling tubes from the ambient temperature, changing the percentage of SF$_6$ gas in the mixture at the gas pressure of 0.24MPa. As shown in the figure, the temperature rise of each part is constant from the ratio of 0/100 to 20/80. When the percentage of SF$_6$ is higher than 20%, the Reynolds number is over the upper limit of laminar flow. Since the Reynolds number for pure SF$_6$ gas is less than the lower limit of turbulent flow, the area from 20/80 to pure SF$_6$ is the transition area to turbulent flow. Since the temperature rise at the coils is highest among them, the ratio of the gas mixture is restricted by the temperature rise 70 °C at the coils which is the upper limit for E-class insulation materials specified in JEC-2200 (1995) "Power Transformers of Japanese Standard". The critical ratio of the gas mixture is SF$_6$ 35%/N$_2$ 65% for this case study.

Case study (2) for the Gas Pressure The calculation on the temperature rise at the coils was performed for the conditions of the gas pressures of 0.1 to 0.7MPa and the ratios of gas mixture 100/0, 50/50, 30/70, 20/80 and 0/100 as shown in Fig.9. The pressure rise shown in Table 1 is necessary to maintain the average temperature rise at the coils below the design temperature for pure SF$_6$ gas insulated transformer.

Case study (3) for Derating the Rated Current Figure 10 shows the temperature rise at the coils for the rated current as a parameter of the mixture ratio. The rated current has to

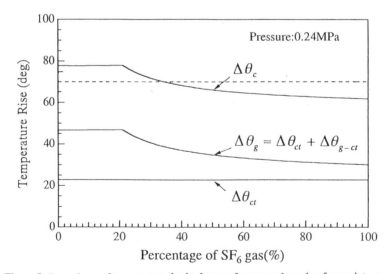

Figure 8. Dependence of temperature rise in the transformer on the ratio of gas mixture $\Delta\theta_c$:temperature rise at the coils from ambient temperature, $\Delta\theta_g$:average gas temperature rise, $\Delta\theta_{ct}$:temperature rise at cooling tubes from ambient temperature.

Table 1. Gas pressures and percentage of current derating to maintain the temperature rise at the coils below the design temperature.

SF_6/N_2	case study(2)	case study(3)
50/50	0.306MPa	95%
30/70	0.362MPa	89%
20/80	0.412MPa	85%
0/100	0.626MPa	85%

be derated for the gas mixtures as shown in Table 1 when the temperature rise is needed to be below the design temperature.

DISCUSSION

We need to examine the design margin of the insulating and thermal aspects when we apply the gas mixture of 50/50 to the transformer without major change of the structure. We consider that an ample margin between the actual breakdown and design stresses at the H-L coils should be taken to secure high reliability. The theoretical estimation of breakdown voltages at the H-L coils suggests that there is some margin left for gas mixture of 50/50. If the decrease from the theoretical curve for gas mixture of 50/50 as shown in Fig.3 takes place at the actual transformer, we would find a little margin at applying the gas mixture to the transformer. Increasing the gas pressure to higher than 0.3MPa results in losing an advantage of exemption from the regulation on pressurized tank, although the breakdown strength equivalent to pure SF_6 at 0.22MPa can be satisfied at the higher pressure than 0.5MPa.

We studied the average temperature rise at several parts in the transformer. The temperature rise at the coils which showed the maximum is needed to maintain below 70 °C (E-class insulation). Applying the gas mixture of 50/50 to the transformer without major change of the structure means to reduce the margin between the design temperature and the upper limit of E-class insulation. The case study conducted in the paper showed that N_2 gas can be filled to 65% into SF_6 gas. However, the case is not always a representative of the gas insulated transformers. The thermal design varies, depending on the structure and the ratings of the transformer. There are the cases in which the design temperature is set closely to the upper limit. For these cases, the percentage of N_2 gas would be much less than 65% in

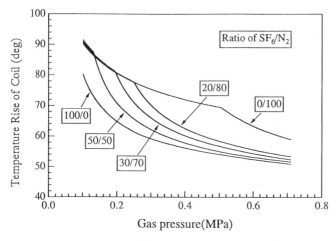

Figure 9. Pressure dependences of temperature rise at the coils on the gas pressure

the mixture. Increase of the gas pressure will not give a solution as mentioned in the above paragraph. Derating the load current may be an alternative way although we need to change the transformer rating and get through the complicated procedures.

CONCLUSIONS

The following conclusions are drawn from the simulation and experimental studies for a case of applying the gas mixture of 50/50 to a natural cooling 10MVA transformer without changing the major structure.

(1) The case study conducted in the paper showed that there are some cases in which the gas mixture of 50/50 can be applied on the condition that adequate margins are confirmed electrically and thermally. Even in such a case, we need further tests using the actual equippment.
(2) Increase of the gas pressure does not indicate prospective development since major structural reinforcement is needed.
(3) Derating the load current may show a possibility although the method generates some disadvantages.

We need to select the transformer to apply the gas mixture, examining the insulating and thermal designs which vary with the transformer.

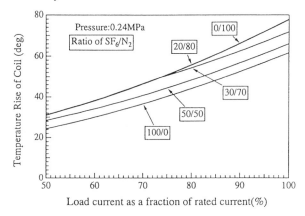

Figure 10. Temperature rise at the coils for the derated current

REFERENCES

1. Task Force 01 of Working Group 23.10 of CIGRE, SF_6 and The Global Atmosphere, ELECTRA, No.164, February, p.121 (1996)
2. Pathfinder Meeting -US-Japan Environmental Leadership Workshop for Climate Change Protection-, April 2-4 (1997)
3. L.G.Christophorou, J.K.Olthoff and D.S.Green, Gases for Electrical Insulation and Arc Interruption:Possible Present and Future Alternatives to Pure SF_6, NIST Technical Note 1425 (1997)
4. S.Okabe, M.Chiba and T.Kono, "Calculation Method of Flashover Voltage of Gas Mixtures", Journal of IEE Japan, Vol.103-B, p.507 (1983)
5. R.Byronbird, W.E.Stewart and E.N.Lightfoot, Transport Phenomena, John Wiley & Sons Inc., (1960)
6. K.Satoh, Estimation Method for Physical Properties, Maruzen Co. Ltd., (1954)
7. M.Thiesen, Verth. deutsch. phys. Ges., 4, p.348, (1902)
8. C.R.Wilke, J.Chem. Phys., 18, p.517, (1950)
9. A.Wassiljewa, Phydik, Z., 5, p.737, (1904)
10. W.Elenbass, "Heat Dissipation of Parallel Plates by Free Convection", Physical IX, No.1, Jan., (1942)
11. E.Siede and G.Tate, "Ind. Eng. Chem.", 28, p.1429 (1942)
12. R.K.Shah and A.L.London, "Advances in Heat Transfer Suppl", 1, Academic Press (1 978)

DISCUSSION

J. CASTONGUAY: What is the typical leak rate of gas (percent per year) in the type of gas transformer you studied? Please comment on the reduction of SF_6 losses by using SF_6/N_2 mixture while increasing the gas pressure, which will probably result in a higher leak rate.

K. NAKANISHI: I do not have exact information concerning the leak rate for the gas-insulated transformer. Considering the fact that the leak rate for GIS is less than 0.1%/year, in some cases less than 0.01%/year, the leak rate for the transformer should be less than that for GIS. It is because the filled pressure of 0.24 MPa in the transformer is lower than the 0.5 MPa in GIS. Increase of pressure of the gas mixture might lead to higher leak rate if we do not strengthen the structure of the transformer.

YICHENG WANG: Your Fig. 4 shows that to achieve the same rating of the breakdown voltage 270 kV, the total pressure required is 0.22 MPa when pure SF_6 is used, while the total pressure is 0.5 MPa when a 50%SF_6/50%N_2 mixture is used. Adding nitrogen makes the whole thing worse, not better, in your case because the absolute amount of SF_6 required is more. Do you have an explanation?

K. NAKANISHI: We have been considering the point you raised. We initially thought that corona stabilization happened in the gas mixtures around a 50%/50% composition. Additional tests to check this possibility showed a negative result. Presently, we do not find a mechanism for the phenomenon. We need further studies to clarify this point.

S. DALE: Are you designing the transformer tank for vacuum in order to recover as much SF_6 as possible after testing and during maintenance? This will reduce the emission of SF_6 into the atmosphere.

K. NAKANISHI: We are actually taking the voluntary action to evacuate the SF_6 gas up to 0.05 MPa from the transformer tank during tests and maintenance.

POWER FREQUENCY AND SIL WITHSTAND PERFORMANCE OF A GIC WITH 5 AND 10 PERCENT SF$_6$/N$_2$ MIXTURES

H. I. Marsden,* M. D. Hopkins,** and C. R. Eck III**

*ABB Power T&D Company Inc., Electric Systems Technology Institute, Raleigh, NC, USA
** ABB Power T&D Company Inc., Substations Division, Westboro, MA, USA

ABSTRACT

This publication presents the results of dielectric tests performed on a scale model GIC with mixtures of N$_2$ and SF$_6$. The purpose of these tests were to establish the allowable design stress levels for the power frequency, and SIL ratings of the 420 GIC system with mixtures of N$_2$ and SF$_6$ where SF$_6$ is the minority constituent. The goal was to optimize a system that would utilize less than 30% SF$_6$. The test model had the same geometry, conductor surface finish and insulators as the 242 kV commercial system currently offered by ABB. The results from the tests were used to establish the dielectric design parameters for the 420 kV GIC system as well as the satisfactory ratios of SF$_6$ in N$_2$.

1. INTRODUCTION

EdF is evaluating the various possibilities of reducing the environmental impact of its grid. In this context, an R&D program was jointly funded by EdF and ABB to demonstrate the technical and economic feasibility of an underground gas insulated cable technology as an alternative to 400 kV overhead lines.

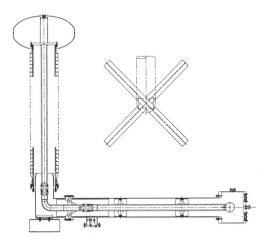

Fig. 1. Detailed view of the 242 kV GIC System Dielectric Test Model

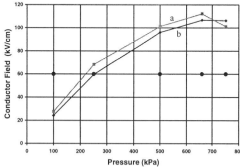

Fig. 2. 60 Hz Conductor Fields for the 242 kV GIC Withstand Test with various Gas Pressures and Gas Mixtures, a) AC$_{rms}$ & 10 % SF$_6$/N$_2$, b) AC$_{rms}$ & 5% SF$_6$/N$_2$, conductor field limit is 60 kV/sm

The technical features and construction of the GIC system should be suitable and cost effective for power transmission lines exceeding 100 km in length, with ampacities up to 6000 A. System losses should be low. The GIC system should have a reliability and availability consistent with that of the main power transmission grid. No special maintenance requirements over that of other cable systems should be required over the life of the system. The life expectancy should be over 40 years. The proposed transmission system should be suitable for underground installation over most of its length, either buried or in tunnels.

In addition, environmental impact of the GIC system should be as small as possible, and a gaseous insulation based upon nitrogen should be considered.

The purpose of this study was to access the feasibility of using nitrogen and SF_6, where SF_6 would be the minor constituent gas as the main dielectric.

The GIC should be rated for the 400 kV grid in France. The rated withstand levels for the equipment on the grid where the GIC will be used are:

-rated lightning withstand strength:	1425 kV (BIL)
-rated switching withstand strength:	1050 kV
-rated 60 Hz withstand strength:	630 kV

2. EXPERIMENTAL

2.1. Test Model and Procedures

All the tests were conducted on a 242 kV GITL system size, with an enclosure inner diameter of 29.21 cm and a conductor outer diameter of 10.16 cm. The test models are shown in Figure 1.

The bushing section of the test model was separated from the coaxial test configuration with a conical insulator. In order to ensure that flashover did not occur on the bushing side of the conical insulator, the bushing and elbow sections were filled with SF_6 to 600 kPa.

To verify the capability of the 242 kV test model, it was tested with SF_6 at 400 kPa to the factory test level for this voltage rating of 450 kV, 60 Hz rms, 1 min. withstand, and with lightning impulses to the BIL level of 1050 kV negative polarity according to IEC 60.1.

The high voltage tests were performed with SF_6/N_2 and pure SF_6. Lightning and switching impulse tests were performed with mixtures of 5 and 10 percent SF_6 in nitrogen. These tests were performed at ABB Electric Systems Technology Institute (ETI) in Raleigh, NC. For each of the gas mixture tested, discrete pressure magnitudes were selected as 100 kPa, 250 kPa, 500 kPa, 750 kPa. Where the available voltage was insufficient, such as was the case for the switching impulse tests at ETI, the maximum mixture pressures tested were determined such that breakdown could be obtained.

The withstand voltage for both of the impulse test wave forms was obtained using a 3X9 pass/fail criterion. In this test criterion, the voltage was increased from approximately 80 % of the anticipated breakdown value in 3% steps. At each test level, the 242 kV model was considered to withstand or pass a given voltage magnitude if no failure was experienced in three consecutive shots, or only one failure in twelve (12) shots.

If two failures occurred (2 failures in 3 shots) in the application of three consecutive shots, the withstand level was reported as the previous step voltage.

If a single failure occurred in the application of the first three shots, then the model was subjected to an additional 9 test shots, any additional failure in this series of 9 shots was considered to fail the applied voltage magnitude. In this case (2 failures in 12 shots) the withstand level was reported as the previous step voltage.

3. POWER FREQUENCY WITHSTAND TEST FOR 242 KV GIC SYSTEM

The results of the 60 Hz dielectric testing are shown in Figure 2. As can be seen in the figure, the dielectric performance of the 242 kV test model varies as a function of both the gas mixture (% SF_6 in N_2) and the gas pressure (100 kPa to 750 kPa). The horizontal line in the Figure 2 represents the minimum conductor field required to meet the EdF over voltage requirement of 630 kV_{rms} and the IEC over voltage requirement of 680 kV_{rms} for a 420 kV GITL. For the 420 kV GITL proposed by ABB this conductor field worked out to be 60 kV/cm with an applied voltage of 680 kV.

For the 5 percent N_2/SF_6 mixtures at 500 kPa the breakdown field was approximately 2 percent higher than the withstand level, while the 10 percent mixture demonstrates approximately 11 percent increase in breakdown over withstand.

As can be seen from the figure, both the withstand and breakdown results are significantly above the conductor stress that will be experienced during IEC 680 kV over voltage, and no problems with power frequency performance are expected for the proposed 420 kV GITL design.

The power frequency results (withstand and breakdown test) showed a steady increase in dielectric performance with gas pressure up to 660 kPa. In addition a net increase in performance was demonstrated with the higher concentration of SF_6 in the SF_6/N_2 gas mixtures.

As can be seen in the Figure 2, the 60 Hz experimental data for breakdown testing in the 145 kV model and for withstand testing in the 242 kV GIC model follows approximately the same shape or trajectory. The dielectric performance of these gas mixtures can be represented by a power law relationship whose exponent is in the neighborhood of 0.65 for pure SF_6. A typical expression may be written as follows.

$$V = A \cdot P^{0.65} \text{ (practical geometry with pure } SF_6)$$

The results shown may be described to demonstrate a reduction in dielectric performance with increasing pressure. A similar trend was also observed during breakdown testing with pure SF_6 in the test model. The observed trends are known to result from the conductor surface roughness and area effects, while the gas mixture plays a secondary role for SF_6/N_2 and the various mixtures thereof [2, 3, 4, and 5].

As can be seen in the Figure 3, the effects of surface roughness in a nitrogen/SF_6 gas mixture does play a significant role in the dielectric performance of a coaxial geometry. However, the magnitude of surface roughness in ABB design is approximately 2 µm.

As a consequence, large reductions in dielectric strength are not expected for N_2/SF_6 gas mixtures at 300 kPa. The family of curves shown in Figure 3 would shift to the left slightly at the higher pressure and then back to the right because of the lower N_2/SF_6 concentrations. The net effect should not be significant in the ABB design.

The 5% and 10% withstand results for 750 kPa show a reduction in dielectric performance from the 660 kPa results. During a tear down of the GIC, the observed reduction was found to be due to heavy marks on the tripost insulators. This kind of insulator marking can be found to occur for repeated high voltage testing at the low SF6 concentrations and the highest pressures, e.g. above 500 kPa.

It is generally accepted that pure N_2 and N_2/SF_6 mixtures are less sensitive than SF_6 to the effects conductor surface roughness and particle contamination, it was not surprising to find experimental results with power law exponents in the range of 0.72 to 0.82. Since the test model had the same materials and surface roughness as are present in commercial available designs with pure SF_6 (~ 2 µm), a large reduction in dielectric strength was not expected.

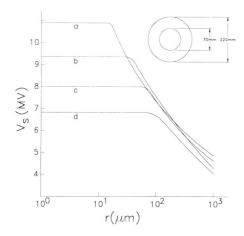

Fig. 3. Calculated breakdown voltage vs. roughness characteristic for coaxial electrodes in SF_6/N_2 mixtures at pressures of about 300 kPa. Curves a, b, c, and d correspond to SF_6 content of 100, 50, 25, and 10 % by vol. respectively. [7]

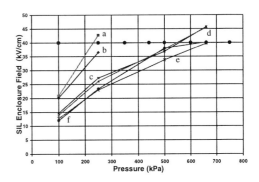

Fig. 4. A comparison (+/−) SIL Enclosure Fields for the 242 kV GIC model. a) +SI SF_6 b) −SI SF_6 c) +SI & 10% SF_6/N_2 d) −SI & 10% SF_6/N_2, e) −SI & 5% SF_6/N_2, f) +SI & 5% SF_6/N_2, enclosure field limit is 40 kV/cm

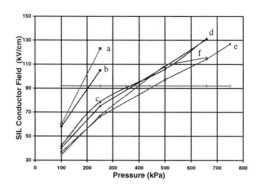

Fig. 5. A comparison (+/-) SIL Conductor Fields for the 242 kV GIC model. a) +SI SF_6 b) –SI SF_6 c) +SI & 10% SF_6/N_2 d) –SI & 10% SF_6/N_2, e) –SI & 5% SF_6/N_2, f) +SI & 5% SF_6/N_2, conductor field limit is 92 kV/cm

4. SWITCHING IMPULSE TEST FOR 242 KV GIC SYSTEM

The switching impulse test results for the conductor and enclosure stresses are shown in Figures 4 and 5 respectively. As can be seen in the figures, the SIL conductor stresses are well behaved in that the positive SI results are in general greater than or approximately equal to the negative SI values for both 5 and 10 percent mixtures at 560 kPa. The data shown indicates that the negative SI for a 5 percent mixture has the lowest dielectric performance over a pressure range from 250 kPa to 600 kPa. The conductor field limit of 92 kV/cm was determined from the 420 kV GIC insulation system requirements (SIL at 1050 kV) and GITL dimensions. The SIL conductor field limit is seen to be approximately 12 percent less that the minimum 242 kV GIC SIL performance at 560 kPa.

In a similar fashion, the SI enclosure fields were established for the 420 kV EdF design. As can be seen in Figure 5, with a 10 percent SF_6 mixture the data indicates 560 kPa may be marginally adequate for both the positive and negative SI wave forms. However, the SI results for the 5 percent SF_6 mixtures required a pressure of nearly 650 kPa to meet the same enclosure stresses. At 560 kPa the 5 percent SF_6 mixtures data did not meet the 420 kV SI enclosure design requirements.

A combination of the increased enclosure stress requirements for the EdF 420 kV GITL and the local stress enhancements of the trap edges have resulted in unacceptably low dielectric performance for SIL wave forms. Similar results were reported by Cookson [3]. To this end there are three possible corrective actions that can be taken to address the higher enclosure stresses required to meet the SIL dielectric performance of the EdF 420 kV GITL.

- Reduce the local field enhancements at the trap edges and conductor transitions
- Increase the gas pressure of the N_2/SF_6 gas mixture
- Slightly increase the N_2/SF_6 mixture concentration

5. CONCLUSIONS

High voltage testing of a model GIC system was performed to verify the results in the literature [2, 3] and to determine the dielectric design parameters for power frequency and SI voltages. Although a considerable amount of work has been reported on SF_6/N_2 dielectric mixtures, these were for the most part intended to reduce the cost of SF_6. Specifically, only a few early studies had been made on practical geometry's with SF_6 concentrations down to 20 percent [2,3,4]. Recent studies have investigated the effects of SF_6/N_2 mixtures with concentrations between zero and 20 percent SF_6 in nitrogen and for practical geometries, [1, 5, 6]

The results presented in this publication are in good agreement with earlier reported work and have shown that the addition of small amounts of SF_6 in nitrogen can dramatically improve the dielectric performance of a GIC by operating the system at marginally higher pressures and/or enlargements to the conductor enclosure geometries of a practical CGIT line.

The high voltage results reported here and in references [1,3,4,7] all demonstrate an increase in withstand performance with pressure up to about 600 kPa. However above 600 kPa the high voltage AC withstand performance suggest an increase in scatter. In a similar manner the SIL withstand performance demonstrates a decrease in benefit for pressures above 600 kPa and particularly for positive polarities. These reductions in the observed withstand performance above 600 kPa were attributed to surface roughness of the electrodes and stress enhancements at the particle traps.

6. ACKNOWLEDGMENTS

The authors are grateful to Electricite de France and ABB for their joint support of this work.

7. REFERENCES

1. Marsden, H.I., et al., "High Voltage Performance of a Gas Insulated Cable with N_2 and N_2/SF_6 Mixtures", 10th International Symposium on High Voltage Engineering, August 1997.
2. EPRI Final Report, "Gases Superior to SF_6 for Insulation and Interruption", November 1981.
3. Cookson, A.H., et al., " High Voltage Performance of Mixtures of SF_6 with N_2, air and CO_2 in Compressed Gas Insulated Equipment", 5th IKE International Conference on Gas Discharges, September 1978.
4. Yingshen et al., "Breakdown Characteristics of Particle Contaminated SF_6 and SF_6/N_2 Mixtures in Large Coaxial Systems", Sixth Int. Symp. High Voltage Engineering, September 1989.
5. Pace, M. O., McCorkle, D. L., and Waymel, X. (1996). "SF_6/N_2 Breakdown in Concentric Cylinders: Consolidation of Data", 1996 IEEE Conf. on Elec. Ins. and Diel. Phen., pp 571-576.
6. Pace, M. O., McCorkle, D. L., and Waymel, X. (1995). "Possible High Pressure Nitrogen-Based Insulation for Compressed Gas Cables", 1995 EKE Conf. on Elec. Ins. and Diel. Phen., pp. 195-198.
7. Christophorou, et al., "SF_6/N_2 Mixtures, Basic and HV Insulation Properties", IEEE Transactions on Dielectrics and Electrical Insulation, Vol. 2, No. 5, pp 952-1003, October, 1995.

DISCUSSION

H. HAMA: Do you consider the area effect of the electrode in your experiments?

H. MARSDEN: Yes, area effects will be finally considered in a 300 m test loop, and then in a 5 km, 3φ test system. The 5 km test system will be part of the EDF grid.

A. BULINSKI: Have you considered the effect of fast front voltage transients on the breakdown field of various gas mixtures?

H. MARSDEN: Yes, initially the past experiences of the open literature were considered, then an insulation coordination was considered for the EDF grid. The combination of the literature search and system overvoltages on the EDF grid suggested that LI overvoltages were more severe than VFTs.

H. OKUBO: What kind of dielectric test will you do on site after laying the GIC?

H. MARSDEN: Although it has not been formally described, it is my expectation that AC testing plus PD detection will be used.

J. CASTONGUAY: What is the order of magnitude of the cost of the GIL compared to overhead lines of the same capacity? Can you comment on the increase cost for increase capacity since GIL can be designed more easily with increase power capacity? Also, what is the overload capability of such GIL (percent overload/duration characteristics)?

H. MARSDEN: ABB has met the economic criteria of EDF in that the installed system should be less than ten times the cost of an equivalent overhead transmission line for 2000 MVA and 3000 MVA. The increase in system cost for a given increase in system capacity would require a detailed analysis, so this part of your question cannot be directly answered. The GIC overload ability again would require knowledge of the interconnected system and a detailed assessment of the soil properties and burial depth.

A. SABOT: Regarding the question of Dr. Okubo, I would like to add that the on site test procedures for GIL is an item which will be discussed at the next CIGRE general session in Paris and that is a key issue for GIL reliability. About the question of VFT, I would like to add that with the chosen BIL of 1425 kV the VFT overvoltages will not be dangerous as their amplitude (2-2.5 p.u.) will remain lower than the expected on site withstand of the GIL installed in the networks (> 0.8 LIWL).

INSULATION CHARACTERISTICS OF GIS

FOR NON-STANDARD LIGHTNING SURGE WAVEFORMS

Masanori KOTO,[1] Shigemitsu OKABE,[1] Takeshi KAWASHIMA,[1]
Tokio YAMAGIWA,[2] Toshio ISHIKAWA[2]

[1]Tokyo Electric Power Company
[2]Hitachi,Ltd.

INTRODUCTION

Recent advances in gas insulated switchgear (GIS) have promoted trends to improve the performance of surge arresters[1] which has allowed lower lightning impulse test voltages for substation equipment[2].

When investigating test voltages, it is important to evaluate the waveforms of actual lightning surges invading substations. There have been few systematic investigations of dielectric characteristics in SF_6 gas, except VFT with metallic particles[3], for waveforms other than standard lightning impulse in the category of lightning surges (hereinafter "non-standard waveforms"). Again, in tests to prove the dielectric performance of equipment, standard lightning impulse waveforms (hereinafter "standard waveforms") are used. Therefore, from the viewpoint of proper insulation design, it is very important to convert non-standard waveforms into equivalent (in terms of minimum breakdown voltages) standard waveforms.

While varying waveform constants, experiments were carried out to obtain dielectric characteristics against three different non-standard waveforms. This report deals with the selection and the production of non-standard waveforms and non-standard waveform discharge characteristics in SF_6 gas gaps which simulate actual ones in GIS.

SELECTED WAVEFORMS AND THEIR GENERATION

Selected waveforms

Results of analyses of lightning surges in GIS at actual substations show that surge waveforms are generally complicated oscillatory waves caused by various reflections in GIS. From their features, oscillatory waves can be classified into two groups: one having a large first wave and the other showing a sustained oscillation. On the basis of these waveforms, three waveforms were selected. They are shown in Fig. 1. Their features are as

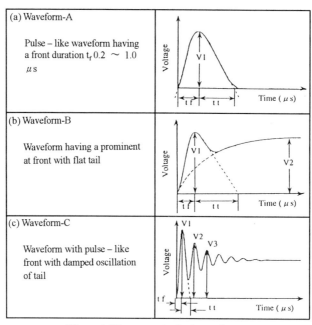

Figure 1. Three non standard waveforms

follows:
- Waveform-A: Only the first wave of the oscillatory wave is extracted. It is the basis of the other two waveforms.
- Waveform-B: Composed by superimposing waveform-A on the wave front of the ordinary lightning impulse waveform.
- Waveform-C: Composed by superimposing a multiple oscillatory wave on waveform-B.

Waveform producing circuits

For waveforms A, B, and C, non-standard waveforms each having a fast rise and an oscillation, a study was carried out to generate voltages. Fig. 2 shows the configuration of the voltage generation circuit. It consists of an ordinary impulse generator (IG) combined externally with air gaps, a resistor, an inductance, and a capacitance. On the test apparatus (C_1=500pF), waveforms are applied via air gaps arrayed in series. Resistor R_d coupled in parallel with C_1 is used to adjust the tails of the non-standard waveforms applied to the test apparatus. The rise time is controlled mainly by the impedance coupled between test apparatus C_1 and capacitor C_s on the power supply side. L_1, L_2, and L_3 in the figure denote the inductance components of the waveform application cables that connect component units.

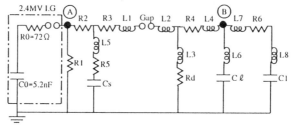

Figure 2. Equivalent circuits of non-standard lightning impulse for generating

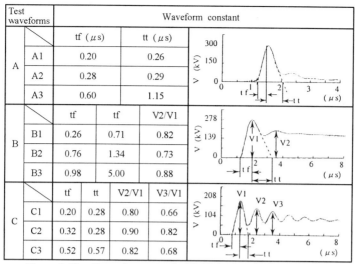

Figure 3. Waveform of non –standard lightning impulse by generating

First, an EMTP analysis was carried out to select circuit constants for generating the required non-standard waveforms. These constants were applied to an actual voltage generating circuit to prove the validity of the waveforms generated. Fig. 3 shows typical waveforms generated.

EXPERIMENTS

Fig. 4 shows the setup for the experiment, including the test apparatus and the electrode. The electrode was a rod-plane, and considering equivalence in actual use, 0.60 was selected for the field utilization factor(η). The gas pressure was constant 0.5MPa (20℃). The applied voltages were negative.

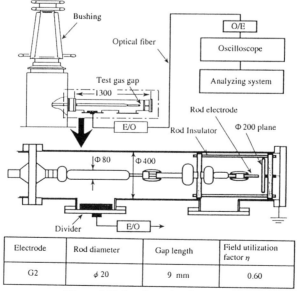

Figure 4. Test equipment and gap

TEST RESULTS

Waveform-A

Voltage-time (V-t) characteristics. Fig. 5 (a) shows the V-t curve of waveform-A1. It rose in the wave crest zone. After the peak, discharges were scarcely observed. With waveform-A2 having polarity reversed, discharges were also observed in the reverse-polarity zone with a trend of slightly declining discharge voltages (Fig. 5 (b)).

Comparison with standard waveforms. Fig. 6 shows the lower limit envelope curves of the V-t characteristics of variations of waveform-A shown in Fig. 5 in comparison with the V-t curve of the standard waveform with the same test electrode.

The discharge zones of the variations of waveform-A were in the short-time zone of the standard V-t curve. In this time zone, the discharge voltages were lower than those of the standard waveform but their minimum (V_{min}) was not below that of the standard waveform.

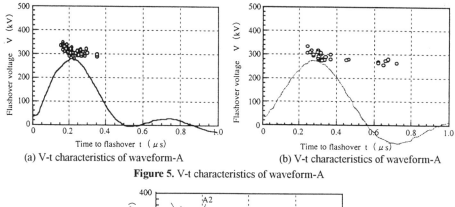

(a) V-t characteristics of waveform-A (b) V-t characteristics of waveform-A

Figure 5. V-t characteristics of waveform-A

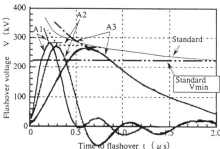

Figure 6. Comparison of the V-t characteristics of between standard and waveform-A

Waveform-B

V-t characteristics. Fig. 7 shows the V-t characteristics of waveform-B1. It gradually rose from the wave crest. The breakdown characteristics of the succeeding zone depended on the ratio of the succeeding voltage level (V_2) to the maximum peak (V_1) of the wave front and if V_2/V_1 was large, discharges were observed in the succeeding zone.

Comparison with standard waveform. Fig. 8 shows the V-t characteristics of the variations of waveform-B in comparison with those of the standard waveform. In the shorter time zone than the wave front of the standard waveform, lower flashover voltages than for the standard waveform are found. Despite the decline of the actual voltage level to 96% of the standard waveform, there were delayed flashovers.

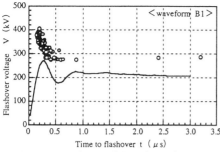

Figure 7. V-t characteristics of waveform-B

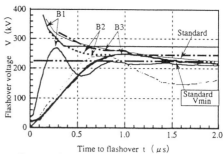

Figure 8. Comparison of the V-t characteristics of between standard and waveform-B

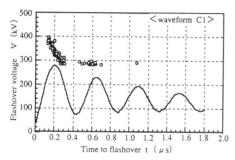

Figure 9. V-t characteristics of waveform-C

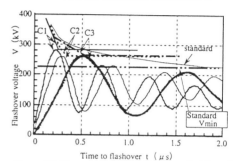

Figure 10. Comparison of the V-t characteristics of between standard and waveform-C

Waveform-C

V-t characteristics. Fig. 9 shows the V-t curve of waveform-C1. Similar to waveform-B, it gradually rises from the wave crest.

Comparison with standard waveform. Fig. 10 shows the V-t characteristics of the variations of waveform-C in comparison with those of the standard waveform. Like the other two waveforms, in the shorter time zone than the wave front of the standard waveform, the flashover voltages are lower than the standard. Flashovers mainly took place at the peak of the first wave, although with quick-cycling and slow-damping waveforms, flashovers were also observed at the peaks of second and third waves.

DIELECTRIC CHARACTERISTICS OF NON-STANDARD AND STANDARD WAVEFORMS

Table 1 summarizes the dielectric characteristics in all cases described above. The figures in parentheses denote values compared to the lower limit value (1.0) of the standard V-t curve. None of the cases in the present study showed lower values than V_{min}, but all of them showed values of 10% or more higher.

With waveform-A, as the wave front time decreased the breakdown voltage rose. In some cases, rises about 27% were observed with a wave front duration of $0.2 \mu s$.

With waveform-B, if the voltage of the succeeding wave exceeded 96% of the standard voltage, flashovers occurred in the succeeding zones.

With waveform-C, if damping was small, flashovers occasionally took place subsequent to second or third waves, and the actual discharge voltages in such cases were generally above 86% of the standard.

Table 1. Flashover characteristics of non-standard waveforms

waveforms (No.)		Waveform constant (μs, etc)				Minimum flashover voltage Vmin (kV)			
		Pulse part (1st wave)		Ratio to V1		Whole range	1st wave (corresponding to A)	B:follow C:2nd wave	3rd wave (C only)
		Front duration tf	Tail duration tt	V2	V3				
Standard		1.50	—	—	—	221 (1.00)	232 (1.05)	—	—
A	A1	0.20	0.26	—	—	280 (1.27)	280 (1.27)	—	—
	A2	0.28	0.29	—	—	273 (1.24)	273 (1.24)	—	—
	A3	0.60	1.15	—	—	265 (1.20)	265 (1.20)	—	—
B	B1	0.26	0.71	82%	—	273 (1.24)	273 (1.24)	273 (1.24)	—
	B2	0.76	1.34	73%	—	243 (1.11)	243 (1.11)	NFO	—
	B3	0.98	5.00	88%	—	244 (1.10)	252 (1.14)	244 (1.10)	—
C	C1	0.20	0.28	80%	66%	280 (1.27)	282 (1.28)	280 (1.27)	289 (1.31)
	C2	0.32	0.28	90%	82%	261 (1.18)	265 (1.20)	261 (1.18)	267 (1.21)
	C3	0.52	0.57	82%	68%	262 (1.19)	262 (1.19)	262 (1.19)	NFO

NFO;No-flashover ( );Ratio to standard wave

CONCLUSION

To determine lightning impulse test voltages for GIS, the dielectric characteristics of typical non-standard waveforms were investigated and compared to those of standard waveforms, revealing the following:

1) In shorter wave front zones than the standard zone, flashover voltages are occasionally lower than the standard, but none of them are below the minimum flashover voltages of the standard waveform.

2) Waveforms having a wave front duration of about 0.2 μs show about 27% improvements above the minimum flashover voltage of the standard waveform.

3) If there is a succeeding portion or wave higher than 86% of the standard waveform, flashovers can occur there.

The present research group is studying methods of calculating dielectrically equivalent conversions of non-standard waveforms into standard ones[4], intending to investigate other non-standard waveforms and electrode configurations to obtain generalizations.

REFERENCES

1. T. Kawamura et al., Pursuing Reduced Insulation Coordination for GIS Substation by application of High Performance Metal Oxide Arrester, CIGRE Paper 33-04 (1988)
2. JEC-0102 "Test Voltages" (1994) (in Japanese)
3. For example, S. Okabe et al., Insulation Characteristics of GIS Spacer for Very Fast Transient Overvoltage, IEEE, PES, 95WM006-7 PWRD (1995)
4. S. Okabe et al., Evaluation of Non-standard Lightning Surge Waveforms in GIS, 10th ISH (1997)

DISCUSSION

W. PFEIFFER: Your results are very interesting. They were achieved in the approximately homogeneous field. Do you think that the results would be similar for a stronger inhomogeneous field?

T. YAMAGIWA: Thank you for your comment. We tested only two cases of the utilization factor, $\eta = 0.45$ and $\eta = 0.6$. We compared the results for these two values. The rising rate of the short time region for the case of $\eta = 0.45$ is a little larger than for $\eta = 0.6$. If the utilization factor becomes smaller (more inhomogeneous field), we think that the rising rate may be higher than this test result.

SECTION 10: GAS-INSULATED EQUIPMENT II

A UTILITY PERSPECTIVE ON SF_6 GAS MANAGEMENT ISSUES

H.D. Morrison, F.Y. Chu, J.-M. Braun, and G.L. Ford
Ontario Hydro Technologies
800 Kipling Ave.
Toronto, Ontario
Canada M8Z 5S4

INTRODUCTION

Ontario Hydro and many other electrical utilities have substantial investments in SF_6 gas-insulated switchgear (GIS). Increasing SF_6 prices and environmental concerns on the effects of releasing SF_6 to the atmosphere are motivating utilities to review their SF_6 management procedures, such as gas handling, recycling, and leak detection. Eventually, solutions to these problems may be offered by equipment manufacturers. In the short to medium term, however, the utilities need to continue to operate their existing GIS, while at the same time reducing maintenance costs and satisfying environmental requirements.

Why, you may ask, should electrical utilities be taking the lead in managing issues related to SF_6? Last year, an inventory of historical production of SF_6 classified by end-use was created by a collaboration of all the major gas manufacturers.[1] The inventory shows that the electrical industry has been and continues to be the chief consumer of SF_6, about 80% of total known production. Thus it is incumbent on electrical utilities and equipment manufacturers to address the issue of SF_6 management, including emissions.

Fortunately, some improvements in SF_6 management are available now. There exists a wealth of scientific knowledge and practical expertise to support recycling of SF_6 gas for reuse. Several technologies are available for detecting and locating SF_6 leaks. Nevertheless there are significant gaps in knowledge in the field of quantifying individual gas compartment leakage rates so that remedial action can be assessed and scheduled, and in the field of preventing leaks with practical and effective technology that can be retro-fitted.

This paper will review all aspects of SF_6 management from a utility perspective and highlight some opportunities for development of improved technology. First we have a closer look at the motivation for SF_6 management, then examine some of Ontario Hydro's policy developments and their practical application, and conclude with a look at future solutions.

MOTIVATION FOR SF_6 MANAGEMENT

Since Ontario Hydro first started using SF_6, the main impetus for the management of SF_6 has been financial: the gas is expensive, so gas carts are used to recover and reuse the gas as much as possible. The financial argument also works the other way: the cost of new gas can be traded off against the cost of fixing leaks, including outages and the repair or replacement of equipment, or against the cost of labour spent waiting for gas to be removed from equipment for construction, commissioning or maintenance. Although SF_6 is only useful to Ontario Hydro when contained in electrical equipment, the SF_6 may have been viewed in some cases more as a consumable than as a capital cost, despite the expense.

Now, the much higher cost of SF_6 coupled with world-wide concern for the environment has increased the value and importance of managing SF_6 to minimize losses. As such, it should be treated neither as a consumable nor as a capital cost, but as an investment. A consumable is easily replaced, a capital cost can be depreciated, but an investment provides value and a return for as long as you maintain it.

The cause of the concern for the environment is based on the theory of global warming by the greenhouse effect and the discovery that SF_6 is the most potent greenhouse gas known, as measured by the Global Warming Potential (GWP)[2]: for a 100-year horizon the GWP for SF_6 is 23,900 on a scale where the GWP for carbon dioxide is 1. In simple terms, one tonne of SF_6 in the atmosphere will produce the same global warming effect as 23,900 tonnes of carbon dioxide. For utilities, such as Ontario Hydro, that generate electricity by burning fossil fuels, releasing carbon dioxide in the process, this conversion factor is important. It allows the utility to place its emissions of SF_6 in context with its primary greenhouse gas emissions of carbon dioxide.

The first formal requirement for limitations on SF_6 emissions as a method to control global warming appears in the Kyoto Protocol of the UN Framework Convention on Climate Change. SF_6 is included in the list of greenhouse gases along with carbon dioxide (CO_2), methane and several other gases. The Kyoto Protocol requires each signatory country to reduce total emissions of the listed gases by at least 5% below 1990 levels in the period 2008-2012; for SF_6 the base year can be 1995. The specific target for Canada is 6% and the target for the USA is 7%. Allowing for the projected growth in energy consumption, the overall reduction required for Canada by 2010 is 21%.

Although national regulations resulting from this Protocol are still a few years off, our discussions with representatives of other utilities and electrical equipment manufacturers indicate that voluntary control of SF_6 emissions now will help to secure the future of SF_6 as a component of electricity transmission and distribution.[3] As there is no candidate gas or gas mixture available now,[4] or in the near future, that can replace SF_6, we must do our part to ensure that SF_6 is used with due diligence for the environment.

SF_6 HANDLING AT THE EQUIPMENT

The current estimate of SF_6 in equipment at Ontario Hydro is 206,000 kg or 206 tonnes. About 85% of the gas is in high-voltage GIS or bus duct distributed among eight indoor stations. The remaining 15% is in free-standing circuit breakers and transformers outdoors, placed throughout the province of Ontario.

The policy of SF_6 handling in GIS stations has always been to recover the gas for reuse, generally into the same equipment from which it was removed. Ontario Hydro has more than 10 gas carts, each with a storage capacity of about 1000 kg, to service the gas handling needs at the stations. Gas handling at sites away from the stations has been a problem because of the cost and difficulty associated with transporting the gas carts. The carts may not be moved if they contain SF_6; they must be empty.

In 1993, our Environment department interpreted the Ontario Environmental Protection Act to recommend against a request to release to atmosphere some surplus SF_6. By the end of 1993, we had drafted "Guidelines for Reporting SF_6 Gas Releases" as an adjunct to our more general policy on spills management. Since then, more than 20 small recovery units have been purchased to handle gas in equipment away from the stations.

SF_6-filled equipment is inspected and maintained regularly, although the trend in recent years has been to reduce the frequency of inspections as a cost-saving measure. These inspections include checking the gas density monitors, and measuring the humidity level. With currently available technology, both procedures result in the release of small amounts of SF_6 from every compartment checked.

The gas density monitors are the principal means of monitoring for leaks in the equipment. The majority of the monitors are designed to detect reductions in density on the order of several per cent. As such they are sufficient to ensure the safe operation of the equipment, but not sensitive enough to track leak rates less than 1% per year for a single compartment.

RECLAIMING AND RECYCLING

Our operating procedures include safe handling and analysis of contaminated SF_6 to reclaim the SF_6 for reuse. The sources of contamination are well known and, in most cases, the gas carts are sufficient for removing the contaminants. When necessary, additional filter units are deployed between the SF_6 compartments and the gas carts to reclaim highly decomposed or humidified gas. We have never used the services of another company to reclaim SF_6.

The practice of reclaiming gas carries with it a requirement on the quality of the gas after reclaiming. This is not so much an issue for removing and refilling gas to the same compartment, but it is a greater issue for refilling different equipment, especially new equipment for which the manufacturer may insist on new gas for the warranty. While there are international standards for the quality of new gas purchased from gas manufacturers, a similar standard for used SF_6 is still being written. Even so, it is possible to obtain agreements with an equipment manufacturer to fill new equipment with used SF_6. For example, in 1993 we renovated one of our major stations, replacing some bus duct and adding several extensions. In the process we recovered and reused about 8,000 kg of gas.

The only contaminants that cannot be easily removed by utilities are air and CF_4, of which air is the more important. Separating SF_6 from air requires specialized equipment, not generally available. One possible solution, provided the gas mixture is otherwise free of contamination, is to dilute the concentration of air by adding pure new SF_6.

MODES OF SF_6 EMISSION

Emissions of SF_6 occur by three mechanisms: 1) leaks due to the failure of mechanical seals, 2) accidental releases due to breaks in the enclosures, and 3) procedural releases, such as during gas transfer operations or maintenance checks.

For new equipment, Ontario Hydro handles type 1 emissions by specifying maximum leak rates at the time of purchase: no more than 0.5% per year by weight of gas for any compartment. As equipment ages, both type 1 and type 2 emissions can occur and be treated by repair, or in extreme cases by total replacement of major components. Some of the leaks Ontario Hydro has experienced are the result of poor design, leading to premature failure of seals or joints. Design problems have been tackled by working with the equipment manufacturers.

In the past, the greatest source of SF_6 emission has probably been type 3, procedural releases. The typical case involves not removing all the gas from a compartment prior to opening, such as for a major overhaul or repair of an internal fault. Smaller releases occur from routine maintenance checks as mentioned before. This type of release will benefit the most from improved procedures and proper training of the technical staff.

GREENHOUSE GAS EMISSIONS

The electrical industry is changing dramatically, moving toward open access, emissions trading, and dismantling of public monopolies. Mergers and acquisitions are creating larger investor-owned companies in the "energy" business, not just electricity. In the last five years, Ontario Hydro has restructured several times. Despite the changes, there has been progress in the development of policies and procedures for environmental protection and more efficient SF_6 management.

In January 1995, the Board of Ontario Hydro approved a "Management Strategy for Greenhouse Gas Emissions." Although the Strategy was mainly for carbon dioxide and did not mention SF_6, the intention was to address the environmental impact of all Ontario Hydro operations. Later that year the Board approved a new "Sustainable Energy Policy and Principles," replacing the 1984 environmental policy that focused on compliance with regulations. The new policy provides a framework for the development of procedures for SF_6 management to go beyond simply controlling emissions.

Last year, SF_6 was still not included in our "Action Plan to Manage Greenhouse Gas Emissions," which serves as a report card on progress against the corporate Strategy. As a major user of fossil fuels, Ontario Hydro is more concerned about the production and release of carbon dioxide. Last year, we had CO_2 emissions of 22.7 Tg, or 22.7 million metric tonnes (MMT). Since the Canadian government has demonstrated its commitment to sign the Kyoto Protocol, senior management at Ontario Hydro has inquired about our emissions of SF_6. The Action Plan will likely be revised to refer to SF_6.

The estimate for 1997, based on what was bought and used for maintenance, is releases totalling 4300 kg of SF_6, equivalent to about 83 cylinders, or just over 2% of our total installed inventory. The estimate was made possible by a new central purchasing system, described in more detail below. Estimates for prior years were based on total purchases of SF_6. They did not reflect either the quantity of gas installed in new equipment or the amount of gas stockpiled for future use.

We can put the SF_6 emissions into context with our CO_2 emissions by using the GWP for SF_6 as the conversion factor. Thus the 4300 kg of SF_6 is equivalent to 0.10 Tg of CO_2, or 0.45% of our CO_2 emissions. In this view, the SF_6 emissions are only a fraction of the total greenhouse gas emissions, but they could be a significant part of the overall reduction plan as the control of SF_6 leaks should be a manageable task compared to CO_2 reduction.

EMISSION TRADING

The concept of emission trading is becoming a reality, and Ontario Hydro has included the possibility of purchasing emission credits for CO_2 in its "Action Plan" mentioned above. While Ontario Hydro remains an integrated company, it can also trade off SF_6 emissions against CO_2 emissions. Obviously, the only real way to reduce CO_2 emissions is to reduce the consumption of fossil fuels with a potential loss of revenue for lost sales of generation. Let us examine this potential loss of revenue as a way of calculating the financial cost of allowing SF_6 emissions.

In 1997, Ontario Hydro generated 133.8 TWh for a gross revenue of 8421 M$, so it received 62.9 M$ per TWh of generation. Generation by fossil fuel was only 24.8 TWh accompanied by CO_2 emissions of 22.7 Tg. Thus the CO_2 emissions can be considered as returning a revenue of 62.9 x 24.8 = 1560 M$, or 68.8 M$ per Tg of emission.

From the previous section, the 2% loss of SF_6 was equivalent to 0.10 Tg of CO_2. A reduction in CO_2 emission to compensate for this loss of SF_6 would result in losing revenue of 6.88 M$. Furthermore, the replacement cost of the SF_6 is about 0.17 M$ (assuming 40$/kg). Clearly, the total cost of SF_6 leaks must also include the manpower cost to replace the gas in the equipment, the cost of repairs, and the cost of equipment outages if taken.

Note that for utilities with lower total inventories of SF_6, the lost revenue to compensate for SF_6 emissions would be proportionately lower, but the same basic messages apply: losing SF_6 is expensive, not losing it may have considerable value.

INVENTORY CONTROL

In response to environmental concerns, Ontario Hydro has continually revised its handling practices for SF_6. Until last year, the most important question was still difficult to answer: how much SF_6 have we released to the atmosphere in any given year? A major improvement in this area was the change to central purchasing for inventory control. Before then, each station or site was responsible for purchasing SF_6 directly from the supplier. An attempt to assess inventory and gas usage at the station level met with great resistance to the paperwork involved.

Now, all gas cylinders are bought by a central unit. As the new cylinders are received, their ID numbers and weight are logged. Gas cylinders are distributed to sites on request, and returned to the central unit when they are finished. Before returning a cylinder to the supplier, it is weighed and any significant amount of residual gas is transferred to other storage containers. In this way, we do not pay for gas returned to the supplier.

With this global tracking system in place, we only need a couple more elements to provide reliable estimates of SF_6 inventory and emissions. These elements are a comprehensive database of all SF_6-filled equipment in service, and an annual inventory of SF_6 in storage containers, including gas carts and spare cylinders. The equipment database should include the year of commissioning, the recommended amount of gas for operation, and whether it was bought filled or partially filled.

The next step is to use the inventory control to identify the causes for emissions. Then we will have a more accurate assessment of the usage patterns for SF_6 at Ontario Hydro, and can modify our management procedures to seek lasting remedies.

WHAT LIES AHEAD?

The preceding discussion paints a broad picture of the available tools for SF_6 management. Let us now examine the opportunities for expanding the tool set and improving SF_6 management.

In some facilities, various types of online monitoring systems are being tested, usually for detecting partial discharge in gas compartments. Online monitoring is a continuous system intended to warn the system owner of problems before a catastrophic failure. The gas density monitors on our GIS now only provide a central alarm facility. They are generally not tied together to a central system for tracking loss of SF_6, and do not provide the necessary precision for a single compartment. Some new types of density monitors have been tested that provide greater precision and reliability, and allow for remote data retrieval.

What would be really useful is an online monitor with remote data access that could measure 1) gas density, either directly or from pressure and temperature, 2) gas humidity (ie water vapour concentration), and 3) the concentration of decomposition byproducts. The latter would be in support of detecting partial discharges and incipient failure. Online monitoring, provided as a retrofit during routine maintenance, would reduce the need for taking gas samples for analysis and minimize the risk of releasing gas.

Just detecting the occurrence of a leak is not sufficient; the leak must also be located. Some advanced methods of leak detection now employ a laser tuned to the main absorption frequency of SF_6 in the infrared. One such system scans the laser beam across the field of view to provide a video image of the equipment and the leak. The image is created from the backscattered laser light, showing gas plumes from leaks as darker areas because the light is absorbed by the gas. Slow movement of the gas plumes can be observed by recording the image on videotape and replaying at higher speed. At the moment, the laser systems are expensive and require highly trained operators. EPRI has sponsored the tests of one unit at several US utilities to encourage further development leading to cheaper, more convenient leak location devices.

The final step after leak location is leak mitigation by repairs to the leaking compartment. For instance, in the case of power transformers, Ontario Hydro has been engaged in a major effort to identify and develop permanent repair technologies capable of field application without oil removal from the transformer. Innovative welding strategies have been implemented on power transformers along with more conventional sealant technologies. A key element for welding performance involves the control of the condition of the steel backing paint as paint flaking and carbonization would adversely affect dielectric performance. Field application to large (35 MVA and up) power transformers was performed successfully without oil removal. These technologies could be applied, with some modifications, to GIS equipment.

Finally, all utilities will eventually need another policy tool: a plan for dealing with the end of life of SF_6-filled equipment. The plan must include methods for reclaiming and storing the SF_6, possibly for use in replacement equipment or perhaps to be returned to the gas supplier. The development of an end of life policy could be facilitated through utility organizations and international standards committees.

SUMMARY

Managing our investment in SF_6 provides direct economic and environmental benefits. Voluntary controls on emissions may be cheaper to implement than regulatory controls, especially if it extends the useful life of SF_6-filled equipment. The simplest form of control is the development of handling procedures to minimize losses, and their careful implementation by proper training to ensure the desired results are achieved. Reclaiming used SF_6 for reuse is a standard practice to prolong the use of SF_6 indefinitely, and minimize expenditures for new gas. A related issue is the treatment of equipment as it reaches end of life, an issue not yet duly considered. A global approach to inventory control is a central purchasing system to minimize paperwork at the stations. This approach may also help to identify systemic problems in SF_6 handling for which solutions may then be implemented. Ultimately, what must be answered is the loss of SF_6, which should be reported as a fraction of installed inventory as well as an absolute quantity. The SF_6 emissions may then be assessed with respect to the total emissions of greenhouse gases.

ACKNOWLEDGEMENTS

The authors gratefully acknowledge informative discussions with Noel Wylie, David Bray and Barb Reuber of Ontario Hydro, with members of the CIGRE WG23.10 Task Force on SF_6, and with members of the IEC Working Group 15 on Revision of IEC 480.

REFERENCES

1. *Sales of Sulfur Hexafluoride (SF6) by End-Use Applications*, report prepared by Science & Policy Services, Washington (1997).
2. *Summary for Policymakers: The Science of Climate Change*, Part of the Second Assessment Report of the Intergovernmental Panel on Climate Change (1995).
3. G. Mauthe, B.M. Pryor, L. Niemeyer, R. Probst, J. Poblotzki, H.D. Morrison, P. Bolin, P. O'Connell, and J. Henriot. *SF_6 Recycling Guide: Re-Use of SF_6 Gas in Electrical Power Equipment and Final Disposal*, CIGRE (1997).
4. L.G. Christophorou, J.K. Olthoff, and D.S. Green, Gases for electrical insulation and arc interruption: possible present and future alternatives to pure SF_6, *NIST Technical Note 1425* (1997).

DISCUSSION

J. CASTONGUAY: I have been monitoring the SF_6 concentration in the air of two Hydro Quebec GIS indoor substations. This has been carried on by now for 26 months. The average leak rates measured were typically ~ 0.54%/year and ~ 0.62%/year. This corresponds to 1 to 1.5 kg/month depending on the amount of SF_6 in the corresponding GIS. These stations are now 20 and 17 years old.

H. MORRISON: This is an interesting result.

B. DAMSKY: EPRI has been testing a "laser camera" in the field for several months. We find it quick and accurate to locate SF_6 leaks down to a rate of a few pounds per year. This technology has the advantage of allowing a leak service to be made while the equipment is energized.

J. CASTONGUAY: Out of your 4300 kg of SF_6 lost in 1997, can you tell the relative proportion of "natural" leaks from the "procedural" releases?

H. MORRISON: Unfortunately not. Our procedures are not yet fully implemented to provide answers on all the sources of emission.

S. DALE: Has Ontario Hydro decided that 1990 is the base year for SF_6 reporting? Both 1990 and 1995 were choices in the Kyoto protocol. Why did Canada decide on 1990?

H. MORRISON: I believe we have not yet decided on the year. The Kyoto protocol insists on 1990 for CO_2 but allows 1995 for SF_6. We may find it easier to use 1995 for SF_6 because the records are better than for 1990.

THE COMPARISON OF ARC-EXTINGUISHING CAPABILITY OF SULFUR HEXAFLUORIDE (SF$_6$) WITH ALTERNATIVE GASES IN HIGH-VOLTAGE CIRCUIT-BREAKERS

Hartmut Knobloch

Siemens AG
Berlin, Germany

ABSTRACT

This paper describes the influence of the switchgear geometry, the priming pressure of the SF$_6$ switchgear and the magnitude of the short-circuit current on the dielectric recovery of the arc gap, using the example of a high-voltage circuit-breaker for a 145 kV network. The results of other tests document the influence of the arc-quenching gases nitrogen (N$_2$), carbon dioxide (CO$_2$) and a nitrogen/SF$_6$ mixture on the arc gap recovery in comparison with SF$_6$. The results show that the arc control capability of these gases is very limited compared with SF$_6$. Consequently, it will not be possible totally to replace the SF$_6$ in high-voltage switchgear in the short to medium-term.

INTRODUCTION

Fluoride compounds make a direct contribution to the greenhouse effect. They are currently the subject of political discussions and negotiations aimed at stabilising greenhouse gas concentrations in the Earth's atmosphere. The fluoride compounds under consideration include sulphur hexafluoride (SF$_6$), one of the uses of which is as an arc quenching and insulating gas in high-voltage switchgear and substations. Although worldwide SF$_6$ contributes just 0.08% to the greenhouse effect[1], and only around a half of this originates from switchgear and substations[2], manufacturers and operators are working hard nevertheless to reduce this level. They are already recycling the gas[3] and reducing leaks and losses due to operation, but the discussion has now moved on to consider how all or at least part of the SF$_6$ can be replaced by other gases[4,5]. The work and results described in this paper may help to clarify the options available for replacing SF$_6$ with other gases or mixtures of gases. In all this work, the emphasis has been placed on the influence of these gases and gas mixtures on the arc control capability in high-voltage power switchgear.

The demand for electricity is increasing world-wide, thus requiring power supply networks to be interlinked to an increasing extent. When a short-circuit occurs, this leads to increasing short-circuit currents within the switching device. SF_6 used as the arc-quenching medium now assures that it is possible to switch short-circuit currents of up to 63 kA in either a 245 kV network with one interrupter unit per pole, or in a 550 kV network with two interrupter units connected in series. The switching device requires both thermal and dielectric quenching capability, i.e. specific dielectric recovery of the arc gap after the short-circuit current.

INFLUENCE OF GEOMETRIC PARAMETERS AND CURRENT INTENSITY ON ARC GAP RECOVERY

Before discussing the influence of different switching gases on the arc gap recovery of a high voltage circuit-breaker in greater detail, we should first consider how this variable is affected if the arc gap geometry and short-circuit current are used as parameters.

To test the recovery, a high voltage circuit-breaker[6] designed for 123 kV and 40 kA was exposed to a damped, oscillating AC voltage with a peak value of 100 kV and a frequency of 100 kHz. This voltage was switched to the interrupter unit exactly as a 36 kA short-circuit current with an arcing time of 17 ms passed through current zero. When we used a commercially available circuit-breaker filled with SF_6 at 6 bar, we recorded no limit values in the recovery voltage range under consideration (up to 100 kV) or in the period up to 5 µs after the short-circuit current passed through zero, since the recovery occurred faster than the rise in the test voltage. To be able to record any recovery limit value curves at all, we had to carry out the tests using a high-voltage circuit-breaker with part of the puffer function[6] disabled. This resulted in reduced compression and blow-out. The arc gap was intentionally further reduced by decreasing the SF_6 filling pressure from 6 bar to 1 bar.

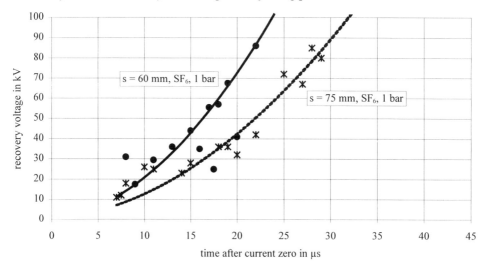

Figure 1. Influence of the arc gap geometry on the dielectric recovery of a weakened high-voltage circuit-breaker after a short-circuit current.

Figure 1 shows the extent to which dielectric recovery after a short-circuit current is dependent on the geometry of the arc gap (in this case on the distance between the fixed contacts) of a weakened high voltage circuit-breaker designed for 123 kV and 40 kA. The recovery becomes slower as the distance between the contacts increases. This is probably because the arc occurring during the short-circuit current phase increases in length as the contact gap widens, resulting in increased energy conversion in the arc gap. In turn, this causes slower recombination of the residual arc plasma in the arc gap, leading directly to the slower increase in dielectric strength.

The same effect occurs if a higher short-circuit current flows across the breaking contacts before recovery. In this case, the increased energy conversion due to the higher current again indirectly results in a slower increase in the dielectric strength. Figure 2 shows test results for short-circuit currents of 36 kA and 50 kA which confirm these hypotheses.

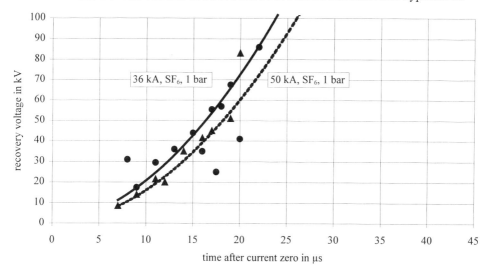

Figure 2. Influence of the short-circuit current on the dielectric recovery of a weakened high-voltage circuit-breaker.

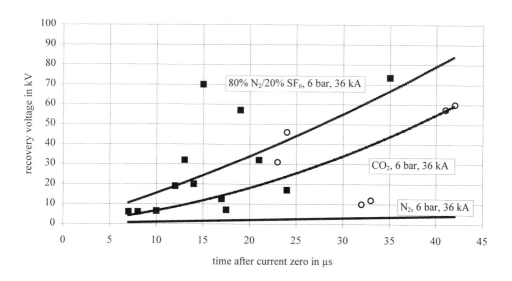

Figure 3. Influence of different arc-quenching gases on the dielectric recovery of a weakened high-voltage circuit-breaker.

INFLUENCE OF DIFFERENT QUENCHING GASES ON THE ARC GAP RECOVERY

Figure 3 shows the arc gap recovery after application of a 36 kA short-circuit current with an arcing time of 17 ms for different arc-quenching gases at 6 bar. Pure nitrogen exhibits the slowest recovery. In this case, voltages of just 15 to 20 kV were recorded approximately 100 µs after current zero, i.e. values which fall a long way outside the scale selected for Figure 3.

Carbon dioxide (CO_2) exhibits faster recovery than nitrogen (N_2), although the values are very disperse. Slightly faster dielectric recovery on average can be obtained with a mixture of 20% SF_6 and 80% N_2. Here, again, the values are scattered over a wide range. However, the dielectric recovery of all three arc quenching gases is still much lower than that of pure SF_6, even if the pressure is reduced to 1 bar (compare with Figure 1). This may be explained by the higher arcing time constants of the gases compared to SF_6. These constants influence the arc gap recovery after the short-circuit current. Table 1 lists the values[7] which correctly describe the measured dielectric recovery characteristics, even though they are qualitative, rather than absolute, since they were determined under different marginal conditions.

Table 1. Arcing time constant τ for different gases for a 1 A arc[7].

Type of gas	τ in µs
SF_6	0.8
CO_2	1.5
N_2	210

INFLUENCE OF QUENCHING GAS PRESSURE ON ARC GAP RECOVERY

The speed of arc gap recovery can be increased by raising the gas pressure, as shown by the measurements for the CO_2 example in Figure 4. At a pressure of 8 bar, for example, CO_2 achieved an average recovery speed roughly corresponding to that of the N_2/SF_6 gas mixture tested in Figure 3. In physical terms, the higher gas priming pressure improves the dielectric strength of the arc gap.

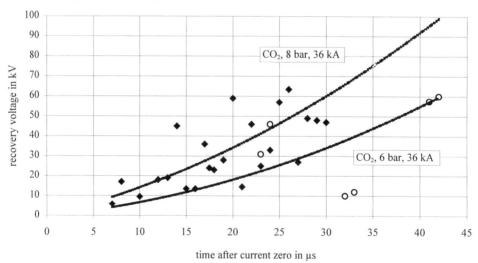

Figure 4. Influence of the gas priming pressure on the dielectric recovery of a weakened high-voltage circuit-breaker.

CONCLUSIONS

The arc gap recovery obtained with pure nitrogen was poor and unacceptable. At the same pressure, a nitrogen/SF$_6$ mixture containing 80% N$_2$ and 20% SF$_6$ would also be unable to assure the necessary dielectric recovery fast enough after a 40 kA current in a 145 kV network. The same applies to pure carbon dioxide (CO$_2$), even if the switchgear priming pressure is increased by a further 2 bar above 6 bar.

This observation is supported by test results obtained with a high-voltage circuit-breaker based on an insulating nozzle system with a self-compression device. This is designed to be filled with SF$_6$ for 40 kA and 145 kV applications[8]. If this device is operated with CO$_2$ at 8 bar, rather than with SF$_6$ at 6 bar, then it is still able to switch a 25 kA short-circuit current in a 38 kV network. In this case it is the dielectric, rather than the thermal switching capacity, which is limited. Since the geometry of the tested interrupter unit for SF$_6$ is optimum with respect to dielectric stress and pressure build-up (and thus gas temperature, blow-out and contact speed), the dimensions of the device would have to be modified in order to be able to use CO$_2$ as the arc quenching gas. Even this would not exceed the quenching capability of an 80% N$_2$ / 20% SF$_6$ gas mixture, as demonstrated by the results in Figure 3.

The current practice of reducing the SF$_6$ operating pressure or using N$_2$ / SF$_6$ gas mixtures in circuit-breakers for low temperature applications always implies a concomitant reduction in the switching capacity requirements, as confirmed by the results discussed here.

However, if existing switchgear is used with gases or gas mixtures other than pure SF$_6$, the operating limits will always have to be recalculated and reconfirmed by further type testing. This is not a practicable option with older devices, given that the service life of switchgear can be 30 or even 50 years.

These disappointing results show that there is certainly no possibility of fully replacing SF$_6$ in switchgear equipment in the short to medium-term, since the arc control capability of other gases or gas mixtures is extremely limited. We therefore need to find alternative solutions, based either on the use of other gases at comparatively high pressures, as it was the case decades ago when nitrogen at several 10 bar was used as the arc quenching gas, or totally different physical arc control mechanisms, such as vacuum switching.

SUMMARY

Increasing the size of the arc gap delays the dielectric recovery of high-voltage circuit-breakers after a short-circuit current, as does reducing the switchgear priming pressure or increasing the short-circuit current. Pure nitrogen (N$_2$), pure carbon dioxide (CO$_2$) and an 80% N$_2$ / 20% SF$_6$ gas mixture all exhibit unacceptably slow recovery compared to SF$_6$. With pure carbon dioxide (CO$_2$) as the quenching gas, an equivalent switching device with SF$_6$ designed for 145 kV and 40 kA would still be able to switch a 25 kA short-circuit current in a 38 kV network, the only proviso being that the device's priming pressure would have to be raised from 6 to 8 bar.

REFERENCES

1. International Electrotechnical Commission. *High-Voltage Switchgear and Controlgear - Use and Handling of Sulphur Hexafluoride (SF$_6$) in High-Voltage Switchgear and Controlgear,* IEC 1634, Technical Report, 1995-04.

2. CAPIEL. *The Influence of SF_6 Switchgear on the Greenhouse Effect,* Joint position paper of UNIPEDE (International Union of Producers and Distributors of Electrical Energy) and CAPIEL (Co-ordinating Committee for Common Market Associations of Manufacturers of Industrial Electrical Switchgear and Controlgear), Paris (1997).
3. CIGRE, SF_6 recycling guide - re-use of SF_6 gas in electrical power equipment and final disposal, *Electra* 173:43 (1997).
4. L.G. Christophorou et al., Gases for electrical insulation and arc interruption: possible present and future alternatives to pure SF_6, *NIST Technical Note 1425,* US Dept. of Commerce, Gaithersburg, Maryland (1997).
5. M. Zhou and J.P. Reynders, Synergy between SF_6 and other gases to enhance dielectric strength, *10th International Symposium on High Voltage Engineering*, Montreal, Quebec, Canada (1997).
6. Siemens, SF_6-circuit-breaker 3AQ1-E, 72,5 kV-245 kV, *Technical Data Sheet, Siemens,* Power Transmission and Distribution, Berlin.
7. E. Philippow. *Taschenbuch der Elektrotechnik, Band 5,* Carl Hanser Verlag München, Wien (1981), 658.
8. Siemens, 3AP1-FG high-voltage circuit-breaker, 72,5 kV to 145 kV, *Technical Data Sheet, Siemens,* Power Transmission and Distribution, Berlin.

DISCUSSION

L. NIEMEYER: Could you estimate how many non-SF_6 circuit breaker units you would have to put in series in order to reach the same performance as one SF_6 breaker unit?

H. KNOBLOCH: At least the connection of four interrupter units in series filled with 8 bar of CO_2 is needed to master the test voltage after short-circuit currents for a 145 kV network. This very rough estimation does not take into consideration the need of additionally grading capacitors to realize a nearly linear voltage division across the interrupter units. It has to be considered also that the maximum imterruptable short-circuit current is lowered from 40 kA for SF_6 to 25 kA for CO_2 at the same time.

A. GLEIZES: First, a comment. The interrupting capability of the 20%SF_6/80%N_2 mixture at a filling pressure of 6 bars is lower than that of pure SF_6 at one bar, in spite of the fact that the partial pressure of SF_6 in the mixture is higher than 1 bar. This corresponds to a very negative effect of nitrogen that tends to confirm one conclusion of my paper indicating that the proportion of SF_6 in the SF_6/N_2 mixture (for HV circuit breaker) should be of the order of or higher than 50%. Second, a question. What is the definition of the arcing time constants given in your paper?

H. KNOBLOCH: The arcing time constants, given in Table 1 of the paper, are from reference 7. The values are valid only for 1 A currents and they are probably different for higher currents like short-circuit currents of several 10 kA. Nevertheless, they indicate the direction: the time constant for CO_2 is higher than for SF_6, and for N_2 it is much higher than for SF_6. The relatively small difference between the arcing time constants of SF_6 and CO_2 was the reason for carrying out the tests with CO_2.

L. G. CHRISTOPHOROU: Have you tried these measurements on a 50%SF_6/50%N_2 mixture?

H. KNOBLOCH: Tests with 50%SF_6/50%N_2 mixtures on the weakened double-nozzle circuit-breaker described in the paper have not been done. But commercially available circuit-breakers operate with these mixtures for very low temperature applications. In this case the short-line fault and terminal-fault switching capability has to be lowered by at least one or two steps, from 40 kA, for instance, to 31.5 kA - 25 kA. It has also to be mentioned that the capacitive switching capability is like the SF_6 partial pressure of the mixture in that case.

B. DAMSKY: Did you measure any of the upstream pressures at the time of interruption? I expect that there would be differences depending on the test gas.

H. KNOBLOCH: Pressures in the compression cylinder of the weakened circuit-breaker were measured during the short-circuit current tests. If I remember right, the highest pressures were recorded for N_2, because of the highest temperature of this gas, followed by CO_2 and SF_6.

INSULATION CHARACTERISTICS OF DC500kV GIS

M. Shikata,[1] K. Yamaji,[2] M. Hatano,[3] R. Shinohara,[4] F. Endo[4] and T. Yamagiwa[4]

[1]The Kansai Electric Power Co., Inc.
[2]Shikoku Electric Power Co., Inc.
[3]Electric Power Development Co., Ltd.
[4]Hitachi, Ltd.

INTRODUCTION

A DC500kV transmission system will be operated at the beginning of the 21st century in Japan, and DC GIS will be used in this system. As DC500kV GIS is the first application of its type in the world, many insulation problems had to be solved.

Under DC voltages, particle movement, electric field distribution, charge accumulation and insulation strength in SF_6 gas are quite different from those under AC voltages. Many investigations were carried out to clarify these characteristics and special new insulation concepts and techniques have been developed. DC500kV GIS incorporates these concepts and technologies, and it has successfully passed a factory test, a one-year site test, and a breakdown test after the site test.

This paper describes the basic, practical characteristics of DC-GIS, new insulation concepts, and the test results of DC GIS.

CONCEPT OF DC500kV GIS

Fig. 1 shows the problems that had to be solved in DC GIS. In particular, consideration had to be given to particles and change accumulation. In the DC electric field, particles go across the gap when they start to float, and collide with a high voltage conductor, and in the case of negative polarity, they float around the high voltage electrode, which is called the firefly phenomenon. Therefore, they sometimes adhere to an insulator and greatly reduce the dielectric strength.

The insulator may be charged when DC voltage is applied for a long time, and the charge accumulation affects the dielectric strength. Therefore, optimization of spacer shape is important to enhance insulation reliability. Furthermore, as shown in Fig. 2, the DC electric field distribution depends on not only the dielectric constant of an insulator, but its resistivity. Therefore, special consideration must be paid to insulators, especially around the operating rod used in GIS.

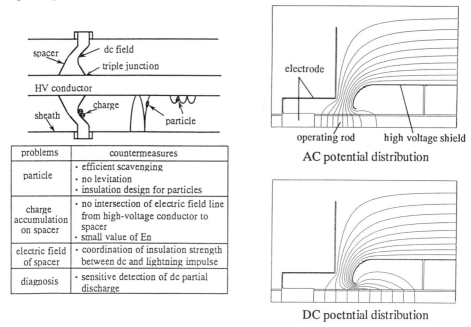

Fig. 1 Problems in DC GIS   Fig. 2 Potential distribution of operating rod

Fig. 3 shows the structure of the DC500kV GIS bus bar. The bus bar consists of a spacer area, particle scavenging area, and non-levitating area. In the spacer region, the spacer shape is optimized for DC and AC voltages. A special particle driver is placed adjacent to the spacer, which drives particles into the trap. At distances from the spacer, the electric field strength is designed to be lower than that in the particle-lifting field. Moreover, the inside of the sheath is coated with insulation materials.

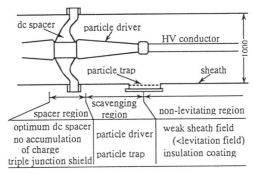

Fig. 3 Internal structure of DC500kV GIS (bus-bar unit)

TEST OF DC500kV GIS

Factory Test and Long-term Site test

DC500kV GIS developed with this design concept successfully passed the factory test shown in Table 1, and then underwent the long-term site test at DC ±625 kV (1.25 p.u) in the Yamazaki Experiment Center for one year starting from September 1995. The site test voltage value of 1.25 p.u for one year is equivalent to a working voltage of 1.0 p.u for 30 years. Fig. 4 shows DC500kV GIS under the long-term site test. Prior to the long-term site test, a conditioning test to remove particles, a polarity reversal test, and a withstand voltage test (DC ±625 kV : duration 2 hours each) were conducted to check the GIS was not damaged during transportation and assembly at the site. After that, the long-term site test of voltage DC -625 kV was started. After half a year, voltage polarity was reversed from negative to positive. Then, DC +625 kV was applied for half a year. Finally, a polarity reversal test (positive-to-negative) and a withstand voltage test were conducted. To ensure no damage owing to polarity reversal test, DC ±625 kV was applied in this withstand voltage test.

Partial discharges, leakage currents of the lightning arrester and the potential divider output voltages were monitored during the long-term site test. The results were acceptable and no degradation was found.

Factory Test after Long-term Site test

The DC500kV GIS was transported to the factory to check the insulation performance. The insulation test in Table 1 was carried out. It was confirmed that the GIS had the same performance as during the initial phase of manufacturing.

BREAKDOWN TEST

Purpose of Breakdown Test and Testing Method

The insulation level of DC500kV GIS after the long-term site test was checked with a breakdown test. DC electric field distribution depends on not only the dielectric constant of the insulator but its resistivity. As resistivity is generally nonlinear to the electric field strength, DC electric field distribution varies with applied voltages. DC breakdown voltages of the GIS was calculated under consideration of these characteristics, and their values were over 2000 kV. This value exceeded the capacity of the testing equipment, so a breakdown test was conducted with lightning impulse voltages.

Fig. 4 DC500kV GIS under long-term site test at Yamazaki Experiment Center
(The Kansai Electric Power Co., Inc.)

Table 1 Insulation test procedure and results

Voltages	Specifications	Results
Lightning impulse	±1300 kV (3 times)	±1560 kV withstood (3 times)
Switching impulse	±1175 kV (3 times)	±1410 kV withstood (3 times)
AC	645kV(5min.) / 559 kV (30 min.) / 559 kV (30 min.)	Withstood no partial discharge up to 774 kV
DC	±900 kV(1 min.) / ±750 kV (2 hrs.) / ±750 kV (2 hrs.)	Withstood no partial discharge up to ±900 kV
Polarity reversal	±625 kV (2 hrs.) / ∓625 kV(30 min.)	Withstood

Calculation Procedure of Breakdown Voltage

Prior to this test, breakdown voltages were estimated in consideration of the breakdown probability distribution and area effect. Breakdown voltages are influenced by the electrode surface roughness and electrode area. Therefore, accurate knowledge of probability distribution of the first breakdown is necessary to estimate minimum breakdown voltages. Electrode area was weighted with this probability, and breakdown voltages were calculated at each portion of the GIS.

Breakdown Test Results

Fig. 5 shows an applied voltage pattern and Fig. 6 shows DC500kV GIS under breakdown test in the factory. The applied voltage was increased from 1800 kV to 2250 kV, which was the limitation of the test bushing, in increments of 150 kV. Lightning impulses of positive and negative polarity were applied 3 times at each step. When the GIS withstood ±2250 kV, the gas pressure was reduced in 0.05 MPa increments, and similar tests were repeated.

This GIS consists of 6 units and breakdown tests were conducted on each unit. Fig. 6 shows the estimated breakdown voltages of the bus bar unit. The hatched range is the calculated breakdown voltages. This range was obtained with characteristic probability. The breakdown voltage of the bus bar unit was -2030 kV at a gas pressure of 0.3 MPa, and this value was equal to 2640 kV at designed gas pressure.

The test was conducted for all units. Breakdown voltages were almost equal to the estimated voltages and exceed the specification value of 1300 kV for every unit. Therefore, these results shows the insulation performance had not been degraded by the long-term site test.

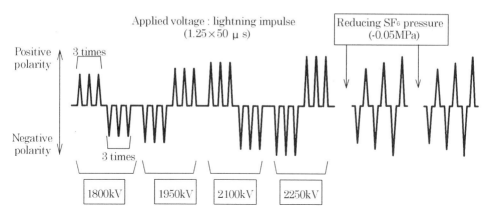

Fig. 5 Voltage application pattern

Fig. 6 DC500kV GIS under breakdown test

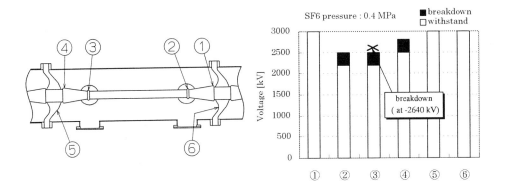

Fig. 7 Estimated breakdown portions and breakdown voltages

CONCLUSION

The insulation structure of DC GIS was investigated on the manufactured DC500kV GIS. The results are summarized as follows.

(1) The insulation structure consisting of a spacer specially designed for DC voltages, a particle scavenging region, and a non-levitating region, attained high insulation reliability.
(2) DC500kV GIS has successfully passed the factory test, the one-year site test and the factory test after the site test.
(3) Breakdown voltages were calculated with first breakdown probability and electrode area effect coincided with measured ones, and this means DC500kV GIS maintained high insulation strength after the long-term site test.
(4) The results of the factory test after the site test and breakdown test showed the insulation characteristics of DC500kV GIS had not been degraded by the long-term site test and the GIS has high insulation reliability for HVDC transmission system.

REFERENCES

1. Hasegawa et al., *Control of particle motion and reliability improvement in High-Voltage DC GIS*, 11th International Conference on Gas Discharges and Their Applications (1995)
2. Yoshida et al., *Enhancing insulation reliability towards particles in DC GIS*, 10th International Symposium on High Voltage Engineering(1997)
3. Hasegawa et al., *Development of Insulation Structure and Enhancement of Insulation Reliability of 500kV DC GIS*, IEEE Trans. on Power Deliverly, vol. 2(1997), pp.194-202
4. Hasegawa et al., *Charging and Breakdown Characteristics of Various Surface-Treatment Spacers under DC Voltage in SF_6*, in:"Gaseous Dielectrics VII", Loucas G. Christophorou and David R. James, ed., Pergamon Press, New York (1994), pp.503-510

DISCUSSION

B. DAMSKY: EPRI, working with GE and Consolidated Edison of New York, installed a pair of 400 kV DC converter stations with gas bus in 1983. While our prototypes passed laboratory tests similar to those described here, our installation was found to have converter valve supports not made to design which caused poor performance. We discovered that the voltage distribution involving different materials was markedly different at different temperatures.

R. SHINOHARA: Thank you for your comment. The AC potential distribution differs according to the temperature. We already considered those characteristics in the calculation.

S. DALE: Did you perform tests where the impulse voltage was superimposed on the DC bias voltage? Is this an important test to perform?

R. SHINOHARA: As you mentioned, we have to consider the impulse superimposed on the AC bias voltage. We then conduct tests between the contact of disconnecting switch. However, it is difficult to conduct tests between the H.V. conductor and ground. We have to calculate the field strength when superimposed LI on the DC voltage, when the DC potential distribution is greatly different compared with that of LI (for example, around the insulator, operating rod, etc.).

W. BOECK: Have you ever investigated the improvement by a conductive layer on the spacer with respect to field distribution and charge accumulation?

R. SHINOHARA: That idea is very good. However, it is very difficult to control the resistivity distribution. We use the same material used in AC-GIS, and control the potential distribution by optimizing the shape.

IMPROVEMENT OF WITHSTAND VOLTAGE AT PARTICLE CONTAMINATION IN DC-GIS DUE TO DIELECTRIC COATING ON CONDUCTOR

T. Hasegawa[1], A. Kawahara[2], M. Hatano[3],
M. Yoshimura[4], H. Fujii[4], K. Inami[4], H. Hama[4], and K. Nakanishi[4]

[1]The Kansai Electric Power Co., Inc.
Osaka, Osaka, Japan

[3]Electric Power Development Co., Ltd.
Tokyo, Japan

[2]Shikoku Electric Power Co.,Inc.
Takamatsu, Kagawa, Japan

[4]Mitsubishi Electric Corporation
Amagasaki, Hyogo, Japan

INTRODUCTION

±500kV DC transmission between Honshu island and Shikoku island in Japan will be operated at the beginning of the next century. High voltage DC-Gas Insulated Switchgear (DC-GIS) equipped in the DC-AC converter station has been developed. One of considerable items for designing the DC-GIS is the effect of the particle contamination to electrical properties more crucial than conventional GIS. The particles floating around the high voltage conductor with negative DC voltage give the most detrimental effect to the insulation properties at application of positive lightning impulse. Therefore it is essential to study and elucidate the phenomena for the realization of compact and highly reliable DC-GIS.

However there are only several papers on the investigation for the discharge characteristics at particle contamination in DC-GIS[1-3]. Diessner et al[1] proposed a model of particle motion near the negative conductor. On the other hand, Anis and et al[2] estimated breakdown voltages in the case that a particle was floating at a certain position.

In this paper, we first studied the behaviors of a particle floating around a negative conductor in both cases of bare and dielectric coated electrodes. And the effect of dielectric coating on high voltage conductor was investigated using a practical GIS bus model. We will discuss a possible mechanism in which the charge quantity of the floating particle much influences to breakdown properties of the system.

EXPERIMENTAL PROCEDURES

Experimental setups are shown in Fig.1. Observation of metallic particle motion was carried out by using the setup in Fig.1 (a). Gap distance between hemispherical electrode in radius of 25mm and plane electrode was 20mm. The high voltage hemispherical electrode was bare or dielectric coated. As the dielectric coating material, Al_2O_3 (dielectric constant :8.6) was used and the thickness was 30 μ m. As metallic particle aluminum wire in diameter of 0.2mm and in length of 3mm was used.

The particle was set on the plane electrode in a gas chamber, with a SF_6 pressure of 0.4MPa. Negative dc voltage was applied to the hemispherical electrode. Particle motion during dc voltage application was observed with a high-speed camera (FASTCAM-ULTIMA:Photoron). The camera was focused near bottom of hemispherical electrode.

In the experiment shown in Fig. 2(b), the breakdown voltage at particle contamination was measured using practical GIS bus model with ID/OD of 240mm/900mm and 5m length at the gas pressure of 0.4MPa. The effect of dielectric coating on high voltage conductor to the breakdown voltage was investigated. The coating was Al_2O_3. The aluminum particle had the same size with that used in the experiment shown in Fig. 1(a)

Applied voltage to the high voltage conductor had negative polarity. When metallic particles were floating around the conductor, a positive lightning impulse voltage was superimposed.

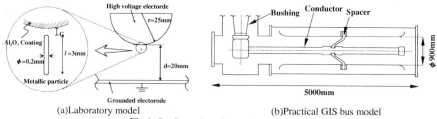

(a)Laboratory model (b)Practical GIS bus model
Fig.1. Configuration of experimental setup

EXPERIMENTAL RESULTS IN LABORATORY MODEL

An example of the particle motion observed with the high-speed camera is shown in Fig. 2. In this case the shutter speed of the camera was 18000 frames/sec. From the figure, the metallic particle returns small gaps (G in Fig. 1(a)) of a few hundreds μ m and in a few ms. Return times in the case of dielectric coated electrode was the same in the case of bare electrode as dielectric electrode.

The dependence of small gap G on applied voltage is shown in Fig. 3. The error bars in the figure show maximum and minimum values. In the case of bare electrode the gap lengths becomes long with increase of applied voltages. On the other hand, in the case of the dielectric coated electrode gap length is almost constant and the smaller than that of bare electrode.

We calculated average velocity $\bar{v}$ (=2G/T) of particle motion using the gap lengths G and the return time T obtained from this experiment. The dependence on applied voltages of the velocity $\bar{v}$ is shown in Fig. 4. The velocity $\bar{v}$ in the case of bare electrode is larger than that of dielectric coated electrode. The velocity increases with applied voltage.

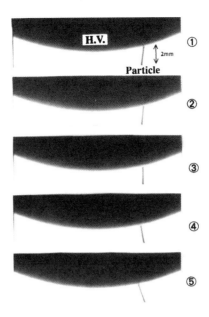

Fig.2. Example of metallic particle motion obtained by high-speed camera
(The intervals of about 220 μs between pictures)

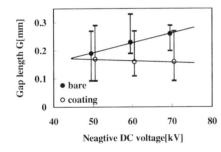

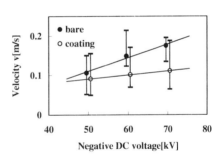

Fig.3. Gap length G of particles as a function of DC voltage

Fig.4. Average velocity of particle as a function of DC voltage

DISCUSSIONS

Characteristic of Metallic Particle Motion near High Voltage Electrode

At first, we discuss about charge quantity of a particle floating around the electrode at this session. We calculated the charge quantity of the metallic particle by solving of the following motion equation,

$$m\frac{dv(t)}{dt} = qE - mg - F_a \quad \cdots(1)$$

, where m is the mass of the particle, E is electric field, g is the acceleration of gravity and F_a is viscous force. In this case, we supposed, for simplification of calculation, that the particle had point charge.

The velocity v(t) is expressed by the equation;

$$v(t) = \frac{qE - mg - F_a}{m}t - \frac{qE - mg - F_a}{2m}T \quad \cdots(2)$$

G is given by the next equation ;

$$G = \int_0^{T/2} v(t)dt \quad \text{(down)} \quad \cdots(3) \qquad G = \int_{T/2}^{T} v(t)dt \quad \text{(up)} \quad \cdots(4)$$

From the equations (2)-(4), q is given by,

$$q = \frac{-4m\bar{v}/T + mg + F_a}{E} \quad \text{(down)} \quad \cdots(5) \qquad q = \frac{4m\bar{v}/T + mg + F_a}{E} \quad \text{(up)} \quad \cdots(6)$$

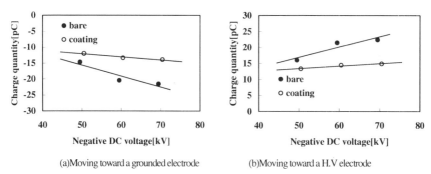

(a) Moving toward a grounded electrode (b) Moving toward a H.V electrode

Fig.5. Charge quantity of particle on DC voltage

The charge quantity q calculated by eq.(5) and (6) is shown in Fig. 5. The charge in the case of the bare electrode is larger than dielectric coated electrode. Metallic particles can be charged much by electrostatic induction as a time of contact to the bare electrode than by partial discharge[3].

On the other hand, in the case of dielectric coated electrode the charge of metallic particle is supplied by partial discharge only in a gap between the particle and the dielectric[3].

Effect of Dielectric Coating on Electrode to Flashover Voltage Initiated by Particle

In this secsion, we discuss about the characteristics of partial discharge onset voltage in the gaps G in the case that metallic particle had charge the quantity shown in Fig. 5. The possibility of breakdown is higher under the application of positive lightning impulse voltage than under that of negative lightning impulse voltage. Therefore we calculated the onset voltage only when metallic particle has negative charge (Fig. 5(a)). The calculation method of the partial discharge onset voltages is as follows. We at first calculated the electric field near the particle by means of the numerical method, and then the streamer onset voltage was calculated by the equation

$$\ln N = \int (\alpha - \eta)/p\,dx = 10^8 \quad \cdots(7)$$

where N is the number of electron in avalanche, α is the ionization coefficient, η is recombination coefficient and p is gas pressure. The result is shown in Fig. 6. It is found that the partial discharge onset voltage in the case of the dielectric coated electrode is a few percents higher than that in the case of bare electrode.

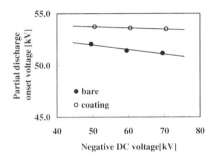

Fig.6. Calculated partial discharge onset voltage   Fig.7. Breakdown process initiated by metallic particle

In the breakdown process initiated by the metallic particle, as shown in Fig. 7, the partial discharge occurs at a small gap between the metallic particle and the high voltage electrode. After that main discharge occurs between the metallic particle and the grounded electrode. Therefore the partial discharge at a small gap triggers main breakdown.. It is dependent of electric fields at the tip of the metallic particle toward the grounded electrode. The electric field at the tip in the case of the bare electrode is almost equal to that in the case of dielectric coated electrode. Therefore breakdown voltage in the case of the dielectric coated electrode is higher than that in the case of bare electrode because the partial discharge onset voltage in the case of the bare electrode is higher than that in the case of bare electrode.

EXPERIMENT IN PRACTICAL GIS BUS

In order to investigate the effect of dielectric coating on the high voltage conductor, we measured breakdown voltage with the model of GIS bus shown in Fig. 2(b). Relative flashover voltage initiated by the metallic particle is shown in Fig. 8. Flashover voltage is normalized with voltage V50 of the breakdown probability of 50% in the case of bare electrode. In the case of dielectric coated conductor, the minimum voltage is about 16% higher than that in the case of the bare conductor and V50 is about 8% percents larger. Therefore dielectric coating on conductor is effective for improvement of the withstand voltage of DC-GIS.

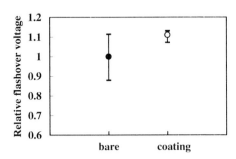

Fig.8. Relative flashover voltage initiated by metallic particle in GIS bus model

CONCLUSIONS

We studied metallic particle motion and breakdown characteristics in DC-GIS from the viewpoint of improvement of withstand voltage. In result, we have drawn the following conclusions.
1. The charge quantity of particle floating around high voltage conductor in the case of bare electrode is smaller than that in the case of dielectric coated electrode.
2. The breakdown voltage is improved by dielectric coating on conductor because partial discharge onset voltage is high as the particle has small charge quantity.

REFERENCES

1. A.Diessner et al, "Free Conducting Particles in a Coaxial Compressed-Gas-Insulated System", IEEE Trans.PAS Vol.89 No.8,1970(1970)
2. H.Anis et al, "Corona Stabilized Breakdown in Particle Contaminated Compressed SF_6 Systems", Gaseous Dielectrics IV, Pergamon, 387(1984)
3. T.Ooishi et al, "Charging Mechanisms of a Conducting Particle on Dielectric Coated Electrode at AC And DC Electric Fields", Gaseous Dielectrics VII, No.73,(1994)

SECTION 11: DISCUSSION GROUPS

DISCUSSION GROUP A: OTHER INDUSTRIAL APPLICATIONS OF GASEOUS DIELECTRICS AND DATABASES

PANELISTS: *A. Garscadden* (Chairman, WPAFB, USA), *D. S. Green* (NIST, USA), *P. Haaland* (Mobium Co., USA), and *E. Marode* (CNRS/ESE, France).

A. GARSCADDEN: The Gaseous Dielectrics Symposia have concentrated on high pressure discharges and on the gas SF_6 as the primary gas insulator. However, Loucas Christophorou astutely built into his symposia programs sessions on the basic physics of gaseous dielectrics, studies on cross sections, simulation and modeling, measurement techniques, surface discharges and environmental aspects of gaseous dielectrics and recycling. Also consider that negative ion mass spectrometry is much more unique in its identification of pollutants, pesticides and explosives. The current environmental interest in halons and in their applications to microelectronics processing, MEMS fabrication, materials processing and modifications ranging from turbine blades to textiles and catheters, means that the data base for gaseous dielectrics in the broadest sense (including data, experimental techniques and understanding of the fundamental physics) has many uses and applications in other technology areas.

Plasma technologies contribute to a cleaner environment in two ways. First, the use of plasma methods often permits materials processing to be achieved with much smaller amounts of reactant and less waste products than other methods, and secondly, the waste of traditional methods can be made more innocuous using plasma based treatments.

France has recognized the technical benefits of a vigorous electrical-based economy and one of the leaders in the use of plasma technologies for pollution control is Dr. Emmanuel Marode, Director of the CNRS, Plateau de Moulon, who will illustrate some of his research in this area.

E. MARODE: I would like to emphasize the uses of plasma technologies for pollution treatments.

Figure 1 illustrates two principal approaches for providing the plasmas involved in these treatments, according to the control of current density in the discharge. When the current density is permitted to increase to the limit of the power supply and the gas properties, the discharge becomes an electric arc whose properties are closely described by 'local thermal equilibrium', viz. a thermal plasma where $T_e = T_i = T_g$ = about 1 eV (where T_e is the electron temperature, T_i is the ion temperature, and T_g is the gas temperature). When the current density is limited to values below a few mA/cm^2, the discharge remains in the glow mode, a 'cold' or non-equilibrium plasma where T_e = several electron volts and $T_i = T_g$ = approximately room temperature. The enhanced chemistry promoted by these two different plasmas is quite distinct. The thermal plasma works mainly through the high translational energy of the heavy particles producing dissociation and faster reactions. The glow discharge activates the chemistry through the production of excited species and radicals by electron impact.

The most important reactions for situations involving combustion or air as a carrier gas inherently include OH, NH_3, O_3, O, N, H, etc. The data base of the gaseous dielectrics community relates to these studies because the species to be treated are such gases as NO_x, SO_x and volatile organic compounds that contain halogen atoms.

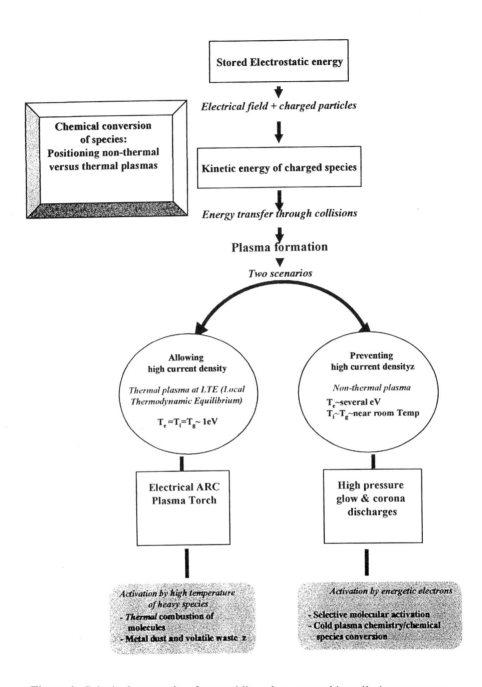

Figure 1. Principal approaches for providing plasmas used in pollution treatments.

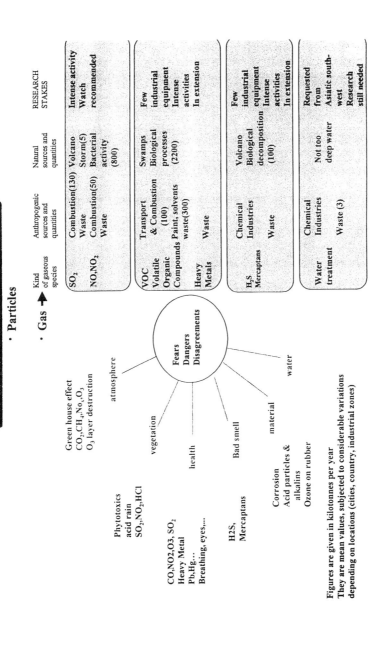

Figure 2. Annual waste production for various kinds of gaseous species.

PRODUCING A NON-THERMAL PLASMA OR HOW TO AVOID THE SPARK FORMATION?

- *Primary species and radicals are produced by a shower of fast electrons*

    E-Beam plasma

- *Non-uniform gaps develop non-sparking discharges in a wide range of voltage*

    - DC applied voltage

        Corona (point-plane, wire cylinder)

    - Pulsed or AC applied voltage

        Pulsed corona
        Low frequency corona

    - Surface discharges

        Surface discharges along dielectrics
        Packed bed (pellets of high ε)

- *Dielectric plates between electrodes + alternating voltage*

    Dielectric Barrier Discharge
    Silent discharges

- *A rapid gas flow prevent the spark formation*

    Fast-flow corona

- Controlled spark phase

    Glidarc
    Prevented spark regime

Figure 3. Schemes used to maintain non-equilibrium plasmas at higher pressures.

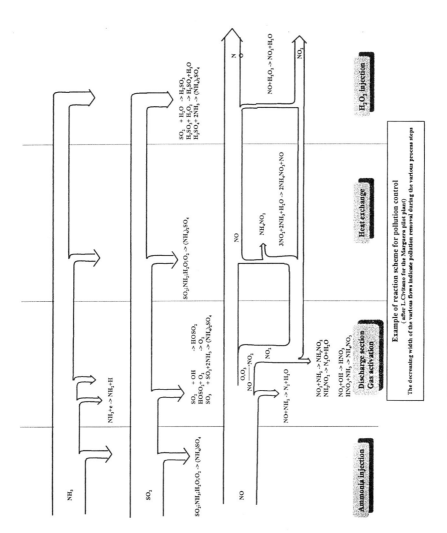

Figure 4. Example of plasma enhanced treatment process.

The other aspect of the gaseous dielectrics technologies that relate to plasma pollution abatement processes is the fact that the scale of plasma circuit interrupters and switches is one of the few applications close to the scale of the installations that will be required to treat the enormous amounts of waste that is produced annually. Figure 2 shows waste production quantities in kilotons/year.

As there is an approximate exponential dependence of radical production on the electron temperature, there is a considerable advantage in having a non-equilibrium plasma. This is difficult to achieve at higher pressures. The constriction occurs when the electron temperature relaxes within the diffusion length and therefore is promoted in molecular gases and at higher pressures. Figure 3 shows various schemes that are used to maintain nonequilibrium plasmas at higher pressures.

One example of a plasma enhanced treatment process (including the reaction scheme) is shown for the Marguera pilot plant in Fig. 4, as given by L. Civitano.

D. SCHWEICKART: What percentage of power may reasonably be expected to be used for pollution abatement processes in a power plant?

E. MARODE: A good target would be 2% of the overall power plant energy.

A. GARSCADDEN: Our second speaker is Dr. David Green from the US National Institute of Standards and Technology (NIST). Electronics and communications is now a trillion-dollar-plus industry employing more people than the metals, automobile and aerospace industries combined. Critical to the computer and electronics revolution, and to its spin-offs in electro-optics, micro electronic mechanical systems (MEMS) and micro optical mechanical systems (MOMS) is plasma-enhanced processing. It is appropriate that NIST has initiated experimental and theoretical programs to support this nationally important area.

D. GREEN: I would like to briefly discuss the use of dielectric gases in the semiconductor industry, with emphasis on applications and databases.

Dielectric gases are used in semiconductor chip manufacturing for patterned wafer etching and chemical vapor deposition (CVD) chamber cleaning. Fluorinated gases are the most widely used to generate *in-situ* fluorine radicals and ions by plasma-assisted reactions. These species are required to etch materials such as nitrides, oxides, or polysilicon from wafers or to remove silica from tool chambers. Halocarbon (HC) based species include tetrafluoromethane (HC-14, CF_4), hexafluoroethane (HC-116, C_2F_6), octafluoropropane (HC-218, C_3F_8), and trifluoromethane (HC-23, CHF_3). Non-carbon containing gases include nitrogen trifluoride (NF_3) and sulfur hexafluoride (SF_6). Chlorine gas (Cl_2) is often used in combination with fluorinated gases for modification of hard coatings, and metallic and organometallic surfaces.

Modeling and simulation of plasma processing requires understanding the nature of the decomposition mechanisms and interactions among precursors and byproducts in reactive flows at moderate to low pressures. Fundamental data are used to characterize the interactions among species and between plasma species and bounding surfaces. Selection of reactions, cross sections and rate coefficients is critical to making predictions on deposition and etching processes. Improved semiconductor manufacturing and environmental management require knowledge of accurate physical and chemical data of gas-phase and surface processes.

Material Processing

Low-pressure plasmas are used for semiconductor processing and for thin film and surface modification applications [P.F. Williams (1997), *Plasma Processing of Semiconductors*, NATO ASI Series Vol. 336, Dordrecht, The Netherlands: Kluwer Academic Publishers; National Research Council (1991) *Plasma Processing of Materials: Scientific Opportunities and Technological Challenges*, Washington DC: National Academy Press]. These partially ionized gases are sustained by the application of electromagnetic power. They consist of complex mixtures of neutral molecules, free radicals, ions, and electrons. Interactions between species take place in the gas phase and at the surfaces bounding the plasma. Often these species exist and interact under conditions far from thermal equilibrium. It is typical for mean electron energies to be on the order of several electron volts in the plasma, but neutral species may be no more than twice room temperature. Ions often impact surfaces at energies from several tens to many hundreds of electron volts.

As an example of the applications of dielectric gases, SF_6 is used by the semiconductor industry for plasma etching prior to CVD. These processes include oxide etching, nitride etching, and cleaning. The fluoride ions generated in the SF_6 plasma are excellent for etching tungsten and tungsten silicide films. Volatile reaction products formed include tungsten hexafluoride and silicon tetrafluoride. Oxygen additions to the plasma or gas stream form volatile sulfur/oxygen byproducts.

It is clear that a large resource for cross sections and rate coefficients is currently available for plasma chemistry. Unfortunately, this resource was developed largely for fields other than material processing, and therefore any relevant data is scattered and difficult to assemble. The most serious challenge is data related to processing new thin film materials including hard coatings and low-k dielectrics which use unique gases and plasma chemistries. Similarly there is a need for data relevant to the plasma surface modification of new metallic, organometallic, and organic materials.

Roadmap

Modeling and simulation is identified in *The National Technology Roadmap for Semiconductors* [Semiconductor Industry Association (1997) *The National Technology Roadmap for Semiconductors*, Austin, TX: SEMATECH] as a requirement cross-cutting all aspects of semiconductor manufacture. Plasma modeling and simulation is used to study plasma processes on various length scales ranging from the 10's of centimeters typical of a plasma tool or reactor to less than a micron in the case of critical etching dimensions [National Research Council (1996) *Database Needs for Modeling and Simulation of Plasma Processing*, Washington DC: National Academy Press].

The drivers for modeling include equipment design, process control, sensor design, process design, environmental safety & health, and process integration. Obtaining the required detailed chemistry and physics information is cited as one of the biggest hurdles. In particular, equipment modeling calls for experimental and computational data on gas and surface/wall reactions and rates as well as transport and thermal constants.

Environment

The dielectric gases used by the semiconductor industry are recognized as perfluorinated gases (PFCs), a class of greenhouse gases with global warming potentials (GWPs) orders of magnitude greater than carbon dioxide (CO_2) [U.S. Environmental Protection Agency (1998) *Proceedings of the Global Semiconductor Industry Conference on Perfluorocarbon Emissions Control, Monterey, CA. April 7 and 8, 1998*,Washington, D.C.: U.S. EPA]. Moreover, these gases are extremely stable resulting in long atmospheric lifetimes. The utilization efficiency of feedgases in semiconductor process tools varies widely depending on the specific gas and process used. Even though the semiconductor industry is not the largest source of PFC emissions it is actively advancing technologies for process optimization, alternative chemistries, recover/recycle, reuse, and abatement.

Plasma etch processes are difficult to optimize or find replacement chemistries, since equipment design, etch profile development and device performance are intimately related. Furthermore, there are significant risks associated with process changes in existing manufacturing. Therefore, abatement may present a lower risk way of reducing or eliminating remissions without process impact. Traditional abatement systems have relied on combustion but recent advances in technology have allowed non-thermal plasma and catalytic oxidation to be applied to the destruction of halogenated compounds. Both these processes lower the operating temperatures relative to the thermal process, thereby reducing energy consumption, generation of thermal NO_x, and some of the products from partial oxidation.

The National Technology Roadmap for Semiconductors [Semiconductor Industry Association (1997) *The National Technology Roadmap for Semiconductors*, Austin, TX: SEMATECH] calls for PFC emissions minimization and eventual elimination. This is to be achieved by improved byproduct use and management and the availability of data for chemical assessment. Modeling of reduction technologies requires similar data to that of process equipment modeling but extends over broader temperature and pressure ranges.

Data

There is a large number of potentially important collisional processes and reactions involved in a comprehensive plasma chemistry and physics database [National Research Council (1996) *Database Needs for Modeling and Simulation of Plasma Processing*, Washington DC: National Academy Press]. Those interested in modeling such systems are faced with deciding which cross-sections, rate coefficients, or thermodynamic parameters are to be included. It is a further challenge to locate such data and be assured of their accuracy and functional form.

In the best approach, data evaluation and compilation, including estimation, is accompanied by modeling and sensitivity analysis. This requires a collaborative effort that does not currently exist. Instead, most individuals rely on data compilations assembled for modeling other complex chemical systems such as combustion, lasers, atmospheric chemistry, and discharge physics. Measured or theoretically derived numbers are the best data sources. Yet many of the compilations contain a significant number of values derived under a completely different set of physical conditions (e.g., pressure) from those where the application demands.

The reactive species in the fluorocompound discharge include electrons, neutral atoms and molecules, radicals, and ions. Both ground state and excited state atoms and molecules will be present. The situation is further complicated by the fact the feedgases are used with additives such as oxygen, nitrogen, and hydrogen, in a carrier gas such as argon. Therefore data may be needed on the interactions involving many byproducts.

The approach emerging in database development is one of targeting critical species and intermediates as well as core sets of mechanisms. Modeling then evolves from considering primary reactive plasma species consisting of the feedgas and additives in a carrier gas to those involving intermediates and byproducts. Surface reaction data and validated mechanisms are considerably sparse.

In many cases these mechanisms may consist of reduced reaction sets that address the conditions of interest but may not be applicable to, for example, higher pressures. In constructing them, sensitivity analyses will identify key reaction sets or sequences that are particularly important and for which data may be lacking.

The summary from the 1996 NRC [National Research Council (1996*) Database Needs for Modeling and Simulation of Plasma Processing*, Washington DC: National Academy Press] report on database needs for plasma processing remain relevant today:

1. The database for ion-molecule and neutral-neutral chemistry varies considerably. For some species and reactions, the data are good. This is especially true for cases in which there is overlap with processes occurring in the upper atmosphere or in some cases in chemical vapor deposition processes. In other cases, however, most notably for etching processes, few data are available.

2. Thermochemical data are sketchy for many species of interest in plasma processing. These data are important in helping to establish boundaries for reaction pathways and in estimating reaction rate coefficients. Techniques, both experimental and computational, are generally available to obtain these quantities, but few efforts are under way at present to meet these needs.

Summary

Dielectric gases, especially the perfluorocompounds, are used by the semiconductor industry for plasma processing of wafers. The industry is faced with rapid advances in process technology and new materials accompanied by the need to address environmental sustainability. Advances are being made in process optimization, alternative chemistries, recapture/recover, reuse, and abatement. Databases are available or being developed to address the need for improved modeling, diagnostics and process control.

E. MARODE: Using recombination coefficients, given in the literature as a function of neutral densities – negative-positive ion recombination coefficients in air at atmospheric pressure, more specifically – we found out that the recombination is so large that the recombination frequency was larger than the moment collision frequency. How to do we get out from this drawback?

D. GREEN: Recombination rates are faster than collision frequency. Ion-ion is typically 20-50 times faster while electron-ion can be 1000 times faster. There are several issues to appreciate: i) free electrons are light and move quickly; ii) for charged particles there are additional interaction forces not considered with neutrals; and, iii) ion-ion collisions can have harpoon mechanisms especially with clusters and there species do not have to collide. Binary rate coefficients for ion-ion mutual neutralization are generally independent of the complexity of the reacting ions , only varying over a limited range of typically $4\text{-}10 \times 10^{-8}$ cm^{3}s^{-1} at 300 K.

One source of ion-ion recombination data is the atmospheric chemistry and physics literature. Versatile techniques are available for measuring the rate coefficients of dissociative electron-ion recombination yet there are a limited number of complete studies. See for example D.Smith, N.C.Adams, *Topics Curr. Chem.*, Vol 89 pp.1-43 (1980), N.G.Adams, *Int. J. Mass*

Spectrom. Ion Proc. Vol 132 pp.1-27 (1994).

At NIST, the recently completed a paper by L.W. Sieck, J. T. Herron and D.S.Green on the *Chemical Kinetics Database And Predictive Schemes For Humid Air Plasma Chemistry. Part I. Positive Ion-Molecule Reactions* addresses some of these issues. The second in the series of papers focuses on the neutral species reactions.

M. SCHMIDT: How are we to understand the change of data for ion-molecule reactions with changing pressure?

D. GREEN: Estimation schemes for rate coefficients in the "fall-off" pressure region describe the transition from third order (termolecular) to pseudo second order kinetics and enable one to fit or predict rate data from the low pressure limit to one atmosphere.

At low pressure association reaction rates are proportional to the third body concentration [M] and the reaction is kinetically third order. As the pressure in increases, the probability for three-body interaction increases and collision-stabilized association reactions can proceed at rapid rates. When this occurs there is a transition from third-order to pseudo second-order kinetics. At high-pressure, where all quenching collisions are effective, the association reaction rate reaches a value independent of the third-body concentration. The second-order constant k_∞ can be related to the capture rate and predicted (to within ± 10 %) using a variety of theoretical means. At low pressure, the limiting low-pressure (third-order) rate constant k_0 can be determined experimentally. Having extrapolated the high-pressure second-order limiting rate constant k_∞ and knowing k_0[M], it remains to describe the functional form of the rate constant in the transitional or pressure "fall-off" regime. For predictive purposes it is useful to adopt the same approximations applied to neutral reactions. Consequently, the "effective" rate constant across the fall-off range and at low to moderate temperatures is expressed as

$$k(M,T) = \{k_0[M]/(1+k_0[M]/k_\infty)\} F,$$

where F is a general broadening factor (based on the derivation by Troe *et al.*) given by $\log F \simeq \log F_c /\{1+[\log(k_0[M]/k_\infty)]^2\}$ and F_c is the parameter by which the rate constant of a given reaction is less than the limiting value. See for example: J. Troe, *Ber. Bunsenges. Phys. Chem.* Vol. **87**, p.161 (1983); R. G. Gilbert, K. Luther, and J. Troe, *Ber. Bunsenges. Phys. Chem.* Vol. **87**, p.169 (1983).

A. GARSCADDEN: Can you comment on database needs for new semiconductor materials and processing like silicon carbide or silicon on copper systems?

D. GREEN: There are expanding data needs in all areas of material processing including semiconductor, thin film and surface modification. The data that are needed include thermodynamics, kinetics, and electron collision. Data exist in related areas that can be useful such as combustion, atmospheric chemistry, and lasers. Some of these data are captured in databases that are simply compilations or bibliographic libraries. In some cases the databases include evaluated data where additional critical analysis and documentation has been provided.

The data needs are being driven by modelers and the expanding capabilities of process simulations. The availability of data from experimentation is likely limited because the materials are new and the experiments themselves challenging. This might be one area where a combination of experimental and computational data will make the greatest impact.

Among the evaluated databases are the NIST chemical kinetics and ion-molecule databases. The data relevant to many of the new semiconductor materials and processes have not been considered. Recently, a new database effort at NIST by S. Lias and E. Hunter has compiled all the literature references to ion-molecule reactions involving F, Cl, and Si species. This project was targeted at the needs of the semiconductor community and should be available shortly. A number of reactions may be relevant to mechanisms associated with silicon carbide for example. In the case of new etch and deposition schemes dealing with copper, other metals, and various low-k and high-k materials there are currently no electron-molecule, ion-molecule, kinetic or thermodynamic databases and the literature is fragmentary.

See for example: National Research Council (1991) *Plasma Processing of Materials: Scientific Opportunities and Technological Challenges.* Washington DC: National Academy Press; Semiconductor Industry Association (1997) *The National Technology Roadmap for Semiconductors.* Austin, TX: SEMATECH.; National Research Council (1996*). Database Needs for Modeling and Simulation of Plasma Processing.* Washington DC: National Academy Press.

A. GARSCADDEN: In many situations in addition to microelectronics, the engineer wishes to have a material that combines certain structural properties with selected surface characteristics. Nature showed the way in composites and also demonstrates the advantages of thin films and distributed structures that have amazing mechanical strengths, and optical, catalytic and sensory characteristics. A colleague who has great insight in these areas and has made valuable contributions to diverse technologies is Dr. Peter Haaland. Peter is nominally a chemist but is a renaissance person currently working in chemicals for fire-fighting, optical filters, spin coating, materials surface processing and nano-materials, and just how materials grow and self organize. Peter will share some of his recent studies that involve plasmas and their data bases.

P. HAALAND: I would like to discuss applications of gaseous dielectrics to solids. The homogeneous collisional and transport properties of gaseous dielectrics have been extensively studied to better explain their insulative and switching characteristics. The generation of reactive neutral and charged radicals constrains the performance of these materials as insulators but provides opportunities for the development of new materials and processing strategies.

I shall discuss a synopsis of three domains for the transition of gaseous dielectric data and understanding to solid materials. The first application involves design and control of interfaces to adhere inexpensive alkyd paints to aluminum surfaces. The formation of cationic precursors through selection of gas composition and electrical excitation produces a functionally gradient interface SiC_x interface less than 100 nm thick that inhibits corrosion and promotes adhesion on a macroscopic level.

A second area for exploitation of gaseous dielectric data is the quest for lower k interlayer dielectric films used in microelectronic components. Conventional interlayer dielectrics such as SiO_x and TiN_y make the rate limiting contribution to chip speed through their capacitance as feature sizes diminish below 0.2 µm. A combination of low k and adequate insulative properties is found in fluoropolymers, but the auxiliary requirements (adhesion, homogeneity, mechanical strength, oxidative stability, etc.) are demanding. Porous media such as aerogels, thermally cured polymers such as benzocyclobutene, and photopolymers compete with plasma enhanced chemical vapor deposition, so that knowledge about the chemistry of fluorine-bearing film precursors such as CHF_3, $c-C_4F_8$, and CF_4 is needed to develop plasma enhanced chemical vapor deposition of interlayer dielectrics.

The third area for transition of gaseous to solid dielectric materials involves synthesis of nanoparticles in electron attaching gases. The role of negative ions is crucial to the production of nanoparticles since anions and negatively charged clusters are kept apart from each other and from reactor walls by electrostatic sheaths. Whether these particles grow by accretion of charged or neutral radicals, one can envision control of the particle microstructure, dispersity, and composition through the same means that are used in thin film manufacturing, namely tuning the excitation and gas composition. Appropriate scaling of nanoparticle synthesis will find applications in catalytic, microcomposite, and optical materials.

These three examples involve different couplings between homogeneous kinetics and surface development. The energy and composition of cationic fluxes governs SiC_x film formation. Neutral and charged precursors are both important to rapid production of interlayer dielectrics, while anions play a central role in nanoparticle production. Improved definition of plasma chemistry in discharges that contain gaseous dielectrics provides insight and illuminates new corridors for the evolution of plasma processing.

A. GARSCADDEN: This session has demonstrated that there are many challenges and consequently many opportunities for the uses of electronegative plasmas, their data bases and their associated technologies. Many of the applications actually depend on scaling the discoveries and techniques to improve the reliability of the process and to increase the throughput of material. The modeling of such processes is mostly multidisciplinary. While there is a temptation to take advantage of the large parallel computing resources now available and to resort immediately to systems of equations with many species and reactions, I would urge that reduced reaction sets and understanding of the physics and chemistry of the process be kept in mind. The sensitivities of the product to the boundary conditions cannot be too extreme as current electronic devices often go through hundreds of processing steps. Many of the actual reaction rates will never be exactly determined. Some process sensitivities need to be macroscopically resolved. Dr. Haaland illustrated that the relative roles of ions and neutrals in thin film deposition may be quickly determined using a small magnet and that the morphology of materials is (self-consistently) important in surface reactions, especially in a plasma. Another exciting feature about these technology areas is their dynamic range. Dr. Marode described systems that may eventually rival an oil refinery in their size and power requirements. Drs. Green and Haaland were mostly concerned with the technology of the very small. As well as electrical and optical properties, we do need to devise standards and methods to provide data bases for thermal and mechanical properties of nano-materials, which translate into adsorption and accommodation coefficients, coefficients of expansion, elasticity, etc., for these materials which now have dimensions such that the classical definition of a solid has to be modified. So we have challenges and neat scientific problems with applications for scientists ranging from chemical engineers to quantum physicists. In turn, their results eventually will feed back to benefit conventional gaseous dielectrics.

DISCUSSION GROUP B: SF_6 SUBSTITUTES

PANELISTS: *A. Cookson* (Chairman, NIST, USA), *W. Boeck* (Technical University of Munich, Germany), *M. Frechette* (IREQ, Canada), *T. Yamagiwa* (Hitachi, Japan), and *A. Sabot* (EDF, France).

A. COOKSON: The question of SF_6 substitutes has already been a "hot" issue at this conference, and there have been several papers discussing the high voltage breakdown characteristics of alternative gases. In this panel, we are fortunate to have several experts from around the world to discuss what the impact will be on the design and operation of the gas-insulated substations (GIS) and other equipment, and what we need to do for future research. Each speaker will make their presentation, and then we will open the floor for questions and discussions.

W. BOECK: I would like to address the subject of SF_6 emission into the atmosphere by GIS and other technologies.

At the Kyoto Conference, SF_6 has been classified as one of six greenhouse gases due to its high relative thermal absorption and its extreme lifetime. Today its extremely low concentration in the atmosphere results in a negligible contribution of less than 0.1% to the greenhouse effect. But the SF_6 content in the atmosphere has increased almost constantly about 7% per year since 1970, from 0.25 ppt to 3.5 ppt in 1995.

Sixty five percent (65%) of the SF_6 production went in 1995 into electric power industry and to utilities. But it is banked in GIS. Only GIS leakages and the handling of SF_6 during filling, repair work, and testing result in a release into the atmosphere. Therefore, the contribution of this technology to the SF_6 concentration in the atmosphere is only 35%.

SF_6 is in use for many other, mostly open, applications as, for instance, magnesium production or semiconductor production. Investigations of CAPIEL (Co-ordinating Committee for Common Market Associations of Manufacturers of Electrical Switchgear and Controlgear) and Unipede (International Union of Producers and Distributors of Electrical Energy), further studies (M. Maiss and C.A.M. Brenninkmeijer, "Atmospheric SF_6, trends, sources, and prospects," paper accepted for publication in Environmental Science and Technology) based on a compilation of world wide SF_6 sales data by end-use markets by an independent consulting firm in Washington D.C., and measurements of the SF_6 concentration in the atmosphere presently result in the situation presented in Fig. 1.

The total global emission of 6200 t/a in 1995 has been calculated from the measured SF_6 concentration in the atmosphere. Its distribution to different regions is estimated by market studies. The SF_6 emission from GIS includes losses by leakage and gas handling. The usual leakage rate guaranteed today is less than 1%/a. In Europe and Japan there are few old GIS designs in operation with higher leakage rates. Furthermore the main GIS manufacturers are located there. They cause additional losses by gas handling during tests. Altogether the losses are estimated to be 6.6%/a or 9.7%/a respectively. Surprising is the extremely high emission rate of 26%/a in North America. This figure needs further investigation. Only partially it can be explained by old dead tank switchgear designs which are known to be high leakers with leakage rates of 100%/a or more. Extremely low loss values are given for the Near East (2.8%/a) and the Southern Hemisphere (2.1%/a) where mainly modern very tight enclosure designs are in use.

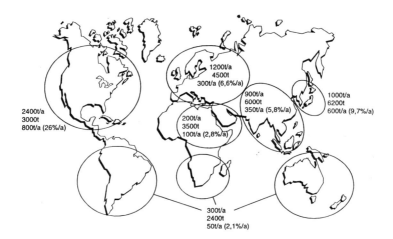

Figure 1. Estimated regional distribution of SF_6 emission and SF_6 installed in switchgear in 1995. The three values shown in the figure are: total SF_6 emission by all applications (t/a); SF6 banked in GIS (t); and SF_6 emission from GIS (t/a) in (%/a).

Altogether, GIS technology contributes with 2200 t/a only 35% to the total SF_6 emission of 6200 t/a. According to the global marked study about 1800 t/a are released by other known applications as;
- blanketing of molten magnesium in the magnesium industry,
- car tires and sport shoes,
- sound insulated windows,
- degassing of aluminium in the aluminium industry,
- semiconductor production in the electronic industry,
- and other known applications.

There is a remaining difference of 2200 t/a due to unknown applications which makes further studies necessary.

According to the Kyoto protocol, a total emission reduction of 5.2% is demanded for the period 2008-2012 based on 1990 or 1995 levels. This can easily be realized if the following measures are taken in the electric power industry:
- a reduction of the leakage rate from the usual 1%/a today to 0.5%/a,
- a reduction of the gas handling losses to 2.5%/a by the new CIGRE recommendation (Task Force 23.10.01, CIGRE, "SF_6 recycling guide - Re-use of SF_6 gas in electrical power equipment and final disposal," Electra, No 173,1997, p. 43), and
- improvement of the tightness in old designs in North America, Europe, and Japan or their exchange by modern equipment.

All this together would result in a reduction of the SF_6 release from 2200 t/a to 1400 t/a. With respect to further SF_6 applications the following reductions are considered as possible. Instead of SF_6 other gases can be applied for the magnesium and aluminium production, alternative technologies are available for sound insulating windows, car tires and sport shoes. Thus, the SF_6 release by these open SF_6 applications will be reduced from 1800 t/a to 1000 t/a. Altogether there would be a reduction of 28 % of the SF_6 emission.

Each year approximately 2000 t SF_6 is sold and banked in new GIS. Under this assumption the total amount of SF_6 banked in GIS of 25,600 t in 1995 will become approximately 75,000 t in 2020 and the total release will be again 2250 t/a as today if the total emission rate of 3%, including leakages and gas handling, is realized and all sold GIS are kept in operation. But further improvements are possible. A nearly total tightness can be achieved at higher costs. In parallel for gas-insulated cables (GIL), where no arc quenching properties are demanded, gas mixtures (SF_6-N_2) will get in use. Of course any open application of SF_6 should be exchanged by new technologies. There is sufficient time to solve these problems and to consider which further steps can be taken to ensure that in the future, as today, SF_6 will not contribute to the greenhouse effect.

T. YAMAGIWA: I would like to make some comments on the topic of SF_6 substitutes from the viewpoint of manufacture of Gas Insulated Substations (GIS). I will present these in the following sequence:
STEP I: Minimize leakage and handling loss.
STEP II: New GIS / minimize using SF_6 gas;
- use only for switchgear [Circuit Breaker (CB), Disconnect Switches (DS), and Earthing Switches (ES)],
- use only for circuit breakers.
STEP III: Stop use of SF_6 gas.

STEP I: MINIMIZE LEAKAGE AND HANDLING LOSS

Figure 2 shows an example of the reduction in the volume of SF_6 for 550kV Gas Circuit Breakers (GCB). In the 1970s, 550kV GCB consisted of four interrupters and the tank length was about 9 m. Now, 550kV GCB with one interrupter have been developed, and the SF_6 gas volume has been halved in 20 years.

Figure 3 shows the example of GIS. Conventional 550kV GIS used two-interrupter GCB. GIS now use one-interrupter GCB, and the gas volume has been reduced by 30%.

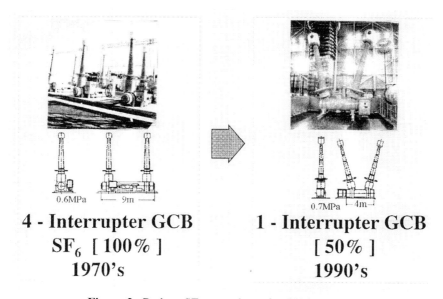

4 - Interrupter GCB
SF_6 [100%]
1970's

1 - Interrupter GCB
[50%]
1990's

Figure 2. Reduce SF_6 gas volume for 550 kV GCB

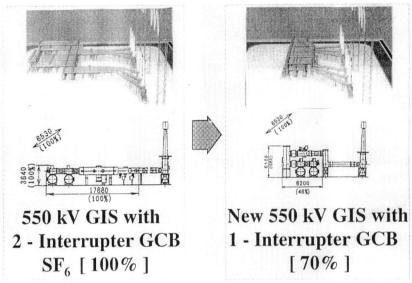

Figure 3. Reduce SF$_6$ gas volume for GIS

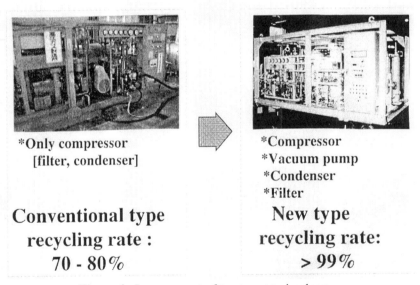

Figure 4. Improvement of treatment technology

Figure 4 shows the improvement of treatment technology for SF$_6$ gas. The conventional type consisted of only a compressor with a filter and condenser. The recycling effectiveness of this type was 70% to 80%, such that 20% to 30% of the filled gas was emitted into the air. The new type now has a vacuum pump. The recycling effectiveness has increased to about 99%, and therefore the emission gas level has become extremely low.

Figure 5 shows the example of preventative maintenance technology for gas leakage. That is, a gas monitoring system for GIS. A pressure sensor is put on the GIS tank and the signal goes to the local panel for the system. There, the analog signal is changed to digital format, and the real-time gas pressure and gas leakage are then diagnosed continuously.

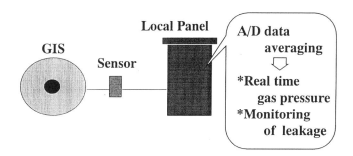

Figure 5. Preventive maintenance technology

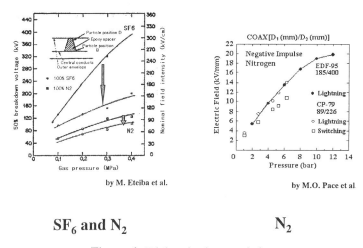

Figure 6. Dielectric characteristics

STEP II: NEW GIS MINIMIZE USING SF_6 GAS

Figure 6 shows the dielectric breakdown characteristics of SF_6 and N_2. The left-side figure shows surface flashover characteristics with and without particles. For clean conditions, the ratio of the flashover voltage of N_2 to SF_6 is about 1:3. On the other hand, with particles present, the ratio is about 1:2. In N_2, particles are less effective on surface flashover. The right-side figure shows the pure N_2 dielectric characteristics. The flashover field strength tends to saturate at about 10 bars, so we need to use an operating pressure less than 10 bars.

Figure 7 shows the new type of hybrid insulation GIS. The concept is as follows. SF_6 gas is used for switchgear (CB, DS and ES). The other parts are filled with pure N_2 or a 5% SF_6 mixture. The total volume of SF_6 gas is reduced by about 50% compared with conventional GIS. In both cases, the gas pressure must be increased from 0.5 MPa to 0.8 MPa, and the size of the GIS diameter increased by 110% to 150%. In the case of mixtures, the big problem is how to recycle. There are two counter measures to reduce these increases in pressure and size. One is the reduction of lightning surge level. The other is reduction of DS switching surge level.

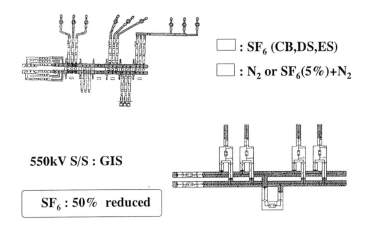

Figure 7. Hybrid insulation GIS [Part 1]

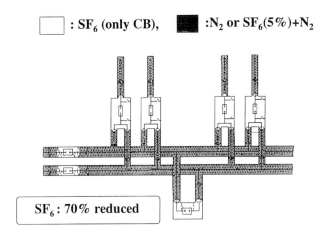

Figure 8. Hybrid insulation GIS [Part 2]

Figure 8 shows the other type of hybrid insulation GIS. In this case, SF_6 gas is used only for the CB. The volume of SF_6 reduced is 70%. In this case, the most important point is the interrupting capability of DS and ES. For bus DS, loop current interruption is needed. For line ES, inductive current interruption is needed. My idea is as follows. We use Air Blast Breaker (ABB) or other gases and mixtures.

Figure 9 shows the application of high performance lightning arrestor (LA). The V-I characteristics of the new ZnO element are double that of the conventional element. Therefore, the size of LA is about half the conventional one, so we can put more LAs in GIS for surge reduction. As a result of this counter measure, we can reduce the Lightning Impulse Waveform Voltage (LIWV).

Figure 10 shows the application of DS with resistors. The Very Fast Transient (VFT) voltage is produced by DS operation. Theoretically, the maximum level of VFT is 3.0 pu. This level can be reduced by inserting resistors in the switching circuit. Figure 9 shows the actual movement process. We insert 500 Ω resistors into the circuit, and the VFT level decreases to about 1.2 pu.

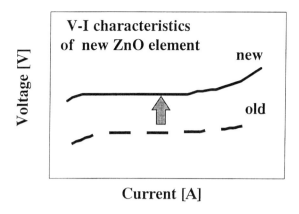

Figure 9. Application of high performance LA

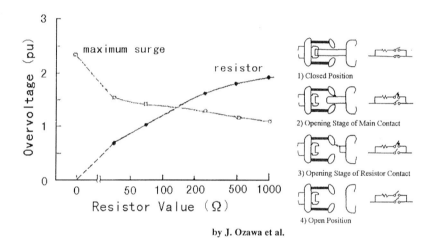

Figure 10. Application of DS with resistors

STEP III: STOP USE SF_6 GAS

Fig. 11 gives the interruption ability of a puffer type GCB with various SF_6 gas mixtures. This result shows that the SF_6 content is important in determining the interrupting ability of the GCB. How to interrupt effectively is a very important problem that needs to be addressed.

CONCLUSIONS

1. The recycling rate of SF_6 gas must be above 99%.
2. The SF_6 gas may be reduced 50% to 70% by application of new Hybrid-Insulation-GIS (e.g., N_2 or 5% SF_6+N_2).
    - First step: SF_6 gas is used for switchgear (CB, DS, ES).
    - Second step: SF_6 gas is used only for circuit breakers.
3. The following counter measures must be realized.

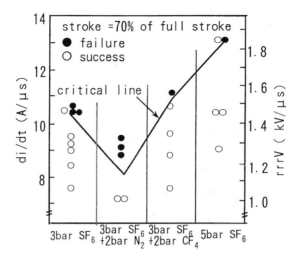

Figure 11. Interruption ability (puffer type GCB; current ~15 kA)

- Reduction of LIWV (e.g., application of new LA).
- Reduction of DS surge (e.g., application of new DS with resistors), etc.

4. A large amount of research must still be performed for any new gas or gas mixture to be used in switchgear (for adequate interrupting capability).

A. SABOT: In Table 1, I summarize a utility perspective on SF_6 substitutes.

Table 1: SF_6 Substitutes: A Utility Point-of-View

1. Interest of industry for SF_6

- Electric equipment
- Metallurgy manufacture process: Aluminium, Magnesium
- Medical equipment: X ray, ...
- Scientific equipment: electronic microscope, electrostatic particle accelerators
- Domestic house: double glazing windows
- Meteorology

2. Interest for power equipment: from 20 kV up to 1000 kV

GIS, Circuit breaker, Gas Insulated Transformer (power & instrument) needs insulating medium with:

- High dielectric strength
- High thermal conductivity: quality for arc extinction and energy evacuation
- Chemical stability and non toxicity (except by-products)
- Fast dielectric and thermal recovery characteristics that SF_6 enables for equipment:
- Size reduction: GIS
- Improvement of the availability and life cycle cost of circuit breakers (no maintenance)

- Long life expectation
- Some environmental aspects superior to oil: e.g., flammability (transformers).

3. Environment, Life cycle, and Health Issues

- Non toxic chemical gas at "normal" temperature and pressure conditions
- Handling, risks and health:
    - SF_6 pure no effect on health
    - SF_6 leaks no risk for health
    - No influence on ozone layer but contribution to greenhouse effect if no control of leakage
- Recycling of the by-products
    - No physical difficulty
    - Handling procedures can be mastered
- End of life
    - Planned end of life
    - Recovering and reuse of the SF_6 with by-products

4. Rules and standards

National rules:
- Pressure vessel: safety rules for workers
- Work atmosphere: safety rules for workers
- Transportation: special material transportation rules

European rules:
- EU 94/909: SF_6 by-products due to arc decomposition considered as dangerous

First standard answer:
- IEC 1634: Solution for mastering the handling and life cycle of SF_6 in power equipment

5. Consequence for power equipment: more stringent specifications

- Minimizing leakage rates: 3% down to 1% and more severe monitoring and management
- As far as possible reduction of the amount of SF_6: N_2 / SF_6 mixtures in equipment where it is possible like GIL (N_2/SF_6 at 10%).
- As far as possible use of substitute insulating medium. Need to satisfy the real needs with less environmental impacts than SF_6 and be technically feasible and economically viable: for example GIL with pure N_2 as the insulating medium is not acceptable from the cost point of view.

6. Costs

Installation costs vary considerably depending on relief, soil type and the technical characteristics of the structure.

Overhead lines are naturally much less dependent on the first two parameters.

Table 2 below gives a rough indication of minimum costs/additional costs. Installation cost of a single-circuit underground line (1) (excluding project management costs and capitalized interest).

Table 2. Costs

Voltage level	Cross-section of underground cable (mm^2)	Underground line over flat land in rural area (2) (excluding terminations) (FF million/km $ million/km)	Additional cost over flat land in rural area compared with line mounted on lattice towers (3) (FF million/km $ million/km)
high voltage	400 aluminum	1.4 0.23	0.9 0.15
	1200 aluminum	2.1 0.35	1.5 0.25
245 kV	1200 copper	4 0.67	3.2 0.53
400 kV gas and synthetic insulation cables: 2,000 MW		23 3.8	21 3.5

(1) A double-circuit underground line costs almost twice as much as a single-circuit line; a double-circuit overhead line costs approximately 1.6 times more than a single-circuit line.

(2) The costs given correspond to projects undertaken in rural areas with no specific difficulties; they are between 1.5 and 2 times less than the costs for urban areas.

(3) Excluding compensation and accompanying costs.

A. COOKSON: Our speakers had raised many issues. It has been clearly pointed out that we need to significantly reduce leakage, which does appear to be practicable. Using alternative to SF_6 may be feasible, but there is the critical question of is there loss in performance. We have other experts here in the audience, and at this point, I would like to invite comments.

S. DALE: At the present we are suffering from very poor data when it comes to regional SF_6 use. What is the source of the regional data shown? Who supplied the data on the North American region? You mention that a source of unknown emission in North America comes from utilities owned by magnesium companies. Contrary to the European situation for magnesium production where the major producer also owns significant electricity production and transmission, North American utilities do not own magnesium plants or vice versa. Thus, there is no "hidden" source for SF_6 emission in North America as claimed.

A. COOKSON: Concerning the quantity of SF_6 associated with different equipment in different parts of the world reported in the study, if there are errors they need to be corrected before they get into print or they will then be accepted as true.

H. MORRISON: Can we clarify the data presented by Professor Boeck? Science Policy published a survey of SF_6 by end-use and Allied Signal contributed to that survey. The categories in the survey were not as detailed as the breakdown provided by Professor Boeck. Perhaps we need to examine how he obtained his data.

A. COOKSON: I understand Figure 1, which includes the data in question, is from a paper by Maiss and Brenninkmeijer to be published in Environmental Science and

Technology, so we should then be able to get the details that Dr. Dale and Dr. Morrison have been requesting.

B. HAMPTON: The advantage of using SF_6/N_2 mixtures are minimal, especially when taking into account the probable higher working pressure and less efficient reclamation plant. There is actually no disadvantage to using pure SF_6 in transmission plant, provided that its loss be at an absolute minimum. In the consumer society it is increasingly the case that goods from refrigerators to car engines are being made maintenance free sealed for life. The same approach should be taken in circuit breakers and other items, which should be sealed for life and then returned to a purpose made factory for reprocessing. This would be entirely in line with the utilities ideal of zero maintenance, and allow the use of pure SF_6 to be continued.

L. NIEMEYER: Sealed-for-life equipment already exists at the distribution level. It has to be taken back by the manufacturer at end of life (at least in Europe). Thus, the recycling loop is already now closed for this type of equipment, both technically and legally. Of course the development of transmission equipment is going in this same direction.

H. MORRISON: The concept by Dr. Yamagiwa of N_2 or low SF_6 concentration mixtures for compartments that are not circuit breakers will require at least two considerations for gas monitoring and handling. First, we would need to monitor for gas leaks between adjacent compartments to prevent contamination of N_2 into pure SF_6 compartments. Second, the gas connections and gas handling equipment must be completely separate and distinct so that no error can be made to mix the gases, for example, by connecting the N_2 gas equipment to the SF_6 compartment.

L. NIEMEYER: The Standard ISO 14040 defines how to determine the environmental impact of a functional unit of equipment including not only the insulation gas but all other materials (and their environmental impact). This standard makes sure that the design changes demanded by an alternative insulation medium are quantitatively accounted for. As an example, when a less-performing gas is used, its pressure or the dimensions of the equipment have to be increased calling for more enclosure material. The environmental impact of the latter may "eat up" the impact saved by not using SF_6.

A. COOKSON: Are there alternatives to SF_6 mixtures or are people only considering using nitrogen?

S. DALE: The important issue with SF_6 and global warming is to release as little SF_6 as possible into the atmosphere. We tend to equate the use of SF_6 in power equipment with leakage to the atmosphere. We discuss using mixtures to reduce the amount of SF_6 used in equipment. However, it is not obvious that this translates into reduced emission. The issue of reclamation of pure SF_6 in a gas mixture (e.g., with nitrogen) must be looked at to ensure that by going to mixtures we are not releasing more SF_6 into the atmosphere. The reclamation of SF_6 from a mixture with N_2 is very difficult and expensive. We may be able to reduce emission more by improving SF_6 handling procedures than going to mixtures.

M. MISAKIAN: The suggestion was made that utilities could go to sealed units and when the units failed, the equipment would go back to the manufacturer for decommissioning. Another member of audience noted that the approach was being used in some distribution systems. There were no objections raised and then we went back to earlier thoughts about mixtures, etc. Would the panel indicate the problems associated with the "sealed system" approach?

L. G. CHRISTOPHOROU: Can you really take all gas-insulated equipment back to the manufacturer?

B. HAMPTON: In response to the comment regarding standardization, may I point out that there are far more manufacturers of motor cars than GIS. And in principle there is no reason why a switchgear manufacturer should not be required to produce units to a standard size, which could be interchanged with others at the end of their lives.

A. COOKSON: In the automotive industry there is a lot more standardization compared to HV circuit breaker industry, so the "sealed-for-life" units are more economically difficult for GIS.

L. G. CHRISTOPHOROU: SF_6 is difficult to destruct in the troposphere. It is not destructed photolytically. Also, the number of free electrons present in the troposphere with suitable energies to cause destruction via negative ion reactions is too small to be of any consequence.

K. D. SRIVASTAVA: Can we reduce the current interrupting requirement for HV circuit breakers, which in turn might permit new technologies?

K. NAKANISHI: In Japan, we are studying the current limiter using the superconducting technology. But, it is at too early a stage to state the possibility of application to power systems.

H. KNOBLOCH: Current limiters can be used to connect separate networks to keep the short circuits low. But it has to be mentioned that low temperature superconductors need a lot of cooling energy and in addition a disconnector is still needed.

A. COOKSON: This has been a most stimulating discussion, which has confirmed that substitution for the SF_6 is indeed a complex challenge. There are alternative gases and design considerations that look promising purely for insulation, but the situation is far more difficult when there is arc interruption. The point was also made in the discussion when looking at the environmental impact of the SF_6, that we should also begin to consider the environmental impact of the complete system from manufacture through its operational life to eventual decommissioning.

PARTICIPANTS

PARTICIPANTS

R. Arora
IIT Kanpur
Dept. of Elec. Eng.
Kanpur, 208016
INDIA

K. H. Becker
Dept. of Phys. & Eng. Phys.
Stevens Institute of Technology
Castle Point Station
Hoboken, NJ 07030

W. Boeck
Tech. University of Munich
Inst. of High Volt. Eng.
Electric Power Trans.
D-80290 Munich
GERMANY

M.-C. Bordage
CPAT
Univ. Paul Sabatier
118 Rt. de Nerbonne
31062 Toulouse Cedex
FRANCE

A.F. Borghesani
Unita INFM di Padova
Dipartimento di Fisica
Via Marzolo 8
35131 Padova
ITALY

A. Bulinski
National Research Council
Montreal Rd.
Bldg. M50
Ottawa, Ontario, K1A 0R6
CANADA

J. Castonguay
IREQ, Hydro-Quebec
1800 Boul. Lionel-Boulet, P89B
Varennes, Quebec, J3X 1S1
CANADA

R. Champion
College of William & Mary
Dept. of Physics
Williamsburg, VA 23187

S. Chigusa
Nagoya University
Dept. of Elec. Eng.
Furo-cho Chikusa-ku
Nagoya, 464-8603
JAPAN

L. G. Christophorou
NIST
Bldg. 220, Rm. B344
Gaithersburg, MD 20899

A. Clark
NIST
Bldg. 220, Rm. B258
Gaithersburg, MD 20899

J. D. Clark
Wright State University
Dept. of Physics
Dayton, OH 45435

I. Coll
Universite Paul Sabatier
C.P.A.T. 118
Route de Narbonne
31062 Toulouse Cedex
FRANCE

A. H. Cookson
NIST
Bldg. 220, Rm. B358
Gaithersburg, MD 20899

S. Dale
ABB Power T&D Company
1021 Main Campus Dr.
Raleigh, NC 27606

B. Damsky
EPRI
3412 Hillview Ave.
Palo Alto, CA 94304

J. de Urquijo
Instituto de Fisica
P.O. Box 48-3
Cuernavaca, Mor, 62251
MEXICO

C. T. Dervos
National Technical Univ.
Dept. Elec. Eng.
Zografou Campus, 15773
GREECE

J. Desormeaux
Old Dominion University
Dept. of Elec. and Comp. Eng.
Norfolk, VA 23529

E. J. Dolin
US EPA
6202J 401 M Street, SW
Washington, DC 20460

K. Ellerton
Allied Signal
P.O. Box 1053
Morristown, NJ 07962

M. Forys
Agric. & Teachers Univ.
3 Maja 54
08-110
Siedlce
POLAND

R. Fouracre
University of Strathclyde
200 George St.
Glasgow G1 1XW
U.K.

M. F. Frechette
Hydro-Quebec
1800 Boul. Lionel-Boulet, P89B
Varennes, Quebec, J3X 1S1
CANADA

H. Fujii
Mitsubishi Electric Corp
8-1-1 Tsukaguchi-Honmachi
Amagasaki, Hyogo, 661-811
JAPAN

C. Gaillac
Schneider Electric
38050 Cedex 9
Grenoble
FRANCE

A. Garscadden
WL/CA
2130 Eighth Street
Wright-Patterson AFB, OH 45433

H. Gaxiola
Eindhoven Univ. of Tech.
P.O. Box 513
Einohoven 5600 MB
NETHERLANDS

A. Ghazala
Old Dominion University
Dept. of Elec. & Comp. Eng.
Norfolk, VA 23529

A. Gleizes
CPAT Universite Paul Sabatier
118 Route de Norbonn
F31062 Toulure Codex
FRANCE

T. Goda
Toshiba Corporation
2-1 Ukishima
Kawasaki-shi, Kanaga
JAPAN

A. Goldman
CNRS
Sunelec
91192 Gif Wax
FRANCE

M. Goldman
CNRS
Sunelec
91192 Gif Wax
FRANCE

D. Green
NIST
Bldg. 221, Rm. A303
Gaithersburg, MD 20899

P. Haaland
Mobium Industries
518 W. Linden Street
Louisville, CO 80027-3124

H. Hama
Mitsubishi Electric Corp.
1-1 Isukaguchi
Honmachi 8 Chome
Amagasaki City, Hyogo, 661-8661
JAPAN

B. Hampton
University of Strathclyde
204 George St.
G1 1XW, Glasgow
U.K.

X. Han
NIST
Bldg. 220, Rm. B344
Gaithersburg, MD 20899

B. Hankla
NSWCDD
17320 Dahlgren Road
Code B20
Dahlgren, VA 22448

R. Hegerberg
Sintef Energy Research
S. Selands VE1 11
Trondheim, N-7034
NORWAY

M. Hikita
Kyushu Inst. of Technology
1-1 Sensui, Tobata
Dept. of Elec. Eng.
Kitakyushu, 804-8550
JAPAN

J. Horwath
Air Force Research Lab.
2645 Fifth St.
AFRL/PRPS
Wright Patterson AFB, OH 45433

C. Jiro
Wright Laboratory
AFRL/PRPS, Bldg. 450
E125
Dayton, OH 45433

S. Kacherski
Soluay Fluorides, Inc.
41 W. Putnam Ave.
Greenwich, CT 06830

T. Kawamura
Dept. of Elect. Systems
  for Urban Eng.
Shibaura Inst. of Tech.
3-9-14 Shibura
Minato-ku
Tokyo 108-8548
JAPAN

H. Knobloch
Siemens AG
EV HS 23
D-13623 Berlin
GERMANY

T. Kobayashi
Tokyo Electric Power Co.
Trans. & Substation Construction
1-1-3 Chiyodaku
Tokyo 100
JAPAN

D. Koenig
Darmstadt Univ. of Tech.
Landgiaf-George
64283 Darmstadt
GERMANY

M. Kosakada
Toshiba Corporation
1-1-6 Uchisaiwai-cho
Chiyoda-ku, Tokyo, 100
JAPAN

E. Kuffel
Univ. of Manitoba
Elec. & Comp. Eng. Dept.
Winnipeg, Manatoba
CANADA

V. Lakdawala
Old Dominion University
Dept. of Elec. & Comp. Engineering
KDH 231N
Norfolk, VA 23529

S. Mahon
NIST
Bldg. 220, Rm. B344
Gaithersburg, MD 20899

E. Marode
CNRS
LPGP-EDEE-Supelec
Plateau de Moulon
91192 Gif-Cedex
FRANCE

H. Marsden
ABB Power T&D Co. Inc.
1021 Main Campus Drive
Raleigh, NC 27606

S. Matsumoto
Toshiba Corporation
2-1 Ukishima
Kawasaki-ku
Kawasaki-shi, 210-0862
JAPAN

I. W. McAllister
Tech. Univ. of Denmark
Dept. of Elec. Power
Building 325
DK-2800 Lyngby
DENMARK

A. Meier
Solvay Fluorides, Inc.
41 W. Putnam Ave.
Greenwich, CT 06830

M. Misakian
NIST
Bldg. 220, Rm. B344
Gaithersburg, MD 20899

J. H. Moore
University of Maryland
Dept. of Chemistry
College Park, MD 20742

M. M. Morcos
Kansas State University
289 Rathbone Hall
Manhattan, KS 66506

H. Morrison
Ontario Hydro Tech.
800 Kipling Ave.
KR151
Toronto, Ontario, M8Z 5S4
CANADA

A. Mueller
Old Dominion University
Dept. of Elec. & Comp. Engineering
Norfolk, VA 23529

A. H. Mufti
King Abdul Aziz Univ.
P.O. Box 9027
E&CE Dept.
21413
SAUDI ARABIA

M. S. Naidu
Indian Inst. of Science
High Voltage Eng.
Bangalore 560 012
INDIA

K. Nakanishi
Mitsubishi Electric Corp.
Advanced Technologies R&D Center
8-1-1 Tsukaguchi-Honmachi
Amagasaki, 661-8661
JAPAN

L. Niemeyer
ABB Research Center
CH 5405 Braden
SWITZERLND

H. Okubo
Nagoya University
Dept. of Elec. Eng.
Furo-cho Chikusa-ku
Nagoya, 464-8603
JAPAN

J. K. Olthoff
NIST
Bldg. 220, Rm. B344
Gaithersburg, MD 20899

B. L. Peko
College of William & Mary
Dept. of Physics
Williamsburg, VA 23187

W. Pfeiffer
Darmstadt Univ. of Tech.
Fachbereich 17
Landgraf-George
64283 Darmstadt
GERMANY

M. Pierontesi
ABB Research Center
CH 5405 Braden
SWITZERLAND

M. Pittroff
Solvay Fluorides, Inc.
3002 Hanover-Hans
Bockler-Allee 20
GERMANY

Y. Qiu
Xi'an Jiaotang Univ.
High Voltage Division
28 Xianning Rd.
Xi'an, Shaanxi, 710049
CHINA

X. Q. Qiu
Univ. of Strathclyde
204 George Street
Glasgow, G1 1XW
U.K.

G. Raju
University of Windsor
2435 Delmar St.
Windsor, Ontario, N9H 1L4
CANADA

M. V. V. S. Rao
Thermosciences Institute
NASA-Ames Res. Center
M.S. 229-1
Moffett Field, CA 94035

A. Sabot
EDF/DER/LGE
Les Renardieres
F-77818 Moret/Loing
FRANCE

A. Salinas
S. California Edison
501 S. Merengo
Bldg C
Alhambra, CA 91802

M. Schmidt
Inst. for Low Temperature Plasma
Physics
Robert-Blum-str. 8-10
17489 Greifswald
Germany

D. Schoen
Univ. of Tech. Darmstadt
Landgraf Georg Str 4
64287 Darmstadt
GERMANY

D. Schweickart
Air Force Research Lab.
2645 Fifth St.
AFRL/PRPS
Wright Patterson AFB, OH 45433

R. Shinohara
Power Systems Dept.
Hitachi, Ltd.
4-6 Surugadai
101-8010
Kanda, Chiyoda-ku, Tokyo
JAPAN

H. Slowikowska
Electrotechnical Inst.
Pozaryskiego 28
04-703 Warsaw
POLAND

S. J. Slowikowski
Electrotechnical Inst.
Pozaryskiego 28
04-703 Warsaw
POLAND

K. D. Srivastava
Univ. of British Columbia
2356, Main Mall
McLeod Building
Vancouver, BC, V6T 1Z4
CANADA

S. Stangherlin
ABB Corporate Research
Viale Edison 50
Sesto S.G., MI 20099
ITALY

I. Szamrej
Agricultural & Teachers Univ.
Chemistry Dept.
3 Maja 54/08-110
Siedlce
POLAND

T. Takahashi
Nagoya University
Dept. of Elec. Eng.
Furo-cho Chikusa-ku
Nagoya 464-8603
JAPAN

R. J. Van Brunt
NIST
Bldg. 220, Rm. B344
Gaithersburg, MD 20899

P. Vasiliou
National Tech. Univ.
Dept. Elec. Eng.
Zografou Campus, 15773
GREECE

L. Vial
Universite Paul Sabatier
C.P.A.T. 118
Route de Narbonne
31062 Toulouse Cedex
FRANCE

L. Vuskovic
Old Dominion Univeristy
4600 Elkhorn Ave.
Rm. 306
Norfolk, VA 23505

L. Walfridsson
ABB Corporate Research
72178 Vasteras
SWEDEN

S. Walton
College of William & Mary
Dept. of Physics
Williamsburg, VA 23187

Yicheng Wang
NIST
Bldg. 220, Rm. B344
Gaithersburg, MD 20899

Yan Wang
Old Dominion University
4600 Elkhorn Ave.
Rm. 306
Norfolk, VA 23529

X. Waymel
EDF-DER
Route de Sens
Ecuelle
F-77818
FRANCE

Z. Wei
Old Dominion University
Dept. of Physics
Norfolk, VA 23539

T. Yamada
Nagoya University
Dept. of Elec. Eng.
Furo-cho Chikusa-ku
Nagoya 464-8603
JAPAN

T. Yamagiwa
Hitachi, Ltd.
Kokubu Works
1-1-1 Kokubu-cho
316-8501
Hitachi-shi, Ibaraki-ken
JAPAN

PHOTOGRAPHS OF PARTICIPANTS

G. Raju   A. Ghazala   M. Piemontesi   J. Desormeaux
R. Arora   H. Marsden   E. H. Gaxiola
  M. Forys   M. Morcos   A. Mueller
V. Lakdawala   K. Srivastava I. Szamrej   A. Gleizes
  M. Frechette   M.-C. Bordage
Z. Wei   I. Coll   L. Vial

J. Clark   M. Rao   B. Hampton   R. Hegerberg
E. Kuffel   J. Horwath   M. Schmidt   M. Pittroff
A. Garscadden   C. Jiro   T. Takahashi   H. Knobloch
Y. Wang   C. Dervos   M. Naidu
H. Okubo   S. Slowikowski   A. Meier
H. Hikita   H. Slowikowska   A. Mufti   C. Gaillac

D. Green
R. Fouracre S. Dale J. Moore B. Damsky M. Misakian
  B. Hankla A. Cookson D. Schweickart
    M. Rao R. Van Brunt
      V. Lakdawala

AUTHOR INDEX

AUTHOR INDEX

W. M. Al-Baiz, 319
I. Alvarez, 51
H. Aoyagi, 473
A. O. Arafa, 319
R. Arora, 275, 325
R. Basner, 3, 161
E. Basurto, 51
K. H. Becker, 3, 13, 21, 37, 161
W. Boeck, 37, 195, 367, 527, 579, 601
M.-C. Bordage, 29, 141
A. F. Borghesani, 73, 79, 85, 91
J.-M. Braun, 557
A. Bulinski, 111, 123, 262, 343, 351, 367, 457, 487, 495, 545
A.-M. Casanovas, 379, 387
J. Casanovas, 379, 387
J. Castonguay, 385, 393, 402, 443, 450, 539, 545, 564
I. D. Chalmers, 189
H. Champain, 113
R. L. Champion, 45
L. G. Christophorou, 21, 39, 139, 179, 195, 307, 343, 361, 367, 375, 386, 571, 612
F. Y. Chu, 557
C. Cisneros, 51
J. D. Clark, 23, 29
R. Clavreul, 379
A. Cookson, 601, 610, 611, 612
I. Coll, 379, 387, 393
P. Coventry, 189
G. C. Crichton, 239, 283
S. Dale, 539, 564, 579, 610, 611
B. Damsky, 430, 442, 564, 571, 579
M. G. Danikas, 489
J. de Urquijo, 51
C. T. Dervos, 417
H. Deutsch, 3
E. J. Dolin, 343, 425, 430
I. V. Dyakov, 45
C. R. Eck, III, 541
F. Endo, 573
R. Foest, 161
G. L. Ford, 557
M. Foryś, 69, 402
R. A. Fouracre, 481, 487
M. Frechette, 601
U. Fromm, 497
H. Fujii, 529, 581

A. Garscadden, 23, 37, 57, 589, 594, 598, 599, 600
J.-M. Gauthier, 443
E. H. R. Gaxiola, 105, 111, 131, 252, 511, 527
A. Gleizes, 169, 179, 571
T. Goda, 263, 267, 301, 505
A. Goldman, 113, 123, 147, 275
M. Goldman, 111, 147, 195, 450, 487, 495
D. S. Green, 361, 589, 594, 597, 598
W. Gu, 181
S. Gubanski, 403
P. D. Haaland, 57, 450, 487, 589, 599
H. Hama, 211, 353, 359, 409, 545, 581
S. Hamano, 353
B. Hampton, 211, 253, 262, 351, 479, 611, 612
X. Han, 307
M. Hanai, 263, 301, 333, 473
T. Hasegawa, 581
M. Hatano, 573, 581
C. Heitz, 369
F. Hempel, 161
M. Hikita, 269, 295
M. Holmberg, 403
M. D. Hopkins, 541
J. Horwarth, 313
Y. Hoshina, 473, 479
T. Hoshino, 529
K. Inami, 353, 581
T. Inoue, 263, 301, 505
A. Inui, 505
M. Ishikawa, 263, 301, 505, 547
C. Q. Jiao, 57
E. Kaneko, 473
A. Kawahara, 581
T. Kawamura, 457
T. Kawashima, 333, 547
H.-J. Kloes, 489
H. Knobloch, 343, 565, 571, 612
T. Kobayashi, 263, 301, 505
S. Koch, 147
D. Koenig, 489, 495
F. Koenig, 369
M. Koto, 547
M. Krishnamurthi, 411
A. Kron, 497
E. Kuffel, 179, 195, 197, 203, 211, 519

M. Lalmas, 113
T. Łaś, 277
R. Liu, 497
S. J. MacGregor, 481
T. D. Märk, 3
E. Marode, 13, 375, 589, 594, 597
H. I. Marsden, 267, 541, 545
F. Massines, 141
S. Matsumoto, 333, 343, 503
S. Matt, 3
I. W. McAllister, 239, 252, 283
M. Meguro, 505
A. Meier, 465
L. Ming, 497
M. Misakian, 451, 457, 612
J. H. Moore, 13, 15, 21
M. M. Morcos, 403, 409
H. D. Morrison, 385, 393, 430, 442, 557, 564, 610, 611
S. Motlagh, 15
A. H. Mufti, 131, 139, 195, 319, 343, 367, 479
H. Murase, 473
Y. Murayama, 333
M. S. Naidu, 155, 203, 205, 211, 213, 219, 495
K. Nakanishi, 131, 393, 409, 495, 503, 527, 529, 539, 581, 612
L. Niemeyer, 179, 359, 367, 369, 431, 442, 450, 459, 571, 611
E. Odic, 147
S. Okabe, 547
H. Okubo, 125, 187, 262, 269, 275, 289, 295, 351, 457, 479, 527, 545
J. K. Olthoff, 31, 39, 361
L. Parissi, 147
B. L. Peko, 45
W. Pfeiffer, 111, 133, 139, 479, 553
M. Piemontesi, 369, 375
M. Pittroff, 465
S. Popović, 97
C. Pradayrol, 387
S. Prem, 325
A. Qiu, 225
X. Q. Qiu, 189, 195, 267
Y. Qiu, 111, 181, 187, 225, 519, 527
A. Rabehi, 141
D. Raghavender, 205, 219
G. R. G. Raju, 13, 29, 231
M. V. V. S. Rao, 31, 37
A. Rosa, 69
A. Sabot, 545, 601, 608

M. Santini, 73, 79, 85, 91
R. Sarathi, 411
J. Sato, 473
M. Schmidt, 3, 161, 598
D. Schoen, 133
T. Schütte, 465
D. L. Schweickart, 313, 487, 594
P. Ségur, 141
M. Shikata, 573
R. Shinohara, 573, 579
H. Słowikowska, 277, 385, 527
J. Słowikowska, 277
K. D. Srivastava, 403, 479, 612
M. Storer, 379
H. N. Suresh, 213
I. Szamrej, 63, 69
T. Takahashi, 125, 269, 275, 289, 295
V. Tarnovsky, 3
T. Teranishi, 263, 301, 505
T. Toda, 125, 289
K. Toufani, 197
K. Tsuji, 529
S. M.Turnbull, 481
R. J. Van Brunt, 301
P. Vassiliou, 417
S. K. Venkatesh, 155
J. Verbrugge, 39
L. Vial, 379, 385, 386
L. Vušković, 97
L. Walfridsson, 497, 503
Yan Wang, 97
Yicheng Wang, 13, 29, 37, 39, 307, 313, 539
X. Waymel, 345, 351, 359
J. M. Wetzer, 105, 511
J. D. Wrbanek, 23
B. W. Wright, 23
T. Yamada, 125, 131, 269, 289, 295
T. Yamagiwa, 547, 553, 573, 601, 603
K. Yamaji, 573
C. H. Ying, 97
T. Yoshikawa, 529
M. Yoshimura, 353, 529, 581
C. Zender, 133
Z. Zeng, 225
Q. Zhang, 181

SUBJECT INDEX

SUBJECT INDEX

Air, 147, 181, 192, 325, 335, 336, 369, 511

Anion formation (see electron attachment)

Arcs, 169, 361, 365

Barrier discharge, 147

Basic mechanisms, 103

Breakdown
- in gas mixtures, 125, 133, 167, 181, 189, 197, 205, 219, 289, 333, 345, 353, 361, 369
- in pure gases, 113, 125, 155, 213, 333, 345, 369

c-C_4F_8, 57, 335, 395, 600

CF_4, 19, 45, 126, 289, 335, 336, 387, 395, 594, 600

CHF_3, 23, 39, 45, 395, 594, 600

CH_4, 18, 51, 425

Circuit breakers, 361, 169, 565, 603-608

CO_2, 126, 289, 425, 565, 596

Coatings, 335, 403, 581

Collision-induced dissociation, 48

Collision cross sections
- for CHF_3, 23, 26

Corona
- discharges, 379
- stabilization, 113, 136

Creeping discharges, 497

Databases, 589, 594, 598

Decomposition, 379, 387, 395, 443, 465

Diagnostics, 237, 253

Dielectric barrier, 141

Dissociative charge transfer, 48

Drift velocity
- of electrons in CHF_3, 42

Electric field measurements, 451

Electron
- attachment
  to CHF_3, 27, 39
  to O_2 in dense He and Ar, 73
  dependence on polarizability, 63
- collisions with excited atoms, 97
- detachment, 47
- drift velocity in
  Ar/CHF_3, 23, 39
  CHF_3, 39
- impact dissociation, 3
  of fluorocarbons, 15
  CH_4, 18
  CF_4, 19
- impact ionization, 3, 57
  of c-C_4F_8, 57
  SF_x, 4
  H_2O, 8
  OH, 8
- mobility in
  dense He, 79
  maximum in dense Ar, 91
-swarm, 68
- transport, 23
  in CHF_3, 23

Environmental aspects, 425, 459, 557, 595, 601, 602, 609

Epoxy, 277

Flashover, 473

Fluorocarbons, 15, 395, 425, 459, 597

Gas decomposition, 379, 387, 395

Gas-insulated bus (GIB), 353

Gas-insulated lines (GIL), 345, 541, 609

Gas-insulated switchgear (GIS), 253, 269, 295, 319, 333, 473, 547, 557, 573, 581, 601, 603-608

Gas-insulated transformer (GIT), 263, 301, 505, 519, 529, 608

Gas Mixtures
- decomposition of, 379, 387, 395
- breakdown in, 189, 197, 289, 333, 519
- impulse breakdown, 181, 189, 205, 219
- partial discharges in, 125, 269, 289, 307
- thermodynamic properties, 169

Gaseous Dielectrics
- basic physics of, 1
- basic mechanisms in, 103
- other industrial applications, 589

Glow discharge, 141

Greenhouse gases, 425, 595

Hybrid insulation GIS, 606

Interruption, 169, 364, 608

Ion kinetics, 31, 57

Ion mobility
- of O_2^- in dense Ar, 85

Ion-molecule reactions, 31, 51

Kyoto protocol, 426, 430, 497, 529, 558, 602

Magnetic field measurements, 451

Material processing, 595

Nitrogen, 126, 335, 336, 346, 369, 565, 605

Partial discharge(s), 125, 237, 269, 289, 295, 301, 307, 313, 319
- in gas mixtures, 125, 289, 307
- in pure gases, 125, 269, 277, 295, 301

- transients, 239, 283

Particle(s), 403, 411, 581

Plasma
- enhanced treatment process, 593
- non-equilibrium, 592
- opening switch, 225
- technologies, 589
- thermal, 589, 590

Polyethylene, 313

Polymer (s), 489

Prebreakdown, 105, 511

Radicals, 4, 15

Recycling, 339, 431, 442

RF discharge, 161, 231

Semiconductor industry, 594

SF_6
- as a greenhouse gas, 425, 431
- breakdown, 113, 155, 189, 213, 289, 295, 333, 605
- corona stabilization, 113
- decomposition, 379, 387, 395, 443, 465
- handling, 417, 431, 442, 557, 603
- impulse breakdown, 213, 547
- ion transport in discharges of, 33
- partial discharges, 125, 269, 277, 289, 295, 301
- physical properties, 335
- prebreakdown, 105
- recycling, 431
- reuse of, 465
- substitutes, 361, 459, 473, 601
- time lags, 155

SF_6/CF_4 mixtures, 125, 197, 289, 387, 395

SF_6/CO_2 mixtures, 125, 181, 189, 219, 289, 521, 527

SF_6/N_2 mixtures, 333, 345, 353, 361
- breakdown, 189, 289, 519, 605

- decomposition, 379
- dielectric characteristics for VFTV, 133
- electrical properties, 169, 263
- for circuit breakers, 361, 565, 571, 606
- for GIL(S), 333, 345, 353, 361, 473, 541, 603, 609, 611
- for GIT, 263, 361, 519, 529
- impulse breakdown, 182, 189, 205, 219
- partial discharges, 125, 269, 289, 307
- recycling, 442
- thermodynamic properties, 169
- VFT, 133

Simulations, 167, 225, 601

Spacer(s), 473, 511

Static electrification, 505

Streamer/leader discrimination, 269

Substitutes, 334, 361, 459, 473, 601-608

Surface
- discharges, 481, 489
- flashover, 473

Tetraethoxylsilane, 161

Townsend discharges, 31
- in O_2, 31
- in SF_6, 31

Time lags, 155

UHF, 133, 253

Universal application gas mixtures, 361

Voids, 283

Water, 8, 147

Waste, 591